Lecture Notes in Computer Science 14506

Founding Editors

Gerhard Goos
Juris Hartmanis

Editorial Board Members

The series Lecture Notes in Computer Science (LNCS), including its subseries Lecture Notes in Artificial Intelligence (LNAI) and Lecture Notes in Bioinformatics (LNBI), has established itself as a medium for the publication of new developments in computer science and information technology research, teaching, and education.

LNCS enjoys close cooperation with the computer science R & D community, the series counts many renowned academics among its volume editors and paper authors, and collaborates with prestigious societies. Its mission is to serve this international community by providing an invaluable service, mainly focused on the publication of conference and workshop proceedings and postproceedings. LNCS commenced publication in 1973.

Giuseppe Nicosia · Varun Ojha ·
Emanuele La Malfa · Gabriele La Malfa ·
Panos M. Pardalos · Renato Umeton
Editors

Machine Learning, Optimization, and Data Science

9th International Conference, LOD 2023
Grasmere, UK, September 22–26, 2023
Revised Selected Papers, Part II

Springer

Editors
Giuseppe Nicosia ⓘ
University of Catania
Catania, Catania, Italy

Varun Ojha ⓘ
Newcastle University
Newcastle upon Tyne, UK

Emanuele La Malfa ⓘ
University of Oxford
Oxford, UK

Gabriele La Malfa ⓘ
University of Cambridge
Cambridge, UK

Panos M. Pardalos ⓘ
University of Florida
Gainesville, FL, USA

Renato Umeton ⓘ
Dana-Farber Cancer Institute
Boston, MA, USA

ISSN 0302-9743 ISSN 1611-3349 (electronic)
Lecture Notes in Computer Science
ISBN 978-3-031-53965-7 ISBN 978-3-031-53966-4 (eBook)
https://doi.org/10.1007/978-3-031-53966-4

This Springer imprint is published by the registered company Springer Nature Switzerland AG
The registered company address is: Gewerbestrasse 11, 6330 Cham, Switzerland

Paper in this product is recyclable.

Preface

LOD is the international conference embracing the fields of machine learning, deep learning, optimization, and data science. The ninth edition, LOD 2023, was organized on September 22–26, 2023, in Grasmere, Lake District, England Like the previous edition, LOD 2023 hosted the Advanced Course and Symposium on Artificial Intelligence & Neuroscience – ACAIN 2023. In fact, this year, in the LOD Proceedings we decided to also include the papers of the third edition of the Symposium on Artificial Intelligence and Neuroscience (ACAIN 2023). The ACAIN 2023 chairs were:

Giuseppe Nicosia, University of Catania, Italy
Panos Pardalos, University of Florida, USA

The review process of 18 submissions to ACAIN 2023 was double-blind, performed rigorously by an international program committee consisting of leading experts in the field. Therefore, the last nine articles in the LOD Table of Contents (volume 14506) are the articles accepted to ACAIN 2023.

Since 2015, the LOD conference has brought together academics, researchers and industrial researchers in a unique *pandisciplinary community* to discuss the state of the art and the latest advances in the integration of machine learning, deep learning, nonlinear optimization and data science to provide and support the scientific and technological foundations for interpretable, explainable, and trustworthy AI. Since 2017, LOD adopted the *Asilomar AI Principles*.

The annual conference on machine Learning, Optimization and Data science (LOD) is an international conference on machine learning, deep learning, AI, computational optimization and big data that includes invited talks, tutorial talks, special sessions, industrial tracks, demonstrations, and oral and poster presentations of refereed papers.

LOD has established itself as a premier interdisciplinary conference in machine learning, computational optimization, and data science. It provides an international forum for presentation of original multidisciplinary research results, as well as exchange and dissemination of innovative and practical development experiences.

LOD 2023 attracted leading experts from industry and the academic world with the aim of strengthening the connection between these institutions. The 2023 edition of LOD represented a great opportunity for professors, scientists, industry experts, and research students to learn about recent developments in their own research areas and to learn about research in contiguous research areas, with the aim of creating an environment to share ideas and trigger new collaborations.

As chairs, it was an honour to organize a premier conference in these areas and to have received a large variety of innovative and original scientific contributions.

During LOD 2023, 3 plenary talks were presented by leading experts:

LOD 2023 Keynote Speakers:

Gabriel Barth-Maron DeepMind, London, UK
Anthony G. Cohn University of Leeds, UK; The Alan Turing Institute, UK
Sven Giesselbach Fraunhofer Institute - IAIS, Germany

ACAIN 2023 Keynote Lecturers:

Karl Friston, University College London, UK
Kenneth Harris University College London, UK
Rosalyn Moran King's College London, UK
Panos Pardalos, University of Florida, USA
Edmund T. Rolls University of Warwick, UK

LOD 2022 received 119 submissions from 64 countries in five continents, and each manuscript was independently double-blind reviewed by a committee formed by at least five members. These proceedings contain 71 research articles written by leading scientists in the fields of machine learning, artificial intelligence, reinforcement learning, computational optimization, neuroscience, and data science presenting a substantial array of ideas, technologies, algorithms, methods, and applications.

At LOD 2023, Springer LNCS generously sponsored the LOD Best Paper Award. This year, the paper by *Moritz Lange, Noah Krystiniak, Raphael C. Engelhardt, Wolfgang Konen, and Laurenz Wiskott* titled *"Improving Reinforcement Learning Efficiency with Auxiliary Tasks in Non-Visual Environments: A Comparison"*, received the LOD 2023 Best Paper Award.

This conference could not have been organized without the contributions of exceptional researchers and visionary industry experts, so we thank them all for participating. A sincere thank you goes also to the 20 sub-reviewers and to the Program Committee of more than 210 scientists from academia and industry, for their valuable and essential work of selecting the scientific contributions.

Finally, we would like to express our appreciation to the keynote speakers who accepted our invitation, and to all the authors who submitted their research papers to LOD 2023.

October 2023

Giuseppe Nicosia
Varun Ojha
Emanuele La Malfa
Gabriele La Malfa
Panos M. Pardalos
Renato Umeton

Organization

General Chairs

Renato Umeton   Dana-Farber Cancer Institute, MIT, Harvard T.H.
Chan School of Public Health & Weill Cornell
Medicine, USA

Conference and Technical Program Committee Co-chairs

Giuseppe Nicosia   University of Catania, Italy
Varun Ojha   Newcastle University, UK
Panos Pardalos   University of Florida, USA

Special Sessions Chairs

Gabriele La Malfa   University of Cambridge, UK
Emanuele La Malfa   University of Oxford, UK

Steering Committee

Giuseppe Nicosia   University of Catania, Italy
Panos Pardalos   University of Florida, USA

Program Committee Members

Jason Adair   University of Stirling, UK
Agostinho Agra   University of Aveiro, Portugal
Massimiliano Altieri   University of Bari, Italy
Vincenzo Arceri   University of Parma, Italy
Roberto Aringhieri   University of Turin, Italy
Akhila Atmakuru   University of Reading, UK
Roberto Bagnara   University of Parma, Italy
Artem Baklanov   Int Inst for Applied Systems Analysis, Austria
Avner Bar-Hen   CNAM, France

Bernhard Bauer	University of Augsburg, Germany
Peter Baumann	Constructor University, Germany
Sven Beckmann	University of Augsburg, Germany
Roman Belavkin	Middlesex University London, UK
Nicolo Bellarmino	Politecnico di Torino, Italy
Heder Bernardino	Universidade Federal de Juiz de Fora, Brazil
Daniel Berrar	Tokyo Institute of Technology, Japan
Martin Berzins	University of Utah, USA
Hans-Georg Beyer	Vorarlberg University of Applied Sciences, Austria
Sandjai Bhulai	Vrije Universiteit Amsterdam, The Netherlands
Francesco Biancalani	IMT Lucca, Italy
Martin Boldt	Blekinge Institute of Technology, Sweden
Vincenzo Bonnici	University of Parma, Italy
Anton Borg	Blekinge Institute of Technology, Sweden
Jose Borges	University of Porto, Portugal
Matteo Borrotti	University of Milano-Bicocca, Italy
Alfio Borzi	Universität Würzburg, Germany
Goetz Botterweck	Trinity College Dublin, Ireland
Will Browne	Queensland University of Technology, Australia
Kevin Bui	University of California, Irvine, USA
Luca Cagliero	Politecnico di Torino, Italy
Antonio Candelieri	University of Milano-Bicocca, Italy
Paolo Cazzaniga	University of Bergamo, Italy
Adelaide Cerveira	UTAD and INESC-TEC, Portugal
Sara Ceschia	University of Udine, Italy
Keke Chen	Marquette University, USA
Ying-Ping Chen	National Yang Ming Chiao Tung University, Taiwan
Kiran Chhatre	KTH Royal Institute of Technolgy, Sweden
John Chinneck	Carleton University, Canada
Miroslav Chlebik	University of Sussex, UK
Eva Chondrodima	University of Piraeus, Greece
Pedro H. Costa Avelar	King's College London, UK
Chiara Damiani	University of Milano-Bicocca, Italy
Thomas Dandekar	University of Würzburg, Germany
Renato De Leone	University of Camerino, Italy
Roy de Winter	Leiden Institue of Advanced Computer Science, The Netherlands
Nicoletta Del Buono	University of Bari, Italy
Mauro Dell'Amico	University of Modena and Reggio Emilia, Italy
Clarisse Dhaenens	University of Lille, France

Giuseppe Di Fatta — Free University of Bozen-Bolzano, Italy
João Dionísio — FCUP, Portugal
Stephan Doerfel — Kiel University of Applied Sciences, Germany
Vedat Dogan — University College Cork, Ireland
Rafal Drezewski — AGH University of Science and Technology, Poland
Juan J. Durillo — Leibniz Supercomputing Centre, Germany
Nelson Ebecken — COPPE/UFRJ, Brazil
Andries Engelbrecht — University of Stellenbosch, South Africa
Des Fagan — Lancaster University, UK
Fabrizio Falchi — ISTI-CNR, Italy
Louis Falissard — Université Paris-Saclay, France
Giovanni Fasano — University of Venice, Italy
Hannes Fassold — Joanneum Research, Austria
Carlo Ferrari — University of Padova, Italy
Lionel Fillatre — Université Côte d'Azur, France
Steffen Finck — Vorarlberg University of Applied Sciences, Austria
Enrico Formenti — Université Cote d'Azur, France
Fabrizio Fornari — University of Camerino, Italy
Valerio Freschi — University of Urbino, Italy
Simon Frieder — University of Oxford, UK
Nikolaus Frohner — TU Wien, Austria
Carola Gajek — University of Augsburg, Germany
Claudio Gallicchio — University of Pisa, Italy
Bruno Galuzzi — University of Milano-Bicocca, Italy
Paolo Garza — Politecnico di Torino, Italy
Bertrand Gauthier — Cardiff University, UK
Tobias Gemmeke — IDS, RWTH Aachen University, Germany
Claudio Gennaro — ISTI - CNR, Italy
Kyriakos Giannakoglou — National Technical University of Athens, Greece
Giorgio Gnecco — IMT - School for Advanced Studies, Lucca, Italy
Lio Goncalves — UTAD, Portugal
Denise Gorse — University College London, UK
Vladimir Grishagin — Nizhni Novgorod State University, Russia
Alessandro Guazzo — University of Padua, Italy
Vijay Gurbani — Illinois Institute of Technology, USA
Selini Hadjidimitriou — University of Modena and Reggio Emilia, Italy
Joy He-Yueya — University of Washington, USA
Verena Heidrich-Meisner — CAU Kiel, Germany
David Heik — HTW Dresden, Germany

Michael Hellwig	Vorarlberg University of Applied Sciences, Austria
Carlos Henggeler Antunes	University of Coimbra, Portugal
Alfredo G. Hernndez-Daz	Pablo de Olavide University, Spain
J. Michael Herrmann	University of Edinburgh, UK
Vinh Thanh Ho	University of Limoges, France
Colin Johnson	University of Nottingham, UK
Sahib Julka	University of Passau, Germany
Robert Jungnickel	RWTH Aachen University, Germany
Vera Kalinichenko	UCLA, USA
George Karakostas	McMaster University, Canada
Emil Karlsson	Saab AB, Sweden
Branko Kavek	University of Primorska, Slovenia
Marco Kemmerling	RWTH Aachen, Germany
Aditya Khant	Harvey Mudd College, USA
Zeynep Kiziltan	University of Bologna, Italy
Wolfgang Konen	Cologne University of Applied Sciences, Germany
Jan Kronqvist	KTH Royal Institute of Technology, Sweden
T. K. Satish Kumar	University of Southern California, USA
Nikolaos A. Kyriakakis	Technical University of Crete, Greece
Alessio La Bella	Politecnico di Milano, Italy
Gabriele Lagani	University of Pisa, Italy
Dario Landa-Silva	University of Nottingham, UK
Cecilia Latotzke	RWTH Aachen University, Germany
Ang Li	University of Southern California, USA
Zhijian Li	University of California, Irvine, USA
Johnson Loh	RWTH Aachen University, Germany
Gianfranco Lombardo	University of Parma, Italy
Enrico Longato	University of Padua, Italy
Angelo Lucia	University of Rhode Island, USA
Hoang Phuc Hau Luu	University of Lorraine, France
Eliane Maalouf	University of Neuchatel, Switzerland
Hichem Maaref	Université d'Evry Val d'Essonne, France
Antonio Macaluso	German Research Center for Artificial Intelligence, Germany
Francesca Maggioni	University of Bergamo, Italy
Silviu Marc	Middlesex University London, UK
Nuno Marques	FEUP, Portugal
Rafael Martins de Moraes	New York University, USA
Moreno Marzolla	University of Bologna, Italy
Shun Matsuura	Keio University, Japan

Sally McClean	Ulster University, UK
James McDermott	University of Galway, Ireland
Silja Meyer-Nieberg	Universität der Bundeswehr München, Germany
Atta Muhammad	Politecnico di Torino, Italy
Kateryna Muts	Hamburg University of Technology, Germany
Hidemoto Nakada	National Institute of Advanced Industrial Science and Technology, Japan
Mirco Nanni	KDD-Lab ISTI-CNR, Italy
Alexander Nasuta	RWTH Aachen University, Germany
Nikolaos Nikolaou	University College London, UK
Leonardo Nunes	Microsoft, USA
Federico Nutarelli	IMT Alti Studi Lucca, Italy
Beatriz Otero	Universidad Politécnica de Cataluña, Spain
Mathias Pacher	Goethe-Universität Frankfurt am Main, Germany
George Papastefanatos	ATHENA Research Center, Greece
Eric Paquet	National Research Council, Canada
Seonho Park	Georgia Institute of Technology, USA
Konstantinos Parsopoulos	University of Ioannina, Greece
Nicos Pavlidis	Lancaster University, UK
Antonio Pellicani	University of Bari, Italy
David Pelta	University of Granada, Spain
Guido Perboli	Politecnico di Torino, Italy
Gianvito Pio	University of Bari, Italy
Nikolaos Ploskas	University of Western Macedonia, Greece
Alessandro Sebastian Podda	University of Cagliari, Italy
Agoritsa Polyzou	Florida International University, USA
Denys Pommeret	Aix-Marseille University, France
Piotr Pomorski	University College London, UK
Steve Prestwich	Insight Centre for Data Analytics, Ireland
Buyue Qian	Xi'an Jiaotong University, China
Michela Quadrini	University of Camerino, Italy
Günther Raidl	Vienna University of Technology, Austria
Reshawn Ramjattan	University of the West Indies, Jamaica
Jan Rauch	Prague University of Economics and Business, Czech Republic
Ralitsa Raycheva	Medical University Plovdiv, Bulgaria
Steffen Rebennack	Karlsruhe Institute of Technology, Germany
Dirk Reichelt	HTW Dresden, Germany
Wolfgang Reif	University of Augsburg, Germany
Cristina Requejo	University of Aveiro, Portugal
Simone Riva	University of Oxford, UK
Humberto Rocha	University of Coimbra, Portugal

Jose Gilvan Rodrigues Maia	Universidade Federal do Ceará, Brazil
Jan Rolfes	KTH Royal Institute of Technology, Sweden
Roberto Maria Rosati	University of Udine, Italy
Hang Ruan	University of Edinburgh, UK
Chafik Samir	University of Clermont, France
Marcello Sanguineti	Universita di Genova, Italy
Giorgio Sartor	SINTEF, Norway
Claudio Sartori	University of Bologna, Italy
Frederic Saubion	University of Angers, France
Robert Schaefer	AGH University of Science and Technology, Poland
Robin Schiewer	Ruhr-University Bochum, Germany
Joerg Schlatterer	University of Mannheim, Germany
Bryan Scotney	University of Ulster, UK
Natalya Selitskaya	IEEE, USA
Stanislav Selitskiy	University of Bedfordshire, UK
Marc Sevaux	Université Bretagne-Sud, France
Hyunjung Shin	Ajou University, South Korea
Zeren Shui	University of Minnesota, USA
Surabhi Sinha	University of Southern California, USA
Cole Smith	New York University, USA
Andrew Starkey	University of Aberdeen, UK
Julian Stier	University of Passau, Germany
Chi Wan Sung	City University of Hong Kong, China
Johan Suykens	Katholieke Universiteit Leuven, Belgium
Sotiris Tasoulis	University of Thessaly, Greece
Tatiana Tchemisova	University of Aveiro, Portugal
Michele Tomaiuolo	University of Parma, Italy
Bruce Kwong-Bun Tong	Hong Kong Metropolitan University, China
Elisa Tosetti	Ca Foscari University of Venice, Italy
Sophia Tsoka	King's College London, UK
Gabriel Turinici	Université Paris Dauphine - PSL, France
Gregor Ulm	Fraunhofer-Chalmers Research Centre for Industrial Mathematics, Sweden
Joy Upton-Azzam	Susquehanna University, USA
Werner Van Geit	EPFL, Switzerland
Johannes Varga	TU Wien, Austria
Filippo Velardocchia	Politecnico di Torino, Italy
Vincent Vigneron	Université d'Evry, France
Herna Viktor	University of Ottawa, Canada
Marco Villani	University of Modena and Reggio Emilia, Italy
Dean Vucinic	Vrije Universiteit Brussel, Belgium

Contents – Part II

Artificial Intelligence and Neuroscience (ACAIN 2023)

Contents – Part I

Machine Learning, Optimization, and Data Science (LOD 2023)

Integrated Human-AI Forecasting for Preventive Maintenance Task Duration Estimation

Jiye Li[1]([⊠]) [iD], Yun Yin[2], Daniel Lafond[1] [iD], Alireza Ghasemi[2] [iD], Claver Diallo[2] [iD], and Eric Bertrand[3]

[1] Thales Research and Technology, Quebec, Canada
{jiye.li,daniel.lafond}@thalesgroup.com
[2] Dalhousie University, Halifax, Canada
{yn987224,alireza.ghasemi,claver.diallo}@dal.ca
[3] Royal Canadian Navy, Halifax, Canada
eric.bertrand@dal.ca

Abstract. Maintenance task duration estimations help manage shipyard resource usage and allow planners to decide on maintenance priorities within a limited time frame. Better estimated task durations help produce more robust resource schedules, perform more tasks in facilities such as shipyards, reduce resource idling time and increase ship operational availability. However, task duration estimations have until now been historically performed by human experts with essentially no artificial intelligence-based forecasting for shipyard operations. The analysis of historical data is also not a common practice to complement any expert-driven forecasting. To explore opportunities for using AI in this work domain, and to improve on human estimations for task durations, we propose a novel hybrid Human-AI approach that involves integrating human forecasts with data-driven models. Our empirical data comes from two fleet maintenance facilities in Canada, containing more than 13,000 anonymized historical ship work orders (WO) ranging from 2017 to 2022. We used supervised learning algorithms to forecast the preventive maintenance task duration on this data, with and without expert task duration estimates and the results demonstrate that hybrid models perform better than both human expert model and historical data alone. An average of 8.6% improvement from Hybrid human-AI model over human expert model is observed based on R2 evaluation metric. Results suggest that human forecasts, which tend to rely on a broader contextual knowledge than the inputs captured in a historical database, remain key for effective task duration estimation, yet can be fine-tuned by the pattern recognition capabilities of machine learning algorithms.

Keywords: Hybrid AI · Human-AI collaboration · planning · task duration estimation · ship refit · preventive maintenance · machine learning

1 Introduction

Task duration estimation is an important and challenging aspect of project scheduling and optimization in many industries. This problem is significantly more acute for projects with a very large number of activities such as ship refits or when historical data is lacking.

G. Nicosia et al. (Eds.): LOD 2023, LNCS 14506, pp. 3–18, 2024.
https://doi.org/10.1007/978-3-031-53966-4_1

This scarcity of historical data makes activity scheduling and planning very challenging, which can lead to unplanned conflicts, insufficient resources being planned, project delays and increased costs. Human experts can provide estimations on task duration; however, their availability is limited, and they cannot provide estimates when there are thousands of activities to deal with. Thus, there is a need to better integrate the expert opinions with the limited data available from historical records or from similar equipment or projects.

We discuss our studies in the context of navy ship maintenance and refit operations domain. In this area, ship refit has become one of the most important tasks in a shipyard [1]. Refit tasks of ships or vessels include repairing, fixing, restoring, mending, and renovating. A given refit task may require multiple activities as well as equipment and human resources for its completion. It is important to have an accurate task duration estimate for ship refit [2]. Schedules built with accurate duration estimates enable more tasks to be completed in the shipyard as well as reducing the utilization of resources such as equipment and personnel, lowering idling time, costs, and increasing ship operational availability. Such necessity applies to other related domains such as manufacturing, pharmaceutics, health care and the food industry [3, 4].

One challenge in the ship refit industry is that human experts perform most of the task duration estimations manually. It also requires an expert in the task-related refit domain to be on board the ship and physically evaluate the relevant ship components to forecast likely durations in hours or days. Such demand, despite being reliable and trustworthy, often is a tedious and expensive process, and may introduce involuntary human errors. On the other hand, data-driven forecasting models have not been seen to be widely adopted in this industry. Historical data is seldom available or exists in limited quantities in the ship refit industry. Classic machine learning forecasting algorithms often require significant amounts of historical data to train accurate predictive models. However, in certain use cases, accessing historical data is very difficult. New tasks might not have any historical data available, and existing tasks may only have a very limited number of recorded historical cases.

In this paper, we investigate and propose a hybrid human-AI forecasting approach for task duration estimation. We demonstrate how to elicit and model expert forecasts to compensate for the lack of historical data, and how to augment the data-driven method by integrating human forecasts to create a hybrid approach. The rest of the paper is organized as follows. In Sect. 2, we present related works in the human expert forecasting, and machine learning domains. In Sect. 3, we describe the fleet maintenance facilities data source used to conduct our studies. In Sect. 4, we show the data pre-processing and feature construction work on the data set. We discuss experiments and results in Sect. 5. We conclude our work and discuss directions for future work on hybrid human-AI forecasting in Sect. 6.

2   Related Work

Below, we review the background work for task duration estimation in the fields of project management, scheduling, as well as AI and machine learning.

A project can be defined as a "one-time endeavour that consists of a set of activities, whose executions take time, require resources and incur costs or induce cash flows" [5].

The time, resources and performance elements can further be complicated by additional factors such as due date penalties, precedence constraints and multiple execution modes. The task of effectively coordinating all these characteristics from start to finish is known as project management. Project managers are usually required to "perform the project within time and cost estimates at the desired performance level in accordance with the client, while utilizing the required resources effectively and efficiently" [5].

Mathematical formulations using network models [6] such as the one shown in Fig. 1 have been successfully applied to project scheduling optimization using the early critical path method (CPM) [7] and the program evaluation and review technique (PERT) [8] to the recent advances in Resource Constrained Project Scheduling Problem (RCPSP) models [9–12]. Interrelated activities on the critical path will impact project completion time if delayed and require most accurate task duration estimations.

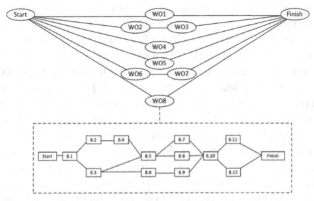

Fig. 1. Resource-dominated network topology common to maintenance and refit operations [6], where "WO" stands for work orders.

2.1 Demand for Precise Input for Planning

An accurate baseline schedule is needed prior to the project start date to assign specific high value or scarce resources - such as cranes, highly specialized technicians, and dry-dock in the naval maintenance environment - to specific periods during the project. Time is one of the most critical resources in any project. The impact of poorly allocating scarce or high value resources is significant to overall organizational performance. A slight delay affecting these crucial resources has the potential to disrupt the current project as well as the schedule and budget of other ongoing and subsequent projects.

The ability to accurately and reliably define the duration of activities is key to the effective use of scheduling and optimization models in project management. This has been an "Achilles' heel" in establishing the usefulness of networking models [13]. Activity durations can sometimes be very difficult to estimate because of a lack of historical data or a large number of activities that an automated procedure would need. Human experts can provide estimates on task durations. However, their availability is limited, and it is also not obvious how one can best integrate human expert's estimations into forecasting models of task durations.

2.2 Expert Input for Planning

Because of the complex, subjective nature of subject matter expert (SME) opinion, many studies contributed to establishing a more systematic approach to elicit expert opinion [14]. Cooke and Goossens [15] provide formal protocols, comprehensive procedures and guidelines on the elicitation process and handling of such data in uncertainty analysis. Clemen and Winkler [16] classify the elicitation and aggregation processes of SME assessments in two groups: mathematical and behavioral approaches [17]. In mathematical approaches, SME individual assessments on an uncertain quantity are expressed as subjective probabilities. They are combined through various mathematical methods by the decision-maker after their elicitation. Some of these mathematical methods are non-Bayesian axiomatic models, Bayesian models, and psychological scaling models (paired comparisons) [18]. Behavioral approaches aim at producing some type of group consensus among SMEs, who are typically encouraged to interact with one another and share their assessments. Some of the most well-known behavioral approaches are the Delphi method and the Nominal Group method [19–21]. It is generally accepted that mathematical approaches yield more accurate results than do behavioral approaches in aggregating expert opinions [15, 22]. Regardless of the approach used, the goal is to find the most efficient and reliable ways to aggregate quantitative and qualitative estimations.

With recent developments in artificial intelligence (AI) and machine learning (ML), new models have been proposed to integrate expert opinions into predictions. For example, Gennatas et al. [23] used a random forest classifier to predict categories of mortality. Human experts (clinicians) were asked for evaluations on the selected rules containing significant features while not exposing the decisions. This hybrid approach of using experts and ML showed better results than the ML approach alone. In addition, N., Madhusudanan et al. studied time duration analysis for maintenance work order data using machine learning models [24]. Regression models for maintenance work hour predictions were discussed in [25].

Recently, a study on hybrid human-AI forecasting classification models [26] for task duration estimation was proposed. Synthetic data was used to simulate both historical data and human expert data with simulated human errors. The results demonstrated that the hybrid method performs better than either the human experts' model and historical data-based model alone.

2.3 Shipyard-Related Task Estimations for Planning

A study for overtime budget predictions for naval fleet maintenance facilities was discussed in [27]. The authors stressed the importance of accurate estimations for task scheduling and management. Task durations are estimated manually by human experts most of the time. Approaches on how to aggregate expert judgements were surveyed in [29] to combine expert judgments into forecasting models to form an augmented single predictive model.

There have been multiple research efforts on analyzing the ship refit historical data such as the DRDC fleet maintenance facility (FMF) data set for various prediction tasks. A cost model has been investigated on this FMF data set to estimate the maintenance costs of the vessels [30]. A regression tree, gradient boost model, and hidden Markov model (HMM) based time-series analysis were explored as modeling algorithms on this data set [28]. Task duration predictions on individual ship level were performed, and overall results indicate regression tree (tree-based prediction model) is suitable for such task duration predictions. An ML model for overtime budget estimation was presented in [27]. The authors mentioned that overtime happened when the tasks required longer than planned hours to finish due to multiple factors such as age, types of the vessels, maintenance policies, and supply chain constraints. Human experts were not involved in the empirical studies mentioned above.

3 Data Description

In this section, we describe the data source used in this paper. Our data is historical data from two Royal Canadian Navy Fleet Maintenance Facilities (FMF) in Canada, which maintain their fleet of ships, submarines and other formation vessels using stand-alone or shared services. Different types of maintenance tasks are performed at a FMF such as corrective maintenance (CM) and preventive maintenance (PM) [31]. Each maintenance task contains many work orders (WO) including the type of the work, the main work center, the location as well as the compartment where the work is carried, the scheduled starting date and scheduled end date, the actual starting date and the actual end date, the status of the work and so on. The 20MB data was shared with us in excel format, ranging from 2017 to 2022, containing more than 13,000 different WO. The names of WOs are anonymized in our study.

4 Data Preprocessing and Feature Construction

We discuss the process of data cleaning, identifying, and creating features for forecasting WO duration.

4.1 Preprocessing

Since the data contains historical maintenance records that were maintained manually, there exists inconsistencies and discrepancies. We cannot use such erroneous data directly into data analytics. After consulting with subject matter experts, we first removed data columns that contain no values or missing values. We removed columns that contained the same value for all rows, since such columns did not bring discernibility for the data and thus were not considered informative. We removed WOs with actual duration less than or equal to zero, which indicated a canceled WO. We also removed orders whose scheduled durations (in days) were greater than the planned hours. The reasoning was that if the gap between scheduled hours and planned hours was too large, then the task was probably postponed or rescheduled due to a lack of resources.

Each WO contains a field "status" indicating whether this work is still ongoing, completed, waiting for predecessor tasks to finish, or being cancelled. We only considered the WO that were completed in our experiments.

We focused on PM WO in this data set as they represented the majority of reoccurring historical records, which enabled the use of ML algorithms to identify patterns in order to perform forecasting tasks. We obtained 1601 PM WO of more than 6000 rows of data after preprocessing the historical data covering 5 years of maintenance work an naval vessels. The average number of data points per WO was 3. The maximum number of historical records of any WO is 129, and the minimum number of records is 1. We identified 13 WO that contains more than 50 historical records in the data, and carried out our experiments based on these 13 WOs. A WO[1] may be related to deck painting, engineering repair, chiller set repair, etc.

4.2 Feature Construction

The final data was presented in an Excel file format. The columns of the original data files were considered as original features. These include features such as the maintenance plant, planned hours, actual hours, order scheduled start date and end date.

To enrich the feature set, we also created features indicating the locations of the WO, time related features such as the year and month of the WO starting date, weather related features such as the amount of the rain and snow, high and low temperatures, wind speed as well as features representing the age of the ship and the class of the ships. The weather information was obtained from an external source[2].

Table 1 lists the selected features and their meanings. "Actual hours" records the historical work hours spent on this task and it is inputted manually after the work is completed. "Planned hours" indicates the estimated hours that a human expert provides for a given WO before the job is started. Note that in this study, we consider that only one human expert estimates the task duration. In situations with multiple human experts providing estimated planned hours, an aggregated value representing groups of estimations will be used as human experts' input. Strategies such as best estimations, majority-voted estimations, and weighted estimations will need to be explored. (For confidential reasons, the value ranges for the features are not shown in Table 1). We applied one hot encoding for categorical features, i.e., "Source", "Ship classes", "Season".

In the following experiment, for each WO, we considered the "Actual hours" as the *target attribute* to predict, and the rest of the features were considered as *conditional attributes*.

---

[1] For confidential reasons, WO identification numbers and names have been anonymized.

[2] External resources for weather information is from https://weatherspark.com/y/466/Average-Weather-in-Victoria-Canada-Year-Round, and https://weatherspark.com/y/28434/Average-Weather-in-Halifax-Canada-Year-Round.

Table 1. A list of features for task duration estimations

Feature	Description	Type
Actual hours	Number of hours the WO actual takes	float
Planned hours	Number of hours planned / estimated	float
Scheduled days	Number of days previously scheduled	int
Age	Age of the vessels	int
Source	Indicating data from certain maintenance facilities	categorical
Maintenance plant	Vessels where the WOs are carried out	string
Order scheduled start date	Scheduled start date for the WO	DateTime
Ship classes	Classes of ships	categorical
Start year	Indicating order scheduled start year	int
Start month	Indicating order scheduled start month	int
Season	Weather related to scheduled start time	categorical
Compartment key	Compartment of WO	string
Order functional location key	Location information	string
Rain, snow, mixed precipitation, temperature, wind speed	Weather related to scheduled start time	float

5 Experiments

In this section, we describe the experiments and demonstrate the advantages of hybrid models. Given that the "actual" task duration is a float type data and that this value is predicted based on the previously defined features, we used sklearn [38] regression algorithms as the prediction algorithms. The purpose of this experiment was not to compare between multiple regression algorithms, but to compare the hybrid human-AI model with any data-driven models and with human expert only model. In the following experiment, we compared four regression algorithms' performance, which are decision tree (DT), random forest (RF), support vector regression (SVR) and K-Nearest Neighbors (KNN) regressors [10]. The experiment framework is shown in Fig. 2. For data-driven AI model and hybrid human-AI model, we applied four regression algorithms to perform the prediction, and selected the best performing algorithm from each model to compare with the other models. Hyper-parameter tunings were performed for all model training processes. We compared the following three models in the experiments.

- Data-driven AI model (trained based on features from Table 1 except "Planned hours") which relies purely on historical data to forecast
- Human expert (HE) model which relies on the manually estimated task duration hours. There is no AI used in this model. "Planned hours", as a single feature inputted by the human expert, is used for predictions.
- Hybrid human-AI model (trained based on all features from Table 1, including "Planned hours") which combines the above two data sources fusing into one model to forecast for task durations.

Our first experiment indicated that prediction models trained with weather related features perform very similar to models without the weather-related features. One possible explanation is that all the WO used in the experiments were performed indoors, thus the durations were less impacted by weather. For the rest of the experiments, we used the features from Table 1 excluding the weather-related features. The evaluation metrics were Mean Absolute Error (MAE), Mean Absolute Percentage Errors (MAPE) and the coefficient of determination R2[3] for regression algorithms.

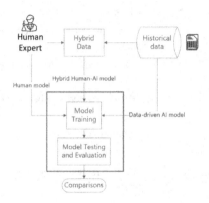

Fig. 2. Structure of the methodology used to compare three predictive models

5.1   Experiment on Hybrid Model vs. Data Only Model

Two sets of experiments with and without "planned hours", which represents human expert's input were carried out to find the impact of the human expert (HE) model, and whether historical data-driven AI model alone was effective for forecasting task durations. For each set of experiments, we ordered the data according to the "Order scheduled start date", and took 80% for training, and test on 20% of the most recent data

---

[3] These metrics are from Python Sklearn library (version 0.24.2). https://scikit-learn.org/stable/modules/generated/sklearn.metrics.mean_absolute_percentage_error.html and https://scikit-learn.org/stable/modules/generated/sklearn.metrics.r2_score.html#sklearn.metrics.r2_score.

set. We performed hyper-parameter tuning for each of the prediction algorithms, and for reach WO predictions[4].

The advantage of hybrid model is shown in Figs. 3 and 4, using a sample WO from our data set. The x-axis shows the date when the WO was performed. The y-axis shows the task duration in hours. The predicted hours are plotted for each regression model as well as the ground truth - "actual hours", which is plotted in purple color. The closer the other curves are to the actual hours, the better performing the regression model is.

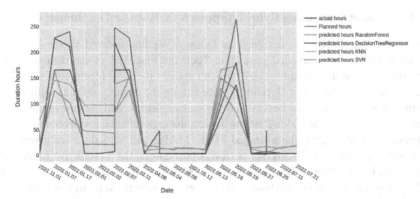

Fig. 3. Results of predictive models without human expert's input (Color figure online)

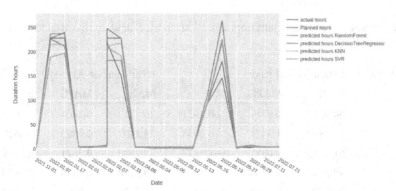

Fig. 4. Results of predictive models with human expert's input (Hybrid human-AI) (Color figure online)

Historical data only prediction models are shown on Fig. 3. Hybrid human-AI model performance is shown in Fig. 4. These experiments demonstrate that human expert's input

---

[4] As an example, we list the hyper-parameters for each algorithm for the WO in Fig. 3. The following set of parameters are used for the decision tree algorithm, DecisionTreeRegressor(criterion='friedman_mse', max_depth=12.0, max_features='auto', min_samples_leaf=4, min_samples_split=3, splitter='random'). For random forest algorithm, RandomForestRegressor(criterion='mae', max_depth=25, min_samples_split=20, n_estimators=3). For support vector regressor and KNN regressor, the default parameters are found to be the best.

is essential on guiding the forecast. Without this resource, historical only data-driven model does not capture the patterns. From the ground truth curve, we observe two work cycles of peak hours vs. minimum hours. Such pattern was not successfully modeled by historical data alone. Hybrid human-AI model (orange curve) with human experts' input is shown to perform better than human (red curve) alone and historical data model (purple curve) alone. Thus, for the remaining experiments, the human expert's input is included as a feature for predictions.

5.2   Experiment Comparing Hybrid Approach vs. Historical or Human Only

This second experiment was designed as follows. We first ordered all the historical data according to the scheduled start date. Then the first 80% of the data was used as training set, and the remaining 20% which contains the most recent data were kept as testing set. The regression algorithms were run on the training set data with hyper-parameter tuning to obtain the best models for each algorithm, and then the models were tested against the test sets. Table 2 below depicts the experimental results obtained. For each regression algorithm, the averaged MAE, MAPE and R2 values are given. The results indicate that the DT regressor performs best among the 5 predictive models for MAPE, and KNN performs better according to MAE and R2 measures. An average of 8.6% improvement from Hybrid human-AI model over human expert model is observed based on R2 evaluation metric. The Hybrid human-AI model performs better than the data-driven AI models.

Table 2.   Averaged prediction results from regression models comparison based on 13 WOs (80% Training 20% Test)

Algorithms	Data-driven AI model			Hybrid human-AI model		
	MAE	MAPE	R2	MAE	MAPE	R2
RF	54.9276	0.9911	−0.2921	26.5217	0.5404	0.7014
SVR	40.9840	1.3475	0.2058	23.1492	0.6823	0.8527
DT	61.7243	1.0387	−1.3872	25.7589	**0.3821**	0.5589
KNN	49.2387	1.5365	0.1371	**20.9750**	0.5444	**0.8940**
HE	26.2670	0.9205	0.8232	26.2670	0.9205	0.8232

RF: Random Forest; SVR: Support Vector Regression; DT: Decision Trees; KNN: K-Nearest Neighbors; HE: Human Expert

We further performed 10-fold cross validation tests per WO with the test conducted on shuffled complete data set. The results in Table 3 show the overall performance of regressors on predictions averaged from 13 WOs. For the overall 10-fold cross validation analysis, hybrid human-AI model always performs better than human expert only, and always better than historical data-driven model.

5.3   Experiment on Outliers as Human Errors

From the data set, we observed that for some WO, human expert's input as "planned hours" varied a lot from the "actual hours". As an example, a given task may have an average of 20 h as actual recorded hours, however, due to unknown reasons, the estimated hours by human expert were sometimes 2 or 3 times longer than the actual working hours. Such information was captured in the database; however, we do not have information on how the expert manually evaluated the task. Would it be due to external reasons that affected experts' judgment? Would it be typos since the numbers were written down manually? Would it be possible that an inexperienced expert performed the estimation instead of the regular skilled expert? It is shown that including highly erroneous human estimations into forecasting may cause worse performance than historical data models [26]. One approach to tackle such discrepancies is to consider outliers as human errors. We experiment on modeling data with and without human prediction outliers and compare the results obtained.

Table 3. Averaged prediction results based on 13 WOs (10-fold cross validation)

Algorithms	Data-driven AI model			Hybrid human-AI model		
	MAE	MAPE	R2	MAE	MAPE	R2
RF	54.5935	1.1133	0.2797	29.3069	0.6390	0.7048
SVR	57.2337	1.1577	0.1770	**24.1499**	0.7363	**0.9301**
DT	61.8340	0.8101	0.0188	31.8800	**0.5117**	0.5966
KNN	52.2757	1.2834	0.4486	25.3037	0.6215	0.8290
HE	33.9370	1.7540	0.9142	33.9370	1.7540	0.9142

In this experiment, we define human estimation outliers as follows. For a given WO, consider all the absolute difference between actual hours and planned hours. Data beyond a certain number of standard deviations from the mean was considered as outliers. In Table 4, we vary the limits from 1.5-sigma to 3-sigma and run the models with and without filtering any outliers. Experiments were performed on 4 regression algorithms for hybrid human-AI model, and the HE forecasting model, through hyper-parameter tuning with 5-fold cross validation and 10-fold cross validation. The experiments were performed over 13 WOs and the averaged MAPE and R2 results for each regression algorithm are shown in the Table 4 below.

Table 4. Averaged prediction results on comparing outlier threshold

	Algo.	1.5 sigma		2 sigma		2.5 sigma		3 sigma		No filter	
		5-fold	10-fold	5-fold	10-fold	5-fold	10-fold	5-fold	10-fold	5-fold	10-fold
MAPE	RF	0.64	0.58	0.63	0.59	0.69	0.61	0.79	0.72	0.91	0.78
	SVR	0.68	0.57	0.75	0.61	0.90	0.73	1.00	0.83	1.12	0.97
	DT	0.80	0.68	0.52	0.62	0.60	0.56	0.77	**0.60**	0.85	0.72
	KNN	0.96	0.93	0.97	0.93	0.95	0.98	1.00	1.02	1.08	1.00
	HE	0.84	0.84	1.09	1.09	1.22	1.22	1.42	1.42	1.75	1.75
R2	RF	0.85	0.86	0.85	0.85	0.83	0.84	0.83	0.84	0.72	0.83
	SVR	0.96	0.97	0.94	0.96	0.92	0.94	0.92	**0.94**	0.90	0.94
	DT	0.76	0.81	0.74	0.82	0.76	0.77	0.75	0.74	0.70	0.66
	KNN	0.78	0.81	0.82	0.83	0.80	0.82	0.80	0.82	0.71	0.82
	HE	0.96	0.96	0.95	0.95	0.94	0.94	0.93	0.93	0.91	0.91

Results indicate that using 2.5-sigma or 3-sigma as outlier removing criteria would further enhance the hybrid models, while models trained from 10-fold cross validation perform similar to 5-fold cross validation.

5.4 Experiment on Individual Prediction Model vs. All-in-One Model

As we stressed earlier, in the ship refit domain, historical data is very scarce. To address such frugal learning problem, in addition to creating enriched data resources, we are also interested to explore richer predictive models. Models trained with larger data sets are normally more stable and perform better. The following experiment was designed to test if a single multi-WO model would outperform WO-specific models. First, for each WO, we removed the outliers with the 3-sigma limit from the above experiment and obtained the processed data for each of the 13 WOs. We conducted a 10-fold cross validation on each of the WOs, and averaged the performance results for each algorithm. After the outlier removal, we concatenated all the WOs into one data, added an extra flag indicating the WO ID, and conducted 10-fold cross validation with the 4 algorithms. The results shown in Table 5 indicate that individual prediction models perform better than all-in-one models, and the individual models provide a more personalized model per WO.

Table 5. Averaged prediction results of comparing individual models and all-in-one model

Algorithms	Individual prediction models			All-in-one prediction model		
	MAE	MAPE	R2	MAE	MAPE	R2
RF	23.6817	0.5711	0.7986	23.9657	0.6379	0.8192
SVR	**19.2026**	0.6836	**0.9591**	22.4593	1.2904	0.9535
DT	23.7528	**0.5464**	0.8261	26.9376	0.7906	0.7932
KNN	22.3044	0.5783	0.8251	21.5941	0.6304	0.8335
HE	27.8922	1.4162	0.9330	27.8922	1.4162	0.9330

6 Conclusion

Task duration estimation is an important element of scheduling and optimization in various industries. However, work completion times vary based on endogenous and exogenous factors. The estimates for completion times are typically made by human experts and often entail assumptions and uncertainty. In this paper, we investigated the performance of hybrid human-AI model and its advantages through experimentations on a real-world maintenance data set in the ship refit domain.

Hybrid human-AI forecasting uses data augmentation to address task estimation problems by integrating both historical data and human expert data. This novel approach is designed to assist experts on task estimation and planning rather than replacing them, and is for application domains that rely on human experts for activity estimations, but contain insufficient historical data for traditional data-driven model for forecasting (thus cannot use deep learning algorithms). The basic hybrid model architecture tested herein could also be extended into a two-stage hierarchical forecasting method. In stage I, we create a model based on historical data and a separate model based on human expert forecasts. Then, in stage II we train a meta-model on the same dataset while adding in the feature list for each case the prediction of each stage I model. This meta-model would have an added flexibility to learn under which circumstances it should rely more or less on the stage I models' outputs.

Results demonstrate that although human forecasts are still an essential resource for task duration estimation, hybrid human-AI forecasting model serves as an add-on to compensate for the limitations of both human only model and historical data-driven model. This fused model can be applied to various domains such as ship refit, aerospace, drone mission planning, mining, and construction industries, to help program managers/users better estimate task durations, produce more robust resource schedules by incorporating uncertainty, thus performing more tasks, reducing resources idling time, costs, and increasing asset operational availability.

Next research goals are to collaborate with experts on integrating their inputs into the forecasting model, to use multi-criteria modeling software such as Myriad [32–35] to model experts' forecasts, and to study how to better integrate estimates from multiple experts. Issues about trust-ability of human experts and utility functions are worth considering when aggregating human expert related feature values into decisions. Process tracing [36] as a feature selection method, and data augmentation method for expert forecast modeling will be studied as ways to improve the hybrid model's performance. Active learning as an intelligent querying method with continuously updated models will also be considered to improve model accuracy through better case selection when collecting forecasts from experts [37].

References

1. Canadian Defence, Aerospace and Commercial and Civil Marine Sectors Survey (2014). https://ised-isde.canada.ca/site/shipbuilding-industrial-marine/en/shipbuilding-repair-maintenance-and-refit. Accessed 26 Apr 2023
2. Lafond, D., Couture, D., Delaney, J., Cahill, J., Corbett, C., Lamontagne, G.: Multi-objective schedule optimization for ship refit projects: toward geospatial constraints management. In: Ahram, T., Taiar, R., Groff, F. (eds.) IHIET-AI 2021. AISC, vol. 1378, pp. 662–669. Springer, Cham (2021). https://doi.org/10.1007/978-3-030-74009-2_84
3. Torres, I.C., Armas-Aguirre, J.: Technological solution to improve outpatient medical care services using routing techniques and medical appointment scheduling. In: IEEE 1st International Conference on Advanced Learning Technologies on Education & Research, pp. 1–4 (2021)
4. Yeung, W., Choi, T., Cheng, T.C.E.: Optimal scheduling of a single-supplier single-manufacturer supply chain with common due windows. IEEE Trans. Automatic Control 55(12), 2767–2777 (2010)
5. Schwindt, C., Zimmermann, J.: Handbook on Project Management and Scheduling, vol. 1. Springer, Cham (2015). https://doi.org/10.1007/978-3-319-05443-8
6. Bertrand, E.: Optimization of the naval surface ship resource-constrained project scheduling problem. Master's thesis, Dalhousie University (2020)
7. Kelley, J.E., Jr., Walker, M.R.: Critical-path planning and scheduling. Papers Presented at the 1–3 December 1959, Eastern Joint IRE-AIEE-ACM Computer Conference, pp. 160–173 (1959)
8. Malcolm, D.G., Roseboom, J.H., Clark, C.E., Fazar, W.: Application of a technique for research and development program evaluation. Oper. Res. 7(5), 646–669 (1959)
9. Węglarz, J., Józefowska, J., Mika, M., Waligóra, G.: Project scheduling with finite or infinite number of activity processing modes - a survey. Eur. J. Oper. Res. 208(3), 177–205 (2011)
10. Deblaere, F., Demeulemeester, E., Herroelen, W.: Proactive policies for the stochastic resource-constrained project scheduling problem. Eur. J. Oper. Res. 214(2), 308–316 (2011). https://doi.org/10.1016/j.ejor.2011.04.019
11. Pellerin, R., Perrier, N., Berthaut, F.: A survey of hybrid metaheuristics for the resource-constrained project scheduling problem. Eur. J. Oper. Res. 280, 395–416 (2020)
12. Van Den Eeckhout, M., Maenhout, B., Vanhoucke, M.: A heuristic procedure to solve the project staffing problem with discrete time/resource trade-offs and personnel scheduling constraints. Comput. Oper. Res. 101, 144–161 (2019)
13. Halpin, D.W.: Subjective and interactive duration estimation: discussion. Can. J. Civ. Eng. 20(4), 719–721 (1993)
14. Winkler, R.L.: Expert resolution. Manag. Sci. 32(3), 298–303 (1986)
15. Cooke, R.M., Goossens, L.H.J.: Procedures guide for structured expert judgment in accident consequence modeling. Radiat. Prot. Dosim. 90(3), 303–309 (2000)
16. Clemen, R.T., Winkler, R.L.: Combining probability distributions from experts in risk analysis. Risk Anal. 19(2), 187–203 (1999)

17. Ouchi, F.: A literature review on the use of expert opinion in probabilistic risk analysis. World Bank Policy Research Working Paper 3201 (2004)
18. Bedford, T., Cooke, R.T.: Probabilistic Risk Analysis: Foundations and Methods. Cambridge University Press, Cambridge (2001)
19. Cooke, R.M.: Experts in Uncertainty: Opinion and Subjective Probability in Science. Oxford University Press, Oxford (1991)
20. Parenté, F.J., Anderson-Parenté, J.K.: Delphi inquiry systems. Judgmental Forecasting (1987)
21. Delbecq, A., Van de Ven, A., Gusstafson, D.: Group Techniques for Program Planning, Glenview, III, Scott-Foresman (1975)
22. Mosleh, A., Bier, V.M., Apostolakis, G.: A critique of current practice for the use of expert opinions in probabilistic risk assessment. Reliab. Eng. Syst. Saf. **20**, 63–85 (1988)
23. Gennatas, E.D., et al.: Expert-augmented machine learning. Proc. Natl. Acad. Sci. **117**(9), 4571–4577 (2020)
24. Navinchandran, M., Sharp, M.E., Brundage, M.P., Sexton, T.B.: Studies to predict maintenance time duration and important factors from maintenance workorder data. In: Annual Conference of the PHM Society, vol. 11 (2019). https://doi.org/10.36001/phmconf.2019.v11 i1.792
25. Khalid, W., Albrechtsen, S., Sigsgaard, K., Mortensen, N.H., Hansen, K., Soleymani, I.: Predicting maintenance work hours in maintenance planning. J. Qual. Maintenance Eng. (2020). https://doi.org/10.1108/JQME-06-2019-0058
26. Li, J., Lafond, D.: Hybrid human-AI forecasting for task duration estimation in ship refit. In: The 8th International Online & Onsite Conference on Machine Learning, Optimization, and Data Science (2022)
27. Eisler, C., Holmes, M.: Applying automated machine learning to improve budget estimates for a naval fleet maintenance facility. In: International Conference on Pattern Recognition Applications and Methods (2021)
28. Maybury, D.: Predictive analytics for the Royal Canadian Navy Fleet Maintenance Facilities. DRDC – Centre for Operational Research and Analysis. Reference Document, DRDC-RDDC-2018-R150 December (2018)
29. McAndrew, T., Wattanachit, N., Gibson, G.C., Reich, N.G.: Aggregating predictions from experts: a review of statistical methods, experiments, and applications. Wiley Interdiscip. Rev. Comput. Stat. **13**(2), e1514 (2021)
30. Bouayed, Z., Penney, Ch.E., Sokri, A., Yazeck, T.: Estimating Maintenance Costs for Royal Canadian Navy Ships. Scientific Report DRDC-RDDC-2017-R147. https://cradpdf.drdc-rddc.gc.ca/PDFS/unc307/p805887_A1b.pdf. Accessed 26 Apr 2023
31. Syamsundar, A., Naikan, V.N.A.: Assessment of maintenance effectiveness for repairable systems: PM and CM case studies. In: Asia-Pacific International Symposium on Advanced Reliability and Maintenance Modeling (APARM), Vancouver, BC, Canada, pp. 1–5 (2020). https://doi.org/10.1109/APARM49247.2020.9209529
32. Labreuche, C., Le Huédé, F.: Myriad: a tool for preference modeling application to multi-objective optimization. In: 7th International Workshop on Preferences and Soft Constraints, Spain, 1 October (2005)
33. Le Huédé, F., Grabisch, M., Labreuche, C., et al.: Integration and propagation of a multi-criteria decision making model in constraint programming. J. Heuristics **12**, 329–346 (2006)
34. Lafond, D., Gagnon, J., Tremblay, S., Derbentseva, D., Lizotte, M.: Multi-criteria assessment of a whole-of-government planning methodology using MYRIAD. In: IEEE International Multi-disciplinary Conference on Cognitive Methods in Situation Awareness and Decision, pp. 49–55 (2015)
35. Barbaresco, F., Deltour, J.C., Desodt, G., Durand, G., Guenais, T., Labreuche, C.: Intelligent M3R radar time resources management: advanced cognition, agility & autonomy capabilities. In: International Radar Conference "Surveillance for a Safer World" (RADAR), pp. 1–6 (2009)

36. Labonté, K., et al.: Combining process tracing and policy capturing techniques for judgment analysis in an anti-submarine warfare simulation. In: 65th International Annual Meeting of the Human Factors and Ergonomics Society, pp. 1557–1561 (2021)
37. Chatelais, B., Lafond, D., Hains, A., Gagné, C.: Improving policy-capturing with active learning for real-time decision support. In: Ahram, T., Karwowski, W., Vergnano, A., Leali, F., Taiar, R. (eds.) IHSI 2020. AISC, vol. 1131, pp. 177–182. Springer, Cham (2020). https://doi.org/10.1007/978-3-030-39512-4_28
38. Pedregosa, F., et al.: Scikit-learn: machine learning in Python. J. Mach. Learn. Res. **12**, 2825–2830 (2011)

Exploring Image Transformations with Diffusion Models: A Survey of Applications and Implementation Code

Silvia Arellano, Beatriz Otero$^{(\boxtimes)}$ (ID), and Ruben Tous (ID)

Universitat Politècnica de Catalunya, Barcelona, Spain
silvia.arellano@estudiantat.upc.edu, {beatriz.otero,ruben.tous}@upc.edu

Abstract. Diffusion Models have become increasingly popular in recent years and their applications span a wide range of fields. This survey focuses on the use of diffusion models in computer vision, specially in the branch of image transformations. The objective of this survey is to provide an overview of state-of-the-art applications of diffusion models in image transformations, including image inpainting, super-resolution, restoration, translation, and editing. This survey presents a selection of notable papers and repositories including practical applications of diffusion models for image transformations. The applications are presented in a practical and concise manner, facilitating the understanding of concepts behind diffusion models and how they function. Additionally, it includes a curated collection of GitHub repositories featuring popular examples of these subjects.

Keywords: Diffusion Models · Image Transformations · Applications · Computer Vision · Inpainting · Restoration · Translation · Editing · Super-resolution

1 Introduction

Denoising Diffusion Probabilistic Models (DDPMs), also known as diffusion models, were introduced by Ho et al. [3]. A DDPM is a generative model used to produce samples that match a given dataset within a finite time. The diffusion model consists of two processes: a forward diffusion process that gradually adds Gaussian noise during several timesteps to the input data until the signal has become completely noisy, and a reverse denoising process that learns to generate data by removing the noise added in the forward process.

The diffusion process uses small amounts of Gaussian noise, allowing for a simple neural network parameterization. Diffusion models are known for their efficiency in training and their ability to capture the full diversity of the data distribution. However, generating new data samples with a diffusion model requires the model to be forward-passed through the diffusion process several times.

To gain a more comprehensive understanding of diffusion models, there are multiple accessible resources available that provide in-depth explanations of diffusion models in a simplified manner [26,28].

G. Nicosia et al. (Eds.): LOD 2023, LNCS 14506, pp. 19–33, 2024.
https://doi.org/10.1007/978-3-031-53966-4_2

The popularity of diffusion models has risen significantly in the 2020s, and their applications include a wide amount of subjects, such is that there is still not a clear classification of them. However, this survey covers a line of research of diffusion models in the field of computer vision, according to the classification stablished by Yang et al. [30]. This survey aims to gather information on state-of-the-art (SOTA) applications of diffusion models in the field of computer vision, specifically in image transformations. It includes highlights of research in image inpainting, super-resolution, restoration, translation and editing. Figure 1 shows a synthesis of the content of this survey along with the applications and codes implemented in each area of the image transformation considered.

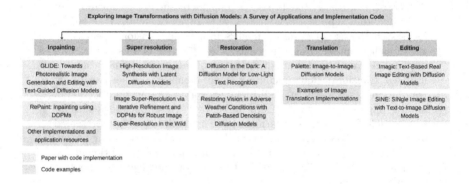

Fig. 1. Breakdown of the survey's content.

In this way, the main contributions of this survey are the following:

– A comprehensive review of SOTA diffusion models in the field of image transformations, covering relevant papers and providing an overview of the current research landscape.
– A study of diverse models and applications developed by researchers and amateurs, highlighting the potential of diffusion models.
– A practical and concise introduction to the concepts of diffusion models, to facilitate a better understanding of the topic.

The remainder of this paper is organized as follows. Section 2 provides the main methods and resources for applications in inpainting. Section 3 presents relevant works in image super-resolution. Section 4 shows methods and applications in restoration. Section 5 illustrates the applications of translation, and Sect. 6 describes the methods and applications of editing. Additionally, each section provides a collection of popular GitHub repositories with examples related to the respective subject. Finally, the main conclusions are presented in Sect. 7.

2 Image Inpainting

Image inpainting is the task of filling in masked regions of an image with new content, either to restore missing or corrupted parts of the image, or to remove unwanted content from the image. The objective of the inpainting process is to create a completed image that looks coherent and balanced, with the new content seamlessly integrated into the original image. Initially, this generated content is uniquely based on the rest of the pixels of the image. However, recent approaches to image inpainting involve text-conditional image synthesis. In such cases, the model uses its understanding of the image context and incorporates external knowledge or prior assumptions to generate new content.

This section features two prominent papers in the field of image inpainting: Lugmayr's method [10] and Nichol's approach [14], both of which achieve SOTA results. Additionally, it provides a set of resources to better understand the code behind image inpainting approaches.

2.1 RePaint: Inpainting Using DDPMs

In this paper, Lugmayr et al. [10] argue that current inpainting methods are limited by their training data, which typically consists of images with masks from a specific distribution. This can negatively impact the model's ability to generate accurate results for new, previously unseen mask types. To overcome this issue, the authors propose a technique that avoids mask-specific training.

Their approach uses an unconditionally trained DDPM, which conditions the generation process by sampling from the given pixels during the reverse diffusion iterations. By not being mask-conditioned, the model can adapt to any type of mask and achieve a better understanding of the image. Unlike typical DDPMs, this approach goes both forward and backward during all the inference process, ensuring that the image is both texturally and semantically coherent.

One characteristic of DDPMs is that each denoising step depends only on the previous step. This paper takes advantage of this, and during each denoising step, the model combines the unmasked parts of the noisy input image from that timestep with what the denoised sample has generated for the masked pixels, in order to generate the denoised output of that specific step. For better understanding, find this process illustrated in Fig. 2.

The authors note that there may be issues with harmonization when integrating the inpainted section with the rest of the image because the known pixels are sampled without considering the generated parts of the image. To address this problem, they propose a technique called resampling, where the output of the previous step is diffused back to the current step. This approach allows the model to have new information at each step, which helps move towards a more accurate output while preserving information from the known region. This method has been shown to be more effective than reducing the sampling steps.

This method is compared to other SOTA inpainting approaches using four different masks with varying characteristics and evaluated using the LPIPS metric. The RePaint method outperforms the other methods. The results show that

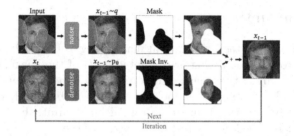

Fig. 2. Diagram of the modified denoising process used in RePainting [10].

when the inference image has a large mask, RePaint can produce a semantically more meaningful image compared to the other methods, due to the fact that it was trained without masks. However, this image may differ significantly from the ground truth.

It is important to note that as RePaint aims to generate data within the distribution of the training set, there may be biases towards certain factors such as ethnicity or age, which could be perceivable in the output.

More information about the implementation of RePaint and examples of this application can be found on their GitHub repository [11].

2.2 GLIDE: Towards Photorealistic Image Generation and Editing with Text-Guided Diffusion Models

This paper, written by Nichol et al. [14], presents a new method for photorealistic image generation and editing using text-guided diffusion models. The code used can be found in the project's official GitHub repository [15]. The main contribution of the paper is the development of a new training procedure for diffusion models that enables the use of text prompts to guide image synthesis. The proposed method, called GLIDE, achieves SOTA results in image synthesis and editing tasks, such as super-resolution, inpainting, and style transfer.

This approach differs from others in the field of text-to-image generation due to the use of the classifier-free guidance for achieving photorealism, a technique introduced by Ho and Salimans [4]. To understand this technique, we first have to refer to classifier guidance, that uses a classifier to guide the diffusion model to generate images of a desired class. Then, the classifier-free guidance aims to do the same but only using a pure generative model, without the need to use a classifier. The classifier-free guidance has the advantage of being able to use its own knowledge instead of relying on a separate classifier model to guide the generation process. This is useful when conditioning on information that is difficult to predict with a classifier, like it's this case with text. Apart from the classifier-free guidance, this paper also studies how does the system work using CLIP guidance instead. However, after making the study, the first method gives better results, producing realistic outputs with new objects, and correct shadowing and lighting.

To tackle the problem of text-conditional image synthesis, they apply guided diffusion. The text prompts are converted into latent vectors that are used to guide the diffusion process, allowing the model to generate images that represent the text input.

To improve the quality of the results, the model's architecture is modified by adding a second set of RGB channels containing the guidance image and a mask channel. During training, arbitrary parts of the input images are erased, and the remaining parts are used as the input for the model. This enables the model to learn to fill in the missing regions based on the remaining image content and the guidance provided by the extra RGB channels and mask. To avoid deepfakes and unwanted social biases, the training datasets have been filtered and reviewed.

The final results, as seen in Fig. 3 are realistic to humans, and seem to reflect properly the statements of the input. It obtains a competitive Fréchet inception distance (FID) score, and it achieves SOTA performance in terms of image quality and diversity.

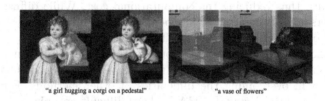

"a girl hugging a corgi on a pedestal!"   "a vase of flowers"

Fig. 3. GLIDE output given the written prompt and the masked source image [14].

2.3   Other Implementation and Application Resources

There are various image inpainting tools available on the internet, including the popular one provided by Runwayml in Hugging Face [21]. The tool provides an inpainting pipeline using Stable Diffusion and the Hugging Face library and can be accessed via a Google Colab file.

Another tool available on the web is *Paint by Example: Exemplar-based Image Editing with Diffusion Models* [1,29]. This code aims to integrate a reference image into a drawn mask of a source image. Unlike simply pasting the reference image, the code adapts it to fit the mask correctly for a more realistic result.

3   Image Super-Resolution

Image super-resolution is a set of techniques used to enhance the resolution of an image, video or other digital content beyond its original quality. It involves taking a low-resolution image and using algorithms to generate additional pixels, resulting in an output with higher-resolution and more detail. Super-resolution has important applications in fields such as medical imaging, remote sensing, and digital entertainment amongst others.

This section of the survey examines a highly regarded paper by Rombach et al. [18] that achieves SOTA results in image synthesis. Additionally, it presents a recent model by Saharia et al. [22] that performs blind super-resolution with outstanding results, even on data that was not part of its original training set.

3.1   High-Resolution Image Synthesis with Latent Diffusion Models

The paper written by Rombach et al. [18] proposes a new approach for generating high-quality images using Latent Diffusion Models (LDMs). They also provide a GitHub repository with pretrained models, demos and examples [19]. The goal is to find a more computationally efficient representation that is still perceptually equivalent. The approach involves doing the learning and training phase in the latent space and training the autoencoder only once.

One of the main challenges with text-to-image models is the requirement for large batch sizes, which is computationally expensive and can restrict access to these models. The proposed solution is to ignore bits allocated to nearly imperceptible details, thus reducing the computational cost. While diffusion models already undersample steps that aren't perceptually relevant, they still need to be evaluated on a pixel level, which can be expensive.

This method stands out from previous approaches as it does not require a delicate weighting of reconstruction and generative abilities. LDMs have proven to be effective in overcoming the difficulties of other methods. For instance, Generative Adversarial Networks (GANs) have difficult optimization processes, whereas LDMs work in a compressed latent space with lower dimensionality, making them quicker without compromising the image synthesis quality. DDPMs produce great results of synthesis quality, but when evaluated and optimized in pixel space, their cost is still high. On the other hand, LDMs have been found to be effective in overcoming this limitation. Additionally, when it comes to autoregressive models (ARM), achieving less compression can result in a high computational cost. LDMs work in higher dimensional latent spaces to avoid this problem.

The proposed approach involves training an autoencoder to learn the semantic compression of data, with conditioning implemented using a semantic map, low-resolution images, or text prompts. The network includes self-attention, cross-attention, and feed-forward layers.

Several important results can be inferred from this study. First, it was found that small downsampling factors for LDMs lead to a slow training process, while overly large values have limitations in terms of image quality. When concatenating the mask and the mask image to the noisy representation, the proposed method achieved the SOTA FID score among inpainting models in 512×512 resolution. Applying classifier-free diffusion guidance was also found to greatly boost sample quality, achieving the SOTA of the FID score. Finally, when using this technique for semantic synthesis, such as using a segmentation mask as conditioning, it was found to be effective on larger resolutions than it was trained on.

3.2 Image Super-Resolution via Iterative Refinement and DDPMs for Robust Image Super-Resolution in the Wild

Early work in the super-resolution field focused on learning a mapping from low dimensional images to higher-resolution outputs. However, this approach did not add any new information. Therefore, Saharia et al. [25] present an alternative approach by treating super-resolution as a conditional generation task, where the output is conditioned on the low-resolution input image. They achieve this by modifying the U-Net architecture. An unofficial implementation can be found on GitHub [8].

The main contributions of this paper are the use of iterative refinement steps to reduce the computational power needed and the superiority of their approach over regression-based models and other approaches with GANs. This paper also inspires the use of unconditional image synthesis.

The model workflow starts by upsampling the input image to the target resolution using bicubic interpolation, which is then concatenated with the noisy image. Next, the image is passed through a modified U-Net, which has learned how to denoise the image.

To make this model work, the authors have replaced the DDPMs residual blocks with the ones used in BigGAN, with appropriate modifications. This helps achieve more detailed and refined textures. During training, noise is added in a scheduled way to avoid overfitting and achieve acceptable output. This also helps in selecting the best noise scheduling during inference, requiring only one-time training of the model.

However, it has been demonstrated that the model does not perform as expected when it is presented with data that significantly differs from the data it was trained on. This type of data is referred to as out-of-distribution (OOD) data. Therefore, [22] presents a new improved system, SR3+, based on the previous one.

High-resolution images are downsampled to obtain the training low-resolution inputs. During this resizing process, some types of degradation like blur or JPEG compression are applied, as it helps achieve better results in OOD data. Once these alterations are applied, and the output has the target resolution, noise-conditioning augmentation is applied, adding noise to the image. This helps the model learn to manage images with different noise levels.

The model is tested on the task of blind super-resolution with a 4× magnification factor. The results are compared with SR3 [25] and Real-ESRGAN [27]. The SR3+ approach generates textures that are more realistic while avoiding significant oversmoothing or saturation. However, for certain image types where precise high-frequency details like text are crucial, it may not perform as well. Considering metrics, SR3+ outperforms the other blind super-resolution methods in FID, but obtains worse scores in SSIM and PSNR. All in all, SR3+ achieves a SOTA performance for blind super-resolution in OOD data.

4   Image Restoration

Image restoration is the process of recovering an image from a degraded version, which may contain noise, artifacts, or blur. Diffusion models can be used for image restoration by smoothing out noise or artifacts while preserving important image features. This section will focus on two studies that address specific image restoration tasks: improving the quality of dark text images and enhancing images that have been degraded by adverse weather conditions.

In this section, we present two recent and innovative approaches in the field of image restoration. Firstly, the paper by Nguyen et al. [12] focuses on recovering text from dark images. Secondly, the study by Özdenizci and Legenstein [17] presents a technique for removing weather-induced artifacts from images of all sizes.

4.1   Diffusion in the Dark: A Diffusion Model for Low-Light Text Recognition

This paper, written by Nguyen et al. [12], proposes a method called Diffusion in the Dark (DiD) to reconstruct images with poor lighting, in order to make the text on them legible. The authors claim that addressing this issue is crucial for advancing towards the future of automation, as real-world images are subject to various types of artifacts that can cause the loss of high-frequency information. High frequencies in images typically represent fine details and sharp edges that can be important for certain image analysis tasks.

DiD surpasses other SOTA approaches in several ways. It has been demonstrated to achieve accurate results in most dark and noisy environments. Moreover, it reduces computational costs and training time, as it operates on patches rather than the whole image.

However, there are still some issues that need to be addressed. For instance, DiD only works with resolutions that are multiples of the patch size, and if this is not the case, interpolation methods need to be used, which can affect negatively the quality of the output image. Additionally, the method can take a long time to produce results, which may not be suitable for some applications.

This method trains on images of various resolutions, dividing them into patches of 32×32 px (which is called *patchifying*). The model uses three conditioning inputs: low-light, well-light, and low-estimate conditions. Knowing this, the image is upsampled to achieve a higher-resolution and Gaussian noise is added. Then, the patch is passed through a DDPM, resulting in a reconstructed version. The DDPM is based on a U-Net architecture and the loss used is a linear combination of L2 loss (also known as Mean Squared Error) and LPIPS. Once all the patches have been processed, they are "stitched" together to obtain the resulting image. Some example results can be seen in Fig. 4.

The inference phase begins by computing the conditioning input values. Once these values are known, the reverse denoising process is applied until the final resolution of 256×256 px is achieved. Due to the patching procedure, the final result may have some inconsistencies with respect to exposure levels and white balancing. To address this issue, a technique called Iterative Latent Variable Refinement is applied.

At the time of writing this paper, the code for this project was not yet available, but the authors have indicated that it will be made accessible through the project's webpage [13].

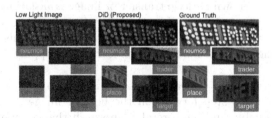

Fig. 4. Results of applying DiD to the low light images [12].

4.2 Restoring Vision in Adverse Weather Conditions with Patch-Based Denoising Diffusion Models

The paper authored by Özdenizci and Legenstein [17] presents a novel image restoration approach that is independent of image size. It achieves SOTA performance on both weather-specific and multi-weather image restoration tasks on benchmark datasets for snow, raindrops, rain, and haze. The code corresponding to this paper is available in its GitHub repository [16]. This method divides the image into patches to improve the efficiency of the process and applies the diffusion process to each one of them. Finally, it blends them together again.

The model starts by dividing the image into squared patches of a fixed size. The patch size and the stride between patches can be changed. It is important to bear in mind that a smaller stride will improve the smoothness of the result, but it will also increase the computational power required.

These patches overlap with each other. As there is no unique solution in such situations, different solutions could have been chosen in the overlapping regions. Therefore, to ensure consistency, the denoising reverse sampling process is guided towards producing smooth outputs across neighboring patches, which helps to improve the overall quality and fidelity of the results.

The diffusion model used for this approach is based on a U-Net architecture and includes techniques such as group normalization, self-attention blocks, and sinusoidal positional encoding to share parameters across time. Additionally, the image patches of the weather-affected image and the ground truth are concatenated channel-wise to provide more information to the network.

This approach restores the image generating the most likely background without being told which weather issue was going on. The experimental results demonstrate that the developed model has the capability to learn the data distribution for various types of adverse weather corruption tasks and outperforms other state-of-the-art approaches. The future work on this paper is teaching the model how to generalize to other unseen scenarios (like low-lighting) and reducing the time of inference.

5   Image Translation

Image transfer, also known as style transfer or image translation, is a technique that involves converting an image from one domain to another while preserving its content and applying the visual style of another image. This process can be accomplished using diffusion models, which learn the mapping between the two domains.

One of most renowned papers in image translation is Palette [23], which represents a significant step forward in the field of image editing by providing a more controllable and interpretable approach than previous methods. It also provides a versatile framework for diverse image editing tasks, including colorization, inpainting, and style transfer.

Moreover, researchers and other individuals interested in the field have developed various innovative versions of image transfer using diffusion models. Examples of these can be found in Sect. 5.2.

5.1   Palette: Image-to-Image Diffusion Models

The paper written by Saharia et al. [23] introduces Palette, a multi-task model trained for colorization, inpainting, uncropping, and JPEG restoration. However, this part will focus on Palette's colorization capabilities.

Palette is a conditional diffusion model, composed by a U-Net based on previous works but with several modifications: it isn't class conditioned, and it uses the grayscale source image as the conditioning factor using concatenation. Some of the remarked works in image colorization work in the YCbCr or LAB color space. However, Palette works in RGB to maintain consistency with the rest of the task it can perform.

This model outperforms the other SOTA approaches, achieving a fool rate of almost 50% and high scores for IS, Classification Accuracy (CA), and FID, indicating its output is nearly identical to the ground truth image. Some example results can be seen in Fig. 5.

Several unofficial GitHub repositories have implemented Palette. One particularly popular implementation is by Jiang and Belousov [9].

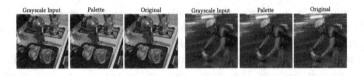

Fig. 5. Palette's colorization output given the grayscale input [23].

5.2   Examples of Image Translation Implementations

Diffusion models on the internet have gained popularity for their ability to transform users' photos into cartoons or famous paintings. These models are typically fine-tuned versions of Stable Diffusion, trained with a dataset of images with the desired style to be transferred. Some of these style-transfer approaches work with text prompts.

For instance, Mackay proposed some Stable Diffusion fine-tuned versions to turn well-known people into characters from diverse films. This approach uses Dreambooth [20], a deep learning generation model that fine-tunes existing text-to-image models. Mackay's image transfer models [2] can style people into characters from *Cats, the Musical* or *Tron: Legacy*.

Pinkney [6] developed a popular model that enables users to convert a famous person's photo into a Pokémon. The tool uses a fine-tuned Stable Diffusion model and exponential moving average (EMA) weights to avoid overfitting and control the amount of tuning.

6   Image Editing

Image editing refers to the process of modifying an image to change its appearance or content. Diffusion models can be used for image editing by enabling the user to manipulate the image in a controlled manner. In this section, we present two recent papers which are becoming a reference in the field of image editing. The first paper we selected is the recent research from Zhang et al. [31], which presents an innovative approach to single-image editing that requires only an image and a text prompt. The second paper considered is the study by Kawar and Zada et al. [7], which stands out for its ability to overcome significant limitations in the field of image editing.

6.1   SINE: SINgle Image Editing with Text-to-Image Diffusion Models

In their paper, Zhang et al. [31] present a novel approach to single-image editing that allows for changes to be introduced using only the source image and a text prompt that explains the desired output image. More information about the code of the project can be found in the project's webpage [32]. The key advantage of this approach over others in the field is its ability to avoid issues associated with overfitting while providing precise control over both the content and geometry of the edited image. Furthermore, the method is capable of producing multiple edits simultaneously. The pillar of this method are Latent Diffusion Models (LDMs), which are a specific class of DDPMs.

This approach uses two techniques: test-time model-based classifier-free guidance and patch-based fine-tuning. The first technique implies that the model generates guidance information that can be directly used in the editing process instead of relying on a separate classifier. The second technique is a way to further train a pretrained model on a specific dataset to improve its performance in that domain. However, the latter technique aims to do so using smaller parts of the image (patches) rather than using the whole image at once.

SINE first takes the source image and randomly crops it into patches, which are then encoded into corresponding latent codes that represent the source image, using a pre-trained encoder.

The next step is fine-tuning the model. For this, a denoising model that takes three inputs is used: a noisy latent code (which results from adding Gaussian Noise to the latent code obtained previously), a language condition that describes the target domain, and a positional embedding for the area where the noisy latent code is obtained.

During sampling, the method takes additional language guidance about the target domain to edit the image, helping the model better understand the kind of image that should be generated. It also samples a noisy latent code with the desired output resolution and uses a pretrained language encoder to "condition" the generation process to produce a more accurate output.

In addition, the method employs the diffusion model that was fine-tuned before, taking as input the noisy latent code, a language conditioning, and a positional embedding for the entire image. The method uses a linear combination of the scores calculated by each model for a defined number of steps, and then inference is done only on the pretrained model.

The authors of the paper state that SINE can be applied over different existing text-to-image frameworks. However, the evaluation has been done using Stable Diffusion [18].

This method has demonstrated the ability to preserve the geometry and important features of an image while accurately editing the source image according to the input text prompt. Furthermore, it has proven to be useful in other specific tasks, such as face manipulation, content removal, style generation, and style transfer. All of these results make SINE a SOTA approach in image editing.

6.2   Imagic: Text-Based Real Image Editing with Diffusion Models

Imagic, an approach developed by Kawar and Zada et al. [7], presents a method to edit images using complex text prompts and diffusion models. Imagic overcomes limitations of other image editing techniques, such as being restricted to certain types of edits or images. Furthermore, it does not require additional inputs like masks to indicate where to apply the modification. Instead, it only needs the source image and a text prompt indicating the desired edit.

The Imagic process has three main steps. First, the text prompt is encoded, obtaining an embedding (eTGT) which is optimized with the aim to recover the original image (eOPT). Next, a pre-trained generative model is fine-tuned to recover the input image accurately given the text embedding eOPT and the input image with noise. Finally, a noisy image and an interpolation of both text embeddings are passed through a fine-tuned diffusion process to obtain the modified image. The interpolation step ensures fidelity between the input image and the text prompt. The Imagic authors tested two SOTA generative models for the second step of the process: Stable Diffusion [18] and Imagen [24].

Imagic produces an output image that preserves the background, important structure, and maximum detail of the original image, while accurately reflecting the desired modification indicated by the text prompt, as it can be seen in Fig. 6. This method outperforms other techniques in the field, although there is still room for improvement. Some limitations include imperceptible edits and changes to zoom or camera angle that can alter the viewer's perception. These limitations can be addressed by optimizing text embeddings differently. Additionally, it is important to avoid societal biases in text-generative models in future implementations.

An unofficial implementation is available in Pinkey's GitHub repository [5].

Fig. 6. Imagic's edited output given the written prompt and the input image [7].

7 Conclusions

This survey describes the most recent and notable research on the use of diffusion models applied in the field of computer vision for image transformation. The article includes applications in image processing related to: inpainting, super-resolution, restoration, translation and editing. The field of computer vision is rapidly advancing, and new methods and techniques for image transformation are being developed every day. While the results of the commented approaches are impressive, there is still more work to be done in the field, particularly to address unwanted social biases. Unwanted social biases in computer vision can have harmful consequences, such as perpetuating discrimination and bias in decision-making processes.

Acknowledgements. This work is partially supported by the Spanish Ministry of Science and Innovation under contract PID2019-107255GB and PID2021-124463OB-IOO, by the Generalitat de Catalunya under grants 2021-SGR-00478 and 2021-SGR-00326. Finally, the research leading to these results also has received funding from the European Union's Horizon 2020 research and innovation programme under the HORIZON-EU VITAMIN-V (101093062) project.

References

1. Yang, B., et al.: Paint by Example: Exemplar-Based Image Editing with Diffusion Models. GitHub repository. https://github.com/Fantasy-Studio/Paint-by-Example. Accessed 20 Apr 2023
2. Mackay, D.: Dallin Mackay's Hugging Face repository. https://huggingface.co/dallinmackay. Accessed 22 Apr 2023
3. Ho, J., Jain, A., Abbeel, P.: Denoising Diffusion Probabilistic Models (2020). https://doi.org/10.48550/arXiv.2006.11239
4. Ho, J., Salimans, T.: Classifier-free diffusion guidance. arXiv preprint arXiv:2207.12598 (2022)
5. Pinkney, J.: Implementation of Imagic: Text-Based Real Image Editing with Diffusion Models using Stable Diffusion. https://github.com/justinpinkney/stable-diffusion/blob/main/notebooks/imagic.ipynb. Accessed 22 Apr 2023
6. Pinkney, J.: Text to Pokemon Generator. https://www.justinpinkney.com/pokemon-generator/. Accessed 22 Apr 2023
7. Kawar, B., et al.: Imagic: text-based real image editing with diffusion models. In: Conference on Computer Vision and Pattern Recognition 2023 (2023). https://doi.org/10.48550/arXiv.2210.09276
8. Jiang, L.: Image Super-Resolution via Iterative Refinement. GitHub repository. https://github.com/Janspiry/Image-Super-Resolution-via-Iterative-Refinement. Accessed 20 Apr 2023
9. Jiang, L., Belousov, Y.: Palette: Image-to-Image Diffusion Models. GitHub repository. https://github.com/Janspiry/Palette-Image-to-Image-Diffusion-Models. Accessed 22 Apr 2023
10. Lugmayr, A., Danelljan, M., Romero, A., Yu, F., Timofte, R., Van Gool, L.: Repaint: inpainting using denoising diffusion probabilistic models. In: Proceedings of the IEEE/CVF Conference on Computer Vision and Pattern Recognition, pp. 11461–11471 (2022). https://doi.org/10.48550/arXiv.2201.09865
11. Lugmayr, A., Danelljan, M., Romero, A., Yu, F., Timofte, R., Van Gool, L.: Repaint. GitHub repository. https://github.com/andreas128/RePaint. Accessed 20 Apr 2023
12. Nguyen, C.M., Chan, E.R., Bergman, A.W., Wetzstein, G.: Diffusion in the Dark: A Diffusion Model for Low-Light Text Recognition (2023). https://doi.org/10.48550/arXiv.2303.04291
13. Nguyen, C.M., Chan, E.R., Bergman, A.W., Wetzstein, G.: Diffusion in the Dark: A Diffusion Model for Low-Light Text Recognition. Project web. https://ccnguyen.github.io/diffusion-in-the-dark/. Accessed 21 Apr 2023
14. Nichol, A., et al.: Glide: towards photorealistic image generation and editing with text-guided diffusion models. arXiv preprint arXiv:2112.10741 (2021)
15. Nichol, A., et al.: GLIDE. GitHub repository. https://github.com/openai/glide-text2im. Accessed 25 Apr 2023
16. Özdenizci, O., Legenstein, R.: Restoring vision in adverse weather conditions with patch-based denoising diffusion models. GitHub repository. https://github.com/IGITUGraz/WeatherDiffusion. Accessed 21 Apr 2023
17. Özdenizci, O., Legenstein, R.: Restoring vision in adverse weather conditions with patch-based denoising diffusion models. IEEE Trans. Pattern Anal. Mach. Intell. 1–12 (2023). https://doi.org/10.1109/TPAMI.2023.3238179
18. Rombach, R., Blattmann, A., Lorenz, D., Esser, P., Ommer, B.: High-resolution image synthesis with latent diffusion models. In: Proceedings of the IEEE/CVF

Conference on Computer Vision and Pattern Recognition, pp. 10684–10695 (2022). https://doi.org/10.48550/arXiv.2112.10752

19. Rombach, R., Blattmann, A., Lorenz, D., Esser, P., Ommer, B.: High-resolution image synthesis with latent diffusion models. GitHub repository. https://github.com/CompVis/latent-diffusion. Accessed 20 Apr 2023

20. Ruiz, N., Li, Y., Jampani, V., Pritch, Y., Rubinstein, M., Aberman, K.: Dream-Booth: fine tuning text-to-image diffusion models for subject-driven generation. arXiv preprint arXiv:2208.12242 (2022)

21. Runwayml: Stable-Diffusion-Inpainting. https://huggingface.co/runwayml/stable-diffusion-inpainting. Accessed 20 Apr 2023

22. Sahak, H., Watson, D., Saharia, C., Fleet, D.: Denoising Diffusion Probabilistic Models for Robust Image Super-Resolution in the Wild (2023). https://doi.org/10.48550/arXiv.2302.07864

23. Saharia, C., et al.: Palette: image-to-image diffusion models. In: ACM SIGGRAPH 2022 Conference Proceedings, pp. 1–10 (2022). https://doi.org/10.48550/arXiv.2111.05826

24. Saharia, C., et al.: Photorealistic text-to-image diffusion models with deep language understanding. In: Advances in Neural Information Processing Systems, vol. 35, pp. 36479–36494 (2022). https://doi.org/10.48550/arXiv.2205.11487

25. Saharia, C., Ho, J., Chan, W., Salimans, T., Fleet, D.J., Norouzi, M.: Image super-resolution via iterative refinement. IEEE Trans. Pattern Anal. Mach. Intell. (2022). https://doi.org/10.1109/TPAMI.2022.3204461

26. Seff, A.: What are Diffusion Models? (2022). https://www.youtube.com/watch?v=fbLgFrlTnGU

27. Wang, X., Xie, L., Dong, C., Shan, Y.: Real-ESRGAN: Training Real-World Blind Super-Resolution with Pure Synthetic Data (2021)

28. Weng, L.: What are diffusion models? lilianweng.github.io (2021). https://lilianweng.github.io/posts/2021-07-11-diffusion-models/

29. Yang, B., et al.: Paint by Example: Exemplar-Based Image Editing with Diffusion Models (2022). https://doi.org/10.48550/arXiv.2211.13227

30. Yang, L., et al.: Diffusion models: a comprehensive survey of methods and applications (2022). https://doi.org/10.48550/arXiv.2209.00796

31. Zhang, Z., Han, L., Ghosh, A., Metaxas, D., Ren, J.: SINE: SINgle Image Editing with Text-to-Image Diffusion Models. arXiv preprint arXiv:2212.04489 (2022)

32. Zhang, Z., Han, L., Ghosh, A., Metaxas, D., Ren, J.: SINE: SINgle Image Editing with Text-to-Image Diffusion Models. https://zhang-zx.github.io/SINE/. Accessed 25 Apr 2023

Geolocation Risk Scores for Credit Scoring Models

Erdem Ünal[1], Uğur Aydın[1], Murat Koraş[1], Barış Akgün[2],
and Mehmet Gönen[2(✉)] (iD)

[1] QNB Finansbank R&D Center, İstanbul, Turkey
{erdem.unal,ugur.aydin,murat.koras}@qnbfinansbank.com
[2] Koç University, İstanbul, Turkey
{baakgun,mehmetgonen}@ku.edu.tr

Abstract. Customer location is considered as one of the most informative demographic data for predictive modeling. It has been widely used in various sectors including finance. Commercial banks use this information in the evaluation of their credit scoring systems. Generally, customer city and district are used as demographic features. Even if these features are quite informative, they are not fully capable of capturing socio-economical heterogeneity of customers within cities or districts. In this study, we introduced a micro-region approach alternative to this district or city approach. We created features based on characteristics of micro-regions and developed predictive credit risk models. Since models only used micro-region specific data, we were able to apply it to all possible locations and calculate risk scores of each micro-region. We showed their positive contribution to our regular credit risk models.

Keywords: Geolocation models · Micro-regions · Credit scoring models

1 Introduction

Credit risk models are advanced tools to evaluate the creditworthiness of credit applicants. They show the probability of default (non-payment of debt) of customers so that banks utilize this info at credit evaluation processes [1]. The predictive power of credit risk models is very crucial for the banks. Even a slight increase at the model power can help banks to decrease the amount of credit loss considerably. Therefore, banks want to make sure that they take full advantage of available data.

In this study, we focused on location data of customers. Instead of using high-level features such as city or district, we created features at a very granular level (micro-region) so that our new feature set was able to capture the heterogeneity within the large regions. By leveraging these feature sets, we developed two location-based credit risk models; one for customer residence address and one for customer work address. Then, we applied these models to all geolocations

G. Nicosia et al. (Eds.): LOD 2023, LNCS 14506, pp. 34–44, 2024.
https://doi.org/10.1007/978-3-031-53966-4_3

for which we can collect required data to calculate features. So, we got residence and work address credit risk predictions of all micro-regions, which we call risk level scores of each geolocation.

As a result, we have prepared and made available a table in lookup format which contains these two risk level scores. So, these scores can be used in all model development processes where customer geolocations are available. We have evaluated the effectiveness of the scores by expert validation and performance contribution to a recently developed credit application risk model. Since we achieved satisfactory results, we decided to use this output in our future models.

1.1 Related Work

Geodemography has a broad range of applications, particularly in the retail sector. For instance, a study was conducted to examine the demand location, customer residential features, and population demographics, and based on this analysis, suitable retail locations were identified [2].

Domschke and Drexl [3] identified facilities and the characteristics of their customer base as two of the four crucial elements in location problems. Moreover, Geographic Information Systems (GIS) have been widely utilized in the literature on facility location problems. For instance, in a research project focusing on ambulance placement, GIS was utilized to ensure that emergency medical services could reach accident scenes rapidly [4]. Similarly, a GIS-based spatial model was employed in a study on bank-branch location problems [5]. Another project investigating geo-marketing revealed that the socio-demographic features of a supermarket's trading region have an impact on firms' location tactics [6].

It is not uncommon to use some geographical features, such as customer city, in the credit risk evaluation systems [7,8]. However, most of the time, these features are not granular enough to represent the actual demographics of customers. For example, in Turkey, the population of big cities are very heterogeneous so that socio-economic status changes by district, by town and even by street. There are some famous places where only roads divide ultra-rich and poor populations [9,10]. Therefore, it is required to use more granular data to utilize customers' info.

1.2 Our Contributions

Using cities or districts of customers is not enough to capture this heterogeneity, so that credit scoring models cannot utilize the customer location data well. To overcome this problem, we worked with very granular data, i.e., geolocation data (latitude & longitude). Working with such granular data has some downsides. For example, generally there are not enough data points to calculate a statistic with high confidence. Similarly, doing interpolations or extrapolations where data is not available is not much reasonable. Therefore, we divided the map of Turkey into micro-regions with a length of 100 m. Then, we calculated region-centric features. These features include not only geodemographics such as child rates, education level, socio-economic category, voting rates, immigration rates but also

POIs (point of interest) and banking data such as average credit score, wealth of customers who live in those micro-regions. Then, we developed two credit risk models by using only these region-centric features (no individual data).

To the best of our knowledge, in credit scoring, this is the first work combining these alternative data sources at a micro-region level.

2   Approach

We planned to develop two separate supervised machine learning models, one for customer residence address and one for customer work address. So that, these models consist of features related with geolocations rather than customer demographics, financial indicators, etc. Models predict the payment performances of credit customers. Although models were developed using customer-based data, no customer specific data was used, so we could apply these models to all geolocations in Turkey. At the end, we had a data set that shows payment performance predictions of all geolocations. We also used these predictions as new features in credit risk models to boost their predictive performances. The stages of the modeling process are explained in detail below.

2.1   Preparation of Modeling Data

We created two data sets, one for each model. Our model data sets consist of approved credit card applications as instances and debt payment performances as target variables. Target variable is in binary form and takes values with the following rule:

$$\begin{cases} 1 & \text{if the customer is defaulted or does not pay his/her debt at} \\ & \text{least 90 days in the first 18 months following card usage,} \\ 0 & \text{otherwise.} \end{cases}$$

The reason behind choosing credit cards is that our bank has the most market share at cards among all credit products. Also, credit card applications have the most reliable address collection process, since plastic cards and card statements are delivered to these addresses.

As we can see from Tables 1 and 2, both data sets have the same instances but they differ from each other in locations and in the features related to the locations. Detailed information about the data fields can be seen from Table 3.

Table 1. Data format for residence address risk model.

Card ID	Customer ID	Target	Residence location	Residence address related features
10000001	80	1	(41.025171, 28.868498)	$\cdots$
10000002	92	0	(43.033902, 27.048473)	$\cdots$
$\vdots$	$\vdots$	$\vdots$	$\vdots$	$\vdots$

Table 2. Data format for residence work risk model.

Card ID	Customer ID	Target	Work location	Work address related features
10000001	80	1	(41.082807, 29.008156)	···
10000002	92	0	(40.747484, 28.772720)	···
⋮	⋮	⋮	⋮	⋮

Table 3. Data field explanations for risk models.

Data field	Explanation
Card ID	Card number
Customer ID	Customer number
Target	Binary target; showing payment performance
Residence/Work location	(Latitude, Longitude) of customer's residence/work address
Card ID	Card number
Feature 1	Feature related to customer's residence/work address
Feature 2	Feature related to customer's residence/work address
⋮	⋮

Extracted features belong to the residence/work locations of customers and specific to the related geolocations, so there is no additional feature that represents other characteristics of customers. You can find a detailed explanation about the creation of features in the following section.

Using these two data sets, we have developed two separate supervised models. Both of them predict the default probability (non-payment of debt) of customers. Since the prediction shows the default probability, it gets continuous values between 0 and 1. Please check the following parts for more detailed information about model architectures, selected features, etc.

Note that all extracted features belong to the locations of customers, so actually no customer data (such as age and income) is required to apply these models to new data points. Therefore, we prepared a new data set which consists of a wide range of geolocations in Turkey and the features related to them (without customer IDs or card IDs). We applied "risk model of residence location" to the new data set and we calculated the default probabilities of all locations, and named this output as "risk level of residence address". Similarly, we applied a "risk model of work location" and we calculated the default probabilities of all locations, and named this output as "risk level of work address". Tables 4 and 5 give example outputs of these two location-based risk models.

As it will be detailed later, we do not use geolocation as points with a high precision. Instead, we divided the map of Turkey into small regions (rectangles) with lengths of 80–100 m. So, the output tables do not include every single coordinate in Turkey but the center coordinates of regions. The bigger risk level

Table 4. Residence address risk scores for micro-regions.

Latitude	Longitude	Risk level of residence address
38.411	27.201	0.030
38.705	35.500	0.001
41.509	34.216	0.140
⋮	⋮	⋮

Table 5. Work address risk scores for micro-regions.

Latitude	Longitude	Risk level of work address
38.411	27.201	0.200
38.705	35.500	0.100
41.509	34.216	0.040
⋮	⋮	⋮

indicates that the customers living (or working) in that region have higher credit default probabilities.

Accept/reject decisions for credit applications will not be done directly with these risk level scores. Rather, these scores will be provided as inputs to other machine learning models such as consumer loan application scorecard, over-draft loan behavioral scorecard, etc. Note that we can add these scores to any customer-based data set using customers' latitude and longitude information. Within the scope of this study, we also measured the effect of using these scores as inputs for credit card & overdraft application risk models.

2.2   Feature Engineering

Since each customer has almost a unique geolocation, it is not very effective to work with such a high granular data. Instead, we divided the map of Turkey into non-intersecting small regions (polygons), picked the center points as representatives and calculated all features with respect to these center points. Then, we used the features of center points for the rest of the geolocations included in the same region. Figure 1 illustrates our grid-based micro-regions in İstanbul.

For each center point, we calculated features from different domains such as

- city, district of the center points
- points of interest (POI) information
  - counts of ATMs/schools/shopping malls etc. around center points (200 m-500 m-1000 m)
- demography
  - average kid-ratio/divorced ratio/socio-economic category etc. around center points (200 m-500 m-1000 m)

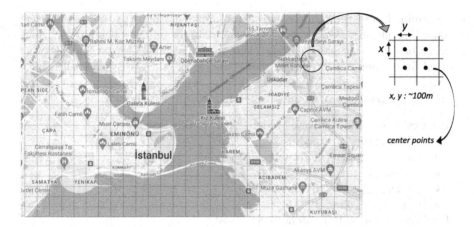

Fig. 1. Representation of grid-based micro-regions.

– neighbors' credentials
  • average credit score of the customers, excluding customers in the model
    data set, around center points (200 m-500 m-1000 m)

POI data and geolocational demographical data provided by third party map
services. At the end of the feature engineering process, 1940 features were created. These features were added to residence address and work address data
sets.

2.3 Modeling Stage

Two separate machine learning models, namely, residence address risk and work
address risk models, were developed using the residence address and work address
data sets. There were about 2000 features and some of them were strongly correlated such as counts of ATMs around 200 m and around 500 m. Our first draft
model was overfitted and we could not try many hyper-parameters because of
the long training time. Several feature elimination techniques (low variance, high
missing rate, low single performance indicator and high multicollinearity) were
used to reduce features to a reasonable number. By doing so, running several
machine learning algorithms with different hyper-parameters had become practical. More than 100 algorithm-hyper-parameter settings were tried.

For each task, logistic regression, decision tree, random forest, gradient boosting, MLP, XGBoost algorithms were tried with several hyper-parameters with
k-fold cross validation technique (100+ algorithm-hyper-parameters were experimented). For logistic regression, ℓ_1 and ℓ_2 regularization; for decision tree, different depths, branches, min observations; for random forest, different number
of estimators, depth, leaf samples, max features; for MLP, different activation
functions, hidden layer sizes, learning rates were tried. The ratio

$$\frac{\text{average cross-validation Gini}}{\text{average train Gini}}$$

was examined, and the models with a ratio smaller than 0.8 were eliminated. Standard deviation of Gini among folds were also examined. The models which perform best on the average were chosen. MLP and XGBoost models were selected for residence address risk model and work address risk model, respectively.

The hyper-parameters of the selected models are listed below.

- For Residence Address Risk Model - MLP (scikit-Learn):
  - `'activation': 'logistic'`
  - `'early_stopping': True`
  - `'hidden_layer_sizes': (60, 60)`
  - `'learning_rate_init': 0.001`
  - `'n_iter_no_change': 5`
- For Work Address Risk Model - XGBoost
  - `'gamma': 10`
  - `'learning_rate': 0.01`
  - `'max_depth': 6`
  - `'min_child_weight': 200`
  - `'n_estimators': 200`
  - `'objective': 'binary:logistic'`

Figures 2 and 3 show feature importance values of top 20 features among 315 and 330 features used in the residence and work address risk models, respectively. It can be seen that feature importance values of residence address risk model are more balanced, while there are dominant features in work address risk model. Top performing feature categories in residence address risk model are residence location, family structure, education level, and socio-economic category. Top performing feature categories in work address risk model are work location, average wealth, population change statistics, POIs, and election statistics.

3   Evaluation

To evaluate the effectiveness of model outputs, we have chosen two methods: (i) submitting results to expert validation and (ii) testing their contributions to risk models.

3.1   Expert Validation

After the development of models, expert validation is done by a focus group using map visualizations (see Fig. 4). This group consists of credit underwriting experts (2 people) and credit analysts (5 people). Focus group validated the consistency of the predictions by analyzing İstanbul map, since İstanbul is well-known with its highly heterogeneous population and socio-economic status. Green regions indicate low risk customers, whereas green and red regions indicate middle and high risk customers, respectively.

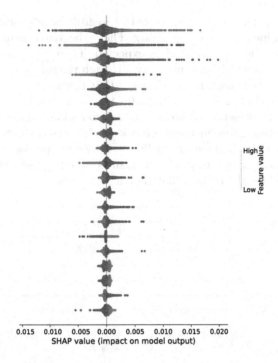

Fig. 2. SHAP values of top 20 features of residence address risk model.

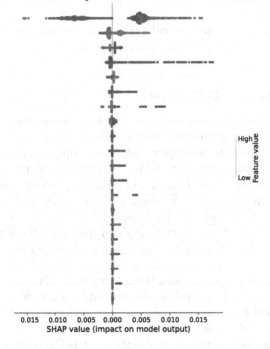

Fig. 3. SHAP values of top 20 features of work address risk model.

At the first stage, experts were asked to identify which locations are most risky and which locations are least risky. Then, the visual map was examined to see if those locations are colored as expected. Then, the experts told us the specific geolocations which are quite risky although the surrounding geolocations are less risky and vice versa. It was observed that the model was able to capture those anomalous micro-regions quite well. At the second stage, we did zoom to some randomly chosen anomalous (the micro-regions which differ from the surrounding micro-regions in terms of risk level) micro-regions together with experts and ask them if these are in line with their expertise. They validated the results for most of the regions, some results were contradictory and experts remained neutral for the regions that they have no idea.

Fig. 4. Visualization of residence and work address risk categories for İstanbul.

3.2   Contribution to Risk Models

The effect of new features, namely, risk level of residence address and risk level of work address, were evaluated by stacking with other models.

We have had a recently developed credit card application risk model. It was developed with a different data set and in a traditional risk modeling fashion. It was also officially validated by related units. Existing model consists of features about the demographics of the customers such as residence cities and districts. Therefore, using location info at high level is not a new practice in risk modeling but using more sensitive location info and features related to the granular locations is. Other details about the existing model are out of the scope of this study.

We prepared a new data set with three features: credit card application risk model score, residence address risk level score, and work address risk level score. Then, we combined these three scores with a stacking method. Figure 5 shows the details of our stacking approach.

Model 1 is the existing model, and Data set 1 is the training data set used in Model 1. The development of Model 1 is out of the scope of this study. The important points are: it was not an old model but newly developed, and it consists of features such as customer city/district etc.

Models 2–3 are residence address and work address risk models developed during this study. So, for each geolocation, we have two corresponding risk scores: residence risk score and work risk score.

Data set 4 is credit based model data. At the initial stage, there were only ids as credit id, targets as payment performance, customer ids, residence and work residence geolocations of customers. First, Model 1 features were created, and Model 1 scores were calculated. Then, by joining with residence and work address risk score tables, these two risk scores were added to the table. Only three features were used (model 1 score, residence address risk score, and work address risk score) during the development of Model 2, which is a regression model that combines aforementioned three features.

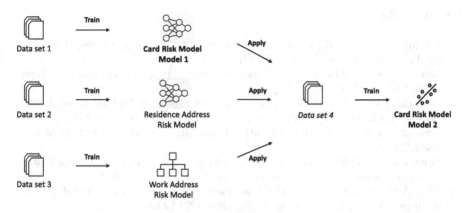

Fig. 5. Stacking process of credit card risk model and address risk models.

To measure the contribution, we compared Gini performances of Model 1 (card risk model without risk level scores) and Model 2 (card risk model with risk level scores). Gini coefficient is used as a performance indicator especially in the evaluation of credit risk models. It has a linear relationship with AUROC, and it is calculated with the formula:

$$\text{Gini} = 2 \times \text{AUROC} - 1$$

where AUROC is the area under the ROC curve obtained using the confusion matrix with changing thresholds.

Gini performance has increased by 1.1% from the credit card application risk model without the addition of risk level scores (Model 1) to the credit card application risk model after the addition of risk level scores (Model 2). Please note that Model 1 was recently developed by choosing the best model among several machine learning techniques, which already included high-level location information such as city & district and has a high predictive performance. The same stacking approach was also applied to the overdraft application risk model and 1.4% Gini increase was obtained.

4 Conclusion

Credit scoring systems, especially credit risk models, are vital for the commercial banking sector. Boosting the predictive power of these models has great financial gain. Traditional data sources are not enough to represent high heterogeneity in customer locations. Our micro-region approaches with alternative data sources improved the predictive performance of classification models considerably. Financial gain has also been quite high compared to alternative approaches such as application of region specific threshold for credit scoring approval.

References

1. Siddiqi, N.: Credit Risk Scorecards: Developing and Implementing Intelligent Credit Scoring. Wiley, Hoboken (2006)
2. Guy, C.: The Retail Development Process: Location, Property, and Planning. Routledge, London (1994)
3. Domschke, W., Drexl, A.: Location and Layout Planning: An International Bibliography. Springer, Cham (2013). https://doi.org/10.1007/978-3-662-02447-8
4. Lindeskov, C.K.: Ambulance Allocation Using GIS. Technical University of Denmark (2002)
5. Morrison, P.S., O'Brien, R.: Bank branch closures in New Zealand: the application of a spatial interaction model. Appl. Geogr. **21**(4), 301–330 (2001)
6. Puig, A.B., Vera, J.B., Perez, C.E.: Geomarketing models in supermarket location strategies. J. Bus. Econ. Manag. **17**(6), 1205–1221 (2016)
7. Kocenda, E., Vojtek, M.: Default Predictors in Retail Credit Scoring: Evidence from Czech Banking Data. William Davidson Institute Working Paper Number 1015 (2021)
8. Basel Committee on Banking Supervision: Credit Risk Modelling: Current Practices and Applications (1999)
9. van Ham, V., Tammaru, T., Ubarevičienė, R., Janssen, H.: Urban Socio-Economic Segregation and Income Inequality. The Urban Book Series, pp. 293–309 (2021)
10. Özaksoy, G.: Urbanization and Social Thought in Turkey, Middle East Technical University (2005)

Social Media Analysis: The Relationship Between Private Investors and Stock Price

Zijun Liu[1]([✉])(iD), Xinxin Wu[1]([✉]), and Wei Yao[2]([✉])(iD)

[1] Columbia University, New York, NY 10025, USA
{zl3031,xw2789}@columbia.edu
[2] The University of Adelaide, Adelaide, SA 5000, Australia
wei.yao@student.adelaide.edu.au

Abstract. Understanding peoples' opinions on social networks, such as Twitter or Reddit, has become easier with the assistance of analysis of users' sentiments on these networks. In our model, social media reveals public opinions and expectations that potentially correlate with stock market price movements. Using natural language processing (NLP), this paper examines reasons for the correlation between public sentiments and stock price fluctuations in the United States. Further, we demonstrate these correlations and provide promising directions for future research.

Keywords: Social Networks · Machine Learning · Natural Language Process · Computational Finance

1 Introduction

By trading stocks, an investor can transact the equity shares of public held companies on the stock market. The primary stock market provides traders who purchase and sell shares with an efficient exchange environment and direct price discovery tool. Due to a variety of factors, it is difficult to predict the stock price. For example, the U.S. economy is greatly affected by political policies, as well as the public's confidence in the system. As a result, investors' sentiments influence the stock price. People in the U.S. can express their opinions freely online, especially on social media. This could have an effect on the stock market. If investors exchanged their daily sentiments and information, the stock market would be severely affected. A compelling example is the General Motors Electric (GME) stock. In a post about GME in August 2020, a prominent investor offered his prediction of a potentially significant price increase for the stock. People were becoming more aware of GME stock as this post spread, creating a positive impression of the stock. Due to positive public sentiment toward the company, many people decided to invest in GME stock, and its price quickly rose. In a short time period, in response to the public's ever-increasing inflow of money, the price rose by nearly 190 times since the initial post, reaching $ 483

Supported by Columbia University.

per share from initial price. As will be discussed, a few U.S. stock prices have been fluctuating dramatically lately. This may indicate both investors' desire for quick wealth and the heightened risk of losing their investments completely. As a result, people pay attention to price changes during business hours. It is hypothesized that these online investors' sentiments, as well as the GME event, affect the subsequent share price according to these analyses. Our natural language processing (NLP) method analyzes the sentiment score of these posts and comments. After analyzing the sentiment score, the Granger causality test is used to determine the correlation between the sentiment score and stock prices on the following trading day.

2   Hypothesis

Stock prices are generally determined by the highest or lowest price someone is willing to pay for stocks. Stock prices are meant to rise in the future so that when investors buy stocks at a given price they then expect to sell them in the future for profit. Eventually, the stock price will reflect the popularity of a stock among investors. As demand increases, new investors will have to pay higher prices for stock.

Stock prices are heavily influenced by investors' sentiments. Optimistic investors are more likely to buy a stock when they feel positive about it. As a result, stock prices and investors' sentiments may be able to be correlated. This phenomenon has been well documented in research. In the last two hours of Wall Street trading, investor sentiment was found to be particularly predictive of market returns [1]. In addition, research in 2017 demonstrated a close relationship between changes in investor sentiments in the first half-hour of a session and market returns in the last half-hour [2]. There is a positive correlation between stock price and dividends. As investors are busy trading during trading hours, their sentiments cannot be expressed during trading hours, so they are more positively related to the price of a stock on the next trading day. Therefore, we hypothesize that private investors' sentiments affect stock prices.

3   Related Work

It is believed that stock prices are a reflection of all available information in the market, according to the Efficient Market Hypothesis, referred as EMH [3]. In other words, stock prices cannot be predicted using historical information and fundamental and technical analysis cannot be used to identify undervalued stocks or predict market trends. It should be noted, however, that the EMH has not always proved correct; for example, a retail investor and an institutional investor primarily use fundamental and technical analysis. Furthermore, investors are increasingly using social media to make investments and recommendations on their own. Researchers, such as Sprenge (2014), have used social media platforms to figure out whether they could gain further insight into stock price trends through online investor discussions. Various studies have investigated psychological factors that influence investors' investment decisions from the perspective of behavioral finance. The investment decisions of irrational investors are likely to be impacted more by external factors than those of rational investors [4]. Irrational investors are heavily influenced through the external factor of social media messages. As a result of technological advancements and when compared with traditional media, it has become easier for users to express their opinions and emotions through online interactive platforms. Thus, social media sentiment can be an important indicator of stock sentiment that can be used for both positive and negative analysis. For example, using machine learning to analyze stock information, we found relationships between stockholder sentiment on Twitter, information volume, and trading volume [5]. Additionally, Reddit comments have been shown to affect GME stock returns based on their tone and volume [6]. The analysis of investor comments and posts on Reddit has been studied only briefly. And, it should be noted, we encountered some research obstacles related to the more emotional nature of Reddit comments than Twitter.

4   Data

Online comments posted by stock investors have been entered into the Reddit database, one of the world's most popular and active discussion forums. Most Reddit users are eager to share their opinions and experiences about investing. Furthermore, investors may be able to communicate with one another through interactive comments under the post. Consequently, those posts, even comments regarding stocks on Reddit, can directly reflect the sentiment of their respective investors.

4.1   Reddit Data

By connecting with external API, Pushshift, mining the data over the time; we have collected more than 5000 Reddit users' posts in Subreddit r/Wall-Streeters about AMC stock and store them into the dataset, ".csv" file. In addition to scraping the user information, we also classified these posts and comments into three categories: time, author, title. The first step is data cleaning which removes the posts and comments that are unrelated to the datasets. Following that, we categorize the users into two categories: Official accounts and individual investors. The official accounts mainly operated by the agency which at the most time are unbiased; that is why the Individual investors' posts and comments are mainly used in the experiment.

4.2   Stock Data

By using the Kaggle dataset, we plot the stock price trend of AMC over the past five years. As the Fig. 1 shows, that AMC stock price has been fluctuating but has consistently remained flat for last five years. It appears that AMC stock prices have increased dramatically in early 2021, which may be related to GME's short squeeze. We compared the stock price and volume of AMC in January 2021 to test this hypothesis. The Fig. 2 below, shows a positive correlation between stock volume and stock price. Scholars have previously demonstrated that the sentiments of the shareholders about GME during the period of the GME Short Squeeze have been positively correlated with changes in GME's stock prices during this period. This study examined comments and posts from that period using keywords like "GME" "AMC" "moon". Figure 3 clearly indicates that the GME Short Squeeze event is directly responsible for the substantial rise in stock prices in January. Based on this historical data, stock prices are positively correlated with stockholders' sentiments. To demonstrate the existence of this positive correlation, the study analyzed the recent AMC stocks data which showed a positive

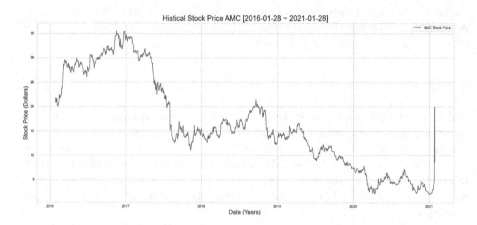

Fig. 1. AMC's stock price over the past five years

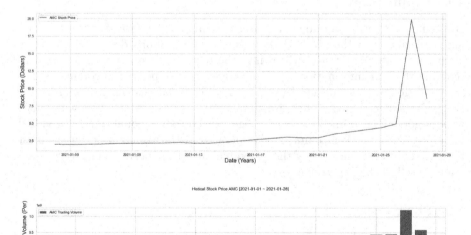

Fig. 2. Hedge funds versus private investments in ACM stock during event

correlation. We used the historical stock exchange data from Dec 10th, 2020, to Jan 6th, 2021 through finance API. The analysis of AMC stocks was conducted on recent data, and then historical data was downloaded through the finance API for the period from Dec 10th, 2021, to Nov 6th, 2021 and compared with the current data on AMC stocks.

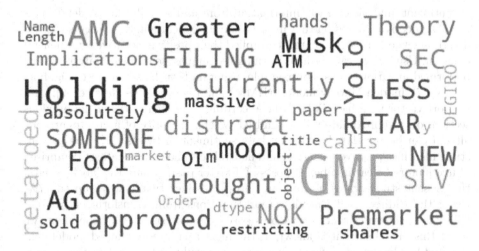

Fig. 3. Frequent words

5   Method

5.1   Data Acquisition and Cleaning

We utilized PushShift API (is also referred to as the pushshift.io Reddit API) to source Reddit comments during trading days. To provide improved functionality for sourcing comments from Reddit, the team behind /r/ datasets mod team developed this tool. However, the API returns extra data columns which are not useful for the data analysis. For example: author flair, collapsed and user background-color. Therefore, it is necessary to filter and keep the data that we need. An analysis of an online comment's content can provide valuable stockholders' sentiments information. The posting time of comments is also required for a time-sequential analysis. The username is also identified for these comments. The effectiveness of sentiment analysis may be reduced by comments containing irrelevant information. The data is therefore cleaned to remove images, comments which included unrelated punctuation marks and other unnecessary information, leaving only valuable data for sentiment analysis. Additionally, we eliminated content that did not contain English words. On every trading day, we collected mainly the closed AMC stock price based on python Yahoo data API called "yfinance". Our data relates to the AMC closed price for the following Monday, since the stock market does not operate on weekends. The data acquisition for AMC stock has been completed by simply checking and recording the closed price from Nov.10 to Dec.06 2021.

5.2 Sentimental Analysis

A sentiment analysis extracts information from user's feedback and determines users' subjectivity based on that information. In conclusion, this process aims to determine whether users' comments implied their positive, negative, or a neutral emotion. Using classical machine learning techniques, sentiment analysis can be performed in various ways. For example, using tools such as k-nearest neighbors (k-NN) [7]; hidden Markov models (HMM) [8]; a Naive Bayes classifier [9]; and, support vector machine (SVM) [10]. According to related works in the previous section, over approximately five hundred tweets, the price of a cryptocurrency altcoin can be correlated with the daily sentiment of tweets [11]. The present study identifies positive, negative, and neutral sentiments in a dataset comprised of five hundred clear number of posts of Twitter posts. Using the labelled data, the present study trains a Random Forest classifier that reaches up to 77%. Using the Bidirectional Encoder Representations from Transformers (BERT) [12], which generates sentiment scores based on analysis of every comment; we select this model since it is known as the most accurate pretrained model constructed by Google researchers. Our next step would be to calculate the average sentiment in a stock after we collected the sentiment scores for that day. We would use this index to analyze the sentiment in the stock further.

5.3 Correlation Analysis

The Granger causality test [13] is a useful method for determining the relationship between two time series. Following are the assumptions that underlie the Granger causality test:

1. A cause occurs before an effect occurs.
2. There is a unique relationship between the cause and the effect, which allows the cause to predict their future values.

Based on the relationship between average sentiment scores and closing prices on the following trading day, the Granger causality test can be applied to these two assumptions. The sentiment can first be calculated before next trading day's closing price is announced. Moreover, sentiment scores provided unique information on the stock's closing price the following day.

The Granger causality test has been applied after analysis of sentiment scores between 10 November and 6 December 2020. The test results revealed what correlations if any exist. We were therefore able to determine whether our hypothesis is accurate.

6   Result

6.1   Sentiment Score vs Stock Price

In our method, we calculate the average sentiment score as an overall indicator of opinions regarding AMC stock by collecting a sentiment score for every comment every day. Reddit's sentiment is intense. The score of sentiment is rated in the range between 0 and 1. Therefore, we choose 0.5 to represent neutral sentiment. On the next trading day, all investors who have a sentiment score higher than 0.5 will tend to buy more stocks related to positive sentiments. For those scores under 0.5 are considered negative sentiment scores. When this occurs, investors tend to sell their stocks the following trading day. In Fig. 4 displays the average sentiment score from 10 November to 6 December in 2021. Figure 5 below shows a chart with the changes in the closing price of AMC stock between 10 November and 6 December. Comparing these two Figures, the sentiments of these investors are clearly affecting the stock price during the period before 19 November and after 29 November. Among Reddit users on 16 November, the average sentiment score was below 0.5, indicating that these users are negatively inclined toward AMC. This resulted in a decrease in the closing stock price the following day. Furthermore, the closing stock price on 13 November was higher than the previous day's and the sentiment score was higher on 12 November. A stock price increases before 19 November and after 29 November when the sentiment score exceeds 0.5; a stock price decreases when the sentiment score drops below 0.5. check the logic It is also important to note that the closing stock price will continue to increase even if the sentiment score decreases to 0.5. There is, however, no agreement between the hypothesis and the behavior of the stock prices between November 19 and November 29. There is no highly positive correlation between the sentiment score and the next day's closing price during this period.

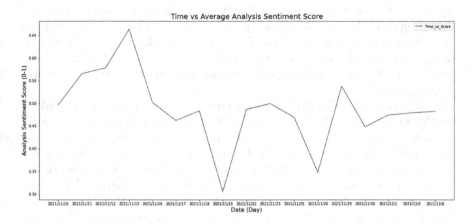

Fig. 4. The movement of the sentimental score between 11/10/2021 and 12/06/2021

Fig. 5. Movement of the stock price of AMC from November 10, 2021 to December 6, 2021

We reviewed some relevant news information in this period to identify the reasons behind this unusual behavior. Following are three more likely possibilities:

1. There was a downturn in shareholder sentiment on November 19th due to a large share sale by AMC's CEO. In spite of this, the stock market will always have many risk takers. According to Warren E. Buffet, an internationally renowned investor [14] "When others are not greedy, you should be greedy; when others are greedy, you should be afraid." A speculative mindset prevailed among many retail investors, who thought that buying a large number of shares when the stock price was low was a good opportunity to earn profit. In this period, many stockholders may still be positive about the AMC stock even though the stock price was dropping dramatically.

2. Tapering and interest rate increases were implemented by the Federal Reserve during this period. According to the U.S. stock market, federal rate hikes are synonymous with regaining dollar liquidity in the market and increasing dollar financing costs. The decline in share prices could be attributed to a decrease in liquidity of U.S. stocks [15]. News of this event was released on 24 November, which caused panic among investors. This sudden news would have caused a drop in stock price even though the sentiment score on November 23rd was higher than 0.5.

3. A new variant of SARS-CoV-2, called Omicron, was reported by the World Health Organization (WHO) during 24 November and 26 November. Public concerns regarding Omicron were exacerbated by conflicting reports on the new variant due to the unpublished results of the study on Omicron. Omicron was the subject of conflicting reports in the press which exacerbated public concerns about the new variant. Omicron study results had not been published, the market experienced general volatility. The stock price of AMC possibly declined as a result of fears about Omicron and uncertainty about immunity against the new variant. Omicron's huge influence cannot be correlated with an average sentiment score higher than 0.5 on 23 November.

To illustrate the relationship between sentiment score and stock price, the Granger causality test was used since these three possibilities, most of which are "black swan" events, do not occur frequently enough to cause a long-term effect.

6.2   Granger Causality Test

As Fig. 4 and Fig. 5 have shown below, stock prices positively correlate with stockholder sentiment scores. Using the Granger causality test, we can further support this conclusion. There is a difference between the stock price the following day and the Reddit sentiment score for the previous day. As a result of stockholders' emotions after business hours, the following day"s stock price will be affected by the stock trading volume.

In the Table 1, Score_X $\geq$ Close_Y means sentimental analysis has correlation next trading day stock price. Close_X $\geq$ Score_Y means stock price on the next trading day has correlation on the sentimental analysis. Since we are focusing on sentiment and stock price, we only need Close_X $\geq$ Score_X, which is 0.1539.

Table 1. Result of Granger Causality Test

	Score_X	Close_X
Score_Y	1.0000	0.1190
Close_Y	0.1539	1.0000

We can conclude that following analysis:

H0: Stock prices are not affected by investor sentiment on the next trading day.
H1: Stock prices are positively correlated with market sentiment of investors on the following day.

Our findings reject the null hypothesis and accept the alternative hypothesis because 0.1539 is greater than the significance level, which is 0.05, as shown in Table 1. During the period November 10th to December 6th year time period, stockholder sentiment scores were positively correlated with stock prices.

7   Conclusion

An analysis of AMC's stock price opinion was conducted in this study. A hypothesis has been formulated based on our research and analysis that the stock price is positively related to public sentiment.

As shown in this paper, using public sentiment as a proxy for AMC's stock price, we conducted an experiment to verify our hypothesis. AMC's closing price was determined by extracting sentiment data from Reddit and Twitter comments and comparing it to the daily average sentiment score. Utilizing the Granger causality test has demonstrated that our hypothesis is accurate, as sentiment scores can be positively correlated with stock prices. Overall, public opinion is a very reliable factor in forecasting stock market fluctuations, however, as previously discussed, factors such as political events, and pandemics need to be accounted for in the analysis.

References

1. Sun, L., Najand, M., Shen, J.: Stock return predictability and investor sentiment: a high-frequency perspective. J. Banking Financ. **73**, 147–164 (2016)
2. Renault, T.: Intraday online investor sentiment and return patterns in the U.S. stock market. J. Banking Financ. **84**, 25–40 (2017)
3. Fama, E.F.: The behavior of stock-market prices. J. Bus. **38**, 34–105 (1965)
4. Lee, C.M., Shleifer, A., Thaler, R.H.: Investor sentiment and the closed-end fund puzzle. J. Financ. **46**, 75–109 (1991)
5. Sprenger, T.O., Tumasjan, A., Sandner, P.G., Welpe, I.M.: Tweets and trades: the information content of stock microblogs. Eur. Financ. Manag. **20**, 926–957 (2013)
6. Long, C., Lucey, B.M., Yarovaya, L.: "I just like the stock" versus "Fear and loathing on main street": the role of reddit sentiment in the GameStop short squeeze. SSRN Electron. J. **31** (2021)
7. Rezwanul, M., Ali, A., Rahman, A.: Sentiment analysis on Twitter data using KNN and SVM. Int. J. Adv. Comput. Sci. Appl. **8**, 19–25 (2017)
8. Soni, S., Sharaff, A.: Sentiment analysis of customer reviews based on Hidden Markov Model. In: Proceedings of the 2015 International Conference on Advanced Research in Computer Science Engineering & Technology (ICARCSET 2015), pp. 1–5 (2015)
9. Gowda, S.R.S., Archana, B.R., Shettigar, P., Satyarthi, K.K.: Sentiment analysis of Twitter data using Naïve Bayes classifier. In: Kumar, A., Senatore, S., Gunjan, V.K. (eds.) ICDSMLA 2020. Lecture Notes in Electrical Engineering, vol. 783, pp. 1227–1234. Springer, Singapore (2022). https://doi.org/10.1007/978-981-16-3690-5_117
10. Firmino Alves, A.L., de Baptista, C., Firmino, A.A., Oliveira, M.G., Paiva, A.C.: A comparison of SVM versus Naïve-Bayes techniques for sentiment analysis in tweets. In: Proceedings of the 20th Brazilian Symposium on Multimedia and the Web, pp. 123–130 (2014)
11. Sasmaz, E., Tek, F.B.: Tweet sentiment analysis for cryptocurrencies. In: 2021 6th International Conference on Computer Science and Engineering (UBMK), pp. 613–618 (2021)
12. Wolf, T., et al.: Transformers: state-of-the-art natural language processing. In: Proceedings of the 2020 Conference on Empirical Methods in Natural Language Processing: System Demonstrations, pp. 38–45 (2020)
13. Granger, C.W.: Investigating causal relations by econometric models and cross-spectral methods. Econometrica 424–438 (1969)
14. Duz Tan, S., Tas, O.: Social media sentiment in international stock returns and trading activity. J. Behav. Financ. **22**, 221–234 (2020)
15. Drobyshevsky, S., Trunin, P., Bozhechkova, A., Sinelnikova-Muryleva, E.: The effect of interest rates on economic growth. Gaidar Institute for Economic Policy, vol. 303 (2017)

Deep Learning Model of Two-Phase Fluid Transport Through Fractured Media: A Real-World Case Study

Leonid Sheremetov[✉], Luis A. Lopez-Peña, Gabriela B. Díaz-Cortes,
Dennys A. Lopez-Falcon, and Erick E. Luna-Rojero

Mexican Petroleum Institute (IMP), 07730 Mexico City, Mexico
{sher,llopezp,gbdiaz,dalopez,eluna}@imp.mx

Abstract. Modelling of fluid flow in well's vicinity in naturally fractured reservoirs is a commonly employed technique used for wells' productivity enhancement, like acid stimulation. Unfortunately, a detailed model reflecting the complex geophysical structure of the porous media is a timely and computationally demanding task. In this paper, a deep learning model is proposed for solving Darcy equation coupled with the transport equation, based on physics-informed neural network (PINN) deep learning technology. Datasets obtained from the 3D numerical simulator are used to train and test our method. We test the sensitivity of our method to the type of optimizer and learning rate, time step size and the number of timesteps, DNN architecture, and spatial resolution. The results of computational experiments on a real-world problem prove a good numerical stability of the solution, its computational efficiency and high precision of the PINN model.

Keywords: Deep learning · Physics-Informed Neural Network · Fluid Transport Model · Acid stimulation

1 Introduction

The analysis of multiphase flow in porous media is of considerable significance in the field of petroleum reservoir simulation, where accurate predictions of fluid flow are important in assessing the performance of oil and gas fields [17]. In the last few years, it has become apparent that the hot topic in reservoir simulation is the application of data analytics or machine learning to numerical simulation including deep learning methods for directly solving the background systems of partial differential equations (PDEs) using automatic differentiation capability of the neural networks [2, 9, 14, 15].

The use of neural network models that approximate the solution to a PDE, replacing traditional numerical discretization started in the 1990s [12]. However, despite the great interest and commercial needs, there is relatively little research that made a step forward from the classical (simplified) fluid flow models to practical applications of data-driven solutions for reservoir engineering, where due to the high degree of uncertainty that characterizes energy systems, most predictive models are parametric and must be updated with observed data [4].

G. Nicosia et al. (Eds.): LOD 2023, LNCS 14506, pp. 55–68, 2024.
https://doi.org/10.1007/978-3-031-53966-4_5

First approaches to solving PDEs with deep learning methods used convolutional deep learning networks, which are data intensive and do not consider the physics of the problem [8]. The main problem of these approaches is that ANNs are incapable of inferring any PDE parameters such as the end-point relative permeability. This is a huge disadvantage especially with the fact that in reservoir engineering, we always deal with large uncertainty due to the complex unknown geology and the complicated fluid flow.

In Raissi et al. 2019, the authors employed automatic differentiation (AD) and proposed the physics-informed neural networks (PINN) approach to approximate PDE solutions, in which the PDE residual is incorporated into the loss function of fully connected neural networks as a regularizer, thus restricting the space of admissible solutions [15]. Using the governing equations to regularize the optimization of parameters in PINNs allows us to train large networks with small datasets. These neural networks are forced to respect any physical principle derived from the laws governing the observed data, as modeled by general time dependent and nonlinear PDEs [11, 14]. Being applied to several classical problems in fluids, quantum mechanics, reaction–diffusion systems, and the propagation of nonlinear shallow-water waves, the method still failed in several cases of fluid-flow models like the Buckley-Leverett equation with a constant initial and boundary condition [6]. Recently, the approach was extended to solve non-Newtonian and complex fluid systems and to estimate parameters in inverse flow and subsurface transport problems [9, 10, 13].

Since the method was first applied to simplified formulations of fluid models (incompressible fluids, one-dimensional horizontal flow, steady flow, immiscible fluids, no capillary pressure), it remains a challenge to apply this approach to the two- and three-dimensional cases, and to tailor the boundary conditions and the parameterization of the heterogeneities, to obtain a physically consistent solution of more complex problems of fluid dynamics [5]. In this paper, we apply the developed framework for solving the practical problem of skin-effect modeling in well's vicinity for acid stimulation design. The main contributions of the paper are the following:

- We extend the PINN approach to the case of complex physics of multiphase flow during acid injection (anisotropic time-dependent porosity and permeability fields).
- We consider 3D geometry of Naturally Fractured Reservoirs (NFR), now, limiting the approach to implicit fracture modeling due to the difficulties of precise static characterization of the reservoirs with triple-porosity structure.
- We test the sensitivity of our model in two different dimensions: PINN-oriented and model-oriented.
- For the PINN dimension, type of optimizer, learning rate, representation of initial and boundary conditions, and the number of iterations were analyzed.
- For the model dimension, we focus on the parameters, which are critical for practical applications: distribution and the number of measurement points, time-step size, number of timestamps[1], and spatial resolution.

---

[1] Timestamps refer to the number of discrete time slices we randomly sample from.

The rest of the paper is organized as follows. In the following section, the theoretical background and the problem formulation are considered. The case under study for experimental validation of the proposed model along with the main experimental results are described in Sect. 3. Lastly, we discuss limitations, areas of improvement of the proposed approach, and future work.

2 Theoretical Background

2.1 Mathematical Basis for Fluid Flow Modeling with Acid Injection

The system of equations governing the flow of two immiscible fluids in the wellbore vicinity through a porous medium consists of the following equations. The mass balance equation describes the transient distribution of mass at every point in space, for a fluid phase α we have:

$$\frac{\partial(\phi \rho_\alpha S_\alpha)}{\partial t} + \nabla \cdot (\rho_\alpha v_\alpha) = \rho_\alpha q_\alpha. \tag{1}$$

In this work we are modeling the fluid of water and oil, therefore $\alpha = $ w and o. Moreover, $\rho_\alpha, S_\alpha, v_\alpha, p_\alpha$ are the density, saturation, velocity, and pressure of the fluid phase α, q_α is the flow rate of the injected fluid and ϕ is the porosity, The velocity, v_α, is described by Darcy's law (without gravity):

$$v_\alpha = -\frac{K_\alpha}{\mu_\alpha} \cdot (\nabla p_\alpha), \tag{2}$$

where K_α and μ_α are the permeability and fluid viscosity.

The saturation of each phase, S_α, is the fraction of space occupied by that phase in the medium; a saturation of zero means that the phase is not present in the reservoir. Fluid density and viscosity, as well as rock porosity can be pressure dependent, i.e., $\rho_\alpha = \rho_\alpha(p)$, $\mu_\alpha = \mu_\alpha(p)$ y $\phi = \phi(p)$.

In this work we use cylindrical coordinates, that means, the vector position is given by $x(r, \theta, z) \in \Omega$, where the computational domain Ω is defined as a cylinder around the well, going from the well radius r_w, to an exterior radius r_e, with a height of $H = z_2 - z_1$, where the z direction (true vertical depth) increases downwards (Fig. 1).

The solute transport is described by the advection-dispersion-reaction equation (ADRE), which describes the change in dissolved acid concentration (C) over time (t) in the flow oriented along the r direction. Additionally, the acid reacts with the rock depending on its concentration:

$$\frac{\partial(\phi C S_w)}{\partial t} = \nabla \cdot (D\phi \nabla(S_w C)) - \nabla \cdot (v_w C) - \alpha_r(1 - \phi)S_w C, \\ x \in \Omega, t \in [0, T] \tag{3}$$

$$\frac{\partial \phi}{\partial t} = \lambda \frac{\beta \alpha_r(1 - \phi)}{\rho_\sigma} S_w C, \tag{4}$$

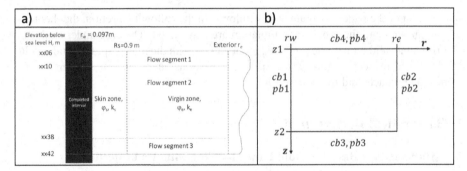

Fig. 1. Geometry of the model: a) dimensions of the model including damaged and virgin zones used for the case study with three flow segments defined by the geostatistical and b) radial projection of the simulated interval with boundary conditions.

where ρ_σ is the density of the rock, β is dissolution power, α_r is mass exchange coefficient (describing that concentration is lost proportionally) and λ is a proportionality constant. Dispersion tensor (without and with diffusion) is defined as follows:

$$D = \alpha\sqrt{v_1^2 + v_2^2}, \tag{5}$$

$$D = D_w \tau I + \alpha\sqrt{v_1^2 + v_2^2}, \tag{6}$$

where α is the dispersivity tensor with principal components α_L α_T. D_w is the diffusion coefficient, τ is the tortuosity, and I is the identity tensor.

Boundary conditions for concentration (C) of the model along with Dirichlet and Von Neumann boundary conditions for pressure (p) are illustrated in Fig. 1a and are defined as follows:

$$C(r_w, \theta, z, t) = C_{in}, \tag{cb1}$$

$$\frac{\partial C(r_e, \theta, z, t)}{\partial r} = 0, \tag{cb2}$$

$$\frac{\partial C(r, \theta, z_2, t)}{\partial z} = 0, \tag{cb3}$$

$$\frac{\partial C(r, \theta, z_1, t)}{\partial z} = 0, \tag{cb4}$$

$$\frac{\partial p(r_w, \theta, z, t)}{\partial r} = q, \tag{pb1}$$

$$p(r_e, \theta, z, t) = P_s, \tag{pb2}$$

$$\frac{\partial p(r, \theta, z_2, t)}{\partial z} = 0, \tag{pb3}$$

$$\frac{\partial p(r, \theta, z_1, t)}{\partial z} = 0. \tag{pb4}$$

where C_{in} is the injected acid concentration and P_s is a static reservoir pressure. It should be mentioned that in the above BCs, a constant flow model is described by Neumann BC (pb1), while for the constant pressure model – by Dirichlet BC (pb2). Finally, the initial condition is given by: $C(x, t = 0) = 0, x \in \Omega$.

2.2 Physics Informed Neural Networks (PINN) Approach

The PINN-based model rests on four pillars:

- The representation of the model variables by means of deep neural networks.
- The encoding of the mathematical model of fluid flow (1)-(4).
- The use of automatic differentiation.
- The training of the model with some reference data (in our case, output data from the numerical simulator).

Let us consider a partial differential equation parameterized by λ given by:

$$\begin{aligned}
f(x, t, \hat{u}, \partial_x \hat{u}, \partial_t \hat{u}, \dots \lambda) &= 0, & x &\in \Omega, t \in [0, T] \\
\hat{u}(x, t_0) &= g_0(x), & x &\in \Omega, \\
\hat{u}(x, t) &= g_\Gamma(t), & x &\in \partial\Omega, t \in [0, T]
\end{aligned} \tag{7}$$

where $x \in \mathbb{R}^d$ is the spatial coordinate and t is the time; f denotes the residual of the PDE containing the differential operators (i.e.,$[\partial_x \hat{u}, \partial_t \hat{u}, \dots]$), $\lambda = [\lambda_1, \lambda_2, \dots]$ are the parameters of the PDE, $\hat{u}(x, t)$ is the solution of the PDE with the initial condition $g_0(x)$ and the boundary condition $g_\Gamma(t)$ (which can be Dirichlet, Neumann or mixed) Ω and $\partial\Omega$ represent the spatial domain and the boundary respectively.

This equation encapsulates a wide range of problems including fluid flow models in porous media. In the PINN approach, neural networks are trained as supervised learning tasks to approximate the solution of the PDE $\hat{u}(x, t)$ following any given laws of physics described by general nonlinear PDE [15]. The derivatives of $\hat{u}$ with respect to the inputs are computed by automatic differentiation and then used to formulate the residuals of the governing equations in the loss function, which is usually composed of multiple terms weighted by different coefficients. The neural network parameters θ and the unknown PDE parameters λ can be learned simultaneously by minimizing the loss function.

Automatic differentiation not only avoids numerical discretization of the original equations, but also is more efficient than finite difference method (FDM) since it only requires one forward pass and one backward pass to compute all the partial derivatives, no matter what the input dimension is. In contrast, using FDM computing each partial derivative $\frac{\partial y}{\partial x_i}$ requires two function valuations $y(x_1, \dots, x_i, \dots x_{d_{in}})$ and $y(x_1, \dots, x_i + \Delta x_i, \dots x_{d_{in}})$ for some small number Δx_i and thus in total $d_{in} + 1$ forward passes are required to evaluate all the partial derivatives. When the input dimension is high, the AD is much more efficient.

If the hidden variable of the k-th hidden layer is denoted by z^k, then the neural network can be expressed as (8).

$$
\begin{aligned}
z^0 &= (x, t), \\
z^k &= \sigma\left(W^k z^{k-1} + b^k\right), \quad 1 \le k \le N_{L-1} \\
z^k &= W^k z^{k-1} + b^k, \quad k = N_L,
\end{aligned}
\tag{8}
$$

where the output of the last layer N_L is used to approximate the true solution, i.e. $\hat{u} \approx z^{N_L}$; W^k and b^k denote the weights and bias vector matrix of the k-th layer respectively; $\sigma(\cdot)$ is a nonlinear activation function. All parameters of the trainable model, i.e. weights and biases, are denoted by θ:

$$
\theta = \left\{W^1, W^2, \ldots, W^{k_i+1}, b^1, b^2, \ldots, b^{k_i+1}\right\}.
\tag{9}
$$

Solving a PDE system (7) becomes an optimization problem by iteratively updating θ with the objective of minimizing the loss function L:

$$
L = \omega_1 L_{PDE} + \omega_2 L_{data} + \omega_3 L_{IC} + \omega_4 L_{BC},
\tag{10}
$$

where ω_{1-4} are the weighting coefficients for the different loss terms. The definition of the loss function shown in (10) depends on the problem, so some terms may disappear for different types of problems.

The first term L_{PDE} penalizes the residual of the governing equations. The other terms are imposed to satisfy the model predictions of the measurements L_{data}, the initial condition L_{IC} and the boundary condition L_{BC}, respectively. In general, the mean square error (MSE) taking the L2 norm of the sampling points is used to calculate the losses in (10). The sampling points are defined as a data set $\{x^i, t^i\}_{i=1}^{N}$, where N the number of points (for different loss terms may be different).

2.3   PINN Model of the Fluid Flow in Well's Vicinity

Individual deep neural networks (DNNs) are used to approximate the state variables of a physical system and then are jointly trained by minimizing the loss function that consists of the governing equations residuals in addition to the error with respect to provided data (measurements or simulation results) [9]. In our case, to solve the system of Eqs. (1)-(4), a set of fully connected feed-forward neural networks (FNN), with space and time coordinates (x, t) as inputs, is used to approximate the solutions $\hat{u} = [v, Sw, p, \phi, k, C]$ (Fig. 2). Seven fully connected neural networks with five hidden layers are used to learn the velocity, water saturation, pressure, porosity, permeability, and concentration fields from training data in four spatial-temporal dimensions. The networks take time t and space polar coordinates r, θ, z as inputs and output the fields. The hyperbolic tangent activation function $\sigma(x) = tanh(x)$ is used. Automatic differentiation is employed to compute the losses according to Eq. (10). An adaptive gradient-based first-order optimization algorithm Adam was used for loss function minimization and optimization of the parameters $\theta = f(W, b)$ of the model during the training of PINN. The limited-memory BFGS with box constraints (L-BFGS-B) method showed to be very time consuming.

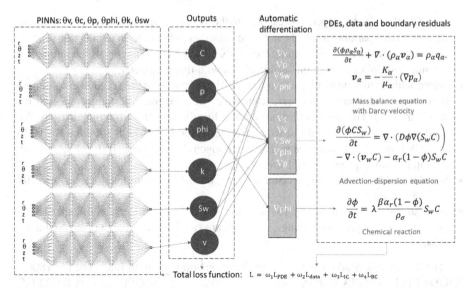

Fig. 2. Schematic diagram of a physics-informed neural network for fluid flow modeling problem (adapted from [3]).

3 Experimental Settings and Results

For experiments, a real-world case-study was selected. The production well from a NFR belonging to the offshore oilfield operated by PEMEX in the Gulf of Mexico was drilled with an approximate deviation of $27°$ and has been producing from the Paleocene Brecha formation. For validation purposes, historical acid stimulation data was selected along with the data from pressure tests before and after the intervention, which were used to calibrate the numerical model.

A well stimulation is defined as the process by which an extensive system of channels is restored (if it was previously damaged) or created in the reservoir, modifying the rock properties to facilitate fluid flow from the formation to the wellbore [1]. Due to the differences in the petrophysical properties of the completed interval, we assume that the damaged zone is non-uniformly distributed in both the vertical and radial directions. For simulation purposes, we can represent it by anisotropic porosity and permeability fields forming several radial layers with different sizes that were determined based on the petrophysical zone analysis during the static modeling (Fig. 1b, Fig. 4). This heterogeneity represents a challenge for PINN prediction model.

The objective of acid stimulation is to remove the damage. So, after the simulation, the skin factor resulting from the production tests should be determined.

3.1 Experimental Settings

The experiments were performed under the following settings:

- Simulation grid spatial dimensions: without loss of generality, the number of cells in the radial coordinate is 100, and for the z coordinate we select 264, 132, 66, and 33 vertical cells. The finest grid corresponds to the available geophysical logs' resolution. Related to this parameter, are the necessary resolution to model acid injection process and the execution time for training the PINN models. The selection of the number of training and collocation points is associated with this parameter.

- Numerical simulation results are saved for different points of time (called timestamps) during the acid injection process, from which 20% of the timestamps are used for model training. The number of timestamps affect the size of the training datasets. We used from 12 to 80 timestamps, which for the $100 \times 1 \times 33$ cells mesh means 1 Mb to 2.5 Mb growth of dataset size. For comparison, for the mesh $100 \times 1 \times 264*80$, the input datafile grows to 20M for each variable.
- We consider the prediction precision as a function of iterations in the range from 10k to 200k iterations.
- Without loss of generality, due to the space limits, we use the results of experiments only for constant pressure settings, which means two Dirichlet and two Von Neumann BCs.

The HP ZBook Studio G8 mobile workstation equipped with Intel core i7 11800H 2.3GHz CPU, NVIDIA GeForce RTX 3070 210 MHz 8 GB GPU, and DDR4 32 GB 532.1 MHz RAM was used for experiments.

3.2 Model Training

For training, the results of numerical simulation of the fluid flow from the house-made 3D simulator were used for homogeneous (Fig. 3a) and heterogeneous distributions (Fig. 3b). The area under study is divided into two parts - the area of interest, where the main effects of acid propagation are present (twice the skin area for 30 min injection, 1.8m for 0.9 m skin zone) and the rest of the simulated area of the well's vicinity. We use more dense distribution of the training points in the skin zone since we are interested in modeling the fluid flow propagation in this zone. In the rest of the space, the process can be considered stationary. All randomly sampled point locations were generated using a Latin Hypercube Sampling (LHS) strategy to fill the space [16].

Apparently, for anisotropic porosity and permeability fields forming vertically different flow units (Fig. 1b), one could expect the PINN model precision depending on the selection of the number of training and collocation points since this selection determines the coverage of the distributions in the training set for both skin and virgin zones (Fig. 3). The total number of training point is usually set to 20%.

At the beginning of the training process, the parameters of the neural networks are randomly initialized using the Xavier scheme [7].

3.3 Deep Neural Network's Meta-Parameters

Below we present an extract from the experiments on the DNN architectures and Adam´s learning rate. In Table 1, we show the comparison of the best three architectures of the DNN with 4 and 5 hidden layers and 60 and 80 neurons for each level. For validation, the following error metrics were used (n is the number of reference data points):

- RMSE (for concentration c): $E_{RMSE} = \sqrt{\frac{1}{n}\sum_{i=1}^{n}(y_i - \hat{y}_i)^2}$,

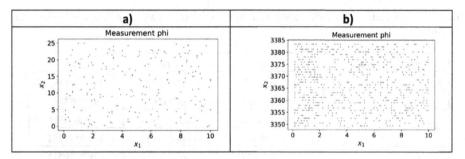

Fig. 3. Distribution of training points: uniform (100 points in total) and heterogeneous (500 points) training sets for porosity neural networks.

- Normalized RMSE by mean and infinite errors,
- MAPE and SMAPE (for all the other variables): $E_{sMAPE} = \sqrt{\frac{1}{n} \sum_{i=1}^{n} |\frac{y_i - \hat{y}_i}{(y_i + \hat{y}_i)/2}|}$.

It was impossible to use MAPE and SMAPE for acid concentration field since it has zero values. As we can see, the best results are obtained for the FNN with 5 hidden layers and 80 neurons in each, improving the SMAPE for almost 3% compared to a reference architecture (5*60). After 50k epochs, the losses converge to similar values for different learning rates in the range of [0.0001, 001] (Table 2).

3.4 Experimental Results

In order to evaluate the prediction capabilities of the developed approach, in Fig. 4 we illustrate the results of the predictions for the concentration field compared to a reference solution (output of the numerical simulation). Though prediction results are generated at the same time points as in the reference model, due to the space limits, we only show the results at the end of the injection period, $t_{max} = 1800$ seg. (half an hour). The above-mentioned accuracy metrics for the maximum time are presented in Table 3.

Prediction results are in very good agreement with the simulation output. The model was able to mimic the main heterogeneities of the concentration, water saturation, porosity and permeability fields, with almost a perfect fit for saturation. The SMAPE error for the illustrated example of 132 cells in the vertical and 100 cells in radial directions, are about 7–8%, mostly due to the high heterogeneity of the modeled fields. For the mesh of 33 vertical cells (less heterogeneity) the errors decrease to 4–5%.

Table 1. Loss and error metrics associated with the DNN architecture (the best results are shadowed).

FNN/loss/error	4*60	5*60	5*80
loss	0.024112128	0.020078912	0.018761199
loss_C	0.005253653	0.002076679	0.000983373
loss_K	0.007804309	0.007690934	0.007018511
loss_h	0.000506361	0.000514122	0.000577764
loss_ϕ	0.007945854	0.007235224	0.007579713
loss_Sw	0.002488048	0.002512732	0.002485585
RMSE-inf c	0.389375928	0.549354578	0.440224129
p	0.000289382	0.000294753	0.000300159
ϕ	0.083125093	0.084978714	0.085183278
S_w	0.070475536	0.074005901	0.075323647
K	0.090212576	0.091342675	0.082465412
Average SMAPE	0.061025647	0.062655511	0.060818124
Relative change, %	**2.60%**	**N/A**	**2.93%**

Table 2. Loss and error metrics as a function of learning rates (the best results are shadowed).

Loss/error	R1	R2	R3	R4	R5
Learning rate	0.0001	0.00031623	0.001	0.00316228	0.01
loss	0.02007891	0.01132999	0.01093152	0.01522937	0.2342281
loss_C	0.00207668	0.00109517	0.00091591	0.00127233	0.09711758
loss_K	0.00769093	0.00434115	0.00388526	0.00605279	0.01991702
loss_h	0.00051412	0.00028133	0.0002905	0.00209528	0.05670343
loss_ϕ	0.00723522	0.0038219	0.00464181	0.00420905	0.03306567
loss_Sw	0.00251273	0.00175844	0.00118667	0.00146396	0.01461687
RMSE-inf C	0.54935458	0.53372323	0.54982321	0.54600906	0.76713179
p	0.00029475	0.000293	0.00029494	0.00032911	0.00040925
ϕ	0.08497871	0.07092853	0.07795473	0.07249022	0.12639518
S_w	0.0740059	0.06808819	0.06559351	0.06928196	0.13783024
K	0.09134268	0.07411157	0.07346865	0.07686339	0.12989804
Average sMAPE	0.06265551	0.05335532	0.05432796	0.05474117	0.09863318

According to (10), the loss function is composed of different components. As we can see from Table 4, the contributions of the equations along with the boundary conditions is much less (in orders of magnitude) than the variables of the model. We tried to change weights for each loss, but the results are similar in general.

Table 3. Error metrics at the end of the injection process (the last timestamp of 1800 s.) and the average errors (N/A – not available).

Var	RMSE	RMSE-norm	RMSE-mean	RMSE-infinite	SMAPE	Average SMAPE
C		0.706407	0.778641	0.474310		0.4402241
p	35086972	N/A	N/A	N/A	0.000353	0.0003001
ϕ	45.30178	N/A	N/A	N/A	0.087520	0.0851832
S_w	140.8673	N/A	N/A	N/A	0.084371	0.0753236
K	1.11E-08	N/A	N/A	N/A	0.084801	0.0824654

We have provided a set of experiments on training time for different experimental settings (mesh dimensions, timestamps, optimizer, number of epochs). In case of timestamps, for 1k iterations, training for 20 timestamps takes 40 s., while for 100 timestamps - 70 s. For the mesh $100 \times 1 \times 33$, for 1k iterations, training takes 45 s., while for $100 \times 1 \times 132$ mesh – 60 s. Total training time for a typical model ($100 \times 1 \times 132$ mesh with 20 timestamps) takes a reasonable time of 1 h 40 min. The prediction is almost instantaneous.

Table 4. Values of loss functions at the end of training (100k epochs)

Variables		Equations		Boundary conditions	
$loss\_C$	1.738e−04	Mass balance ($loss\_f\_pde$)	4.842e−14	Cb1 ($loss\_cN1$)	1.3e−07
$loss\_p$	3.737e−03	Transport ($loss\_fc\_pde$)	3.098e−10	Cb3/cb5 ($loss\_cN2$)	2.356e−06
$loss\_K$	2.199e−03	Chemical reaction ($loss\_fr\_pde$)	2.333e−06	Pb3/pb4 ($loss\_pN2$)	5.756e−05
$loss\_\phi$	8.122e−04			Pb2 ($loss\_pD2$)	9.109e−05
$loss\_Sw$	3.696e−03				

4   Discussion and Concluding Remarks

The main contribution of this paper is that the developed model was able to mimic the distribution and evolution of the main fields of the flow transport model in highly heterogeneous porous media. The proxy simulator even with 100,000 iterations, was able to predict acid injection results within acceptable margins of error. For the case study (half an hour acid injection), the skin reduction was estimated at 9.41 with the numerical simulator and 9.24 with the PINN simulator (of the total damage determined at 17.4^2), representing 1.82% error of the PINN model.

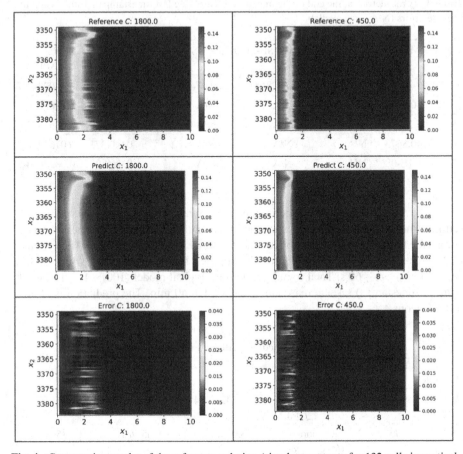

Fig. 4. Comparative results of the reference solution (simulator outputs for 132 cells in vertical dimension) versus the FNN prediction of the concentration field with the respective errors for two timestamps: 450 and 1800 s.

While developing the model, we found out that the PINN model is very sensible to the differences in the equations behind the numeric simulation and the PINN (numeric simulation model, which provides data for training, and the PINN have different parameters). It means, that even small changes in the simulation can cause that the prediction model doesn't converge for some variables. In such cases, behavior patterns of the loss

---

[2] Note, that during the real injection process, two similar phases of acid injection were carried out while we simulated only the 1st one. Thus, the skin reduction is in good agreement with the reference case.

functions also have severe fluctuations when there are discrepancies between equations and training data.

It was observed that the total loss function at the training stage reached values close to 0.01, which allowed training the models with good prediction accuracy. At the same time, it was observed that the model was not able to reproduce some fluctuations of the high frequency variables (although these fluctuations are mainly registered outside the damage zone and do not represent interest for simulation purposes). By increasing the number of points and the number of epochs, one can expect to decrease the insensitivity effect of the model to high frequencies of property change.

In our future work, we will compare the developed model with the pure data-driven approach and with the PDE neural solver (without training data), along with different types of neural operators. One of the main challenges for the application of such type of models in practice is the possibility to use them for ranges of parameters different from those that they have been trained.

Acknowledgements. This work was partially funded by the IMP. The authors would like to thank M.Sc. Humberto Aguilar Cisneros for fluid characterization and Ing. Jorge Javier Vazquez Calderon for his invaluable support of this work.

References

1. Almajid, M.M., Abu-Al-Saud, M.O.: Prediction of porous media fluid flow using physics informed neural networks. J. Pet. Sci. Eng. **208**, 109205 (2022). https://doi.org/10.1016/j.pet rol.2021.109205
2. Baydin, A.G., Pearlmutter, B.A., Radul, A.A., Siskind, J.M.: Automatic differentiation in machine learning: a survey. J. March. Learn. Res. **18**, 1–43 (2018)
3. Cai, S., Mao, Z., Wang, Z., et al.: Physics-informed neural networks (PINNs) for fluid mechanics: a review. Acta Mech. Sin. Mech. Sin. **37**, 1727–1738 (2021). https://doi.org/10.1007/s10 409-021-01148-1
4. Fraces, C.G., Papaioannou, A., Tchelepi, H.: Physics informed deep learning for transport in porous media Buckley Leverett proble (2020)
5. Fraces C.G., Tchelepi, H.: Physics informed deep learning for flow and transport in porous media. (2021). https://doi.org/10.2118/203934-MS
6. Fuks, O., Tchelepi, H.A.: Limitations of physics informed machine learning for nonlinear two-phase transport in porous media. J. Mach. Learn. Model. Comput. **1**(1), 19–37 (2020)
7. Glorot, X., Bengio, Y.: Understanding the difficulty of training deep feedforward neural networks. In: Proceedings of the 13th International Conference on AI and Statistics, vol. 9, pp. 249-256 (2010)
8. Han, J., Jentzen, A., Weinan, E.:.Solving high-dimensional partial differential equations using deep learning. Proc. Natl. Acad. Sci. U.S.A(2018).https://doi.org/10.1073/pnas.1718942115
9. He, Q., Barajas-Solano, D., Tartakovsky, G., Tartakovsky, A.M.: Physics-informed neural networks for multiphysics data assimilation with application to subsurface transport. Adv. Water Res. **141**, 103610 (2020). https://doi.org/10.1016/j.advwatres.2020.103610
10. He, Q., Tartakovsky, A.M.: Physics-informed neural network method for forward and backward advection-dispersion equations. Water Res. Res. **57**, e2020WR029479 (2021). https://doi.org/10.1029/2020WR029479

11. Jagtap, A.D., Karniadakis, G.E.: Extended Physics-Informed Neural Networks (XPINNs): A Generalized Space-Time Domain Decomposition Based Deep Learning Framework for Nonlinear Partial Differential Equations, Commun. Comput. Phys., Vol.28, No.5, 2002–2041 (2020). https://doi.org/10.4208/cicp.OA-2020-0164

12. Lagaris, I.E., Likas, A., Fotiadis, D.I.: Artificial neural networks for solving ordinary and partial differential equations. IEEE Trans. Neural Netw. 9(5), 987–1000 (1998). https://doi.org/10.1109/72.712178

13. Mahmoudabadbozchelou, M., Karniadakis, G.E., Jamali, S.: nn-PINNs: non-Newtonian physics-informed neural networks for complex fluid modeling. Soft Matter, 18(1), 172-185 (2022). https://doi.org/10.1039/d1sm01298c

14. Malik, S., Anwar, U., Ahmed, A., Aghasi, A.: Learning to solve differential equations across initial conditions (2020). https://doi.org/10.48550/arXiv.2003.12159

15. Raissi, M., Perdikaris, P., Karniadakis, G.E.: Physics-informed neural networks: a deep learning framework for solving forward and inverse problems involving nonlinear partial differential equations. J. Comput. Phys. Comput. Phys. 378, 686–707 (2019)

16. Stein, M.: Large sample properties of simulations using Latin hypercube sampling. Technometrics 29, 143–151 (1987)

17. White, I.R., Lewis, R.W., Wood, W.L.: The numerical simulation of multiphase flow through a porous medium and its application to reservoir engineering. Appl. Math. Model. 5(3), 165–172 (1981). https://doi.org/10.1016/0307-904X(81)90039-1

A Proximal Algorithm for Network Slimming

Kevin Bui[1(✉)], Fanghui Xue[1], Fredrick Park[2], Yingyong Qi[1], and Jack Xin[1]

[1] University of California, Irvine 92697, CA, USA
{kevinb3,fanghuix,yqi,jack.xin}@uci.edu
[2] Whittier College, Whittier 90602, CA, USA
fpark1@whittier.edu

Abstract. As a popular channel pruning method for convolutional neural networks (CNNs), network slimming (NS) has a three-stage process: (1) it trains a CNN with ℓ_1 regularization applied to the scaling factors of the batch normalization layers; (2) it removes channels whose scaling factors are below a chosen threshold; and (3) it retrains the pruned model to recover the original accuracy. This time-consuming, three-step process is a result of using subgradient descent to train CNNs. Because subgradient descent does not exactly train CNNs towards sparse, accurate structures, the latter two steps are necessary. Moreover, subgradient descent does not have any convergence guarantee. Therefore, we develop an alternative algorithm called proximal NS. Our proposed algorithm trains CNNs towards sparse, accurate structures, so identifying a scaling factor threshold is unnecessary and fine tuning the pruned CNNs is optional. Using Kurdyka-Łojasiewicz assumptions, we establish global convergence of proximal NS. Lastly, we validate the efficacy of the proposed algorithm on VGGNet, DenseNet and ResNet on CIFAR 10/100. Our experiments demonstrate that after one round of training, proximal NS yields a CNN with competitive accuracy and compression.

Keywords: channel pruning · nonconvex optimization · convolutional neural networks · neural network compression

1 Introduction

In the past decade, convolutional neural networks (CNNs) have revolutionized computer vision in various applications, such as image classification [12,32,37] and object detection [10,16,26]. CNNs are able to internally generate diverse, various features through its multiple hidden layers, totaling millions of weight parameters to train and billions of floating point operations (FLOPs) to execute. Consequently, highly accurate CNNs are impractical to store and implement on resource-constrained devices, such as mobile smartphones.

The work was partially supported by NSF grants DMS-1854434, DMS-1952644, DMS-2151235, and a Qualcomm Faculty Award.

G. Nicosia et al. (Eds.): LOD 2023, LNCS 14506, pp. 69–83, 2024.
https://doi.org/10.1007/978-3-031-53966-4_6

To compress CNNs into lightweight models, several directions, including weight pruning [1,11], have been investigated. Channel pruning [23,33] is currently a popular direction because it can significantly reduce the number of weights needed in a CNN by removing any redundant channels. One straightforward approach to channel pruning is network slimming (NS) [23], which appends an ℓ_1 norm on the scaling factors of the batch normalization layers to the loss function being optimized. Being a sparse regularizer, the ℓ_1 norm pushes the scaling factors corresponding to the channels towards zeroes. The original optimization algorithm used for NS is subgradient descent [31], but it has theoretical and practical issues. Subgradient descent does not necessarily decrease the loss function value after each iteration, even when performed exactly with full batch of data [4]. Moreover, unless with some additional modifications, such as backtracking line search, subgradient descent may not converge to a critical point [25]. When implemented in practice, barely any of the scaling factors have values exactly at zeroes by the end of training, resulting in two issues. First, a threshold value needs to be determined in order to remove channels whose scaling factors are below it. Second, pruning channels with nonzero scaling factors can deteriorate the CNNs' accuracy since these channels are still relevant to the CNN computation. As a result, the pruned CNN needs to be retrained to recover its original accuracy. Therefore, as a suboptimal algorithm, subgradient descent leads to a time-consuming, three-step process.

In this paper, we design an alternative optimization algorithm based on proximal alternating linearized minimization (PALM) [5] for NS. The algorithm has more theoretical and practical advantages than subgradient descent. Under certain conditions, the proposed algorithm does converge to a critical point. When used in practice, the proposed algorithm enforces the scaling factors of insignificant channels to be exactly zero by the end of training. Hence, there is no need to set a scaling factor threshold to identify which channels to remove. Because the proposed algorithm trains a model towards a truly sparse structure, the model accuracy is preserved after the insignificant channels are pruned, so fine tuning is unnecessary. The only trade-off of the proposed algorithm is a slight decrease in accuracy compared to the original baseline model. Overall, the new algorithm reduces the original three-step process of NS to only one round of training with fine tuning as an optional step, thereby saving the time and hassle of obtaining a compressed, accurate CNN.

2   Related Works

Early pruning methods focus on removing redundant weight parameters in CNNs. Han et al.[11] proposed to remove weights if their magnitudes are below a certain threshold. Aghasi et al.[2] formulated a convex optimization problem to determine which weight parameters to retain while preserving model accuracy. Creating irregular sparsity patterns, weight pruning is not implementation friendly since it requires special software and hardware to accelerate inference [20,40].

An alternative to weight pruning is pruning group-wise structures in CNNs. Many works [3,8,19,24,29,33] have imposed group regularization onto various CNN structures, such as filters and channels. Li et al.[20] incorporated a sparsity-inducing matrix corresponding to each feature map and imposed row-wise and column-wise group regularization onto this matrix to determine which filters to remove. Lin et al.[21] pruned filters that generate low-rank feature maps. Hu et al.[13] devised network trimming that iteratively removes zero-activation neurons from the CNN and retrains the compressed CNN. Rather than regularizing the weight parameters, Liu et al.[23] developed NS, where they applied ℓ_1 regularization on the scaling factors in the batch normalization layers in a CNN to determine which of their corresponding channels are redundant to remove and then they retrained the pruned CNN to restore its accuracy. Bui et al.[6,7] investigated nonconvex regularizers as alternatives to the ℓ_1 regularizer for NS. On the other hand, Zhao et al.[40] applied probabilistic learning onto the scaling factors to identify which redundant channels to prune with minimal accuracy loss, making retraining unnecessary. Lin et al.[22] introduced an external soft mask as a set of parameters corresponding to the CNN structures (e.g., filters and channels) and regularized the mask by adversarial learning.

3 Proposed Algorithm

In this section, we develop a novel PALM algorithm [5] for NS that consists of two straightforward, general steps per epoch: stochastic gradient descent on the weight parameters, including the scaling factors of the batch normalization layers, and soft thresholding on the scaling factors.

3.1 Batch Normalization Layer

Most modern CNNs have batch normalization (BN) layers [17] because these layers speed up their convergence and improve their generalization [28]. These benefits are due to normalizing the output feature maps of the preceding convolutional layers using mini-batch statistics. Let $z \in \mathbb{R}^{B \times C \times H \times W}$ denote an output feature map, where B is the mini-batch size, C is the number of channels, and H and W are the height and width of the feature map, respectively. For each channel $i = 1, \ldots, C$, the output of a BN layer on each channel z_i is given by

$$z_i' = \gamma_i \frac{z_i - \mu_B}{\sqrt{\sigma_B^2 + \epsilon}} + \beta_i, \tag{1}$$

where μ_B and σ_B are the mean and standard deviation of the inputs across the mini-batch B, ϵ is a small constant for numerical stability, and γ_i and β_i are trainable weight parameters that help restore the representative power of the input z_i. The weight parameter γ_i is defined to be the scaling factor of channel i. The scaling factor γ_i determines how important channel i is to the CNN computation as it is multiplied to all pixels of the same channel i within the feature map z.

3.2 Numerical Optimization

Let $\{(x_i, y_i)\}_{i=1}^N$ be a given dataset, where each x_i is a training input and y_i is its corresponding label or value. Using the dataset $\{(x_i, y_i)\}_{i=1}^N$, we train a CNN with c total channels, where each of their convolutional layers is followed by a BN layer. Let $\gamma \in \mathbb{R}^c$ be the vector of trainable scaling factors of the CNN, where for $i = 1, \ldots, c$, each entry γ_i is a scaling factor of channel i. Moreover, let $W \in \mathbb{R}^n$ be a vector of all n trainable weight parameters, excluding the scaling factors, in the CNN. NS [23] minimizes the following objective function:

$$\min_{W,\gamma} \frac{1}{N} \sum_{i=1}^N \mathcal{L}(h(x_i, W, \gamma), y_i) + \lambda \|\gamma\|_1, \tag{2}$$

where $h(x_i, W, \gamma)$ is the output of the CNN predicted on the data point x_i; $\mathcal{L}(h(x_i, W, \gamma), y_i)$ is the loss function between the prediction $h(x_i, W, \gamma)$ and ground truth y_i, such as the cross-entropy loss function; and $\lambda > 0$ is the regularization parameter for the ℓ_1 penalty on the scaling factor vector γ. In [23], (2) is solved by a gradient descent scheme with step size δ^t for each epoch t:

$$W^{t+1} = W^t - \delta^t \nabla_W \tilde{\mathcal{L}}(W^t, \gamma^t), \tag{3a}$$

$$\gamma^{t+1} = \gamma^t - \delta^t \left(\nabla_\gamma \tilde{\mathcal{L}}(W^t, \gamma^t) + \lambda \partial \|\gamma^t\|_1 \right), \tag{3b}$$

where $\tilde{\mathcal{L}}(W, \gamma) := \frac{1}{N} \sum_{i=1}^N \mathcal{L}(h(x_i, W, \gamma), y_i)$ and $\partial \| \cdot \|_1$ is the subgradient of the ℓ_1 norm.

By (3), we observe that γ is optimized by subgradient descent, which can lead to practical issues. When $\gamma_i = 0$ for some channel i, the subgradient needs to be chosen precisely. Not all subgradient vectors at a non-differentiable point decrease the value of (2) in each epoch [4], so we need to find one that does among the infinite number of choices. In the numerical implementation of NS[1], the subgradient ζ^t is selected such that $\zeta_i^t = 0$ by default when $\gamma_i^t = 0$, but such selection is not verified to decrease the value of (2) in each epoch t. Lastly, subgradient descent only pushes the scaling factors of irrelevant channels to be near zero in value but not exactly zero. For this reason, when pruning a CNN, the user needs to determine the appropriate scaling factor threshold to remove its channels where no layers have zero channels and then fine tune it to restore its original accuracy. However, if too many channels are pruned that the fine-tuned accuracy is significantly less than the original, the user may waste time and resources by iterating the process of decreasing the threshold and fine tuning until the CNN attains acceptable accuracy and compression.

To develop an alternative algorithm that does not possess the practical issues of subgradient descent, we reformulate (2) as a constrained optimization problem by introducing an auxiliary variable ξ, giving us

$$\min_{W,\gamma,\xi} \tilde{\mathcal{L}}(W, \gamma) + \lambda \|\xi\|_1 \quad \text{s.t.} \quad \xi = \gamma. \tag{4}$$

---

[1] https://github.com/Eric-mingjie/network-slimming.

However, we relax the constraint by a quadratic penalty with parameter $\beta > 0$, leading to a new unconstrained optimization problem:

$$\min_{W,\gamma,\xi} \ \tilde{\mathcal{L}}(W,\gamma) + \lambda\|\xi\|_1 + \frac{\beta}{2}\|\gamma - \xi\|_2^2. \tag{5}$$

In (2), the scaling factor vector γ is optimized for both model accuracy and sparsity, which can be difficult to balance when training a CNN. However, in (5), γ is optimized for only model accuracy because it is a variable of the overall loss function $\tilde{\mathcal{L}}(W,\gamma)$ while ξ is optimized only for sparsity because it is penalized by the ℓ_1 norm. The quadratic penalty enforces γ and ξ to be similar in values, thereby ensuring γ to be sparse.

Let (W,γ) be a concatenated vector of W and γ. We minimize (5) via alternating minimization, so for each epoch t, we solve the following subproblems:

$$(W^{t+1},\gamma^{t+1}) \in \arg\min_{W,\gamma} \tilde{\mathcal{L}}(W,\gamma) + \frac{\beta}{2}\|\gamma - \xi^t\|_2^2 \tag{6a}$$

$$\xi^{t+1} \in \arg\min_{\xi} \lambda\|\xi\|_1 + \frac{\beta}{2}\|\gamma^{t+1} - \xi\|_2^2. \tag{6b}$$

Below, we describe how to solve each subproblem in details.

(W,γ)-subproblem. The (W,γ)-subproblem given in (6a) cannot be solved in closed form because the loss function $\tilde{\mathcal{L}}(W,\gamma)$ is a composition of several nonlinear functions. Typically, when training a CNN, this subproblem would be solved by (stochastic) gradient descent. To formulate (6a) as a gradient descent step, we follow a prox-linear strategy as follows:

$$(W^{t+1},\gamma^{t+1}) \in \arg\min_{W,\gamma} \tilde{\mathcal{L}}(W^t,\gamma^t) + \langle \nabla_W \tilde{\mathcal{L}}(W^t,\gamma^t), W - W^t \rangle$$
$$+ \langle \nabla_\gamma \tilde{\mathcal{L}}(W^t,\gamma^t), \gamma - \gamma^t \rangle + \frac{\alpha}{2}\|W - W^t\|_2^2 + \frac{\alpha}{2}\|\gamma - \gamma^t\|_2^2 + \frac{\beta}{2}\|\gamma - \xi^t\|_2^2, \tag{7}$$

where $\alpha > 0$. By differentiating with respect to each variable, setting the partial derivative equal to zero, and solving for the variable, we have

$$W^{t+1} = W^t - \frac{1}{\alpha}\nabla_W \tilde{\mathcal{L}}(W^t,\gamma^t) \tag{8a}$$

$$\gamma^{t+1} = \frac{\alpha\gamma^t + \beta\xi^t}{\alpha + \beta} - \frac{1}{\alpha + \beta}\nabla_\gamma \tilde{\mathcal{L}}(W^t,\gamma^t). \tag{8b}$$

We see that (8a) is gradient descent on W^t with step size $\frac{1}{\alpha}$ while (8b) is gradient descent on a weighted average of γ^t and ξ^t with step size $\frac{1}{\alpha+\beta}$. These steps are straightforward to implement in practice when training a CNN because the gradient $(\nabla_W \tilde{\mathcal{L}}(W^t,\gamma^t), \nabla_\gamma \tilde{\mathcal{L}}(W^t,\gamma^t))$ can be approximated by backpropagation.

ξ-subproblem. To solve (6b), we perform a proximal update by minimizing the following subproblem:

$$\xi^{t+1} \in \arg\min_{\xi} \lambda\|\xi\|_1 + \frac{\alpha}{2}\|\xi - \xi^t\|_2^2 + \frac{\beta}{2}\|\gamma^{t+1} - \xi\|_2^2. \tag{9}$$

Algorithm 1. Proximal NS: proximal algorithm for minimizing (5)

Input: Regularization parameter λ, proximal parameter α, penalty parameter β
    Initialize W^1, ξ^1 with random values.
    Initialize γ^1 such that $\gamma_i = 0.5$ for each channel i.
1: **for** each epoch $t = 1, \ldots, T$ **do**
2:    $W^{t+1} = W^t - \frac{1}{\alpha} \nabla_W \tilde{\mathcal{L}}(W^t, \gamma^t)$ by stochastic gradient descent or variant.
3:    $\gamma^{t+1} = \frac{\alpha\gamma^t + \beta\xi^t}{\alpha+\beta} - \frac{1}{\alpha+\beta}\nabla_\gamma \tilde{\mathcal{L}}(W^t, \gamma^t)$ by stochastic gradient descent or variant.
4:    $\xi^{t+1} = \mathcal{S}\left(\frac{\alpha\xi^t + \beta\gamma^{t+1}}{\alpha+\beta}, \frac{\lambda}{\beta+\alpha}\right).$
5: **end for**

Expanding it gives

$$\xi^{t+1} = \arg\min_\xi \|\xi\|_1 + \frac{1}{2\left(\frac{\lambda}{\beta+\alpha}\right)}\left\|\xi - \frac{\alpha\xi^t + \beta\gamma^{t+1}}{\alpha+\beta}\right\|_2^2 = \mathcal{S}\left(\frac{\alpha\xi^t + \beta\gamma^{t+1}}{\alpha+\beta}, \frac{\lambda}{\beta+\alpha}\right),$$

where $\mathcal{S}(x, \lambda)$ is the soft-thresholding operator defined by $(\mathcal{S}(x, \lambda))_i = \text{sign}(x_i)\max\{0, |x_i| - \lambda\}$ for each entry i. Therefore, ξ is updated by performing soft thresholding on the weighted average between ξ^t and γ^{t+1}.

We summarize the new algorithm for NS in Algorithm 1 as proximal NS.

4   Convergence Analysis

To establish global convergence of proximal NS, we present relevant definitions and assumptions.

Definition 1 ([5]). *A proper, lower-semicontinuous function* $f : \mathbb{R}^m \to (-\infty, \infty]$ *satisfies the Kurdyka-Łojasiewicz (KL) property at a point* $\bar{x} \in \text{dom}(\partial f) := \{x \in \mathbb{R}^m : \partial f(x) \neq \varnothing\}$ *if there exist* $\eta \in (0, +\infty]$, *a neighborhood* U *of* $\bar{x}$, *and a continuous concave function* $\phi : [0, \eta) \to [0, \infty)$ *with the following properties: (i)* $\phi(0) = 0$; *(ii)* ϕ *is continuously differentiable on* $(0, \eta)$; *(iii)* $\phi'(x) > 0$ *for all* $x \in (0, \eta)$; *and (iv) for any* $x \in U$ *with* $f(\bar{x}) < f(x) < f(\bar{x}) + \eta$, *it holds that* $\phi'(f(x) - f(\bar{x}))\text{dist}(0, \partial f(x)) \geq 1$. *If* f *satisfies the KL property at every point* $x \in \text{dom}(\partial f)$, *then* f *is called a KL function.*

Assumption 1. *Suppose that*

a) $\tilde{\mathcal{L}}(W, \gamma)$ *is a proper, differentiable, and nonnegative function.*
b) $\nabla\tilde{\mathcal{L}}(W, \gamma)$ *is Lipschitz continuous with constant* L.
c) $\tilde{\mathcal{L}}(W, \gamma)$ *is a KL function.*

Remark 1. Assumption 1 (a)-(b) are common in nonconvex analysis (e.g., [5]). For Assumption 1, most commonly used loss functions for CNNs are verified to be KL functions [38]. Some CNN architectures do not satisfy Assumption 1(a) when they contain nonsmooth functions and operations, such as the ReLU activation functions and max poolings. However, these functions and operations can be replaced with their smooth approximations. For example, the smooth approximation of ReLU is the softplus function $\frac{1}{c}\log(1 + \exp(cx))$ for some parameter

$c > 0$ while the smooth approximation of the max function for max pooling is the softmax function $\sum_{i=1}^{n} \frac{x_i e^{cx_i}}{\sum_{i=1}^{n} e^{cx_i}}$ for some parameter $c > 0$. Besides, Fu et al.[9] made a similar assumption to establish convergence for their algorithm designed for weight and filter pruning. Regardless, our numerical experiments demonstrate that our proposed algorithm still converges for CNNs containing ReLU activation functions and max pooling.

For brevity, we denote

$$F(W, \gamma, \xi) := \tilde{\mathcal{L}}(W, \gamma) + \lambda \|\xi\|_1 + \frac{\beta}{2}\|\gamma - \xi\|_2^2.$$

Now, we are ready to present the main theorem:

Theorem 1. *Under Assumption 1, if* $\{(W^t, \gamma^t, \xi^t)\}_{t=1}^{\infty}$ *generated by Algorithm 1 is bounded and we have* $\alpha > L$, *then* $\{(W^t, \gamma^t, \xi^t)\}_{t=1}^{\infty}$ *converges to a critical point* (W^*, γ^*, ξ^*) *of* F.

The proof is delayed to the appendix. It requires satisfying the sufficient decrease property in F and the relative error property of ∂F [5].

5   Numerical Experiments

We evaluate proximal NS on VGG-19 [32], DenseNet-40 [14,15], and ResNet-110/164 [12] trained on CIFAR 10/100 [18]. The CIFAR 10/100 dataset [18] consists of 60,000 natural images of resolution 32×32 with 10/100 categories. The dataset is split into two sets: 50,000 training images and 10,000 test images. As done in recent works [12,23], standard augmentation techniques (e.g., shifting, mirroring, and normalization) are applied to the images before training and testing. The code for proximal NS is available at https://github.com/kbui1993/Official-Proximal-Network-Slimming.

5.1   Implementation Details

For CIFAR 10/100, the implementation is mostly the same as in [23]. Specifically, we train the networks from scratch for 160 epochs using stochastic gradient descent with initial learning rate at 0.1 that reduces by a factor of 10 at the 80th and 120th epochs. Moreover, the models are trained with weight decay 10^{-4} and Nesterov momentum of 0.9 without damping. The training batch size is 64. However, the parameter λ is set differently. In our numerical experiments, using Algorithm 1, we set $\xi \sim \text{Unif}[0.47, 0.50]$ for all networks while $\lambda = 0.0045$ and $\beta = 100$ for VGG-19, $\lambda = 0.004$ and $\beta = 100$ for DenseNet-40, and $\lambda = 0.002$ and $\beta = 1.0, 0.25$ for ResNet-110 and ResNet-164, respectively. We have initially $\alpha = 10$, the reciprocal of the learning rate, and it changes accordingly to the learning rate schedule. A model is trained five times on NVIDIA GeForce RTX 2080 for each network and dataset to obtain the average statistics.

Table 1. The average number of scaling factors equal to zero at the end of training. Each architecture is trained five times per dataset.

Architecture	Total Channels/γ_i	CIFAR 10	CIFAR 100
		Avg. Number of $\gamma_i = 0$	Avg. Number of $\gamma_i = 0$
VGG-19	5504	4105.2	3057.0
DenseNet-40	9360	6936.4	6071.6
ResNet-164	12112	8765.4	7115.8

5.2 Results

We apply proximal NS to train VGG-19, DenseNet-40, and ResNet-164 on CIFAR 10/100. According to Table 1, proximal NS drives a significant number of scaling factors to be exactly zeroes for each trained CNN. In particular, for VGG-19 and DenseNet-40, at least 55% of the scaling factors are zeroes while for ResNet-164, at least 58% are zeroes. We can safely remove the channels with zero scaling factors because they are unnecessary for inference. Unlike the original NS [23], proximal NS does not require us to select a scaling factor threshold based on how many channels to remove and how much accuracy to sacrifice.

We compare proximal NS with the original NS [23] and variational CNN pruning (VCP) [40], a Bayesian version of NS. To evaluate the effect of regularization and pruning on accuracy, we include the baseline accuracy, where the architecture is trained without any regularization on the scaling factors. For completeness, the models trained with original NS and proximal NS are fine tuned with the same setting as the first time training but without ℓ_1 regularization on the scaling factors. The results are reported in Tables 2a–2b.

After the first round of training, proximal NS outperforms both the original NS and VCP in test accuracy while reducing a significant amount of parameters and FLOPs. Because proximal NS trains a model towards a sparse structure, the model accuracy is less than the baseline accuracy by at most 1.56% and it remains the same between before and after pruning, a property that the original NS does not have. Although VCP is designed to preserve test accuracy after pruning, it does not compress as well as proximal NS for all architectures. With about the same proportion of channels pruned as the original NS, proximal NS saves more FLOPs for both VGG-19 and ResNet-164 and generally more parameters for all networks.

To potentially improve test accuracy, the pruned models from the original and proximal NS are fine tuned. For proximal NS, test accuracy of the pruned models improve slightly by at most 0.42% for DenseNet-40 and ResNet-164 while worsen for VGGNet-19. Moreover, proximal NS is outperformed by the original NS in fine-tuned test accuracy for all models trained on CIFAR 100.

A more accurate model from original NS might be preferable. However, the additional fine tuning step requires a few more training hours to obtain an accuracy that is up to 1.5% higher than the accuracy of a pruned model trained once by proximal NS. For example, for ResNet-164 trained on CIFAR 100, proximal NS takes about 7 h to attain an average accuracy of 75.26% while the original

Table 2. Results between the different NS methods on CIFAR 10/100. Average statistics are obtained by training the baseline architectures and original NS five times, while the results for variational NS are originally reported from [40].

(a) CIFAR 10

Architecture	Method	Avg. Training Time per Epoch (s) Pre-Pruned/Fine Tuned	% Channels Pruned	% Param. Pruned	% FLOPS Pruned	Test Accuracy (%) Post Pruned/Fine Tuned
VGG-19	Baseline	38.10/—	N/A	N/A	N/A	93.83/—
	Original NS [23]	40.39/29.40	74.00*	90.22	54.67	10.00/93.81
	Proximal NS (ours)	42.71/30.39	74.59	**91.17**	**57.54**	93.71/93.38
DenseNet-40	Baseline	117.45/—	N/A	N/A	N/A	94.25/—
	Original NS [23]	119.49/74.45	74.01	67.13	**60.46**	41.46/93.94
	VCP [40]	Not Reported	60.00	59.67	44.78	93.16/—
	Proximal NS (ours)	118.86/76.10	74.11	**67.75**	57.35	93.58/93.64
ResNet-164	Baseline	146.41/—	N/A	N/A	N/A	94.75/—
	Original NS [23]	151.62/112.80	71.98	52.95	59.27	16.61/93.21
	VCP [40]	Not Reported	74.00	56.70	49.08	93.16/—
	Proximal NS (ours)	150.13/114.26	72.37	**65.84**	**63.54**	93.19/93.41

(b) CIFAR 100

Architecture	Method	Avg. Training Time per Epoch (s) Pre-Pruned/Fine Tuned	% Channels Pruned	% Param. Pruned	% FLOPS Pruned	Test Accuracy (%) Post Pruned/Fine Tuned
VGG-19	Baseline	37.83	N/A	N/A	N/A	72.73/—
	Original NS [23]	39.98/30.74	55.00	78.53	38.66	1.00/72.91
	Proximal NS (ours)	42.31/30.04	55.54	**79.62**	**41.17**	72.81/72.70
DenseNet-40	Baseline	117.17	N/A	N/A	N/A	74.55/—
	Original NS [23]	119.32/77.95	65.01	**59.29**	**52.61**	25.96/74.50
	VCP [40]	Not Reported	37.00	37.73	22.67	72.19/—
	Proximal NS (ours)	120.89/82.92	64.87	59.15	45.00	73.70/73.98
ResNet-164	Baseline	145.37	N/A	N/A	N/A	76.79/—
	Original NS [23]	150.65/115.95	59.00	26.66	45.17	2.39/76.68
	VCP [40]	Not Reported	47.00	17.59	27.16	73.76/—
	Proximal NS (ours)	149.15/117.88	58.75	**42.28**	**47.93**	75.26/75.68

* This is the maximum possible for all five networks to remain functional for inference.

NS requires about 12 h to achieve 1.42% higher accuracy. Therefore, the amount of time and resources spent training for an incremental improvement may not be worthwhile.

Finally, we compare proximal NS with other pruning methods applied to Densenet-40 and ResNet-110 trained on CIFAR 10. The other pruning methods, which may require fine tuning, are L1 [19], GAL [22], and Hrank [21]. For DenseNet-40, proximal NS prunes the most parameters and the second most FLOPs while having comparable accuracy as the fine-tuned Hrank and post-pruned GAL-0.05. For ResNet-110, proximal NS has better compression than L1, GAL-0.5, and Hrank with its post-pruned accuracy better than GAL-0.5's fine-tuned accuracy and similar to L1's fine-tuned accuracy. Although GAL or Hrank might be advantageous to use to obtain a sparse, accurate CNN, they have additional requirements besides fine tuning. GAL [22] requires an accurate baseline model available for knowledge distillation. For Hrank [21], the compression ratio needs to be specified for each convolutional layer, thereby making hyperparameter tuning more complicated (Table 3).

Table 3. Comparison of Proximal NS with other pruning methods on CIFAR 10.

Architecture	Method	% Param./FLOPs Pruned	Test Accuracy (%) Post Pruned/Fine Tuned
DenseNet-40	Hrank [21]	53.80/61.00	—/93.68
	GAL-0.05 [22]	56.70/54.70	93.53/94.50
	Proximal NS (Ours)	67.75/57.54	93.58/93.64
ResNet-110	L1 [19]	32.60/38.70	—/93.30
	GAL-0.5 [22]	44.80/48.50	92.55/92.74
	Hrank [21]	39.40/41.20	—/94.23
	Proximal NS (Ours)	50.70/48.54	93.25/93.27

Overall, proximal NS is a straightforward algorithm that yields a generally more compressed and accurate model than the other methods in one training round. Although its test accuracy after one round is slightly lower than the baseline accuracy, it is expected because of the sparsity–accuracy trade-off and being a prune-while-training algorithm (which automatically identifies the insignificant channels during training) as discussed in [30]. Lastly, the experiments show that fine tuning the compressed models trained by proximal NS marginally improves the test accuracy, which makes fine tuning wasteful.

6   Conclusion

We develop a channel pruning algorithm called proximal NS with global convergence guarantee. It trains a CNN towards a sparse, accurate structure, making fine tuning optional. In our experiments, proximal NS can effectively compress CNNs with accuracy slightly less than the baseline. Because fine tuning CNNs trained by proximal NS marginally improves test accuracy, we will investigate modifying the algorithm to attain significantly better fine-tuned accuracy.

For future direction, we shall study proximal cooperative neural architecture search [34,35] and include nonconvex, sparse regularizers, such as $\ell_1 - \ell_2$ [36] and transformed ℓ_1 [39].

A   Appendix

First, we introduce important definitions and lemmas from variational analysis.

Definition 2 ([27]). *Let $f : \mathbb{R}^n \to (-\infty, +\infty]$ be a proper and lower semicontinuouous function.*

(a) The Fréchet subdifferential of f at the point $x \in domf := \{x \in \mathbb{R}^n : f(x) < \infty\}$ is the set

$$\hat{\partial}f(x) = \left\{ v \in \mathbb{R}^{n^2} : \liminf_{y \neq x, y \to x} \frac{f(y) - f(x) - \langle v, y - x \rangle}{\|y - x\|} \geq 0 \right\}.$$

(b) The limiting subdifferential of f at the point $x \in \text{dom} f$ is the set

$$\partial f(x) = \left\{ v \in \mathbb{R}^{n^2} : \exists \{(x^t, y^t)\}_{t=1}^{\infty} \ \text{s.t.} \ x^t \to x, f(x^t) \to f(x), \hat{\partial} f(x^t) \ni y^t \to y \right\}.$$

Lemma 1 (Strong Convexity Lemma [4]). A function $f(x)$ is called strongly convex with parameter μ if and only if one of the following conditions holds:

a) $g(x) = f(x) - \frac{\mu}{2}\|x\|_2^2$ is convex.
b) $f(y) \geq f(x) + \langle \nabla f(x), y - x \rangle + \frac{\mu}{2}\|y - x\|_2^2, \ \forall x, y.$

Lemma 2 (Descent Lemma [4]). If $\nabla f(x)$ is Lipschitz continuous with parameter $L > 0$, then

$$f(y) \leq f(x) + \langle \nabla f(x), y - x \rangle + \frac{L}{2}\|x - y\|_2^2, \ \forall x, y.$$

For brevity, denote $\tilde{W} := (W, \gamma)$, the overall set of weights in a CNN, and $Z := (\tilde{W}, \xi) = (W, \gamma, \xi)$. Before proving Theorem 1, we prove some necessary lemmas.

Lemma 3 (Sufficient Decrease). Let $\{Z^t\}_{t=1}^{\infty}$ be a sequence generated by Algorithm 1. Under Assumption 1, we have

$$F(Z^{t+1}) - F(Z^t) \leq \frac{L - \alpha}{2}\|Z^{t+1} - Z^t\|_2^2. \tag{10}$$

for all $t \in \mathbb{N}$. In addition, when $\alpha > L$, we have

$$\sum_{t=1}^{\infty} \|Z^{t+1} - Z^t\|_2^2 < \infty. \tag{11}$$

Proof. First we define the function

$$L_t(\tilde{W}) = \tilde{\mathcal{L}}(\tilde{W}^t) + \langle \nabla \tilde{\mathcal{L}}(\tilde{W}^t), \tilde{W} - \tilde{W}^t \rangle + \frac{\alpha}{2}\|\tilde{W} - \tilde{W}^t\|_2^2 + \frac{\beta}{2}\|\gamma - \xi^t\|_2^2. \tag{12}$$

We observe that L_t is strongly convex with respect to $\tilde{W}$ with parameter α. Because $\nabla L_t(\tilde{W}^{t+1}) = 0$ by (7), we use Lemma 1 to obtain

$$L_t(\tilde{W}^t) \geq L_t(\tilde{W}^{t+1}) + \langle \nabla L_t(\tilde{W}^{t+1}), \tilde{W}^t - \tilde{W}^{t+1} \rangle + \frac{\alpha}{2}\|\tilde{W}^{t+1} - \tilde{W}^t\|_2^2$$
$$\geq L_t(\tilde{W}^{t+1}) + \frac{\alpha}{2}\|\tilde{W}^{t+1} - \tilde{W}^t\|_2^2, \tag{13}$$

which simplifies to

$$\tilde{\mathcal{L}}(\tilde{W}^t) + \frac{\beta}{2}\|\gamma^t - \xi^t\|_2^2 - \alpha\|\tilde{W}^{t+1} - \tilde{W}^t\|_2^2 \geq \tilde{\mathcal{L}}(\tilde{W}^t) + \langle \nabla \tilde{\mathcal{L}}(\tilde{W}^t), \tilde{W}^{t+1} - \tilde{W}^t \rangle$$
$$+ \frac{\beta}{2}\|\gamma^{t+1} - \xi^t\|_2^2. \tag{14}$$

Since $\nabla\tilde{\mathcal{L}}(\tilde{W})$ is Lipschitz continuous with constant L, we have

$$\tilde{\mathcal{L}}(\tilde{W}^{t+1}) \leq \tilde{\mathcal{L}}(\tilde{W}^t) + \langle\nabla\tilde{\mathcal{L}}(\tilde{W}^t), \tilde{W}^{t+1} - \tilde{W}^t\rangle + \frac{L}{2}\|\tilde{W}^{t+1} - \tilde{W}^t\|_2^2 \qquad (15)$$

by Lemma 2. Combining the previous two inequalities gives us

$$\tilde{\mathcal{L}}(\tilde{W}^t) + \frac{\beta}{2}\|\gamma^t - \xi^t\|_2^2 + \frac{L-2\alpha}{2}\|\tilde{W}^{t+1} - \tilde{W}^t\|_2^2 \geq \tilde{\mathcal{L}}(\tilde{W}^{t+1}) + \frac{\beta}{2}\|\gamma^{t+1} - \xi^t\|_2^2.$$

Adding the term $\lambda\|\xi^t\|_1$ on both sides and rearranging the inequality give us

$$F(\tilde{W}^{t+1}, \xi^t) - F(Z^t) \leq \frac{L-2\alpha}{2}\|\tilde{W}^{t+1} - \tilde{W}^t\|_2^2 \qquad (16)$$

By (9), we have

$$\lambda\|\xi^{t+1}\|_1 + \frac{\beta}{2}\|\gamma^{t+1} - \xi^{t+1}\|_2^2 + \frac{\alpha}{2}\|\xi^{t+1} - \xi^t\|_2^2 \leq \lambda\|\xi^t\|_1 + \frac{\beta}{2}\|\gamma^{t+1} - \xi^t\|_2^2.$$

Adding $\tilde{\mathcal{L}}(\tilde{W}^{t+1})$ on both sides and rearranging the inequality give

$$F(Z^{t+1}) - F(\tilde{W}^{t+1}, \xi^t) \leq -\frac{\alpha}{2}\|\xi^{t+1} - \xi^t\|_2^2 \qquad (17)$$

Summing up (16) and (17) and rearranging them, we have

$$F(Z^{t+1}) - F(Z^t) \leq \frac{L-2\alpha}{2}\|\tilde{W}^{t+1} - \tilde{W}^t\|_2^2 - \frac{\alpha}{2}\|\xi^{t+1} - \xi^t\|_2^2 \leq \frac{L-\alpha}{2}\|Z^{t+1} - Z^t\|_2^2. \qquad (18)$$

Summing up the inequality for $t = 1, \ldots, N-1$, we have

$$\sum_{t=1}^{N-1} \frac{\alpha - L}{2}\|Z^{t+1} - Z^t\|_2^2 \leq F(Z^1) - F(Z^N) \leq F(Z^1).$$

Because $\alpha > L$, the left-hand side is nonnegative, so as $N \to \infty$, we have (11).

Lemma 4 (Relative error property). *Let $\{Z^t\}_{t=1}^{\infty}$ be a sequence generated by Algorithm 1. Under Assumption 1, for any $t \in \mathbb{N}$, there exists some $w^{t+1} \in \partial F(Z^{t+1})$ such that*

$$\|w^{t+1}\|_2 \leq (3\alpha + 2L + \beta)\|Z^{t+1} - Z^t\|_2. \qquad (19)$$

Proof. We note that

$$\nabla_W\tilde{\mathcal{L}}(\tilde{W}^{t+1}) \in \partial_W F(Z^{t+1}), \qquad (20a)$$

$$\nabla_\gamma\tilde{\mathcal{L}}(\tilde{W}^{t+1}) + \beta(\gamma^{t+1} - \xi^{t+1}) \in \partial_\gamma F(Z^{t+1}), \qquad (20b)$$

$$\lambda\partial_\xi\|\xi^{t+1}\|_1 - \beta(\gamma^{t+1} - \xi^{t+1}) \in \partial_\xi F(Z^{t+1}). \qquad (20c)$$

By the first-order optimality conditions of (7) and (9), we obtain

$$\nabla_W \tilde{\mathcal{L}}(\tilde{W}^t) + \alpha(W^{t+1} - W^t) = 0, \tag{21a}$$

$$\nabla_\gamma \tilde{\mathcal{L}}(\tilde{W}^t) + \alpha(\gamma^{t+1} - \gamma^t) + \beta(\gamma^{t+1} - \xi^t) = 0, \tag{21b}$$

$$\lambda \partial_\xi \|\xi^{t+1}\|_1 + \alpha(\xi^{t+1} - \xi^t) - \beta(\gamma^{t+1} - \xi^{t+1}) \ni 0. \tag{21c}$$

Combining (20a) and (21a), (20b) and (21b), and (20c) and (21c), we obtain

$$\nabla_W \tilde{\mathcal{L}}(\tilde{W}^{t+1}) - \nabla_W \tilde{\mathcal{L}}(\tilde{W}^t) - \alpha(W^{t+1} - W^t) = w_1^{t+1} \in \partial_W F(Z^{t+1}), \tag{22a}$$

$$\nabla_\gamma \tilde{\mathcal{L}}(\tilde{W}^{t+1}) - \nabla_\gamma \tilde{\mathcal{L}}(\tilde{W}^t) - \alpha(\gamma^{t+1} - \gamma^t) - \beta(\xi^{t+1} - \xi^t) = w_2^{t+1} \in \partial_\gamma F(Z^{t+1}), \tag{22b}$$

$$-\alpha(\xi^{t+1} - \xi^t) = w_3^{t+1} \in \partial_\xi F(Z^{t+1}), \tag{22c}$$

where $w^{t+1} = (w_1^{t+1}, w_2^{t+1}, w_3^{t+1}) \in \partial F(Z^{t+1})$. As a result, by triangle inequality and Lipschitz continuity of $\nabla \tilde{\mathcal{L}}$, we have

$$\|w_1^{t+1}\|_2 \le \alpha \|W^{t+1} - W^t\|_2 + \|\nabla_W \tilde{\mathcal{L}}(\tilde{W}^{t+1}) - \nabla_W \tilde{\mathcal{L}}(\tilde{W}^t)\|_2$$
$$\le \alpha \|W^{t+1} - W^t\| + L\|\tilde{W}^{t+1} - \tilde{W}^t\|_2 \le (\alpha + L)\|Z^{t+1} - Z^t\|_2,$$

$$\|w_2^{t+1}\|_2 \le \alpha \|\gamma^{t+1} - \gamma^t\|_2 + \beta\|\xi^{t+1} - \xi^t\|_2 + \|\nabla_\gamma \tilde{\mathcal{L}}(\tilde{W}^{t+1}) - \nabla_\gamma \tilde{\mathcal{L}}(\tilde{W}^t)\|_2$$
$$\le (\alpha + L)\|\tilde{W}^{t+1} - \tilde{W}^t\|_2 + \beta\|\xi^{t+1} - \xi^t\|_2 \le (\alpha + \beta + L)\|Z^{t+1} - Z^t\|_2,$$

and

$$\|w_3^{t+1}\|_2 \le \alpha\|\xi^{t+1} - \xi^t\|_2 \le \alpha\|Z^{t+1} - Z^t\|_2.$$

Therefore, for all $t \in \mathbb{N}$, we have

$$\|w^{t+1}\|_2 \le \|w_1^{t+1}\|_2 + \|w_2^{t+1}\|_2 + \|w_3^{t+1}\|_2 \le (3\alpha + 2L + \beta)\left\|Z^{t+1} - Z^t\right\|_2.$$

Proof (Proof of Theorem 1). The result follows from Lemmas 3–4 combined with [5, Theorem 1]

References

1. Aghasi, A., Abdi, A., Nguyen, N., Romberg, J.: Net-trim: convex pruning of deep neural networks with performance guarantee. In: Advances in Neural Information Processing Systems, pp. 3177–3186 (2017)
2. Aghasi, A., Abdi, A., Romberg, J.: Fast convex pruning of deep neural networks. SIAM J. Math. Data Sci. **2**(1), 158–188 (2020)
3. Alvarez, J.M., Salzmann, M.: Learning the number of neurons in deep networks. In: Advances in Neural Information Processing Systems, pp. 2270–2278 (2016)
4. Beck, A.: First-order methods in optimization. SIAM (2017)
5. Bolte, J., Sabach, S., Teboulle, M.: Proximal alternating linearized minimization for nonconvex and nonsmooth problems. Math. Program. **146**(1), 459–494 (2014)

6. Bui, K., Park, F., Zhang, S., Qi, Y., Xin, J.: Nonconvex regularization for network slimming: compressing CNNs even more. In: Bebis, G., et al. (eds.) Advances in Visual Computing. ISVC 2020. LNCS, vol. 12509, pp. 39–53. Springer, Cham (2020). https://doi.org/10.1007/978-3-030-64556-4_4

7. Bui, K., Park, F., Zhang, S., Qi, Y., Xin, J.: Improving network slimming with nonconvex regularization. IEEE Access **9**, 115292–115314 (2021)

8. Bui, K., Park, F., Zhang, S., Qi, Y., Xin, J.: Structured sparsity of convolutional neural networks via nonconvex sparse group regularization. Front. Appl. Math. Stat. (2021)

9. Fu, Y., et al.: Exploring structural sparsity of deep networks via inverse scale spaces. IEEE Trans. Pattern Anal. Mach. Intell. (2022)

10. Girshick, R., Donahue, J., Darrell, T., Malik, J.: Rich feature hierarchies for accurate object detection and semantic segmentation. In: Proceedings of the IEEE Conference on Computer Vision and Pattern Recognition, pp. 580–587 (2014)

11. Han, S., Pool, J., Tran, J., Dally, W.: Learning both weights and connections for efficient neural network. In: Advances in Neural Information Processing Systems, pp. 1135–1143 (2015)

12. He, K., Zhang, X., Ren, S., Sun, J.: Deep residual learning for image recognition. In: Proceedings of the IEEE Conference on Computer Vision and Pattern Recognition, pp. 770–778 (2016)

13. Hu, H., Peng, R., Tai, Y.W., Tang, C.K.: Network trimming: a data-driven neuron pruning approach towards efficient deep architectures. arXiv preprint arXiv:1607.03250 (2016)

14. Huang, G., Liu, Z., Pleiss, G., Van Der Maaten, L., Weinberger, K.: Convolutional networks with dense connectivity. IEEE Trans. Pattern Anal. Mach. Intell. (2019)

15. Huang, G., Liu, Z., Van Der Maaten, L., Weinberger, K.Q.: Densely connected convolutional networks. In: Proceedings of the IEEE Conference on Computer Vision and Pattern Recognition, pp. 4700–4708 (2017)

16. Huang, J., et al.: Speed/accuracy trade-offs for modern convolutional object detectors. In: Proceedings of the IEEE Conference on Computer Vision and Pattern Recognition, pp. 7310–7311 (2017)

17. Ioffe, S., Szegedy, C.: Batch normalization: accelerating deep network training by reducing internal covariate shift. In: International Conference on Machine Learning, pp. 448–456 (2015)

18. Krizhevsky, A., Hinton, G., et al.: Learning multiple layers of features from tiny images. Technical report, University of Toronto (2009)

19. Li, H., Kadav, A., Durdanovic, I., Samet, H., Graf, H.P.: Pruning filters for efficient convnets. arXiv preprint arXiv:1608.08710 (2016)

20. Li, Y., Gu, S., Mayer, C., Gool, L.V., Timofte, R.: Group sparsity: the hinge between filter pruning and decomposition for network compression. In: Proceedings of the IEEE/CVF Conference on Computer Vision and Pattern Recognition (CVPR), June 2020

21. Lin, M., et al.: Hrank: filter pruning using high-rank feature map. In: Proceedings of the IEEE/CVF Conference on Computer Vision and Pattern Recognition, pp. 1529–1538 (2020)

22. Lin, S., et al.: Towards optimal structured CNN pruning via generative adversarial learning. In: Proceedings of the IEEE/CVF Conference on Computer Vision and Pattern Recognition, pp. 2790–2799 (2019)

23. Liu, Z., Li, J., Shen, Z., Huang, G., Yan, S., Zhang, C.: Learning efficient convolutional networks through network slimming. In: Proceedings of the IEEE Conference on Computer Vision and Pattern Recognition, pp. 2736–2744 (2017)

24. Meng, F., et al.: Pruning filter in filter. Adv. Neural Inf. Process. Syst. **33**, 17629–17640 (2020)
25. Noll, D.: Convergence of non-smooth descent methods using the Kurdyka-Łojasiewicz inequality. J. Optim. Theory Appl. **160**(2), 553–572 (2014)
26. Ren, S., He, K., Girshick, R., Sun, J.: Faster R-CNN: towards real-time object detection with region proposal networks. In: Advances in Neural Information Processing Systems, pp. 91–99 (2015)
27. Rockafellar, R.T., Wets, R.J.B.: Variational Analysis, vol. 317. Springer Science & Business Media, Berlin, Heidelberg (2009). https://doi.org/10.1007/978-3-642-02431-3
28. Santurkar, S., Tsipras, D., Ilyas, A., Madry, A.: How does batch normalization help optimization? Adv. Neural Inf. Process. Syst. **31** (2018)
29. Scardapane, S., Comminiello, D., Hussain, A., Uncini, A.: Group sparse regularization for deep neural networks. Neurocomputing **241**, 81–89 (2017)
30. Shen, M., Molchanov, P., Yin, H., Alvarez, J.M.: When to prune? A policy towards early structural pruning. In: Proceedings of the IEEE/CVF Conference on Computer Vision and Pattern Recognition, pp. 12247–12256 (2022)
31. Shor, N.Z.: Minimization Methods for Non-Differentiable Functions, vol. 3. Springer Science & Business Media, Berlin, Heidelberg (2012). https://doi.org/10.1007/978-3-642-82118-9
32. Simonyan, K., Zisserman, A.: Very deep convolutional networks for large-scale image recognition. arXiv preprint arXiv:1409.1556 (2014)
33. Wen, W., Wu, C., Wang, Y., Chen, Y., Li, H.: Learning structured sparsity in deep neural networks. In: Advances in Neural Information Processing Systems, pp. 2074–2082 (2016)
34. Xue, F., Qi, Y., Xin, J.: RARTS: an efficient first-order relaxed architecture search method. IEEE Access **10**, 65901–65912 (2022)
35. Xue, F., Xin, J.: Network compression via cooperative architecture search and distillation. In: IEEE International Conference on AI for Industries, pp. 42–43 (2021)
36. Yin, P., Lou, Y., He, Q., Xin, J.: Minimization of ℓ_{1-2} for compressed sensing. SIAM J. Sci. Comput. **37**(1), A536–A563 (2015)
37. Zagoruyko, S., Komodakis, N.: Wide residual networks. arXiv preprint arXiv:1605.07146 (2016)
38. Zeng, J., Lau, T.T.K., Lin, S., Yao, Y.: Global convergence of block coordinate descent in deep learning. In: International Conference on Machine Learning, pp. 7313–7323. PMLR (2019)
39. Zhang, S., Xin, J.: Minimization of transformed l_1 penalty: theory, difference of convex function algorithm, and robust application in compressed sensing. Math. Program. **169**(1), 307–336 (2018)
40. Zhao, C., Ni, B., Zhang, J., Zhao, Q., Zhang, W., Tian, Q.: Variational convolutional neural network pruning. In: Proceedings of the IEEE/CVF Conference on Computer Vision and Pattern Recognition, pp. 2780–2789 (2019)

Diversity in Deep Generative Models and Generative AI

Gabriel Turinici[(✉)][iD]

CEREMADE, Université Paris Dauphine - PSL, CNRS, Paris, France
gabriel.turinici@dauphine.fr
https://turinici.com

Abstract. The decoder-based machine learning generative algorithms such as Generative Adversarial Networks (GAN), Variational Auto-Encoders (VAE), Transformers show impressive results when constructing objects similar to those in a training ensemble. However, the generation of new objects builds mainly on the understanding of the hidden structure of the training dataset followed by a sampling from a multi-dimensional normal variable. In particular each sample is independent from the others and can repeatedly propose same type of objects. To cure this drawback we introduce a kernel-based measure quantization method that can produce new objects from a given target measure by approximating it as a whole and even staying away from elements already drawn from that distribution. This ensures a better diversity of the produced objects. The method is tested on classic machine learning benchmarks.

Keywords: variational auto-encoder · generative models · measure quantization · generative AI · generative neural networks

1 Introduction, Motivation and Literature Review

We investigate in this work an approach to enhance diversity in decoder-based generative AI paradigms i.e., when generating objects (e.g., images) similar to the content of a given (training) dataset. Such procedures received a large audience in the last years, especially after the introduction of several deep neural network architectures widely used today: the Generative Adversarial Networks [3,5,16] (hereafter named GAN), the Variational Auto-Encoders (VAE), see [8,9,15] and the Transformer [20].

All approaches use a small-dimensional set of parameters called *latent space* of dimension L as a companion representation for any object of initial dimension N; for instance, for RGB color pictures N will be 3 times the number of pixels. We will take VAE as an example. At the high level of description, an object e.g., an image, is a vector in $\mathbb{R}^N$; the training dataset becomes a set of points in $\mathbb{R}^N$ and it is hypothesized that it corresponds to some distribution $\mu(dx)$[1] on $\mathbb{R}^N$ of which the dataset is an empirical sampling. The object distribution μ is mapped by the encoder part of the VAE into an empirical distribution μ_L on the latent space $\mathbb{R}^L$; here L is much smaller than N and represents the essential degrees of freedom. The existence of such a L is a crucial hypothesis of most generative AI procedures. The optimization routines of VAE

---

[1] The notation $\mu(dx)$ means that μ is a distribution of objects in $\mathbb{R}^N$ with generic variable x.

© The Author(s), under exclusive license to Springer Nature Switzerland AG 2024
G. Nicosia et al. (Eds.): LOD 2023, LNCS 14506, pp. 84–93, 2024.
https://doi.org/10.1007/978-3-031-53966-4_7

ensure that this mapping of the training dataset as a probability distribution on $\mathbb{R}^L$ will be as close as possible to some target (ideal) latent distribution, chosen usually to be the multi-variate normal distribution $\mathcal{N}(0_L, \mathrm{Id}_L)^2$; here 0_L is the zero vector in $\mathbb{R}^L$ and Id_L is the identity matrix in $\mathbb{R}^{L \times L}$. The GAN and Transformer architectures operate a bit differently but in all cases, generating J new objects resumes to drawing J independent new samples $X_1, ..., X_J$ from $\mathcal{N}(0_L, \mathrm{Id}_L)$. These samples are then "decoded", i.e., passed through a neural network implementing a mapping $D : \mathbb{R}^L \to \mathbb{R}^N$ with $D(X_j)$ being the object e.g., image, corresponding to the latent representation $X_j \in \mathbb{R}^L$ for any $j \leq J$.

Since the J random variables $X_1, ..., X_J$ are independent, some X_j end up being very similar and the decoded objects $D(X_j)$ may lack diversity.

The goal of this work is to propose a method to enforce this diversity. We do this by relating the samples X_j through the requirement that $\frac{1}{J} \sum_{j=1}^{J} \delta_{X_j}$ be as close as possible to the empirical distribution μ_L on the latent space or the ideal latent distribution $\mathcal{N}(0_L, \mathrm{Id}_L)$; here δ_x is a general notation for the Dirac measures centered at the value x. Such a goal will ensure that the objects X_j cover well the (empirical) latent distribution μ_L and the generated objects $D(X_j)$ have adequate diversity.

1.1 Short Literature Review

Representing a target measure μ by the means of a sum of Dirac measures is similar in principle to the "vector quantization" approaches [6, 12] that divide the support of μ in several regions, called Voronoi cells, and replace the values in any Voronoi cell by the value at the center of the cell. It results a "quantized" set of values hence the name of the method.

Our approach is similar to this method, but from a technical point of view we do not work in the Wasserstein norm (which is used by the vector quantization algorithms) but instead our proposal is based on a kernel which gives rise to interesting analytic properties (for instance the distance to a normal can be calculated explicitly, see [17] for details).

From the computational point of view, our contribution is similar to the "energy statistic" (see [13, 14]) but with the modification that we use a kernel which is not exactly $|x|$ but a smooth approximation.

On the other hand, diversity has been evoked in a recent work [2] in the context of GANs when avoiding the "mode collapse". The authors proposes a new GAN framework called diversified GAN (DivGAN) that aims at encouraging the GANs to produce diverse data. The DivGAN module computes a metric called "contrastive loss" that indicates with the level of diversity in the sample. Their approach has objectives aligned with our but instead of the contrastive loss we use state of the art kernel-based statistical distance as in [17]. Note also the approach of [10] that uses uses conditional GANs to avoid mode collapse.

Finally, for a more general discussion on the diversity and fidelity metrics see [11] that propose new metrics to upgrade the standard ones like the Inception Score (IS) and the Frechet Inception Distance (FID).

---

2 Some other choices exist, a popular one being a mixture of normal variables [4].

2   Representation of the Target Distribution

We describe in this section the main part of our procedure. The procedure is based on the minimization, with respect to X_j, of the distance between the Dirac measure sum $\frac{1}{J} \sum_{j=1}^{J} \delta_{X_j}$ and the target distribution μ_L or $\mathcal{N}(0_L, \mathrm{Id}_L)$ (recall that the goal of the VAE is to render μ_L very close to $\mathcal{N}(0_L, \mathrm{Id}_L)$). We use the Adam [7] stochastic optimization algorithm to minimize the distance but one can also choose Nesterov, momentum, SGD etc.

To compute the distance between two sets of Dirac measures we employ the following metric:

$$
d \left(\frac{1}{J} \sum_{j=1}^{J} \delta_{X_j}, \frac{1}{B} \sum_{b=1}^{B} \delta_{z_b} \right)^2 = \frac{\sum_{j,b=1}^{J,B} h(X_j - z_b)}{JB}
$$

$$
- \frac{\sum_{j,j'=1}^{J} h(X_j - X_{j'})}{2\, J^2} - \frac{\sum_{b,b'=1}^{B} h(z_b - z_{b'})}{2 B^2}, \tag{1}
$$

where the kernel h is defined for any $a \geq 0$ by $h(x) = \sqrt{\|x\|^2 + a^2} - a$. The distance $d(\cdot, \cdot)$ is a kernel-based statistical distance; we refer to [17, 18] for considerations its usefulness and properties. Note that in particular it is not obvious that $d \left(\frac{1}{J} \sum_{j=1}^{J} \delta_{X_j}, \frac{1}{B} \sum_{b=1}^{B} \delta_{z_b} \right) \geq 0$ for any choice of vectors X and z, but this can be proven with tools from the theory of Reproducing Kernel Hilbert Spaces (RKHS); in particular in [18] it is proven that such a distance is of Gaussian mixture type. This statistical distance can be extended also to general measures (not only sums of Diracs) by the definition:

$$
d(\mu_1, \mu_2)^2 = -\frac{1}{2} \int_{\mathbb{R}^L} \int_{\mathbb{R}^L} h(x - y)(\mu_1 - \mu_2)(dx)(\mu_1 - \mu_2)(dy). \tag{2}
$$

With these provisions we can introduce our two algorithms below A1 and A2. Both use repeated sampling from the target distribution in order to stochastically minimize the distance from the candidate $\frac{1}{J} \sum_{j=1}^{J} \delta_{X_j}$ to the target distribution. The difference between the two algorithms is the following: A1 samples from the ideal distribution $\mathcal{N}(0_L, \mathrm{Id}_L)$ while A2 samples from the empirical distribution μ_L.

The parameters of the Adam algorithm were set to the defaults. Note that here the unknowns are coordinates of the vectors X_j and the goal of the algorithm is to find the optimal X_j, $j = 1, .., J$. To do so, the stochastic optimization algorithm needs to compute the gradient, with respect to X of the loss function. Such a computation is done with the usual tools of back-propagation even if here the learned parameters do not correspond to neural network layers.

Note that, even if we describe the measure representation for the particular situation that have as target a multi-dimensional normal distribution, the procedure above can be generalized to any other targets. On the other hand, for the normal distribution, the distance from a sum of Dirac measures to the normal distribution can be computed analytically as in [17]; such an analytic formula renders the sampling from the target distribution useless and the whole stochastic optimization in Algorithm A1 can be

Algorithm A1. Diversity sampling algorithm : ideal target case $\mathcal{N}(0_L, \mathrm{Id}_L)$

Inputs : batch size B, parameter $a = 10^{-6}$.
Outputs : quantized points X_j, $j = 1..., J$.
1: **procedure**
2:    initialize points $X = (X_j)_{j=1}^J$ sampled i.i.d from $\mathcal{N}(0_L, \mathrm{Id}_L)$;
3:    **while** (max iteration not reached) **do**
4:        sample i.i.d $z_1, ..., z_B \sim \mathcal{N}(0_L, \mathrm{Id}_L)$;
5:        compute the global loss $L(X) := d\left(\frac{1}{J}\sum_{j=1}^J \delta_{X_j}, \frac{1}{B}\sum_{b=1}^B \delta_{z_b}\right)^2$ as in eq. (1) ;
6:        update X by performing one step of the Adam algorithm to minimize $L(X)$.
7:    **end while**
8: **end procedure**

replaced by a deterministic optimization; we tested the procedure and the results were coherent with what is reported here.

Algorithm A2. Diversity sampling algorithm : empirical target μ_L

Inputs : batch size B, parameter $a = 10^{-6}$, measure μ_L stored previously or computed on the fly.
Outputs : quantized points X_j, $j = 1..., J$.
1: **procedure**
2:    initialize points $X = (X_j)_{j=1}^J$ sampled i.i.d from μ_L;
3:    **while** (max iteration not reached) **do**
4:        sample i.i.d. $z_1, ..., z_B \sim \mu_L$;
5:        compute the global loss $L(X) := d\left(\frac{1}{J}\sum_{j=1}^J \delta_{X_j}, \frac{1}{B}\sum_{b=1}^B \delta_{z_b}\right)^2$ as in eq. (1) ;
6:        update X by performing one step of the Adam algorithm to minimize $L(X)$.
7:    **end while**
8: **end procedure**

The Algorithm A2 is tailored specifically for VAE. It samples from the empirical distribution μ_L in several possible distinct manners:

1. store, during the last epoch of the VAE convergence, the latent points and construct μ_L as a list of Dirac masses. There is no additional computation cost but memory is required to store the data; memory consumption is usually not large because the latent space has very reduced dimension compared to the initial dataset;
2. as previously but the computation is done **after** the last epoch of the GAN/VAE/Transformer; the computational cost increases with less than the cost of one additional epoch of the algorithm (no decoding and no need to compute gradients);
3. on the fly: when sampling from μ_L is required, select at random objects from the initial dataset and encode them into the latent space. The cost is that of the encoding step.

In practice in the numerical tests we selected alternative 2 which gives best quality at a very reasonable cost.

3   Numerical Results for the Ideal Sampling Algorithm A1

All the experiments below are available on the Github site [18] and also as Zenodo repository [19]. In order to test both algorithm we used, even for the Algorithm A1, a VAE setting.

3.1   The VAE Design

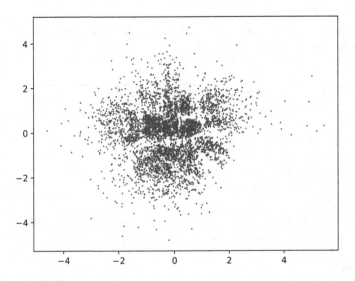

Fig. 1. The latent space representation of the MNIST dataset; we used 5000 images in the dataset. Each image is encoded and its corresponding 2D latent vector is plotted. Compare with Fig. 2 that displays some decoded images; the latent distribution is close to a 2D Gaussian but is not fully so.

We take as example the MNIST dataset (similar results, not shown here, were obtained for the Fashion-MNIST dataset [21]) and generate new images through a VAE; more precisely we use a standard VAE which is the CVAE in the Tensorflow tutorial [1]; however, in order to gain in quality, we replace all convolution Conv2D layers by fully connected (FC) layers (size $28 * 28$) which results in the following encoder/decoder architecture:

Encoder: input 28×28 images; followed by 5 Relu FC layers of dimension $28 * 28$ and a final dense layer of dimension $2L$ (no activation).

Decoder: 4 Relu FC layers of dimension $28 * 28$ and a final dense layer of dimension $28 * 28$ (no activation).

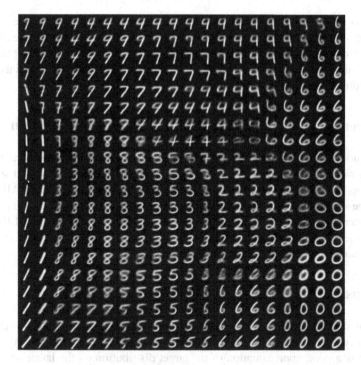

Fig. 2. The latent space representation of the MNIST dataset; we use the same approach as in [1] and sample the distribution with $Q = 20$ equidistant (quantile-wise) points, for instance the point in the lattice at line i_1 and column i_2 corresponds to the i_1/Q-th quantile in the first dimension and i_2/Q-th quantile in the second direction (for the normal distribution). For each such a point we draw the image associated by the decoder $D(\cdot)$ to that point.

The latent space dimension is $L = 2$; the encoding mapping with respect to the image dataset is presented in Figs. 1 and 2 where a good quality is observed, even if some figures, like the 4 and 3 are not well represented (all 4 resemble very much to a 9). Note that although the latent space distribution is close to a 2D Gaussian it is not exactly so. This will affect the quality of the generated images which is not yet optimal.

3.2   Diversity Enforcing Sampling

The Algorithm A1 is used to sample $J = 10$ points from the ideal latent distribution $\mathcal{N}(0_2, \mathrm{Id}_2)$ (recall $L = 2$); these points are then run through the decoder and we compare them with random i.i.d. sampling (plus decoder phase). We see in Fig. 4 that the random i.i.d. sampling has many repetitions (depending on sampling the number of repetitions may vary); on the contrary, the diversity enforcing sampling in the second row images has fewer repetitions (a 9 can be seen as close to a 4 given the latent space in Fig. 2; same a figure which is close to a 3). Of course, the quality of the sampling depends on the initial VAE quality; one component of the VAE quality is the latent distribution which, as illustrated in Fig. 1 can still be improved to match a 2D Gaussian. Since the empirical latent distribution of the dataset, depicted as the blue points in Figs. 1 and 3, does not match perfectly the target $\mathcal{N}(0_2, \mathrm{Id}_2)$ distribution, the diversity enforcing sampling, which use $\mathcal{N}(0_2, \mathrm{Id}_2)$, will not represent an optimal sample for the empirical latent distribution; this is seen in Fig. 3 where the red and black points do not seem to represent optimally the blue points.

4   Numerical Results for the Empirical Sampling Algorithm A2

We move now to the results for the Algorithm A2. As indicated previously, after VAE converged we run a new epoch by asking VAE to encode all the dataset and store the latent points obtained. This was used as input for the Algorithm A2. The VAE setting remains the same. The results are presented in Fig. 4 (third row). The numerical results appear better than those in Sect. 3. This can be explained by the quality of the sampling from μ_L as illustrated in Fig. 5 where the sampling in the latent space appear to represent more accurately the empirical distribution μ_L.

5   Discussion and Final Remarks

We presented a procedure to enforce diversity in the decoder-based generative networks. The diversity is ensured by drawing simultaneously all samples and ensuring that the overall set is a good approximation of the target distribution on the latent space. Two algorithms were proposed and tested numerically on standard learning datasets. Each procedure strikes a different balance between efficiency and quality: Algorithm A1 is very fast and should be used when the generative algorithm converged well and the

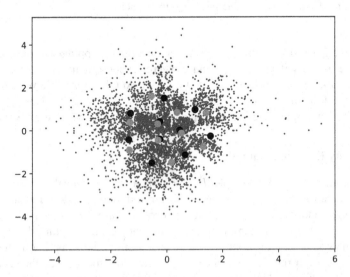

Fig. 3. The latent space representation of the MNIST dataset together with the two sets of latent points corresponding to diversity sampling depicted in the second row of Fig. 4. Blue points are latent distribution points μ_L; red and black points are the two sets of results $X = (X_j)_{j=1}^J$ of the two runs of the Algorithm A1 (red =first run, black= second run).

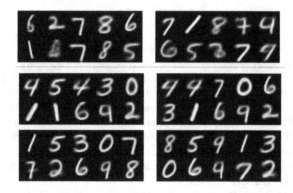

Fig. 4. Diversity sampling results from Algorithms A1 and A2. **First row pictures :** I.i.d. sampling of $J = 10$ points from the target latent distribution (2D normal) and their corresponding images (after decoding); we took two independent samplings in order to show that figure repetition is a common feature of these samplings. The non-figure image in the second line second column is just a VAE artifact due to the fact that the latent distribution is **not** the target 2D Gaussian, so the image is not like images in the dataset. **Second row pictures :** results of Algorithm A1. The repetitions present in the initial i.i.d sampling (e.g. 6, 7, 8, etc.) are much less present; figures never present in the first row (e.g. 3) appear here. **Third row pictures :** results of Algorithm A2. Results improve with respect to the second row (Algorithm A1), only one repetition present.

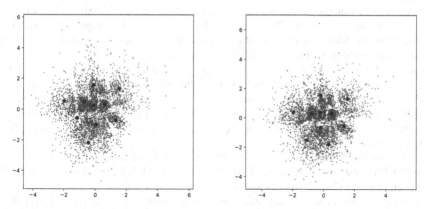

Fig. 5. The latent space representation of the MNIST dataset together with the two sets of latent points corresponding to diversity sampling depicted in the third row of Fig. 4. Blue points are latent distribution points μ_L; orange points are the results $X = (X_j)_{j=1}^{J}$ of the Algorithm A2.

empirical latent distribution μ_L can be supposed close to the ideal value $\mathcal{N}(0_L, \mathrm{Id}_L)$; on the other hand the Algorithm A2 uses the actual latent distribution μ_L and obtains better quality results but at the cost of storing μ_L (or calculating it on the fly) and can be used when the GAN/VAE/Transformer quality is not optimal. This is the one we recommend in routine practice. Nevertheless, both methods confirm the initial expectations as procedures to enhance the diversity of the generative process.

References

1. Cvae, tensorflow documentation. https://www.tensorflow.org/tutorials/generative/cvae. Accessed 30 Jan 2022
2. Allahyani, M.,et al.: DivGAN: a diversity enforcing generative adversarial network for mode collapse reduction. Artif. Intell. **317**, 103863 (2023). https://doi.org/10.1016/j.artint.2023. 103863, https://www.sciencedirect.com/science/article/pii/S0004370223000097
3. Arjovsky, M., Chintala, S., Bottou, L.: Wasserstein generative adversarial networks. In: Precup, D., Teh, Y.W. (eds.) Proceedings of the 34th International Conference on Machine Learning. Proceedings of Machine Learning Research, vol. 70, pp. 214–223. PMLR, International Convention Centre, Sydney, Australia, 06–11 August 2017. http://proceedings.mlr. press/v70/arjovsky17a.html
4. Dilokthanakul, N., et al.: Deep unsupervised clustering with Gaussian mixture variational autoencoders (2017). arXiv:1611.02648
5. Goodfellow, I., et al.: Generative adversarial nets. In: Ghahramani, Z., Welling, M., Cortes, C., Lawrence, N.D., Weinberger, K.Q. (eds.) Advances in Neural Information Processing Systems, vol. 27, pp. 2672–2680. Curran Associates, Inc. (2014). http://papers.nips.cc/paper/ 5423-generative-adversarial-nets.pdf
6. Graf, S., Luschgy, H.: Foundations of Quantization for Probability Distributions. Springer, Berlin, Heidelberg (2007). https://doi.org/10.1007/BFb0103945
7. Kingma, D.P., Ba, J.: Adam: a method for stochastic optimization (2014). arxiv:1412.6980
8. Kingma, D.P., Max, W.: An Introduction to Variational Autoencoders. Now Publishers Inc., Delft, November 2019
9. Kingma, D.P., Welling, M.: Auto-encoding variational Bayes (2013). arxiv:1312.6114
10. Liu, S., Wang, T., Bau, D., Zhu, J.Y., Torralba, A.: Diverse image generation via self-conditioned GANs. In: Proceedings of the IEEE/CVF Conference on Computer Vision and Pattern Recognition (CVPR), pp. 14286–14295, June 2020
11. Naeem, M.F., Oh, S.J., Uh, Y., Choi, Y., Yoo, J.: Reliable fidelity and diversity metrics for generative models. In: III, H.D., Singh, A. (eds.) Proceedings of the 37th International Conference on Machine Learning. Proceedings of Machine Learning Research, vol. 119, pp. 7176–7185. PMLR, 13–18 Jul 2020. https://proceedings.mlr.press/v119/naeem20a.html
12. Gray, R.: Vector quantization. IEEE ASSP Mag. **1**(2), 4–29 (1984). https://doi.org/10.1109/ MASSP.1984.1162229
13. Sriperumbudur, B.K., Fukumizu, K., Lanckriet, G.R.G.: Universality, characteristic kernels and RKHS embedding of measures. J. Mach. Learn. Res. **12**(70), 2389–2410 (2011). http:// jmlr.org/papers/v12/sriperumbudur11a.html
14. Szekely, G.J., Rizzo, M.L.: Energy statistics: a class of statistics based on distances. J. Stat. Plan. Inference **143**(8), 1249–1272 (2013). https://doi.org/10.1016/j.jspi.2013.03.018, http:// www.sciencedirect.com/science/article/pii/S0378375813000633
15. Tabor, J., Knop, S., Spurek, P., Podolak, I.T., Mazur, M., Jastrzebski, S.: Cramer-Wold AutoEncoder. CoRR abs/1805.09235 (2018). http://arxiv.org/abs/1805.09235
16. Tolstikhin, I., Bousquet, O., Gelly, S., Schoelkopf, B.: Wasserstein auto-encoders (2017). arxiv:1711.01558
17. Turinici, G.: Radon-sobolev variational auto-encoders. Neural Netw. **141**, 294–305 (2021). https://doi.org/10.1016/j.neunet.2021.04.018, https://www.sciencedirect.com/ science/article/pii/S0893608021001556
18. Turinici, G.: Huber energy measure quantization, December 2022. https://doi.org/10.5281/ zenodo.7406032. https://github.com/gabriel-turinici/Huber-energy-measure-quantization. original-date: 2022-08-25T14:07:16Z

19. Turinici, G.: Supporting files for the paper Diversity in deep generative models and generative AI, sept 2023 version, September 2023. https://doi.org/10.5281/zenodo.7922519
20. Vaswani, A., et al..: Attention is all you need (2017). https://doi.org/10.48550/ARXIV.1706.03762, https://arxiv.org/abs/1706.03762
21. Xiao, H., Rasul, K., Vollgraf, R.: Fashion-MNIST: a novel image dataset for benchmarking machine learning algorithms. CoRR abs/1708.07747 (2017). http://arxiv.org/abs/1708.07747

Improving Portfolio Performance Using a Novel Method for Predicting Financial Regimes

Piotr Pomorski and Denise Gorse(✉) ⓘ

University College London, London WC1E 6BT, UK
{p.pomorski,d.gorse}@cs.ucl.ac.uk

Abstract. This work extends a previous work in regime detection, which allowed trading positions to be profitably adjusted when a new regime was detected, to ex ante prediction of regimes, leading to substantial performance improvements over the earlier model, over all three asset classes considered (equities, commodities, and foreign exchange), over a test period of four years. The proposed new model is also benchmarked over this same period against a hidden Markov model, the most popular current model for financial regime prediction, and against an appropriate index benchmark for each asset class, in the case of the commodities model having a test period cost-adjusted cumulative return over four times higher than that expected from the index. Notably, the proposed model makes use of a contrarian trading strategy, not uncommon in the financial industry but relatively unexplored in machine learning models. The model also makes use of frequent short positions, something not always desirable to investors due to issues of both financial risk and ethics; however, it is discussed how further work could remove this reliance on shorting and allow the construction of a long-only version of the model.

Keywords: Financial regime prediction · Trading · Random forest

1 Introduction

Systematic methodologies that attempt to exploit patterns in financial time series have been around for a long time, in the US dating back to the late 1800s and the work of Charles Dow (the creator of Dow Jones Industrial Average), and in Japan and the Netherlands even longer. However, this field, referred to in the financial industry as *technical analysis* (TA) fell into disrepute in the 1980s due to the rise of the Efficient Markets Hypothesis (EMH), which claims it is impossible to 'beat the market' using past data, since the stock price already incorporates historical and current information. But the EMH has since itself been challenged by the emerging field of behavioural finance, and the work of Andrew Lo and others [16] has evidenced that there are indeed exploitable signals buried in the noise presented by financial time series data. A further reason for this recovered optimism has been the rise of machine learning (ML) models for

G. Nicosia et al. (Eds.): LOD 2023, LNCS 14506, pp. 94–108, 2024.
https://doi.org/10.1007/978-3-031-53966-4_8

financial time series forecasting, able to discover weaker, and subtler, signals than could otherwise be detected. While some ML models use deep learning to uncover patterns in raw data, the majority use engineered features, including ones derived from TA, to reduce the dimensionality of the data and add value to it, with the strongest-evidenced (for example, in [19]) TA tools being of the moving average type. Moving averages do, however, have issues with the choice of window length, though these can be overcome by the use of adaptive moving averages, in particular *Kaufman's adaptive moving average* (KAMA) [12].

Recently, in [22], KAMA, adept in trend (upward, or 'bullish', versus downward, or 'bearish') detection, was combined with the *two-state Markov switching regression* (MSR) model [13], as a means of distinguishing volatility regimes, in order to separate financial time series into four states: low-variance (LV) bullish, low-variance (LV) bearish, high-variance (HV) bullish, and high-variance (HV) bearish, with the detected regimes then used by the resulting *KAMA+MSR* model as the basis of a profitable trading strategy. However, any strategy based on regime detection will be subject to inevitable losses, due to having remained in a now-disadvantageous trading position after the market has switched regime, or due to not having sufficiently quickly taken advantage of a new trading opportunity. This work therefore builds on the work of [22], by using KAMA+MSR to appropriately label periods of the training data, these labels being then used as targets for a random forest (RF) predictor, resulting in what we term the *KMRF* (KAMA+MSR+RF) model. It will be demonstrated that the KMRF regime predictor does indeed improve substantially on results obtainable using KAMA+MSR (utilising cost-adjusted returns, with realistically-estimated transaction costs), with respect to a range of financial metrics.

2   Machine Learning for Financial Regime Prediction

Regime prediction, like most other areas of financial prediction, is nowadays dominated by machine learning (ML) models, traditional tools such as the Probit model [2], used in the 1990s to predict economic recessions in the US and elsewhere, having now been set aside, as even the simpler ML methods, such as elastic net, have been shown more effective [27]. The dominant ML model for regime prediction in industry and academia is the hidden Markov model (HMM) [4], used for example to infer future price trends (downward and upward) in the oil market [24], Euro/USD and AUD/USD crosses [15], as well as Bitcoin [9]. The HMM however suffers from an inability to predict regime switches without experiencing the onset of the new regime: the HMM can in effect only predict *continuations* of already-changed regimes. Due to this limitation, many other ML models have been explored for regime prediction (though deep learning models have been less-used due to a relative scarcity of data and consequent risk of overfitting these complex models), with random forest (RF) [5] standing out as the most promising model for regime prediction.

RF models have been used successfully in many areas of finance. For example, RF was shown in [3] to be the most effective algorithm for prediction of one-year ahead price direction, for over 5500 European stocks, on the basis of fundamental and macro variables, compared with those made by a number of other algorithms, such as SVM, ANN, and k-nearest neighbours. Furthermore, when compared to SVM and naïve Bayes in [18], RF achieved the highest F-Score in stock price correction (at least 10% upward or downward movement) prediction over the long term. Moving to regime prediction, RF models have been used, for example, to predict bank crises [28], regime turning points [20] (in which RF outperformed gradient-boosted trees and naïve Bayes), US recessions [29] (again outperforming gradient-boosted trees, as well as support vector machine and ANN models), and in the prediction of stock market regimes [26].

It was thus decided to use the RF model to predict the assigned KAMA+MSR labels, with the HMM model, due to its long history of use for regime prediction, as a predictor benchmark (alongside the KAMA+MSR model [22], in order to establish the advantage of regime prediction over detection).

3   Methodology

A summary flowchart of our methodology is given at the conclusion of this section in Fig. 1, with the individual components within this workflow discussed in more detail in the subsections below.

3.1   Data and Feature Engineering

Data Used. Regimes will be predicted for three asset classes: equities, commodities, and foreign exchange (FX) pairs. The component assets for each class are listed online [21], as well as the start dates of the data, after first having gone through scaling process known as *fractional differencing* [10], which balances stationarity and non-stationarity within time series. (It should be noted that it is not possible to release the datasets per se, for commercial reasons; the code used also cannot be shared for these reasons.) Note that each asset class has a separate benchmark used only in out-of-sample model performance comparison, and that all series, including the benchmarks, end on the same date, 29/04/2022.

Candidate Features. A wide range of features, also listed online [21], were considered as a pool from which the feature selection method of Sect. 3.4 could draw. These features fall into three broad categories: *technical* (statistical measures created from price series), *fundamental* (measures related to a company's financial health), and *macroeconomic* (related to wider economic forces), with the technical features, 60 in total, used in the models for all three asset classes, computed using the TA[1] (28 features) and tsfresh [7] (32 features) packages.

3.2   Transaction Costs

Both hyperparameter optimisation and out-of-sample performance assessment will use measures based on cost-adjusted returns (Eq. 1). The costs assumed in this work are listed in Table 1; these are realistic estimates which were adopted after appropriate consultation with industry experts.

---

[1] https://github.com/bukosabino/ta. Last accessed 29 March 2023.

Table 1. Two-way (buy (long), sell (short)) trading costs for each asset class

Asset class	Brokerage commissions (%)	Bid-ask spread (%)	Market impact (%)	Total cost (%)
Equities	0.07	0.065	0.265	0.40
Commodities	0.14	0.13	0	0.27
Currencies (FX)	0	0.13	0	0.13

3.3 Hyperparameter Optimisation

Hyperparameter optimisation (using the Optuna [1] package) for the RF component of the three KMRF models was done in two phases, with the objective of optimising a financially-relevant metric, the Sortino ratio of Sect. 3.7. Search ranges and optimal hyperparameter values are listed in Table 2, other RF hyperparameters, deemed of lesser importance, being set to their default values in scikit-learn[2]. The first phase consisted of a loose optimisation, in order to get an adequately working model for each asset class; feature selection was then done, as will be described below, followed by a second phase of more rigorous and computationally demanding hyperparameter optimisation. 85% of each time series was used for training/validation (except in the case of the benchmarks) and 15% was used for testing. In the case of time series, however, cross-validation needs to respect temporal ordering; in addition, care should be taken to avoid data leakage that might arise due to the computation of lagged statistical features.

The *purged group time-series split* (PGTS) method[3], recommended in [8], is used here. PGTS is a cross-validation technique for time series that avoids data leakage. It uses non-overlapping groups to allow multiple time series as inputs, though these may start on different dates. During cross-validation, PGTS generates multiple splits, with groups remaining distinct across folds, with the training sets progressively including more data, maintaining temporal relationships.

Table 2. Hyperparameter search ranges and optimised values for each of the three RF models (equities, commodities, and currencies (FX)) built.

Hyperparameter	Search space	Equities	Commodities	Currencies
n_estimators	{10, ..., 300}	220	280	240
max_depth	{1, ..., 20}	13	3	7
min_samples_split	{1, ..., 100}	76	18	22
min_samples_leaf	{1, ..., 100}	95	95	60
max_samples	[0.1, ..., 1.0]	0.3649	0.1247	0.3603
min_weight_fraction_leaf	[0.0, ..., 0.05]	0.0454	0.0213	0.0349
max_features	[0.2, ..., 1.0]	0.2481	0.4001	0.3410

---

[2] https://scikit-learn.org/stable/. Last accessed 6 September 2022.

[3] https://www.kaggle.com/code/marketneutral/purged-time-series-cv-xgboost-optuna. Last accessed 12 July 2023.

3.4   Feature Selection

While RF has its own, built-in, feature selection tool, in the form of *mean decrease in impurity* (MDI), it has been argued that this is not ideal as a means to identify the most important features (see, for example, [25] and [14]). Feature importance is therefore investigated here using the *BorutaShap*[4] package. BorutaShap did, however, need adaptation for use with time series data, and the original code was modified to use PGTS cross-validation, described previously, in order to avoid data leakage; note that the same validation splits were used in loose, initial tuning, and in final optimisation.

3.5   Regime Label Generation

As noted in the Introduction, KAMA+MSR [22] creates four regime classes: low-variance (LV) bullish, LV bearish, high-variance (HV) bullish, and HV bearish, though with only the first and the fourth of these being used for trading. The three (one-hot encoded) target labels used by the proposed KMRF model are derived from these four KAMA+MSR classes as below:

– *Bullish* = LV bullish + extension to the peak of next HV bullish regime.
– *Bearish* = HV bearish + extension to the trough of next LV bearish regime.
– *'Other'* = remaining parts of the HV bullish and LV bearish regimes.

The above extensions are carried out because HV bullish and LV bearish states contain periods of up-trending and down-trending markets, respectively. For example, after an LV bullish period the price of a certain asset, such as a commodity, may soar significantly and then suddenly drop, thus effectively entering an HV bullish state due to the volatile movement of the market. Similarly, subsequent to an HV bearish regime, an LV bearish state may occur when the volatility of the asset's price becomes milder, and it initially falls but afterward rebounds (a typical 'bear rally'). Such periods, if classified as 'other', could confuse the learning algorithm, as well as potentially result in lesser profits. Finally, any period during which a bullish or bearish label, even if correctly predicted, could not be usefully exploited because the transaction costs of Table 1 would outweigh the profits made, is converted to the 'other' class.

3.6   Contrarian Interpretation of Predicted Regime Labels

Initially, it was assumed that the models' predictions would be used to establish a trend-following strategy, that is, a high probability of a bullish (prices rising) regime would give a buying signal, while a high probability of a bearish (prices falling) regime would give a shorting signal. (A high probability of the 'other' regime after previously having entered a long or short position in an asset would indicate a point of time to close this position.) However, preliminary results

---

[4] https://github.com/Ekeany/Boruta-Shap. Last accessed 5 April 2023.

indicated that the predictions for the KMRF models in fact produced contrarian signals. In other words, a high probability of a bullish regime here indicated an exhausted and overbought market, which should spur an action to enter a short position in the underlying asset, while a high probability of a bearish regime pointed at oversold market conditions, thus recommending entering a long position in the underlying asset. Contrarian models are commonly used in the industry; many trading strategies act on, and many indicators provide an alarm about, overbought and oversold conditions of the market, and recommend assuming a contrarian position, shorting when overbought or buying when oversold. Following the discovery that a contrarian interpretation of the models' signals was advisable, the following strategy was therefore adopted:

- When a bullish regime is predicted, a short position should be assumed in the underlying asset.
- When a bearish regime is predicted, a long position should be assumed in the underlying asset.
- When the 'other' regime is predicted, an asset should be sold (if it was bought) or bought back (if it was shorted).

3.7   Performance Metrics

Financial Performance Metrics. Three such metrics are used here, the *annualised* (over a trading year of 252 days) *Sortino ratio, adjusted Sharpe ratio,* and *information ratio.* The presented formulae will in each case make reference to the risk-free return r_f, but it should be noted that this quantity will in practice be conventionally set to zero. Furthermore, we note that $\mu(.)$ will denote the mean value of any daily-varying quantity, $\sigma(.)$ its variance, and $\sigma_{neg}(.)$ its negative semi-variance (downside deviation), with each of these calculated over all $t = 1 \ldots T$ trading days in the test period. Each of the three metrics to follow is based on the calculation of a *daily cost-adjusted trading return*, $r_M(t)$,

$$r_M(t) = \sum_{\substack{\text{assets} \\ i}} w_{L,i}(t)r_{L,i}(t) + w_{S,i}(t)r_{S,i}(t), \tag{1}$$

in which $r_{L,i}(t)$ is the return associated with longing (similarly, shorting) asset i on day t, and $w_{L,i}(t)$ is the weight associated with longing (and similarly, for shorting) asset i on day t. The weights are given by

$$w_{L,i}(t) = \begin{cases} \frac{1}{n_{L,t}} & \text{if asset } i \text{ is longed on day } t \\ 0 & \text{otherwise} \end{cases}, \tag{2}$$

where $n_{L,t}$ is the number of assets longed (similarly, shorted) on day t. Note that an asset does not need to be traded, either longed or shorted, on a given day; such a situation is represented by setting $w_{L,i}(t) = w_{S,i}(t) = 0$.

- *Annualised Sortino ratio* (ST). This is our primary metric, used for both hyperparameter tuning and out-of-sample performance assessment:

$$ST = \frac{\mu(r_M - r_f)}{\sigma_{neg}(r_M - r_f)} \times \sqrt{252}. \tag{3}$$

- *Annualised adjusted Sharpe ratio* (ASR) [11]. This is a modification of the Sharpe ratio that treats positive returns in the same manner as the standard Sharpe ratio, but more strongly penalises negative returns:

$$ASR = \frac{\mu(r_M - r_f)}{\sigma(r_M - r_f)^{\mu(r_M-r_f)/\mu|r_M-r_f|}} \times \sqrt{252}. \tag{4}$$

- *Annualised information ratio* (IR). This is the degree to which an actively managed portfolio, typically incurring significant transaction costs, can outperform a buy-and-hold investment in a benchmark index,

$$IR = \frac{\mu(r_M - r_B)}{\sigma(r_M - r_B)} \times \sqrt{252}, \tag{5}$$

where r_B are benchmark returns. A value in excess of 1.0 is desirable (the managed portfolio then being more profitable than the index); over extended periods, such as the test period used here, this can be hard to achieve.

Classification Performance Metric. Additionally to these financial metrics, the *Matthews correlation coefficient* [17] (MCC) has been used to check the robustness of the KMRF model as a classifier. The MCC was chosen for this purpose due to arguments that it is the most informative single-number metric that can be derived from the confusion matrix [6], especially for imbalanced data. We calculate the MCC separately for each class, as below,

$$MCC_i = \frac{(TP_i \times TN_i - FP_i \times FN_i)}{\sqrt{(TP_i + FP_i)(TP_i + FN_i)(TN_i + FP_i)(TN_i + FN_i)}}, \tag{6}$$

in which TP_i, TN_i, FN_i, and FP_i are true positives, true negatives, false negatives (incorrect assignments of a class i example to any $j \neq i$), and false positives (incorrect assignments to class i of any $j \neq i$ example), respectively, and $i \in \{bullish, bearish, other\}$. The calculation of MCCs for each class is useful since the main objective is to predict bullish and bearish regimes and take trading actions based on these predictions; in contrast, the 'other' regime is used only to close an open position, so the value of MCC_{other} is of lesser importance.

4   Results

This section will examine the performance of the proposed KMRF model for each of the asset classes, during the relevant test periods (approximately four years in each case). It will first consider, in Sect. 4.1, the performance of the

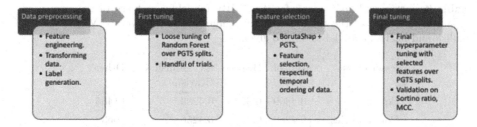

Fig. 1. Summary of the methodology adopted in this work.

model purely as a classifier, making three-way regime predictions of bullish, bearish, or 'other', with performance assessed by the MCC. It will then, in Sect. 4.2, move on to performance in relation to financially-motivated metrics, before finally, in Sect. 4.3, examining which of the many initially considered features listed in Sect. 3.1 were discovered to be the most valuable for regime prediction. It should be noted that the KMRF results presented here are for the models in which BorutaShap feature selection has already been performed (on the basis of training data only) and the RF model then re-trained and re-optimised (again, on the basis of training data only, using the PGTS cross-validation method referenced in Sect. 3.3). This feature selection step reduced the numbers of features to 131 in the case of equities, 63 in the case of commodities, and 24 in the case of FX, the larger number of features used by the equities model being due to the larger number of candidate fundamental features in this case.

4.1 Classification Performance of the Proposed KMRF Model

This section looks at the performance of the model as a classifier, with the Matthews correlation coefficient (MCC) (Eq. 6) used as a metric, for reasons explained in Sect. 3.7. It should be noted that a high MCC does not guarantee a high Sortino ratio, as the MCC is a counting measure that registers only whether a predicted regime label is correct or incorrect; a correct prediction may not be profitable if the market move is insufficient to outweigh the costs of trading. However, conversely, a high Sortino ratio might be achieved during the test period essentially by luck, but an MCC close to zero would expose this. Thus the dual use of MCC and ST is the best guide to the likely reliability of the model outside of its test period, going forward into the future.

Table 3 presents the result of this investigation. It is immediately apparent from the table that the MCCs for both the bullish and bearish regime labels are very high, in the context of financial prediction (for comparison, one can note that a prediction of closing price direction based on previous closing prices would typically not achieve an MCC greater than 0.05), while the MCCs for the 'other' regime label are substantially lower. However, this latter is not a large problem as the 'other' signal is used only to close a position, and hence a misclassification of an actually bullish or bearish regime to 'other' would represent only a loss of profit due to the premature closing of a position, not a potentially much larger loss due to having taken the wrong position.

Table 3. Test period MCC scores for the KMRF model for each asset class, presented separately for each of the three predicted regime labels; the average MCC over the bullish and bearish class labels is also given.

Asset class	Bullish	Bearish	Av. Bullish & Bearish	'Other'
Equities	0.4809	0.3768	0.4289	0.1455
Commodities	0.4260	0.4191	0.4256	0.0632
Currencies (FX)	0.5413	0.3295	0.4354	0.0552

4.2   Trading Performance: Comparison of the Proposed KMRF Model with Competitor Models and Benchmarks

This section considers the result of using the generated regime predictions to trade. The KMRF model was compared to a hidden Markov model (HMM) and the KAMA+MSR model of [22]. The HMM was picked as a comparison model due to its popularity for regime prediction, as evidenced in Sect. 2. The comparison with the KAMA+MSR model aimed to clarify the added value of predicting, as opposed to only detecting, as in [22], financial regimes. The comparison was done in terms of financially-relevant metrics: the Sortino ratio (Eq. 3), used, as noted in Sect. 3.3, for hyperparameter optimisation; but also the adjusted Sharpe ratio (Eq. 4) and information ratio (Eq. 5), neither of which were used in any form of optimisation.

Table 4 shows that the KMRF model performed substantially better than its competitors during the four-year test periods, with respect to all these metrics. Notably, the KMRF model was the only one to achieve an IR > 1.0, i.e., to be able to outperform the relevant index benchmark, for every asset class. It should also be noted that predictions from all the models were both statistically significantly different from random and statistically significantly different from each other (at a significance level of 0.01 or better, in all cases).

Table 4. Test period financial metrics (annualised Sortino and adjusted Sharpe ratios, and information ratio (IR)) for the KMRF model in comparison to its competitor models (HMM, KAMA+MSR), for each asset class.

Asset class	Index benchmark	Model	Sortino ratio (ann.)	Adj. Sharpe ratio (ann.)	IR
Equities	ACWI	KMRF	**24.99**	**1.44**	**1.46**
		HMM	0.32	0.005	−0.11
		KAMA+MSR	−0.29	−0.0004	−0.51
Commodities	BBG_Commodity	KMRF	**12.66**	**1.22**	**1.93**
		HMM	−0.47	−0.002	−1.27
		KAMA+MSR	0.24	0.002	−0.42
Foreign Exchange	BBDXY	KMRF	**4.72**	**0.29**	**1.45**
		HMM	−0.37	−0.0001	-0.88
		KAMA+MSR	0.17	0.1	−0.28

The strength of the KMRF model is further illustrated in Figs. 2, 3, and 4, for each of the asset classes, in terms of cumulative performance,

$$Q(t) = \left[\prod_{s=1}^{t} (1 + r_M(s)) \right] \times 100, \qquad (7)$$

the wealth attained on day $t = 1 \ldots T$ by an investor with an initial ($t = 0$) investment of \$100. This is compared with the performance of the competitor models (HMM and KAMA+MSR), as well as those of the appropriate index benchmarks. Note that the returns are cost-adjusted, apart from the index benchmarks, and that the index benchmarks are not compounded, as they are simply bought at the start of the test period and sold at the end. The best-performing KMRF model was for the commodities asset class, but the outperformance of the KMRF model in the case of equities and FX is also very noticeable.

It may be recalled from Sect. 3.6 that the KMRF model adopts a contrarian interpretation of the generated signals. The HMM and KAMA+MSR models, in contrast, here use conventional interpretations of the trading signals. It should be noted, however, that the competitor models were, as a further test, also run in contrarian mode; this resulted, for both competitors, for all three asset classes, in an even worse performance. Thus we feel confident it is the *combination* of KMRF and contrarian trading that leads to the observed superior results.

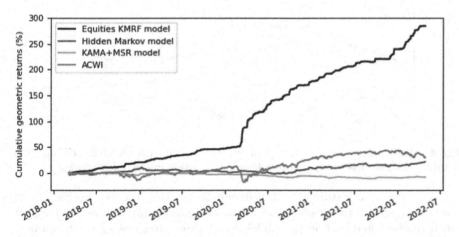

Fig. 2. Test period cumulative performance of the equities KMRF, HMM, and KAMA+MSR models, compared to a long-only position in the index benchmark.

4.3   Feature Importances and Interpretation

Figures 5, 6, and 7 show the top 10 most important features out of those selected by the BorutaShap algorithm, in terms of their absolute impact on the predictions of the equities, commodities, and FX, respectively, KMRF models, with the impact of each feature on prediction being computed using Shapley values [23].

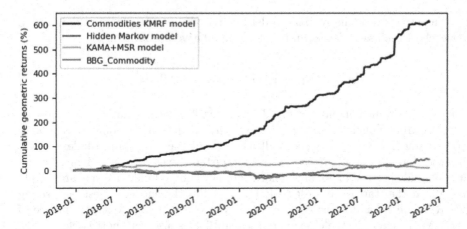

Fig. 3. Test period cumulative performance of the commodities KMRF, HMM, and KAMA+MSR models, compared to a long-only position in the index benchmark.

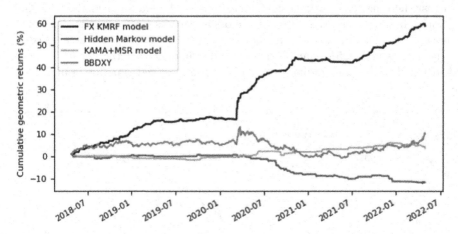

Fig. 4. Test period cumulative performance of the FX KMRF, HMM, and KAMA+MSR models, compared to a long-only position in the index benchmark.

The figures present feature importances only for the 15% test sets, as this is more relevant in relation to out-of-sample use than the same statistics for the training set. It can be noted that all the models for all three asset classes are driven by a similar set of features, though the FX model, unlike the others, is not primarily driven by the 'energy ratio by chunks' feature. The largest impact on all models' output probabilities is from technical momentum variables. Macroeconomic variables have a lesser effect, which is in line with the data frequency (daily); in the financial industry, it is commonly known that technical features are more predictive for higher frequency data, while macroeconomic features conversely tend to be more predictive for lower frequency data, such as monthly. This is because macroeconomic variables tend to change more slowly than technical indicators, as well as usually lagging current market conditions.

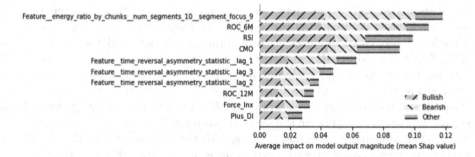

Fig. 5. Feature importances for the equities KMRF model.

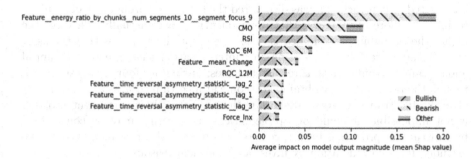

Fig. 6. Feature importances for the commodities KMRF model.

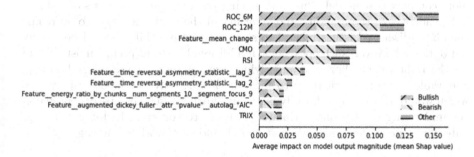

Fig. 7. Feature importances for the FX KMRF model.

5   Discussion and Conclusions

The work of this paper has built upon a previous work [22] which proposed a novel method of detection of regimes, referred to as the KAMA+MSR model, for financial time series. This model combined the Markov-switching regression (MSR) model's ability to detect changes in volatility with the ability of Kauff-man's adaptive moving average (KAMA) to detect an upward or downward trend. However, the KAMA+MSR model suffered from inevitable lag-related losses associated with regime detection, as opposed to prediction, and thus this work has focused on predicting the KAMA+MSR regime classes ex ante, using a random forest classifier, resulting in a regime prediction model we term the *KMRF* (KAMA+MSR+RF) model. Three major asset classes were considered: equities, commodities, and foreign exchange (FX), to increase the robustness of our conclusions. Results over a four-year test period evidenced the success of the proposed methodology, as each of the KMRF models, based on its generated signals, was able to outperform the selected benchmarks, as well as generate substantial risk-adjusted returns, net of costs, with the commodities KMRF model having the highest cumulative geometric returns, while the equities KMRF model achieved the highest annualised Sortino and adjusted Sharpe ratios. However, one major finding of this work was that the KMRF models provide signals best interpreted as contrarian: it was discovered the higher the probability of a bullish regime, the more likely it was that the asset price would actually fall, while the higher the probability of the bearish class, the more likely it was that the asset price would in fact rise. Adopting this interpretation gave rise to substantial and sustained profits for all three asset classes, during our four-year test period, though the profits thereby obtained do come at a certain cost, as a large number of profitable trades are due to short positions. Even though frequent shorting is not impossible, it could be problematical, as shorting may be constrained by institutions providing brokerage services on grounds of financial risk, or be excluded from consideration by investors for ethical reasons.

Turning to future work, the KMRF model could be enhanced with dynamic asset weight allocation (instead of using mean weights, as in Eq. 2), as well as a more complex cost model (this work having assumed the fixed costs of Table 1, independently of the liquidity and volatility of the traded assets). To overcome the above-discussed issues related to shorting, one possibility would be to buy put options; however, such options incur additional costs, as premia must be paid for the right to sell short the underlying asset. A more attractive and widely-adoptable possibility might be to focus on a long-only portfolio strategy, one which would also be capable of dynamically shifting asset weights, depending on the underlying asset return estimates, in order to construct high-Sortino long-only portfolios under a realistic cost model, and such work is ongoing.

References

1. Akiba, T., Sano, S., Yanase, T., Ohta, T., Koyama, M.: Optuna: a next-generation hyperparameter optimization framework. In: Proceedings of the 25th ACM SIGKDD International Conference on Knowledge Discovery & Data Mining, pp. 2623–2631 (2019). https://doi.org/10.1145/3292500.3330701
2. Aldrich, J.H., Nelson, F.D.: Linear Probability, Logit, and Probit Models. Sage (1984)
3. Ballings, M., Van den Poel, D., Hespeels, N., Gryp, R.: Evaluating multiple classifiers for stock price direction prediction. Expert Syst. Appl. **42**(20), 7046–7056 (2015). https://doi.org/10.1016/j.eswa.2015.05.013

4. Baum, L.E., Petrie, T.: Statistical inference for probabilistic functions of finite state Markov chains. Ann. Math. Stat. **37**(6), 1554–1563 (1966). https://www.jstor.org/stable/2238772
5. Breiman, L.: Random forests. Mach. Learn. **45**(1), 5–32 (2001). https://doi.org/10.1023/A:1010933404324
6. Chicco, D., Jurman, G.: The advantages of the Matthews correlation coefficient (MCC) over F1 score and accuracy in binary classification evaluation. BMC Genom. **21**(1), 1–13 (2020). https://doi.org/10.1186/s12864-019-6413-7
7. Christ, M., Kempa-Liehr, A.W., Feindt, M.: Distributed and parallel time series feature extraction for industrial big data applications. arXiv preprint arXiv:1610.07717 (2016). https://doi.org/10.48550/arXiv.1610.07717
8. De Prado, M.L.: Advances in Financial Machine Learning. Wiley, Hoboken (2018)
9. Giudici, P., Abu Hashish, I.: A hidden Markov model to detect regime changes in cryptoasset markets. Qual. Reliab. Eng. Int. **36**(6), 2057–2065 (2020). https://doi.org/10.1002/qre.2673
10. Hosking, J.: Fractional differencing. Biometrika **68**, 165–175 (1981)
11. Israelsen, C.: A refinement to the Sharpe ratio and information ratio. J. Asset Manag. **5**(6), 423–427 (2005). https://doi.org/10.1057/palgrave.jam.2240158
12. Kaufman, P.: Smarter Trading: Improving Performance in Changing Markets. McGraw-Hill, New York (1995)
13. Krolzig, H.: Markov-Switching Vector Autoregressions: Modelling, Statistical Inference, and Application to Business Cycle Analysis. Springer, New York (1997). https://doi.org/10.1007/978-3-642-51684-9
14. Kursa, M.B., Rudnicki, W.R.: Feature selection with the Boruta package. J. Stat. Softw. **36**, 1–13 (2010). https://doi.org/10.18637/jss.v036.i11
15. Lee, Y., Ow, L.T.C., Ling, D.N.C.: Hidden Markov models for forex trends prediction. In: 2014 International Conference on Information Science & Applications (ICISA), pp. 1–4. IEEE (2014). https://doi.org/10.1109/ICISA.2014.6847408
16. Lo, A.W., MacKinlay, A.C.: A Non-random Walk Down Wall Street. Princeton University Press, Princeton (2011)
17. Matthews, B.W.: Comparison of the predicted and observed secondary structure of T4 phage lysozyme. Biochim. Biophys. Acta (BBA)-Protein Struct. **405**(2), 442–451 (1975). https://doi.org/10.1016/0005-2795(75)90109-9
18. Milosevic, N.: Equity forecast: predicting long term stock price movement using machine learning. arXiv preprint arXiv:1603.00751 (2016). https://doi.org/10.48550/arXiv.1603.00751
19. Pavlov, V., Hurn, S.: Testing the profitability of moving-average rules as a portfolio selection strategy. Pac. Basin Financ. J. **20**(5), 825–842 (2012). https://doi.org/10.1016/j.pacfin.2012.04.003
20. Piger, J.: Turning points and classification. In: Fuleky, P. (ed.) Macroeconomic Forecasting in the Era of Big Data. ASTAE, vol. 52, pp. 585–624. Springer, Cham (2020). https://doi.org/10.1007/978-3-030-31150-6_18
21. Pomorski, P.: Features and assets used in this work (2023). https://figshare.com/articles/conference_contribution/LOD2023_appendix/23681205
22. Pomorski, P., Gorse, D.: Improving on the Markov-switching regression model by the use of an adaptive moving average. In: Gartner, W.C. (ed.) New Perspectives and Paradigms in Applied Economics and Business. Springer Proceedings in Business and Economics, pp. 17–30. Springer, Cham (2023). https://doi.org/10.1007/978-3-031-23844-4_2
23. Shapley, L.S., et al.: A Value for N-Person Games (1953)

24. e Silva, E.G.D.S., Legey, L.F.L., e Silva, E.A.D.S.: Forecasting oil price trends using wavelets and hidden Markov models. Energy Econ. **32**(6), 1507–1519 (2010). https://doi.org/10.1016/j.eneco.2010.08.006

25. Strobl, C., Boulesteix, A.L., Zeileis, A., Hothorn, T.: Bias in random forest variable importance measures: illustrations, sources and a solution. BMC Bioinform. **8**(1), 1–21 (2007). https://doi.org/10.1186/1471-2105-8-25

26. Uysal, A.S., Mulvey, J.M.: A machine learning approach in regime-switching risk parity portfolios. J. Financ. Data Sci. **3**(2), 87–108 (2021). https://doi.org/10.3905/jfds.2021.1.057

27. Vrontos, S.D., Galakis, J., Vrontos, I.D.: Modeling and predicting US recessions using machine learning techniques. Int. J. Forecast. **37**(2), 647–671 (2021). https://doi.org/10.1016/j.ijforecast.2020.08.005

28. Ward, F.: Spotting the danger zone: forecasting financial crises with classification tree ensembles and many predictors. J. Appl. Economet. **32**(2), 359–378 (2017). https://doi.org/10.1002/jae.2525

29. Yazdani, A.: Machine learning prediction of recessions: an imbalanced classification approach. J. Financ. Data Sci. **2**(4), 21–32 (2020). https://doi.org/10.3905/jfds.2020.1.040

Ökolopoly: Case Study on Large Action Spaces in Reinforcement Learning

Raphael C. Engelhardt[1]([✉]) [iD], Ralitsa Raycheva[1], Moritz Lange[2] [iD],
Laurenz Wiskott[2] [iD], and Wolfgang Konen[1] [iD]

[1] Faculty of Computer Science and Engineering Science,
Cologne Institute of Computer Science, TH Köln, Gummersbach, Germany
{Raphael.Engelhardt,Wolfgang.Konen}@th-koeln.de
[2] Faculty of Computer Science, Institute for Neural Computation,
Ruhr-University Bochum, Bochum, Germany
{Moritz.Lange,Laurenz.Wiskott}@ini.rub.de

Abstract. Ökolopoly is a serious game developed by biochemist Frederic Vester with the goal to enhance understanding of interactions in complex systems. Due to its vast observation and action spaces, it presents a challenge for Deep Reinforcement Learning (DRL). In this paper, we make the board game available as a reinforcement learning environment and compare different methods of making the large spaces manageable. Our aim is to determine the conditions under which DRL agents are able to learn this game from self-play. To this goal we implement various wrappers to reduce the observation and action spaces, and to change the reward structure. We train PPO, SAC, and TD3 agents on combinations of these wrappers and compare their performance. We analyze the contribution of different representations of observation and action spaces to successful learning and the possibility of steering the DRL agents' gameplay by shaping reward functions.

Keywords: Deep reinforcement learning · Large action space · Cybernetics · Serious games

1 Introduction

Despite the overwhelming success of Deep Reinforcement Learning (DRL) in the last decade, large action spaces can still pose a challenge for DRL algorithms [7]. The serious game *Ökolopoly* [19] is an example of an environment exhibiting such a large action space. The game has its roots in cybernetics as it aims at teaching the players how to steer circular causal processes. The game in its internationalized computer simulation version *Ecopolicy* is cited as an example for training systemic and long-term thinking in complex, interconnected systems as opposed

This research was supported by the research training group "Dataninja" (Trustworthy AI for Seamless Problem Solving: Next Generation Intelligence Joins Robust Data Analysis) funded by the German federal state of North Rhine-Westphalia.

G. Nicosia et al. (Eds.): LOD 2023, LNCS 14506, pp. 109–123, 2024.
https://doi.org/10.1007/978-3-031-53966-4_9

to linear thinking in terms of immediate, simple cause-effect relationships. International competitions in schools were held (*"Ecopoliciade"*) to train pupils the art of thinking holistically. [1, Sect. 2.2] [12, Sect. 3.2]

The combinatorial explosion of choices quickly leads to a large number of possible game states and a very large action space, making this game an interesting test case for DRL algorithms [7].

In this paper we describe how the board game can be formalized as a Reinforcement Learning (RL) environment following OpenAI Gym [2] standards. Given this implementation, we investigate the following research questions:

RQ 1 Is it possible for RL agents to learn to play the game Ökolopoly from self-play?

RQ 2 Which components are essential for successful learning (if any)?

RQ 3 Can the agent learn to propose valid actions or is it necessary that the environment transforms invalid actions into valid ones?

We explain and experimentally test different methods of approaching such large action spaces. We will show which of these methods are essential for a DRL agent to successfully learn to play the game and to what extent the agent can develop an "understanding" of the underlying game mechanics. Our hope is that the results from this Ökolopoly case study are also useful for other RL problems with large action spaces.

The remainder of the paper is structured as follows: Sect. 2 will discuss related work. In Sect. 3 we briefly describe the game of Ökolopoly. Section 4 contains technical information about how the game is translated from a board game to the domain of RL as well as methods to treat the large observation and action spaces, and different reward functions. Section 6 presents the experimental outcomes. In Sect. 7 we answer the research questions and give a short conclusion.

2   Related Work

Large action spaces have been identified as one of the main challenges in RL [7]. Proposed solution techniques may factorize the action space into binary or ternary subspaces [13], embed the discrete action space in a continuous one [6], or use the technique of action elimination [20]. In our work we will use the first two techniques as well. Instead of action elimination we use action normalization (projection onto valid actions, see Sect. 5.2). While the above-mentioned papers investigate action spaces of size 10^2–10^4, the application studied in this work has an action space of size 10^6–10^8 (see Sect. 3.3, depending on whether we use action normalization or not). Huang and Ontañón [11] thoroughly describe and evaluate invalid action masking, a technique to restrict large, discrete action spaces to valid actions only. This is achieved by considering the logits produced by policy gradient algorithms and substituting those corresponding to invalid actions by a large negative number.

Serious games in biology [17] and ecology [12] have a long tradition and are often used for educational purposes or for collective problem-solving in science, as in *Foldit* [3]. The use of RL for serious games is an emerging research topic [5]. Dobrovsky et al. [4] use interactive DRL to balance the transfer of knowledge and the entertainment in serious games based on the context information from gameplay. Another example are rehabilitation serious games [10] where an RL-based approach is used to modify the difficulty of the rehabilitation exercises.

The game of Ökolopoly has—to the best of our knowledge—not been learned successfully by RL methods before.

3   The Game of Ökolopoly

Designed by Frederic Vester and made available as board game in 1984 [18,19], the game of Ökolopoly aims to raise awareness for and deepen the understanding of acting in systems of complex interdependencies.

The game is conceived as a single-player, turn-based strategy game. It models the state of the imaginary country of *Kybernetien* with scores on eight interacting departments or fields (such as Population, Quality of Life or Environment). The player's task is to lead the country to success by cleverly distributing available action points to the fields and developing an understanding of the underlying interdependencies. The fields are described in Table 1 with their minimal, maximal, and starting values.

Table 1. The eight fields, number of rounds played, and available action points determining the state of the game. Five of these fields are directly *actionable*, i.e., they may receive action points.

Field	Min	Max	Start Value	Actionable
Sanitation	1	29	1	yes
Production	1	29	12	yes
Environment	1	29	13	
Education (e)	1	29	4	yes
Quality of Life (q)	1	29	10	yes
Population Growth (g)	1	29	20	yes
Population (b)	1	48	21	
Politics (p)	-10	37	0	
Rounds (r)	1	30	1	
Action Points	1	36	8	

3.1   One Turn of the Game

In each turn (or timestep in RL terms) the player chooses how to distribute the available action points among the five fields Sanitation, Production, Education, Quality of Life, and Population Growth, so that the respective field values are incremented by those action points. Only for the field Production the player may also choose to diminish the value, at the cost of action points, too. There is no minimum value of action points to use, i.e., the player may choose to save action points for the next round.

Once the action points are distributed, certain interdependency functions between the fields, e.g., $G_j(x)$, $j = 1, \ldots, 4$ for the field population growth g, where x is any of the other fields, give rise to a number of automatic adjustments (feedback effects). For example, if Education is $e = 19$, Population Growth changes by $G_1(e) = +3$. The interdependency functions are deterministic, but complex and hard to memorize for a human player. Finally, the action points available for the next round are assigned following the interdependency functions, the round counter is increased by one, and the round ends.[1]

For higher values of Education e, the interdependency function $G_1(e)$ may exhibit a number preceded by $\pm$: in those cases the player can choose to diminish or increase the field Population Growth g in the given range at no additional cost of action points (e.g., if $G_1(e) = \pm 3$ then any choice of $\Delta g \in \{-3, -2, -1, 0, 1, 2, 3\}$ is allowed). The reasoning behind this rule is that a sufficiently educated population is able to steer its growth.

3.2   End of the Game

When one or more fields leave the allowed range (either due to the distribution of action points or due to the automatic adjustments thereafter), or when 30 rounds are played, the game ends. At the end of the game, the balance score B is computed as a function of the field Politics p, the value of the interdepency function $D(q)$, which is monotonically rising with Quality of Life q, and the number of played rounds r:

$$B(p, q, r) = \begin{cases} \frac{10\,[p + 3D(q)]}{r + 3} & \text{if } 10 \leq r \leq 30 \\ 0 & \text{otherwise} \end{cases} \tag{1}$$

This means that a balance score of 0 is given, if the condition $10 \leq r \leq 30$ is not met. The game instructions define a score of over 20 as exceptionally good.

3.3   Observation and Action Spaces

Given Table 1, the observation space allows for $29^6 \cdot 48^2 \cdot 36 \cdot 30 \approx 1.48 \times 10^{15}$ different states.

When distributing $a \in \{1, \ldots, 36\}$ available action points to the five fields, there are in principle $(a + 1)^5$ possible combinations. Since a player cannot distribute more action points than available, the number of valid combinations is much smaller. Counting the number of valid and the number of possible combinations for all values of a, we find that there are 9.7×10^6 valid combinations, which are only 1.1% of all 9.1×10^8 possible combinations.

---

[1] An advanced version of the game provides optional "event cards" to be drawn every five rounds. We ignore this advanced version in our implementation.

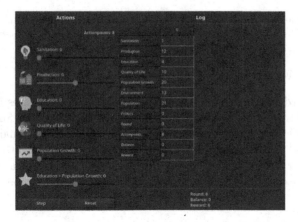

Fig. 1. GUI for the RL environment

This poses two challenges for any DRL agent: Firstly, even when restricting the agent to valid actions (e.g., by sum normalization, see Box action wrapper in Sect. 5.2), there is still a large number of 9.7×10^6 options. Secondly, if we give the agent no information on whether a possible action is valid or not (no action wrapper), it has to learn by reinforcement feedback to suggest valid combinations (otherwise the episode will terminate immediately). This is a demanding task given the small percentage of only 1.1% valid combinations.

4 Implementation of the Game

In this section we briefly describe how the board game was implemented as an OpenAI Gym [2] compatible RL environment. The code and GUI for human-play (Fig. 1) were adapted from [15]. Implementation, experiments, and additional material are available on Github[2].

4.1 Representation of the Observation Space

The observation space is internally represented as a ten-dimensional object of class `MultiDiscrete` containing the values of the eight fields, the number of rounds played, and the currently available action points. The agent has therefore full access to all information visible to the human player of the board game. As different ranges are allowed in the different dimensions of the observation space (the field Politics can even contain negative values which are not supported by `MultiDiscrete`), allowed states are shifted accordingly (see first row of Table 2).

4.2 Representation of the Action Space

In a similar way the action space is encoded as a six-dimensional `MultiDiscrete` object containing the number of action points assigned to each of the five fields.

---

[2] https://github.com/WolfgangKonen/oekolopoly_v1.

The sixth number accounts for the possibility of modifying Population Growth by up to ± 5 points according to the value of $G_1(e)$ (see end of Sect. 3.1).

4.3   Reward Functions

The basic reward function merely implements Eq. 1. This requires the agent to perform a long streak of profitable actions, which is difficult to find by exploration, before receiving any learning signal (only after the terminal step and if it occurs after at least ten rounds). For this reason, we implemented and tested different auxiliary reward structures, which additionally assign a reward after each intermediate step. These dense reward functions are described in detail in Sect. 5.3.

5   Methods

To assess the impact of different ways to handle the large observation and action spaces and different reward structures on the success of training, we performed experiments with different combinations of DRL algorithms and wrappers we describe in the following. A summary of the different spaces and their implementation is given in Table 2.

5.1   Observation Wrappers

We consider three different observation wrapper choices that should enable the algorithms to digest the huge observation space.

None. We treat the observation space as a `MultiDiscrete` object.

Box Observation Wrapper. In the case of `MultiDiscrete` observations, the DRL agent does not have an intrinsic concept of distance between possible values in each dimension of the observation space. To mitigate this problem, the Box observation wrapper represents each of the eight fields, the played rounds, and the available action points as a value in a continuous `Box` reaching from the minimum to the maximum of the respective observation.

Simple Observation Wrapper. This observation wrapper subdivides each dimension into just three possible values: *low, medium,* and *high.* This way each field, the current number of rounds played, and the available action points are represented each by one value in $\{0, 1, 2\}$. The state space is thereby reduced to $3^{10} = 59\,049$ different states. The observation wrapper is implemented as a ten-dimensional `MultiDiscrete` object.

5.2   Action Wrappers

Similarly, we implemented and tested three different choices for the action wrapper to simplify the action space.

None. The action space is the unaltered `MultiDiscrete` object from Sect. 4.2. Since there is no mechanism translating actions into legal moves, the validity of an action is not ensured. In fact, the overwhelming majority of points in this action space do not correspond to valid moves.

Box Action Wrapper. This wrapper transforms the discrete action space into a continuous one of type `Box`. The elements of this six-dimensional vector range from $[0, -1, 0, 0, 0, -1]$ to $[1, 1, 1, 1, 1, 1]$ and have the meaning described in Sect. 4.2. If more than the available action points should be distributed according to the tentative action vector $\mathbf{a'}$, the values are normalized by their absolute sum, multiplied with the currently available action points n, and rounded to the next integer:

$$a_i = \left\lfloor \left| \frac{a'_i}{\sum_{i=0}^{4} |a'_i|} \cdot n + 0.5 \right| \right\rfloor \qquad \text{for} \quad i = 0, \ldots, 4$$

If, due to rounding effects, the sum of action points a_i still exceeds n, the highest element of vector $\mathbf{a}$ is decreased by one[3].

Simple Action Wrapper. Analogous to the Simple observation wrapper, this action wrapper reduces the number of available discrete actions by dividing the number of available action points into three equal or near-equal parts. These blocks of action points can then be distributed among the five different fields in the game. This distribution is encoded as a six-digit string: The first five digits from the left assume values in $\{0, 1, 2, 3\}$ which represent the number of action point partitions assigned to the respective field (the sum of the first five digits may therefore not exceed 3); the rightmost digit encodes whether action points are added (0) or deducted (1) from Production[4]. In total there are 77 such strings to represent legal moves. The action space is implemented as two-dimensional `MultiDiscrete` object whose first element contains the index of one of the 77 legal moves while the second one contains the possibility of modifying Population Growth by up to ±5. The total number of possible actions is therefore reduced to $77 \cdot 11 = 847$. On the other hand, actions are less precise. In certain situations, this can lead to premature episode termination, which could have been avoided by a finer-grained distribution. The Simple action wrapper mimics a possible human strategy of being less precise but actionable in unknown complex environments.

5.3 Reward Functions

We trained DRL agents using the score defined by the rules of the board game (see Sect. 3.2) as well as two other reward functions which alleviate mentioned problems related to this sparse reward structure.

---

[3] The sixth dimension $a'_5 \in [-1, 1]$ (modifier for Population Growth g, Sect. 3.1) is multiplied by 5, rounded, and then appended to $\mathbf{a}$.

[4] As an example the string 020101 encodes the following distribution of action points: one third of the action points are added to Education, two thirds of the action points are deducted (rightmost digit is 1) from Production.

Balance. At intermediate timesteps no reward is given. Only the final step yields a reward B computed according to Eq. 1.

PerRound. This reward wrapper issues an additional constant reward R_c after each intermediate step. After the episode's last step the known reward B from Eq. 1 is added to the return. This should be an incentive for the agent to reach a higher number of rounds, as the rules of the game suggest that between 10 and 30 rounds should be played. Different values for R_c will be investigated. If not stated otherwise, we use $R_c = 0.5$.

Heuristic. A heuristic that keeps Production and Population at healthy, intermediate values (15 and 24 respectively) empirically seemed to be a good strategy. To this end, the following auxiliary reward R is assigned after each intermediate step:

$$R = s \cdot (R_{\text{prod}} + R_{\text{pop}})$$
$$\text{with} \quad R_{\text{prod}} = 14 - |15 - V_{\text{prod}}|$$
$$R_{\text{pop}} = 23 - |24 - V_{\text{pop}}|$$

After the last step of each episode, the usual balance B from Eq. 1 is given as reward. The scaling factor s allows to steer the agent to maximize the auxiliary reward R and therefore the number of rounds (for higher values such as $s = 1$) or to maximize the balance B (for smaller values of s, see Fig. 7). Where not explicitly stated otherwise, we use $s = 1$.

Table 2. Overview of action and observation space wrappers and their representation

Space	Wrapper	Object
Observation	None	MultiDiscrete([29,29,29,29,29,29,48,48,31,37])
	Box	Box(low=[1,1,1,1,1,1,1,-10,0,0], high=[29,29,29,29,29,48,37,30,36])
	Simple	MultiDiscrete([3,3,3,3,3,3,3,3,3,3])
Action	None	MultiDiscrete([29,57,29,29,29,11])
	Box	Box(low=[0,-1,0,0,0,-1], high=[1,1,1,1,1,1])
	Simple	MultiDiscrete([77,11])

5.4   DRL Algorithms

For the experiments we trained agents with the on-policy algorithm PPO [16] and the off-policy algorithms TD3 [8] and SAC [9] using their implementation provided by the Python DRL framework Stable-Baselines3 [14]. We use the default hyperparameters of SB3; for the TD3 agents, Gaussian action noise with $\sigma = 0.1$ is applied. While all mentioned algorithms are suited to handle MultiDiscrete and Box observation spaces, not all are compatible with MultiDiscrete action spaces (see Table 3). The computational effort varies noticeably depending on the algorithm. The elapsed real time to train an agent (no observation wrapper,

Box action wrapper, PerRound reward with $R_c = 0.5$) for $800\,000$ timesteps was $680.0 \pm 2.1\,\mathrm{s}$ for PPO, $13\,216.2 \pm 115.8\,\mathrm{s}$ for SAC, and $11\,786.5 \pm 55.5\,\mathrm{s}$ for TD3 (three repetitions)[5].

Table 3. Compatibility of mentioned algorithms and used action and observation space representations.

Algorithm	Observation Space		Action Space	
	MultiDiscrete	Box	MultiDiscrete	Box
PPO	✓	✓	✓	✓
SAC	✓	✓	✗	✓
TD3	✓	✓	✗	✓

5.5 Experimental Setup

As a consequence of Table 2 and Table 3, not every combination of wrapper and algorithm is possible: we have $3 \cdot 3 \cdot 3 = 27$ combinations for PPO, but only 9 combinations for SAC and TD3 (only Box action wrappers). For our experiments we train DRL agents with all available combinations of observation and action wrappers and reward functions for $800\,000$ timesteps while logging the balance B (Eq. 1), the number of played rounds r, and the reward as seen by the DRL agent. After each completed training episode the agent plays one evaluation episode using deterministic predictions according to its current training state. As a measure of the agents' performance we compute the averages and standard deviations of balance B and played rounds r of the last $1\,000$ evaluation episodes.

We investigate the stability of the training process with respect to initialization of the agents' networks, the role of the different methods in handling observation and action space, the impact of reward functions, and the strategy of single agents.

6 Results

After inspecting the training history we examine our results regarding the role of different representations of observation and action space as well as different reward functions. We touch upon the possibility of steering the agents' focus by suitable choices of rewards and provide more insights regarding the agents' performance in terms of balance and played rounds.

6.1 Stability of Training

To assess how critical the random initialization of the DRL agent's networks is, as an example we trained 5 PPO agents on the combination Box observation

---

[5] Timing experiments were performed on a system with Intel® Core™ i7-1185G7 CPU and 16 GB RAM.

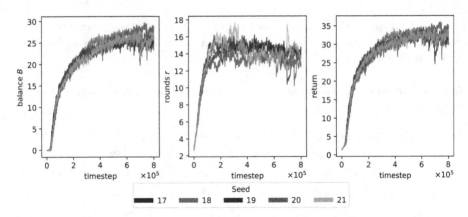

Fig. 2. Training curves of 5 differently seeded PPO agents (Box observation, Box action wrapper, PerRound reward $R_c = 0.5$) encoded by color. (For better visualization a Savitzky-Golay smoothing filter was applied.)

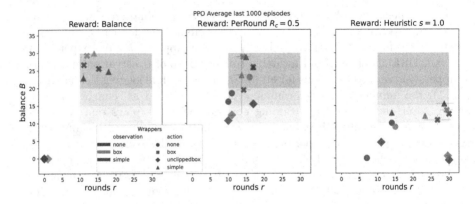

Fig. 3. Average and standard deviation of number of played rounds and balance for the last 1 000 evaluation episodes of **PPO** agents. Different reward functions are shown in separate plots, while observation and action wrapper are encoded by color and marker. The areas marked by colored backgrounds represent scores considered as increasingly good by the rules of the game. (N.b. at $(0,0)$ the combination without observation wrapper and UnclippedBox action wrapper (blue diamond), as well as all combinations without action wrapper (circles) are hidden below the green diamond marker). (Color figure online)

wrapper, Box action wrapper, and PerRound reward $R_c = 0.5$ under identical conditions except for a different seed of the DRL algorithm. As can be seen in Fig. 2, the initialization has at most only marginal effects on the training process and especially on the outcome.

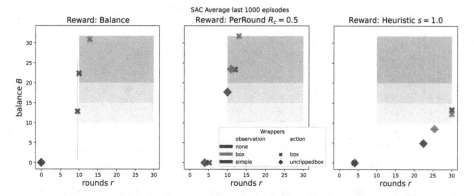

Fig. 4. Average and standard deviation of number of played rounds and balance for the last 1 000 evaluation episodes of **SAC** agents. The plots are structured as in Fig. 3. Tables 2 and 3 show why this plot contains fewer results. (N.b. the three diamond markers at $(0, 0)$ are overlapping in the leftmost plot).

6.2 Impact of Wrappers

To compare the training outcomes of different combination of wrappers, we examine the average and standard deviation of the agents' performance across the last 1 000 evaluation episodes during training in terms of played rounds and balance. The results are shown in Fig. 3 (PPO) and Fig. 4 (SAC).

Without ensuring valid actions by the use of action wrappers and with only sparse reward, the agents do not receive a sufficient learning signal and consequently fail (three overlaying round markers at $(0, 0)$ in the left plot of Fig. 3). With action wrappers intrinsically allowing only valid moves, the agents accumulate in a specific region of the multiobjective plot with exceptionally high score after having played between 10 and 20 rounds. If PPO agents are rewarded with a fixed positive amount R_c after each intermediate step, they are able to handle even large multidiscrete observation and action spaces, learn legal moves, and find a suitable strategy, as shown in the central plot. While a clear clustering by color or marker could not be observed here, the right plot shows the agents' difficulties to find suitable actions by the accumulation of round markers in the lower left side.

While PPO agents exhibit an overall remarkably good performance (Fig. 3), SAC agents using the Simple observation wrapper seem to struggle (Fig. 4). We also trained TD3 agents with default hyperparameters and Gaussian action noise with $\sigma = 1$. Since their results were inferior to PPO and SAC (only some TD3 combinations resulted in a balance above 10; most combinations have a balance $B \approx 0$ and rounds $r < 10$), we do not show these results.

The action wrappers serve a dual purpose: not only do they reduce the number of available actions, they also ensure only valid actions can be performed. To investigate which effect is predominant, all PPO and SAC agents were also trained using a modified Box action wrapper. This *UnclippedBox* action wrap-

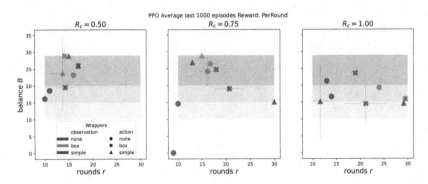

Fig. 5. Effect of different values of constant per-round reward R_c on PPO agents

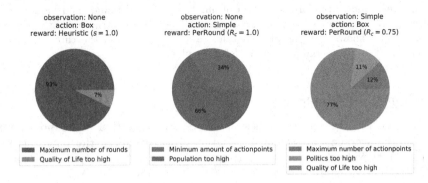

Fig. 6. Reasons for episodes' termination of the last 1 000 evaluation episodes during training. Shown is a small subset of PPO agents from the set of about 140 pie charts.

per still represents the discrete action space as a continuous box space, but it does not ensure that the intended distribution of action points is possible (i.e., does not exceed the number of available action points). Figures 3 and 4 show how the additional task of learning valid distributions of action points negates any success in case of a sparse reward. With dense rewards, *UnclippedBox* leads to somewhat reduced performance but it is noteworthy that the agents learn to perform valid actions.

Investigating the agent's gameplay further, a variety of different reasons for episode's termination emerge, many of which are "positive" in the sense that they leave the range of possible values in a direction that would commonly be considered desirable (e.g., "Quality of Life too high"). For three agents taken as examples, Fig. 6 shows statistics of reasons for termination of the last 1 000 evaluation episodes during training[6].

---

[6] The full set of pie charts is available in the Github repository.

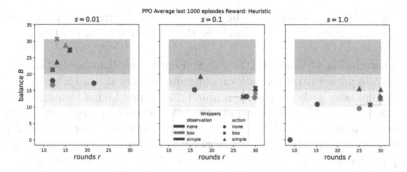

Fig. 7. Effect of different scaling of Heuristic reward on PPO agents.

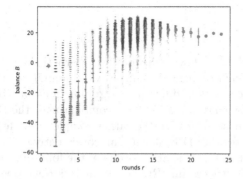

Fig. 8. Correlation of rounds r and balance B. Blue dots mark every pair (r, B) encountered by the PPO agent during training for the combination Box observation wrapper, Box action wrapper, reward Balance. Orange dots mark the average for every number of rounds. (Color figure online)

6.3 Reward Shaping

The importance of a suitable reward function is well-known among RL practitioners. This holds true also in the case of Ökolopoly. Not only does the reward function have arguably the biggest effect on overall performance (see Fig. 3), but also do the parameters of the single reward function itself play an important role. In Fig. 5 we show the results of comparing three different values of the constant per-round reward R_c. By changing this value we can tune the agent towards surviving more rounds r (in case of higher values R_c) on one side or towards prioritizing a higher balance B on the other.

Similarly, Fig. 7 shows the results for different values of the scaling factor s of the Heuristic reward.

6.4 Optimal Balance and Rounds

Looking at Fig. 5, one might ask why in the case $R_c = 0.5$ (where the incentive is more on balance) the high balance results are not possible in combination with a higher number of rounds. The average episode length does not exceed 17 rounds. A similar observation is made in the 'Reward: Balance' plots of Fig. 3 and Fig. 4.

The correlation is visualized in Fig. 8: Firstly, the balance B (for rounds $r < 10$ only hypothetical) shows a local optimum roughly at $r \approx 15$. Secondly, the number of episodes with $r \in [10, 15]$ is about 2 000 times higher than the number of episodes with $r \in [20, 25]$, meaning that the average in Fig. 5, $R_c = 0.5$, is likely to be below $r \approx 15$. The agent has learned that it is not detrimental to stop an episode after $10 - 15$ rounds, because a longer lasting episode would not improve (or even diminish) B. This is why the agent often stops quite early (after surpassing the required 10 rounds) unless a special incentive (e.g., $R_c = 1$) tells it otherwise.

7   Conclusion

In this article we described how we made the cybernetic board game Ökolopoly accessible as an RL environment. Various ways of simplifying the large observation and action space as well as different reward functions have been investigated. We were able to show that in many of our combinations, the DRL agents could successfully learn how to deal with the rather complex interdependencies of the game. Our results show that DRL agents can consistently learn to play the game from self-play (RQ 1). However, at least one of the following conditions have to be met in order to be successful (RQ 2): (i) an action wrapper that inherently maps to valid actions (Box or Simple) or (ii) a dense reward function (PerRound or Heuristic) are applied.

If neither (i) nor (ii) is fulfilled, no learning occurs in this large action space. We have shown that within the Box action wrapper not the mapping to continuous space but the clipping to valid actions is essential. We got the interesting result that even in the absence of action wrappers, the agent can learn by self-play to propose valid actions, given a dense reward structure (RQ 3).

We invite everyone to use our new RL environment Ökolopoly, which is made available as a benchmark for researchers, to test their methods for handling large observation and action spaces.

References

1. Bosch, O., Nguyen, N., Sun, D.: Addressing the critical need for "new ways of thinking" in managing complex issues in a socially responsible way. Bus. Syst. Rev. **2**, 48–70 (2013)
2. Brockman, G., et al.: OpenAI Gym (2016). https://doi.org/10.48550/arXiv.1606.01540
3. Cooper, S., et al.: Predicting protein structures with a multiplayer online game. Nature **466**(7307), 756–760 (2010). https://doi.org/10.1038/nature09304
4. Dobrovsky, A., Borghoff, U.M., Hofmann, M.: Improving adaptive gameplay in serious games through interactive deep reinforcement learning. Cogn. Infocommun. Theory Appl. **13**, 411–432 (2019)

5. Dobrovsky, A., Wilczak, C.W., Hahn, P., Hofmann, M., Borghoff, U.M.: Deep reinforcement learning in serious games: analysis and design of deep neural network architectures. In: Moreno-Díaz, R., Pichler, F., Quesada-Arencibia, A. (eds.) EUROCAST 2017. LNCS, vol. 10672, pp. 314–321. Springer, Cham (2018). https://doi.org/10.1007/978-3-319-74727-9_37

6. Dulac-Arnold, G., et al.: Deep reinforcement learning in large discrete action spaces (2015). https://doi.org/10.48550/arXiv.1512.07679

7. Dulac-Arnold, G., et al.: Challenges of real-world reinforcement learning: definitions, benchmarks and analysis. Mach. Learn. 110(9), 2419–2468 (2021). https://doi.org/10.1007/s10994-021-05961-4

8. Fujimoto, S., van Hoof, H., Meger, D.: Addressing function approximation error in actor-critic methods. In: Dy, J., Krause, A. (eds.) Proceedings of the 35th International Conference on Machine Learning, PMLR, vol. 80, pp. 1587–1596 (2018)

9. Haarnoja, T., Zhou, A., Abbeel, P., Levine, S.: Soft actor-critic: off-policy maximum entropy deep reinforcement learning with a stochastic actor. In: Dy, J., Krause, A. (eds.) Proceedings of the 35th International Conference on Machine Learning, PMLR, vol. 80, pp. 1861–1870 (2018)

10. Hornak, D., Jascur, M., Ferencik, N., Bundzel, M.: Proof of concept: using reinforcement learning agent as an adversary in serious games. In: 2019 IEEE International Work Conference on Bioinspired Intelligence, pp. 111–116 (2019)

11. Huang, S., Ontañón, S.: A closer look at invalid action masking in policy gradient algorithms. In: The International FLAIRS Conference Proceedings, vol. 35 (2022). https://doi.org/10.32473/flairs.v35i.130584

12. Nguyen, N.C., Bosch, O.J.H.: The art of interconnected thinking: starting with the young. Challenges 5(2), 239–259 (2014). https://doi.org/10.3390/challe5020239

13. Pazis, J., Parr, R.: Generalized value functions for large action sets. In: Proceedings of the 28th International Conference on International Conference on Machine Learning, pp. 1185–1192 (2011)

14. Raffin, A., Hill, A., Gleave, A., Kanervisto, A., Ernestus, M., Dormann, N.: Stable-baselines3: reliable reinforcement learning implementations. J. Mach. Learn. Res. 22(268), 1–8 (2021)

15. Raycheva, R.: Erstellung eines custom environments in OpenAI Gym für das Spiel Ökolopoly. Technical report, TH Köln (2021)

16. Schulman, J., Wolski, F., Dhariwal, P., Radford, A., Klimov, O.: Proximal policy optimization algorithms (2017). https://doi.org/10.48550/arXiv.1707.06347

17. Teixeira, J.d.S., Angeluci, A.C.B., Junior, P.P., Martin, J.G.P.: 'Let's play?' A systematic review of board games in biology. J. Biol. Educ. 1–20 (2022). https://doi.org/10.1080/00219266.2022.2041461

18. Vester, F.: Der blaue Planet in der Krise. Gewerkschaftliche Monatshefte 39(12), 713–773 (1988)

19. Vester, F.: Ökolopoly: das kybernetische Umweltspiel. Studiengruppe für Biologie und Umwelt (1989)

20. Zahavy, T., Haroush, M., Merlis, N., Mankowitz, D.J., Mannor, S.: Learn what not to learn: action elimination with deep reinforcement learning. In: Bengio, S., et al. (eds.) Advances in Neural Information Processing Systems, vol. 31 (2018)

Alternating Mixed-Integer Programming and Neural Network Training for Approximating Stochastic Two-Stage Problems

Jan Kronqvist[1] , Boda Li[2] , Jan Rolfes[1(✉)] , and Shudian Zhao[1]

[1] Department of Mathematics, KTH - Royal Institute of Technology,
Lindtstedtsvägen 25, 100 44 Stockholm, Sweden
{jankr,jrolfes,shudian}@kth.se
[2] ABB Corporate Research Center, Wallstadter Str. 59, 68526 Ladenburg, Germany

Abstract. The presented work addresses two-stage stochastic programs (2SPs), a broadly applicable model to capture optimization problems subject to uncertain parameters with adjustable decision variables. In case the adjustable or second-stage variables contain discrete decisions, the corresponding 2SPs are known to be NP-complete. The standard approach of forming a single-stage deterministic equivalent problem can be computationally challenging even for small instances, as the number of variables and constraints scales with the number of scenarios. To avoid forming a potentially huge MILP problem, we build upon an approach of approximating the expected value of the second-stage problem by a neural network (NN) and encoding the resulting NN into the first-stage problem. The proposed algorithm alternates between optimizing the first-stage variables and retraining the NN. We demonstrate the value of our approach with the example of computing operating points in power systems by showing that the alternating approach provides improved first-stage decisions and a tighter approximation between the expected objective and its neural network approximation.

Keywords: Stochastic Optimization · Neural Network · Power Systems

1 Introduction

The area of mathematical optimization often assumes perfect information on the parameters defining the optimization problem. However, in many real-world applications, the input data may be uncertain. The uncertain parameters when optimizing power systems can, e.g., be the power output of wind and solar power units or hourly electricity demands. To address this challenge, two major approaches have been developed in the current literature, namely, robust

Supported by Digital Futures and C3.ai Digital Transformation Institute.

optimization (RO), see, e.g., [1] for further details and stochastic optimization (SO), see, e.g., the seminal surveys [3] and [21]. In RO, only limited information on the underlying distribution of uncertainties is available, e.g., that the uncertain parameter vector belongs to a polyhedron, and the goal is to protect against the worst possible outcome of the uncertainty. In SO, it is typically assumed that the underlying probability distribution of the uncertain parameters is known, or at least that representative samples are available. Here, often the goal is to optimize over the expectation of the uncertain parameters, which results in less conservative solutions compared to RO. The ability to handle uncertainties makes both methods valuable in power-system-related optimization. A comparison between RO and SO for unit commitment with uncertain renewable production is presented in [13]. However, SO tends to be more prevalent in power system operation optimization, as operating scenarios are readily available. Applications of SO include the optimal operation of high-proportion renewable energy systems [23,32,38], pricing of energy and reserves in electricity markets [7,35,39] or day-ahead power scheduling [24,30,36] – to name a few.

Moreover, in some real-world applications, not all the decisions have to be taken before the uncertainty realizes. Here, we distinguish between *first-stage* or here-and-now decisions and *second-stage* or wait-and-see decisions. These *stochastic two-stage* problems or *2SPs* have drawn significant attention over the last years, both due to the practical applicability and computational challenges. Methods for dealing with 2SPs include L-shaped methods [20], Lagrangian relaxation-based methods [6], and the sample average approach of Kleywegt et al. [15]. For more information on 2SP and algorithmic approaches, we refer to [8,19,33] and the references therein.

Here, we focus on 2SPs, where the uncertainties either belong to a bounded discrete distribution or can be approximated sufficiently close by such a distribution. Consequently, we can restrict ourselves to a finite set of scenarios. Furthermore, we are assuming some of the second-stage variables are restricted to integer values and that all constraints are linear. For such problems, it is possible to represent the 2SP as a classical single-stage mixed-integer linear program (MILP) through the so-called deterministic equivalent [3]. The deterministic equivalent formulation creates one copy of the second-stage variables and constraints for each scenario, thus resulting in a potentially huge MILP problem. Especially if the second stage involves many integer variables, it may be computationally intractable to consider a large number of scenarios. To overcome this issue, Dumouchelle et al. [8] proposed the so-called Neur2SP approach ,where the value function, i.e., the optimal objective value of the second stage, is approximated by a neural network (NN). By selecting a MILP representable NN architecture, e.g., ReLU activation functions [9,22,34], the NN approximation of the second-stage problem along with the first-stage problem can then be integrated into a single stage MILP. The advantage of this approach is that the MILP presentation of the NN can require far fewer variables and constraints than forming the deterministic equivalent. The potential drawback is that we only utilize an approximation of the expected value of the second-stage problem.

We build upon the approach presented by Dumouchelle et al. [8] and investigate the proposed algorithm for an application in optimizing power systems. We show that proper training of the NN is crucial, and we propose a dynamic sampling as a training method. Here, we directly train a NN to approximate the expected value function. Our approach can be viewed as a surrogate-based optimization [2,10] approach for 2SPs, as we are iteratively forming a surrogate of the expected value function which is globally optimized to find a new solution and to improve the approximated expected value function. The proposed algorithm is not limited to the specific application in Sect. 3, but applicable to any 2SP of suitable structure, e.g., see the applications used in [8].

For the application of finding optimal operating points of a power grid considered in the present article, the existing approaches in the power system literature usually approximate the integer variables in the second stage, e.g. by ignoring the action variable of energy storage, see [5] or only search for a local optimum of the model, e.g. by applying an augmented Benders decomposition algorithm, see [4]. Thus, the Neur2SP approach keeps the underlying integral structure of the second-level, while at the same time aims for a global optimum.

The outline of the present article is as follows. In Sect. 2, we present the stochastic programming formulation [8] and discuss our methodology. Then, we connect this modeling to achieve data-driven operating states of smart converters in power systems in Sect. 3. Subsequently, we present a numerical comparison between these data-driven results with a baseline algorithm based on the Neur2SP framework in Sect. 4.

2   Replacing the Second-Stage in Stochastic Programming by a Neural Network

The two-stage stochastic programming consists of two types or *levels* of decision variables. The *first-level* variables, denoted by $x \in \mathcal{X}$, describe an initial planning approach. Then, a random vector $\xi \in \Omega$ supported on a compact domain $\Omega \subseteq \mathbb{R}^I$ and distributed by a probability distribution $\mathbb{P} \in \mathcal{P}(\Omega)$ affects the outcome of this initial planning. However, one can adjust the initial planning x after the random vector ξ realizes by choosing the *second-level or recourse* variables $y \in \mathcal{Y}(x, \xi)$, wherein the present article, we suppose that $\mathcal{Y}(x, \xi) \subseteq \mathbb{R}^n \times \{0, 1\}^l$ is assumed to be defined through linear constraints. Thus, we can summarize a *stochastic two-stage problem* as follows:

$$\min_{x \in \mathcal{X}} G(x) + \mathbb{E}_{\mathbb{P}} \left(\min_{y \in \mathcal{Y}(x, \xi)} c^\top y \right), \tag{1}$$

where in this article $\mathcal{X}$ and Ω are assumed to be polytopes, c denotes the objective coefficients, and $\mathcal{Y}(x, \xi)$ a polytope intersected with an integer lattice. Note, that (1) is considered to be a challenging problem as it contains the NP-complete MIP $\min_{y \in \mathcal{Y}(x, \xi)} c^\top y$ as a subproblem. Moreover, stochastic two-stage problems are in general notoriously challenging, see e.g., Sect. 3 in [21] for an overview.

In the present article, we restrict ourselves to the estimated expected value

$$\min_{x \in \mathcal{X}} \ G(x) + \frac{1}{m} \sum_{j=1}^{m} \left(\min_{y \in \mathcal{Y}(x, \xi_j)} c^\top y \right), \tag{2}$$

where ξ_j denote scenarios sampled from $\mathbb{P}$. For the numerical experiments in Sect. 4, we limit the number of scenarios to $m = 400$. We then aim to learn the mapping

$$Q(x) := \frac{1}{m} \sum_{j=1}^{m} \left(\min_{y \in \mathcal{Y}(x, \xi_j)} c^\top y \right)$$

by training a series of neural networks inspired by the Neur2SP framework developed in [8]. However, in contrast to the approach in [8], we aim to learn the mapping $x \mapsto Q(x)$, by training a series of neural networks with iteratively generated training data instead of predefined data, see Algorithm 1.

2.1   Embedding the Neural Network into 2SP

We approximate the map $x \to Q(x)$ by training a fully-connected neural network with 2-hidden layers, where both layers have 40 neurons, denoted as NN 2×40. Hence, the relation between l-th layer input $x^l \in \mathbb{R}^{n_l}$ and output $x^{l+1} \in \mathbb{R}^{n_{l+1}}$ in such neural network with the ReLu activation function is

$$x^{l+1} = \max\{0, W^l x^l + b^l\}, \tag{3}$$

where $W^l \in \mathbb{R}^{n_{l+1} \times n_l}$ is the weight matrix and $b^l \in \mathbb{R}^{n_{l+1}}$ is the bias vector. The ReLu activation function is piece-wise linear and the big-M formulation was the first presented to encode it and is still common [9, 22]. In this way, with binary variable $\sigma^l \in \mathbb{R}^{n_{l+1}}$, (3) is equivalent to

$$
\begin{aligned}
(w_i^l)^\top x^l + b_i^l &\leq x_i^{l+1}, \\
(w_i^l)^\top x^l + b_i^l - (1 - \sigma_i^l) LB_i^{l+1} &\geq x_i^{l+1}, \\
x_i^{l+1} &\leq \sigma_i^l UB_i^{l+1}, \\
\sigma_i^l \in \{0, 1\}, \ x_i^{l+1} &\geq 0, \forall i \in [n_{l+1}],
\end{aligned}
\tag{4}
$$

where w_i^l is the i-th row vector of W^l, b_i^l is the i-th entry of b^l, LB_i^{l+1} and UB_i^{l+1} are upper and lower bounds on the pre-activation function over x_i^{l+1}, such that $LB_i^{l+1} \leq (w_i^l)^\top x^l + b_i^l \leq UB_i^{l+1}$. Encoding ReLu functions has been an active research topic in recent years, as it enables a wide range of applications of mixed-integer programming to analyze and enhance NNs, such as generating adversarial examples [9], selecting optimal inputs of the training data [18, 40], lossless compression [29], and robust training [28].

By encoding the trained NN as (4), we can approximate (1) by the following MIP:

$$\min \ G(x^1) + x^L \tag{5a}$$

$$\text{s.t. } W^l x^l + b^l \leq x^{l+1}, \qquad\qquad \forall l \in [L-1], \tag{5b}$$

$$W^l x^l + b^l - \text{diag}(LB^{l+1})(1 - \sigma^{l+1}) \geq x^{l+1}, \quad \forall l \in [L-1], \tag{5c}$$

$$x^l \leq \text{diag}(UB^l)\sigma^l, \qquad\qquad \forall l \in \{2, \ldots, L\}, \tag{5d}$$

$$x^L = W^{L-1} x^{L-1} + b^{L-1}, \tag{5e}$$

$$\sigma^l \in \{0,1\}^{n_l}, \qquad\qquad \forall l \in \{2, \ldots, L\}, \tag{5f}$$

$$x^l \in \mathbb{R}_+^{n_l}, \qquad\qquad \forall l \in [L-1], \tag{5g}$$

$$x^1 \in \mathcal{X}, x^L \in \mathbb{R}, \tag{5h}$$

where $x^1 := x$ is the input-layer variable that belongs to the polytope $\mathcal{X}$, and x^L is the output of the neural network which approximates the expected value $Q(x^1)$.

2.2 Alternating MIP and NN Training

Since the approximation quality of (5) with regards to (2) highly depends on the accuracy of the neural network, the approach has significant drawbacks for problems with high dimensional first-level decisions, i.e., high-dimensional $\mathcal{X}$. This drawback becomes even more significant as optimal solutions for (2) are often attained at the boundary or even vertices of $\mathcal{X}$. Here, the approximation by the neural network tends to be worse due to a lack of training data around the vertices. Note that, sampling around each vertex is typically computationally intractable as there are potentially exponentially many such vertices.

The approach we propose in the present article uses intermediate solutions of (5) to inform the retraining of neural networks. To this end, after training the NN with an initial uniformly distributed sample of $\mathcal{X}$, we solve (5) and add sample points around the computed optimal solution x_k^* to our training data. Algorithm 1 illustrates our approach.

We would like to stress, that the number of additional datapoints in every iteration k as well as the parameter α towards x_k^* are chosen arbitrarily. Thus, more careful choices of these parameters may give room for significant improvement of the performance of Algorithm 1.

Here, we would like to briefly comment on the relative computational expenses of the different parts of Algorithm 1. We specify the considered application as well as our computational resources in Sect. 4. The most time-consuming computation of Algorithm 1 includes solving optimization problems (5) and $\min_{y \in \mathcal{Y}(x,\xi)} c^\top y$. With the default settings, Gurobi [11] can solve (5) encoding an NN 2×40 within 5 s on a standard notebook. Moreover, the generation of new training data (see line 6–10), i.e., the optimization for each scenario ξ, can be solved in parallel, where solving the problem $\min_{y \in \mathcal{Y}(x,\xi)} c^\top y$ with a pair x and ξ only takes 0.03 s.

Algorithm 1. Alternating MIP-NN Algorithm for 2SP (MIP-NN 2SP)

1: **Input** Initial training data $\bar{\mathcal{X}}$, $\alpha = 0.99$
2: **Output** Approximately optimal solution x^*_{100} for (1)
3: Train a NN 2×40 with training data set $\bar{\mathcal{X}}$
4: **for** $k = 1, \ldots, 100$ **do**
5:    $x^*_k \leftarrow \operatorname{argmin}$ (5)
6:    **for** $i = 1, \ldots, 50$ **do**
7:       Sample $x_i \in X$ uniformly
8:       $x_i \leftarrow \alpha x^*_k + (1 - \alpha)x_i$
9:       $\bar{\mathcal{X}} \leftarrow \bar{\mathcal{X}} \cup \{x_i\}$
10:    **end for**
11:    Retrain NN with $\bar{\mathcal{X}}$
12: **end for**

---

3 Application to Smart Converters in Power System Networks

In order to assess the practical value of the above approach, we apply it to a power system with smart inverters and energy storage. Its effects include reducing power generation costs and improving operational performance are then observed. Since the proposed heuristic given by Algorithm 1 may compute first-stage decisions x that lead to an empty second-stage, i.e., $\mathcal{Y}(x, \xi) = \emptyset$, we assume in the remainder of this article that every feasible $x \in \mathcal{X}$ leads to non-empty $\mathcal{Y}(x, \xi)$ for every scenario ξ. In particular, this assumption is valid whenever, we discuss a value function $Q : \mathcal{X} \times \Omega \to \mathbb{R}, Q(x, \xi) := \min_{y \in \mathcal{Y}(x, \xi)} c^\top y$ as is often the case in the existing literature, see e.g. [8], and also justified as the discussion in Sect. 2.1 in [31] illustrates.

Nevertheless, it is crucial to verify whether this assumption actually holds. For the power system application studied below however, numerical experiments indicate, that a feasible, but potentially costly $y \in \mathcal{Y}(x, \xi)$, exists and thus the use of neural networks will not have an adverse impact on the feasibility of the problem. Moreover, from an engineering perspective, the power system has a slack node, see b_1 in Fig. 1, that ensures stable operation when the system deviates from the set points, although again the operating cost may increase significantly. On the contrary, it can effectively improve the computational efficiency of the solution. The reasons are as follows:

(a) This paper establishes a two-stage optimization model to solve the day-ahead scheduling strategy of the power system (i.e., the first-stage strategy). We only use neural networks to replace the optimization model in the second stage. The first-stage decisions still satisfy the corresponding hard constraints (e.g., power balance).
(b) The second-stage optimization model corresponds to the intra-day operation of the power system. In the cases studied by this paper, the system is connected to the external grid and has energy storage inside as a buffer

to balance the energy. From the perspective of practical engineering experience, when the decisions of the first stage are within a reasonable range, the operation strategy of the second stage is usually feasible. Therefore, it is reasonable to use the first-stage decision variables to estimate the operating cost of the second stage through a neural network.

(c) In practical operations, the day-ahead operation (i.e., the first stage) needs to obtain results within tens of minutes. When there are a large number of integer variables in the second stage of the 2SP (energy storage requires frequent changes in operational states, which introduces a large number of binary variables into the second stage), this runtime requirement is difficult to guarantee. However, since replacing the second stage of (2) by a neural network can significantly reduce the runtime, Algorithm 1 has the potential to be applicable in practice.

A simplified DC power flow model, following Kirchhoff laws, is adopted to describe the energy balance and power flow in the power system. We follow closely the notations by [37] and the presentation in [16] in order to specify the parameters present in (1).

Consider a power grid equipped with a set of buses $\mathcal{B}$ and a set of lines/branches $\mathcal{L}$. Within the grid, power is generated by a set of conventional (fossil fuel) generators denoted by $\mathcal{N}_G$ or by a set of distributed (renewable) generators denoted by $\mathcal{N}_{DG}$. The *system operator* (SysO) may decide, whether to

– store or release power to/from a set of batteries $\mathcal{N}_S$
– purchase or sell power on the day-ahead market or intra-day, this power enters/leaves the power system through a trading node with the main grid

The amount of power traded day-ahead, i.e., on the day-ahead or *first-level market* is denoted by P_{fl} and the corresponding market price by p_{fl}. Similarly, the amount of power and the market price traded intra-day is denoted by P_{sl} and p_{sl}, respectively. A positive value for P_{fl}, P_{sl} is interpreted as a purchase, negative values as a sell P_{fl}, P_{sl} of energy.

The SO's initial planning, refers to deciding the first-level variables $x = (P_G^\top, P_{fl})^\top \in \mathbb{R}^{\mathcal{N}_G} \times \mathbb{R}$ based on the estimated renewable energy production $P_{DG,forecast} \in \mathbb{R}^{\mathcal{N}_{DG}}$ in order to ensure that a given total demand $\sum_{i \in \mathcal{B}} P_{d_i}$ is met. Hence,

$$\mathcal{X} = \{x \in \mathbb{R}^n : \text{(6a) \& (6b)}\},$$

where

$$P_{fl}^t = \sum_{i \in \mathcal{B}} P_{d_i}^t - \sum_{i \in \mathcal{N}_G} P_{G_i}^t - \sum_{i \in \mathcal{N}_{DG}} P_{DG_i,forecast}^t \qquad \forall t \in T, \qquad (6a)$$

$$P_{G_i,\min}^t \leq P_{G_i}^t \leq P_{G_i,\max}^t \qquad \forall i \in \mathcal{N}_G, t \in T, \qquad (6b)$$

with given parameters $P_{G,\min}, P_{G,\max} \in \mathbb{R}^{\mathcal{N}_G \times T}$. Note that the *market-clearing condition* (6a) ensures the active power balance in the whole system. The objective of the first level in (1) is then given by

$$G(x) = x^\top \text{Diag}(c_2)x + c_1^\top x + c_0,$$

where $c_2, c_1 \in \mathbb{R}^{\mathcal{N}_G}, c_0 \in \mathbb{R}$ are given generator cost parameters. Since the uncertainties will impact the initial planning and may jeopardize the power balance, we consider the capacity of the renewable generators as uncertain, i.e., we denote the second-level variable by $\xi = P_{DG,\max} \in \mathbb{R}^{\mathcal{N}_{DG} \times T}$. As these generators are dependent on weather conditions, which are highly uncertain, this is one of the most common uncertainties faced by modern power grids with a high proportion of renewable energies [12,27]. To this end, we draw samples from the following domain

$$\Omega = \left\{ P_{DG,\max} \in \mathbb{R}^{\mathcal{N}_{DG} \times T} : 0 \le P_{DG_i,\max}^t \le P_i^+ \ \forall \ i \in \mathcal{N}_{DG}, t \in T \right\}, \quad (7)$$

where P_i^+ denotes the technical limit of the renewable generator, i.e. its capacity under optimal conditions. In particular, since the capacity of the renewable generators is further limited by the weather conditions.

For our computational results in Sect. 4, we assume that without these limitations $P_{DG_i,\max}$ are independently distributed according to a normal distribution, i.e. $P_{DG_i,\max\text{-}wol} \sim \mathcal{N}(P_{DG_i,forecast}, 0.1 \cdot P_{DG_i,forecast})$. Consequently, every realization of $P_{DG_i,\max\text{-}wol}$ leads naturally to a sample point

$$\xi_{DG_i} = P_{DG_i,\max} = \begin{cases} 0 & \text{if } P_{DG_i,\max\text{-}wol} \le 0 \\ P_{DG_i,\max\text{-}wol} & \text{if } P_{DG_i,\max\text{-}wol} \in (0, P_i^+) \\ P_i^+ & \text{if } P_{DG_i,\max\text{-}wol} \ge P_i^+. \end{cases}$$

On the third level, the SysO adjusts the initial planning according to this realization. In particular, the SysO might regulate the energy output of the conventional generators P_G to $P_{G,\text{reg}}$ by either increasing the production by adding $P_G^+ \ge 0$ at a cost r^+ or decreasing the production by adding $P_G^- \le 0$ at a cost r^-. Similarly, $P_{DG} \in [0, \xi_{DG}]$ is the adjusted energy production that deviates from its forecast by P_{DG}^+ or P_{DG}^- with deviations penalized by f^+, f^- respectively. Moreover, the SysO might also trade power intra-day (P_{sl}) or decide to use the batteries by setting $(\mu_{ch}, \mu_{dch} \in \{0,1\}^{\mathcal{N}_S})$ and change the charging/discharging quantity P_{ch}, P_{dch} in order to balance a potential power deficiency or surplus. Based on these decisions, the following variables vary accordingly: The *state of charge* of a storage, denoted by soc, the power on a line $(k,l) \in \mathcal{L}$, denoted by p_{kl} and the phase angles of the system, denoted by θ. Thus, the SysO adjusts the vector $y = (P_{G_i,\text{reg}}, P_{G_i}^+, P_{G_i}^-, P_{DG_i}, P_{DG_i}^+, P_{DG_i}^-, P_{sl}, P_{ch_i}, P_{dch_i}, p_{kl}, \theta, \text{soc}, \mu_{ch}, \mu_{dch})^\top$ in order to satisfy

$$y \in \mathcal{Y}(x, \xi) := \{ y \in \mathbb{R}^m : (8a)\text{–}(12g) \},$$

where the constraints (8a)–(12g) are given below:

(a) First, we consider the *Generator and DG output constraints*:

$$P^t_{G_i,\text{reg}} = P^t_{G_i} + P^{t,+}_{G_i} + P^{t,-}_{G_i} \qquad\qquad \forall i \in \mathcal{N}_G, t \in T, \quad (8a)$$

$$P^t_{G_i,\text{min}} \leq P^t_{G_i,\text{reg}} \leq P^t_{G_i,\text{max}} \qquad\qquad \forall i \in \mathcal{N}_G, t \in T, \quad (8b)$$

$$P^t_{DG_i} = P^t_{DG_i,\text{forecast}} + P^{t,+}_{DG_i} + P^{t,-}_{DG_i} \qquad \forall i \in \mathcal{N}_{DG}, t \in T, \quad (8c)$$

$$P^t_{DG_i,\text{min}} \leq P^t_{DG_i} \leq P^t_{DG_i,\text{max}} \qquad\qquad \forall i \in \mathcal{N}_{DG}, t \in T, \quad (8d)$$

$$P^t_{sl} - P^t_{fl} = -\mathbf{1}^\top (P^{t,+}_G + P^{t,-}_G) - \mathbf{1}^\top (P^{t,+}_{DG} + P^{t,-}_{DG})$$
$$- \mathbf{1}^\top (P^t_{dch} - P^t_{ch}) \qquad\qquad \forall t \in T. \quad (8e)$$

Here, Constraints (8a)–(8d) describe the output range of the conventional and renewable generators. In particular, (8d) shows that the third-level variables $P^t_{DG_i}$ are restricted by the uncertainties realized in the second level $(P^t_{DG_i,\text{max}})$. The market clearing on the intra-day market, i.e., the actual power demand-supply relations, is reflected by Constraint (8e).

(b) Second, we consider the *operation constraints*:

$$0 \leq P^{t,+}_{G_i} \leq P^{t,+}_{G_i,\text{max}} \qquad\qquad \forall i \in \mathcal{N}_G, t \in T, \quad (9a)$$

$$P^{t,-}_{G_i,\text{min}} \leq P^{t,-}_{G_i} \leq 0 \qquad\qquad \forall i \in \mathcal{N}_G, t \in T, \quad (9b)$$

$$P^{t,+}_{DG_i} \geq 0 \qquad\qquad \forall i \in \mathcal{N}_{DG}, t \in T, \quad (9c)$$

$$P^{t,-}_{DG_i} \leq 0 \qquad\qquad \forall i \in \mathcal{N}_{DG}, t \in T, \quad (9d)$$

where Constraints (9a)–(9d) limit the real output derivations of conventional and renewable generators.

(c) Third, we consider the *power flow constraints*. To this end, we apply a DC approximation for a given line reactance $(x_{ij} > 0)$ and demand in active power $(P^{d,t}_k)$ at every time step t and bus k:

$$\sum_{l \in \delta(k)} p^t_{kl} = \sum_{i \in \mathcal{N}_G : i \sim k} P^t_{G_i,\text{reg}} + \sum_{i \in \mathcal{N}_{DG} : i \sim k} P^t_{DG_i}$$
$$+ \sum_{i \in \mathcal{N}_S : i \sim k} (P^t_{dch_i} - P^t_{ch_i}) - P^{d,t}_k \qquad \forall k \in \mathcal{B} \setminus \{0\}, t \in T,$$
$$(10a)$$

$$\sum_{l \in \delta(k)} p^t_{0l} = \sum_{i \in \mathcal{N}_G : i \sim 0} P^t_{G_i,\text{reg}} + \sum_{i \in \mathcal{N}_{DG} : i \sim 0} P^t_{DG_i}$$
$$+ \sum_{i \in \mathcal{N}_S : i \sim 0} (P^t_{dch_i} - P^t_{ch_i}) + P^t_{sl} - P^{d,t}_0 \qquad \forall t \in T, \quad (10b)$$

$$p^t_{ij} = \frac{1}{x_{ij}}(\theta^t_i - \theta^t_j) \qquad\qquad \forall \{i,j\} \in \mathcal{L}, t \in T, \quad (10c)$$

where (10a) establishes the nodal power flow balance for every node except the root node, which is addressed separately in (10b). The branch power flow is modeled through (10c).

(d) Fourth, we consider the *branch thermal constraints* that guarantee, that the power flow does not exceed the branch's capacities:

$$p_{ij}^t \leq s_{ij,\max} \qquad \forall\{i,j\} \in \mathcal{L}, t \in T. \qquad (11a)$$

(e) Fifth, we consider the *storage constraints*. To this end, note that the storage operation involves two actions, the storage action is represented by two binary variables $\mu_{ch}, \mu_{dch} \in \{0,1\}^{\mathcal{N}_S \times T}$, whereas the quantity of the charging/discharging is represented by continuous variables $P_{ch_i}^t, P_{dch_i}^t$. Thus, we have the following constraints:

$$\text{soc}_{i,\min}^t \leq \text{soc}_i^t \leq \text{soc}_{i,\max}^t \qquad \forall i \in \mathcal{N}_S, t \in T, \qquad (12a)$$

$$\text{soc}_i^t = \text{soc}_i^{t-1} + \frac{(P_{ch_i}^t - P_{dch_i}^t)}{E_i}\Delta T \qquad \forall i \in \mathcal{N}_S, t \in T, \qquad (12b)$$

$$P_{ch_i}^t, P_{dch_i}^t \geq 0 \qquad \forall i \in \mathcal{N}_S, t \in T, \qquad (12c)$$

$$\mu_{ch_i}^t, \mu_{dch_i}^t \in \{0,1\} \qquad \forall i \in \mathcal{N}_S, t \in T, \qquad (12d)$$

$$\mu_{ch_i}^t P_{ch_i,\min}^t \leq P_{ch_i}^t \leq \mu_{ch_i}^t P_{ch_i,\max}^t \qquad \forall i \in \mathcal{N}_S, t \in T, \qquad (12e)$$

$$\mu_{dch_i}^t P_{dch_i,\min}^t \leq P_{dch_i}^t \leq \mu_{dch_i}^t P_{dch_i,\max}^t \qquad \forall i \in \mathcal{N}_S, t \in T, \qquad (12f)$$

$$\mu_{ch_i}^t + \mu_{dch_i}^t \leq 1 \qquad \forall i \in \mathcal{N}_S, t \in T. \qquad (12g)$$

Here, the upper/lower bounds of soc, as well as the relationships between soc and charging/discharging actions, are given by Constraints (12a) and (12b). The connection between the storage actions and the respective quantity is reflected by Constraints (12c)–(12g).

Lastly, the third-level cost function aims to both, minimize the electricity cost as well as reduce the deviation between the intra-day system operation strategy and the day-ahead planning. Thus, the whole adjustment can be summarized as solving

$$\min_{y \in \mathcal{Y}(x,\xi)} c^\top y, \qquad (13)$$

where

$$c^\top y := \sum_{t \in T}\sum_{i \in \mathcal{N}_G}(r_i^{+,t}P_{G_i}^{+,t} + r_i^{-,t}P_{G_i}^{-,t}) + p_{sl}^t(P_{sl}^t - P_{fl}^t) + \sum_{i \in \mathcal{N}_{DG}}(f_i^+ P_{DG_i}^{t,+} + f_i^- P_{DG_i}^{t,-}).$$

After having established the model parameters for (1), we continue with a case study in order to demonstrate the practical value of Algorithm 1.

4   Computational Results

In this section, we present a case study, where we consider the daily power distribution of a 5-bus instance, based on the "case5.m" instance from the matpower library [41]. Here, the day is divided into hourly (24 period) time intervals. The MIP problems in Algorithm 1 are solved by Gurobi [11], which is the state-of-art

MIP solver. The computations were executed via Gurobi 10.0.0 under Python 3.7 on a Macbook Pro (2019) notebook with an Intel Core i7 2,8 GHz Quad-core and 16 GB of RAM. The neural network is trained with PyTorch [25].

The network topology of "case5.m" is illustrated in Fig. 1, where we assume bus 1 to be the slack node, i.e., the trading node connected to other grids. Since the type of generators is not specified in [41] and this study solely serves as an academic example, we chose whether a generator in "case5.m" is a conventional/renewable one or a storage in the following convenient way: The conventional (fossil fuel) generators are connected to the buses 1 and 4, i.e. $\mathcal{N}_G = \{1,4\}$, two distributed generators are connected to buses 1 and 5, i.e., $\mathcal{N}_{DG} = \{1,5\}$ and an energy storage unit is connected to bus 3, i.e. $\mathcal{N}_S = \{3\}$. In order to aid reproducibility and encourage further research on this topic, we provide the underlying system data for public use, see [17].

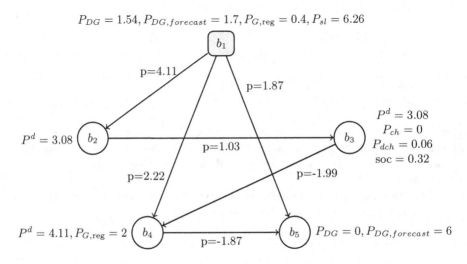

$P_{DG} = 1.54, P_{DG,forecast} = 1.7, P_{G,\mathrm{reg}} = 0.4, P_{sl} = 6.26$

Fig. 1. The case5.m network with its corresponding generators and an exemplary power flow. $P^d = 0$ at buses 1 and 5

Incorporating the instance data closely follows the methodology used in [16]. Hence, both, the day-ahead and intra-day market prices p_{fl}, p_{sl} were taken as averages from the Pecan street database's [26] "miso" data set for March 3rd, June 3rd, October 3rd and December 2nd, 2022. Daily deviations in P^d, denoted by Δ_d^t, or daily deviations in $P_{DG,\mathrm{max}}^t$, denoted by Δ_{DG}^t were taken from the Pecan street database's "california_iso" dataset for the same days, 2022 in the same way. In particular, we also provide the underlying weather data for public use, see [17].

Consequently, we model the daily varying demand by $P^d = P^d \cdot \Delta_d^t$ and the varying potential renewable energy production by

$$P_{DG,\mathrm{forecast}}^t := \min\left\{\frac{P_{DG,\mathrm{min}} + P_{DG,\mathrm{max}}}{2} \cdot \Delta_{DG}^t, P_{i,t}^+\right\}.$$

After incorporating this data, we compare the solutions given by Algorithm 1 to a baseline experiment, where we replace line 6–10 by adding 50 uniformly distributed sample points drawn from $\mathcal{X}$ in each iteration k. This algorithm is inspired by [8] and simply creates the same amount of sample points, but does not incorporate targeted information around x_k^*. We chose 3000 uniformly distributed datapoints from $\mathcal{X}$ as a starting dataset for both algorithms. We use trained neural networks with the same architecture for both experiments, i.e., NN 2×40.

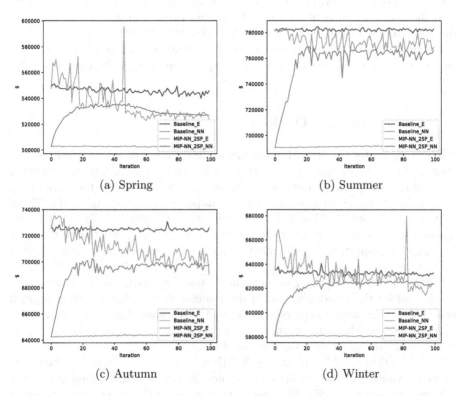

(a) Spring   (b) Summer

(c) Autumn   (d) Winter

Fig. 2. Comparison of Algorithm 1 (in green and red) to successively adding batches of 50 uniformly distributed data points (in blue and orange) with respect to operating costs in $ for weather data of different seasons. (Color figure online)

In Fig. 2, MIP-NN_2SP_NN (resp. Baseline_NN) denotes the objective function value of (5) solved by Algorithm 1 (resp. the baseline experiment) at each iteration. In particular, we include $G(x_k^*) + \frac{1}{400} \sum_{j=1}^{400} Q(x_k^*, \xi_j)$, denoted by MIP-NN_2SP_E (resp. Baseline_E), with optimal solutions x_k^* of line 5 from Algorithm 1 (resp. the baseline experiment). In this way, the difference between MIP-NN_2SP_NN (resp. Baseline_NN) and MIP-NN_2SP_E (resp. Baseline_E) measures the quality of the approximation by the neural networks at each iteration.

Observe, that after 100 iterations, the baseline experiment cannot close the gap between the estimated expected value of the second stage (i.e., Baseline_E) and its neural network approximation (i.e., Baseline_NN) since a gap of 80,000$ remains. In other words, the difference between the NN approximation and the expected value is 80,000$. While including targeted datapoints in the proposed sampling approach significantly reduces the gap between MIP-NN_2SP_NN and MIP-NN_2SP_E. Thus, showing that the dynamic sampling in Algorithm 1 clearly outperforms sampling uniformly distributed and results in a much more accurate approximation of the expected value. Moreover, the results indicate, that the operating state of the conventional generators computed from the baseline experiment is not optimal, and approximately 20,000$ may be saved by using the operating states computed by Algorithm 1. The results, thus, show the importance of obtaining an accurate approximation of the expected value function.

5 Conclusion and Outlook

We have presented an algorithm for solving 2SP problems with integer second-stage variables by iteratively constructing a NN that approximates the expected value of the second-stage problem. This approach is inspired by the work of [8]. In the algorithm, we propose a dynamic sampling and retraining of the NN to improve the approximation in regions of interest. We numerically evaluate the algorithm in a case study of optimizing a power system. The numerical results highlight the importance of obtaining an accurate approximation of the expected value function and show that a poor approximation can lead to a suboptimal solution. The results strongly support the proposed sampling and retraining approach, and by the smart selection of data points we are able to obtain a much better approximation in the regions of interest compared to a simple uniform sampling. For the baseline, using uniform sampling, the accuracy does not seem to improve even after doubling the number of data points. Whereas, the proposed algorithm quickly obtains an NN that seems to closely approximate the expected value function, at least in the neighborhood of the minimizer of the NN approximated 2SP (5). The results, thus, support the idea of approximating the expected value function by an NN and using a MILP encoding of the NN to form a single-stage problem, but clearly show the importance of efficiently training a NN to high accuracy in the regions of interest.

As Algorithm 1 delivers promising numerical results on the considered power system, it is natural to ask, whether these results extend to the benchmark problems given in [8], i.e. Capacitated Facility location, Investment, Stochastic server location and pooling problems. Moreover, also extensions to ACOPF models similarly to the ones in [14] may pose interesting challenges for future research.

Acknowledgements. We would like to thank the anonymous referees for their very valuable comments, that helped to significantly improve the quality of this article.

References

1. Bertsimas, D., Brown, D.B., Caramanis, C.: Theory and applications of robust optimization. SIAM Rev. **53**(3), 464–501 (2011)
2. Bhosekar, A., Ierapetritou, M.: Advances in surrogate based modeling, feasibility analysis, and optimization: a review. Comput. Chem. Eng. **108**, 250–267 (2018)
3. Birge, J., Louveaux, F.: Introduction to Stochastic Programming. Springer Series in Operations Research and Financial Engineering. Springer, New York (2011). https://doi.org/10.1007/978-1-4614-0237-4, https://books.google.se/books?id=Vp0Bp8kjPxUC
4. Cao, X., Sun, X., Xu, Z., Zeng, B., Guan, X.: Hydrogen-based networked microgrids planning through two-stage stochastic programming with mixed-integer conic recourse. IEEE Trans. Autom. Sci. Eng. **19**(4), 3672–3685 (2022). https://doi.org/10.1109/TASE.2021.3130179
5. Cao, Y., Wei, W., Wang, J., Mei, S., Shafie-khah, M., Catalão, J.P.S.: Capacity planning of energy hub in multi-carrier energy networks: a data-driven robust stochastic programming approach. IEEE Trans. Sustain. Energy **11**(1), 3–14 (2020). https://doi.org/10.1109/TSTE.2018.2878230
6. Carøe, C.C., Schultz, R.: Dual decomposition in stochastic integer programming. Oper. Res. Lett. **24**(1–2), 37–45 (1999)
7. Chazarra, M., García-González, J., Pérez-Díaz, J.I., Arteseros, M.: Stochastic optimization model for the weekly scheduling of a hydropower system in day-ahead and secondary regulation reserve markets. Electr. Power Syst. Res. **130**, 67–77 (2016)
8. Dumouchelle, J., Patel, R., Khalil, E.B., Bodur, M.: Neur2SP: neural two-stage stochastic programming (2022). https://doi.org/10.48550/ARXIV.2205.12006
9. Fischetti, M., Jo, J.: Deep neural networks and mixed integer linear optimization. Constraints **23**(3), 296–309 (2018)
10. Forrester, A.I., Keane, A.J.: Recent advances in surrogate-based optimization. Prog. Aerosp. Sci. **45**(1–3), 50–79 (2009)
11. Gurobi Optimization, LLC: Gurobi optimizer reference manual (2022). https://www.gurobi.com
12. Impram, S., Nese, S.V., Oral, B.: Challenges of renewable energy penetration on power system flexibility: a survey. Energ. Strat. Rev. **31**, 100539 (2020)
13. Kazemzadeh, N., Ryan, S.M., Hamzeei, M.: Robust optimization vs. stochastic programming incorporating risk measures for unit commitment with uncertain variable renewable generation. Energy Syst. **10**, 517–541 (2019)
14. Kilwein, Z., et al.: Ac-optimal power flow solutions with security constraints from deep neural network models. In: Türkay, M., Gani, R. (eds.) 31st European Symposium on Computer Aided Process Engineering, Computer Aided Chemical Engineering, vol. 50, pp. 919–925. Elsevier (2021). https://doi.org/10.1016/B978-0-323-88506-5.50142-X
15. Kleywegt, A.J., Shapiro, A., Homem-de Mello, T.: The sample average approximation method for stochastic discrete optimization. SIAM J. Optim. **12**(2), 479–502 (2002). https://doi.org/10.1137/S1052623499363220
16. Kronqvist, J., Li, B., Rolfes, J.: A mixed-integer approximation of robust optimization problems with mixed-integer adjustments. Optim. Eng. (2023). https://doi.org/10.1007/s11081-023-09843-7
17. Kronqvist, J., Li, B., Rolfes, J., Zhao, S.: https://github.com/jhrolfes/alternating_mixed_integer_programming_and_neural_network_training_for_approximating_2SP_data

18. Kronqvist, J., Misener, R., Tsay, C.: P-split formulations: a class of intermediate formulations between big-M and convex hull for disjunctive constraints (2022). arXiv preprint
19. Küçükyavuz, S., Sen, S.: An introduction to two-stage stochastic mixed-integer programming. In: Leading Developments from INFORMS Communities, pp. 1–27. INFORMS (2017)
20. Laporte, G., Louveaux, F.V.: The integer l-shaped method for stochastic integer programs with complete recourse. Oper. Res. Lett. **13**(3), 133–142 (1993)
21. Li, C., Grossmann, I.E.: A review of stochastic programming methods for optimization of process systems under uncertainty. Front. Chem. Eng. **2** (2021). https://doi.org/10.3389/fceng.2020.622241
22. Lomuscio, A., Maganti, L.: An approach to reachability analysis for feed-forward ReLU neural networks (2017). arXiv preprint
23. Meibom, P., Barth, R., Hasche, B., Brand, H., Weber, C., O'Malley, M.: Stochastic optimization model to study the operational impacts of high wind penetrations in Ireland. IEEE Trans. Power Syst. **26**(3), 1367–1379 (2010)
24. Mohseni-Bonab, S.M., Kamwa, I., Moeini, A., Rabiee, A.: Voltage security constrained stochastic programming model for day-ahead BESS schedule in co-optimization of T&D systems. IEEE Trans. Sustain. Energy **11**(1), 391–404 (2019)
25. Paszke, A., et al.: PyTorch: an imperative style, high-performance deep learning library. In: Proceedings of NEURIPS 2019, pp. 8024–8035. Curran Associates, Inc. (2019)
26. pecanstreet.org (2022). https://www.pecanstreet.org/
27. Pfenninger, S., Hawkes, A., Keirstead, J.: Energy systems modeling for twenty-first century energy challenges. Renew. Sustain. Energy Rev. **33**, 74–86 (2014)
28. Raghunathan, A., Steinhardt, J., Liang, P.: Certified defenses against adversarial examples. In: International Conference on Learning Representations (2018). https://openreview.net/forum?id=Bys4ob-Rb
29. Serra, T., Kumar, A., Ramalingam, S.: Lossless compression of deep neural networks. In: Hebrard, E., Musliu, N. (eds.) CPAIOR 2020. LNCS, vol. 12296, pp. 417–430. Springer, Cham (2020). https://doi.org/10.1007/978-3-030-58942-4_27
30. Shams, M.H., Shahabi, M., Khodayar, M.E.: Stochastic day-ahead scheduling of multiple energy carrier microgrids with demand response. Energy **155**, 326–338 (2018)
31. Shapiro, A., Philpott, A.B.: A tutorial on stochastic programming (2007)
32. Sharafi, M., ElMekkawy, T.Y.: Stochastic optimization of hybrid renewable energy systems using sampling average method. Renew. Sustain. Energy Rev. **52**, 1668–1679 (2015)
33. Torres, J.J., Li, C., Apap, R.M., Grossmann, I.E.: A review on the performance of linear and mixed integer two-stage stochastic programming software. Algorithms **15**(4) (2022). https://doi.org/10.3390/a15040103
34. Tsay, C., Kronqvist, J., Thebelt, A., Misener, R.: Partition-based formulations for mixed-integer optimization of trained ReLU neural networks. Adv. Neural. Inf. Process. Syst. **34**, 3068–3080 (2021)
35. Wong, S., Fuller, J.D.: Pricing energy and reserves using stochastic optimization in an alternative electricity market. IEEE Trans. Power Syst. **22**(2), 631–638 (2007). https://doi.org/10.1109/TPWRS.2007.894867
36. Wu, H., Shahidehpour, M., Li, Z., Tian, W.: Chance-constrained day-ahead scheduling in stochastic power system operation. IEEE Trans. Power Syst. **29**(4), 1583–1591 (2014). https://doi.org/10.1109/TPWRS.2013.2296438

37. Yang, Y., Wu, W.: A distributionally robust optimization model for real-time power dispatch in distribution networks. IEEE Trans. Smart Grid **10**(4), 3743–3752 (2019). https://doi.org/10.1109/TSG.2018.2834564
38. Yu, J., Ryu, J.H., Lee, I.B.: A stochastic optimization approach to the design and operation planning of a hybrid renewable energy system. Appl. Energy **247**, 212–220 (2019)
39. Zhang, J., Fuller, J.D., Elhedhli, S.: A stochastic programming model for a day-ahead electricity market with real-time reserve shortage pricing. IEEE Trans. Power Syst. **25**(2), 703–713 (2009)
40. Zhao, S., Tsay, C., Kronqvist, J.: Model-based feature selection for neural networks: a mixed-integer programming approach (2023). arXiv preprint
41. Zimmerman, R.D., Murillo-Sánchez, C.E., Thomas, R.J.: MATPOWER: steady-state operations, planning, and analysis tools for power systems research and education. IEEE Trans. Power Syst. **26**(1), 12–19 (2011). https://doi.org/10.1109/TPWRS.2010.2051168

Heaviest and Densest Subgraph Computation for Binary Classification. A Case Study

Zoltán Tasnádi[(✉)] and Noémi Gaskó

Faculty of Mathematics and Computer Science, Centre for the Study of Complexity, Babeș-Bolyai University, Cluj-Napoca, Romania
{zoltan.tasnadi,noemi.gasko}@ubbcluj.ro

Abstract. This article presents a novel network-based data classification method. The classification problem is discussed as a graph theoretical problem. A real-valued data first is transformed to an undirected graph, and then the heaviest and densest subgraphs are detected based on an ant colony optimization approach. Numerical experiments conducted on a real-valued dataset show the potential of the proposed approach.

1 Introduction

Classification is a key task in machine learning with several application possibilities, such as in image processing [11], in handwriting recognition [7], in data mining [13], etc.

Several algorithms have been proposed for classification problems, such as support vector machines, random forest, k-nearest neighbor, decision trees, and logistic regression (for a detailed description of the algorithms, see surveys [10] and [8]).

Network-based machine learning has gained a lot of attention in recent years. The first question which arises in this type of classification problem is how to build the graph from the dataset. There are two main directions of data transformation: the k-nearest neighbors (kNN) [15], and the ϵ-radius network construction can be considered as uniform methods, while there is no difference between sparse and dense regions of graphs. The other class processes the data adaptively, for example by combining the above presented two network formation algorithms [14], AdaRadius [2], or k-Associated Optimal Graph [3].

The main goal of this article is to use the heaviest and densest subgraph computation problem for network-based data classification. This is a novel approach in which the classification problem is discussed as a graph-theoretical problem. A real-valued dataset is transformed into a graph (considered both weighted and unweighted versions), and then the densest and heaviest subgraphs of size k are searched in the built graph.

This work was supported by a grant of the Romanian Ministry of Education and Research, CNCS - UEFISCDI, project number PN-III-P4-ID-PCE-2020-2360, within PNCDI III.

The article is organized as follows: the next section describes in more detail the binary classification problem, and the heaviest and densest subgraph computation problem. The third section presents the proposed method and the basic idea of the problem formulation; the fourth section describes the numerical experiments, while the article ends up with a conclusion and further work possibilities.

2   Problem Statement and Related Work

In the next section, we will focus on the binary classification problem, which can be described as follows: for a given set of input data $X = (x_1, x_2, \ldots, x_n)^T$, where $x_i \in \mathbf{R}^p, p > 1$ and for a given set of labels $Y = \{y_1, y_2, \ldots, y_n\}$, where $y_i \in \{0, 1\}$ (y_i corresponding to x_i) the problem consists of finding a model that makes a good prediction from the input data X to Y.

Formally, a graph can be described as $G = (V, E, F, W)$, where

- V is the set of nodes
- E is the set of edges
- F is a mapping between the nodes and edges, $F : E \to V \otimes V$
- W is the weight function, $W : E \to \mathcal{R}$

Remark 1. In an abbreviated form, we can refer to a graph as $G = (V, E)$.

Formally, the densest k-subgraph problem can be described as follows:

Definition 1. *Given an unweighted graph $G = (V, E)$, where V is the set of nodes and E is the set of edges, and for a given k (k is less than the number of nodes), to find an $V' \subset V, |V'| = k$ the number of edges should be maximal in the subgraph $G' = (S, E')$, where $E' \subset E$:*

$$\max_{V' \subset V, |V'|=k, E' \subset E} |E'|, \tag{1}$$

where $|\cdot|$ denotes the cardinality of the set.

Remark 2. The densest k-subgraph problem is a special case of the heaviest k-subgraph problem, where every weight is equal to 1.

Definition 2. *The heaviest k-subgraph problem consists in finding $G' = (V', E'), V' \subset V, |V'| = k$ and $E' \subset E$ such that $\sum_{e \in E'} w(e)$ is maximal, where $w(e)$ is the weight of the edge e.*

Example 1. Given the graph $G = (V, E)$, where $V = \{A, B, C, D, E\}$ and $E = \{(A, B), (A, C), (A, D), (B, C), (B, D), (C, D), (D, E)\}$, and the corresponding weights are $(1, 1, 1, 1, 1, 2, 10)$. The graph is visualized in Fig. 1. In this example, the 4-densest subgraph $G' = (V', E'), V' = A, B, C, D$, since there are 6 edges in that subgraph. However, when we take into account the weights of these edges and calculate the 4-heaviest subgraph, it becomes $G' = (V', E'), V' = \{A, B, D, E\}$, because the edge (D, E) has a weight of 10, which is larger as the sum of the weights of the other edges.

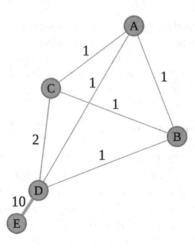

Fig. 1. The visualisation of the example network

Regarding the densest and the heaviest k-subgraph problem, several algorithmic approaches were introduced, for example [12] proposes a semidefinite programming approach. In [9] a simulated annealing and a tabu search approach are proposed to solve a similar variant of the problem. [4] proposed a variable neighborhood search algorithm for the heaviest subgraph problem.

In network-based classification, a key task is how to construct the network from the data. The most common technique is the k-nearest neighbors network (kNN), where each row of the data is represented by a node, and a directed edge exists between nodes v_i and v_j if v_j is between the k most similar nodes of node v_i. Another popular construction method is the ϵ-radius network construction method, where each node is a row, and there is an edge between node v_i and v_j if the distance between the two nodes is less than or equal to ϵ (for distance we can use the Euclidean distance of each attribute). [5] presents a particle swarm optimization-based approach that builds a network while optimizing a quality function based on classification accuracy. In [6] a new measure, called spatio-structural differential efficiency is introduced, which combines the topological and physical properties of the networks.

3   Method

3.1   Network Construction

A straightforward method is to build the ϵ radius network (as described above), where each data row is a node in the network, and there exists an edge between the nodes, if the weights of the edges between the nodes are greater or equal to a certain value of ϵ. The weights can be calculated as the sum of Euclidean distances of the real data values.

3.2   Ant Colony Optimization Algorithm

For the heaviest and densest subgraph detection problem, we use an ant colony optimization algorithm. The algorithm for the detection of the densest k-subgraph was proposed in [17], and for the heaviest k-subgraph in [16].

The algorithm consists of using agents (ants) to generate lists of nodes of k unique elements. The first element is generated randomly, and from there, each next node is selected using a combination of pheromones and a priori knowledge. Calculating the latter is done differently in the two cases, while the former abides by the method described in the MinMaxAntSystem restricting the pheromone levels in order to avoid getting stuck in local maxima. In each iteration, only the highest performing ant places pheromones on the edges that they traversed. The amount placed down is proportional to the fitness of the solution.

The a priori knowledge for the heaviest k-subgraph is calculated for each pair of nodes by summing the weights of edges connected to one of the nodes and adding to it the weight of the edge that connects the pair, if it exists, and in the end transposing the entire matrix in order to counteract negative edges, since the algorithm cannot handle them. This is encapsulated in the following formula:

$$\eta_{i,j} = \sum_{k \in N_j}^{k} w_{j,k} + w_{i,j} - min(w) + 1, \tag{2}$$

where N_j is the neighborhood of node j, $w_{i,j}$ contains the weight of the edge between i and j, and $min(w)$ is the minimal weight.

In the case of the densest k-subgraph we employ a different strategy. For each pair of nodes, we count the number of common neighbors and the number of neighbors of the second node that are not common with the first. Since the latter is less important, it is scaled down accordingly. Both values are also divided by the size of the graph. The final formula looks the following:

$$\eta_{i,j} = \frac{|N_j \cup N_i|}{|G|} + \frac{1}{4} \frac{|N_j \setminus N_i|}{|G|}. \tag{3}$$

After a given number of iterations, the best result from all iterations is returned. The main steps of the algorithm are outlined in Algorithm 1.

4   Numerical Experiments

Dataset. For numerical experiments we used the Algerian forest fires dataset[1] [1]. The dataset has 244 instances and 12 attributes:

---

[1] downloaded from https://archive.ics.uci.edu/ml/datasets/Algerian+Forest+Fires+Dataset++.

Algorithm 1. Ant colony optimization algorithm for the densest and heaviest k-subgraph problem

---

Initialize pheromone trails;
Initialize ants;
$i = 0$
while $i < iter_{max}$ **do**
    $S \longleftarrow \emptyset$
    repeat
        Construct a new solution s
        $S \longleftarrow S \cup \{s\}$
        $i \longleftarrow i + 1$
    until $|S| = k$
    Calculate the iteration-best (s_{ib}), and the global-best (s_{best})
    Compute pheromone trail limits (τ_{min} and τ_{max})
    Update pheromone trail
end while
return s_{best}

---

1. Date - weather data observations, we omitted this data from the graph construction
2. Temperature: 22 to 42
3. Relative Humidity: 21 to 90
4. Wind speed in km/h: 6 to 29
5. quantity of daily rain in mm: 0 to 18
6. Fine Fuel Moisture Code (FFMC) index: 28.6 to 92.5
7. Duff Moisture Code (DMC) index: 1.1 to 65.9
8. Drought Code (DC) index: 7 to 220.4
9. Initial Spread Index (ISI) index: 0 to 18.5
10. Buildup Index (BUI) index: 1.1 to 68
11. Fire Weather Index (FWI) Index: 0 to 31.1
12. Classes: Fire or not Fire

As mentioned above, we constructed the networks (graphs) based on the ϵ-radius network construction method. Two types of network are constructed: using the first eight attributes (omitting the date feature) and using all features (again, omitting the date feature). For ϵ four different values are considered 50, 100, 150 and 200. We consider both weighted and unweighted variants of the graphs, totally eight graphs. Table 2 presents some basic properties of the constructed graphs, the number of nodes ($|V|$), the number of edges ($|E|$), the average degree ($\langle d \rangle$), and the density of the graph (ρ).

Figures 2 and 3 present two networks: the first example presents the constructed network using 8 attributes and $\epsilon = 150$, while the second presents the constructed network from ten attributes and for $\epsilon = 200$ (Table 1).

Table 1. Some basic properties of the constructed networks from the Algerian Forest Fire Dataset.

Graph	ϵ	$|V|$	$|E|$	$\langle d \rangle$	ρ
8 attributes					
G1	50	244	14668	120.23	0.495
G2	100	244	5195	42.582	0.175
G3	150	210	1799	17.133	0.082
G4	200	98	178	3.633	0.037
10 attributes					
G5	50	244	15345	125.779	0.518
G6	100	244	5669	46.467	0.191
G7	150	216	2105	19.491	0.091
G8	200	134	318	4.746	0.036

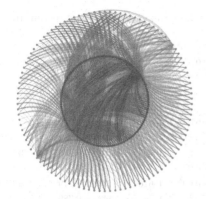

Fig. 2. The constructed network from eight attributes and $\epsilon = 150$. (Color figure online)

Fig. 3. The constructed network from ten attributes and $\epsilon = 200$. (Color figure online)

Remark 3. Figures 2 and 3 represent the constructed network where the nodes in the FIRE group are colored red, while the nodes in the NOT_FIRE group are colored blue.

Performance Evaluation Measure. For the performance evaluation, we use normalised accuracy, the fraction of correctly detected classes over the total number of predictions.

Parameter Settings. The following parameters were chosen for the ACO algorithm: $\alpha = 1$, $\beta = 2$, $\rho = 0.8$, $\epsilon = 0.005$, the number of ants was set to 50, and the number of iterations was set to 300. Furthermore, we chose the maximum length of a solution to be $3k$. Whenever during the generation phase we reach that limit, the solution will be discarded. For k we considered different values $k = 10\%$, $k = 20\%$, $k = 30\%$ and $k = 50\%$ of the total number of nodes.

Results and Discussion. Table 2 presents the proportions obtained for the nodes based on the heaviest and densest subgraph problem. 50 independent runs were conducted and the k nodes with the most frequent appearances were selected. If k is relatively small, no significant conclusion can be drawn, about half and half of the nodes are from the two different classes. The same situation can be observed if k is set to half of the nodes ($k = 50\%$). Regarding the setting of the ϵ value, a small value results in a dense graph with no useful information. Table 3 presents the accuracy values obtained for the 'NOT FIRE' class. The best result is 0.830, obtained considering only 8 attributes and for $k = 122$.

Table 2. Proportion of the detected classes in the constructed graphs, considering both weighted and unweighted networks

Classes	k	graph type	8 attr. $\epsilon = 50$	8 attr. $\epsilon = 100$	8 attr. $\epsilon = 150$	8 attr. $\epsilon = 200$	10 attr. $\epsilon = 50$	10 attr. $\epsilon = 100$	10 attr. $\epsilon = 150$	10 attr. $\epsilon = 200$
FIRE	24	W	0.583	0.5	0.416	0.125	0.5	0.5	0.416	0.208
NOT_FIRE	24	W	0.416	0.5	0.583	0.875	0.5	0.5	0.583	0.791
FIRE	24	UW	0.75	0.791	0.5	0.125	0.75	0.833	0.583	0.208
NOT_FIRE	24	UW	0.25	0.208	0.5	0.875	0.25	0.166	0.416	0.791
FIRE	49	W	0.673	0.448	0.265	0.081	0.632	0.469	0.265	0.081
NOT_FIRE	49	W	0.326	0.551	0.734	0.918	0.367	0.530	0.734	0.918
FIRE	49	UW	0.795	0.469	0.265	0.081	0.795	0.489	0.285	0.163
NOT_FIRE	49	UW	0.204	0.530	0.734	0.918	0.204	0.510	0.714	0.836
FIRE	73	W	0.657	0.342	0.205	0.095	0.616	0.356	0.219	0.136
NOT_FIRE	73	W	0.342	0.657	0.794	0.904	0.383	0.643	0.780	0.863
FIRE	73	UW	0.712	0.356	0.219	0.095	0.726	0.356	0.191	0.123
NOT_FIRE	73	UW	0.287	0.643	0.780	0.904	0.273	0.643	0.808	0.876
FIRE	122	W	0.557	0.336	0.303	0.483	0.549	0.368	0.303	0.483
NOT_FIRE	122	W	0.442	0.663	0.696	0.516	0.450	0.631	0.696	0.516
FIRE	122	UW	0.549	0.352	0.278	0.295	0.581	0.368	0.311	0.286
NOT_FIRE	122	UW	0.450	0.647	0.721	0.704	0.418	0.631	0.688	0.713

Table 3. Normalized accuracy values in the constructed graphs, considering both weighted and unweighted networks

k	graph type	8 attr. $\epsilon = 50$	8 attr. $\epsilon = 100$	8 attr. $\epsilon = 150$	8 attr. $\epsilon = 200$	10 attr. $\epsilon = 50$	10 attr. $\epsilon = 100$	10 attr. $\epsilon = 150$	10 attr. $\epsilon = 200$
24	W	0.094	0.113	0.132	0.198	0.113	0.113	0.132	0.179
24	UW	0.057	0.047	0.113	0.198	0.057	0.038	0.094	0.179
49	W	0.151	0.255	0.340	0.425	0.170	0.245	0.340	0.425
49	UW	0.094	0.245	0.340	0.425	0.094	0.236	0.330	0.387
73	W	0.236	0.453	0.547	0.623	0.264	0.443	0.538	0.594
73	UW	0.198	0.443	0.538	0.623	0.189	0.443	0.557	0.604
122	W	0.509	0.764	0.802	0.594	0.519	0.726	0.802	0.594
122	UW	0.519	0.745	0.830	0.811	0.481	0.726	0.792	0.821

5   Conclusions and Further Work

This article presents a novel network-based data classification method. A real-valued data is first transformed to an undirected graph, and then the heaviest and densest subgraphs are detected based on an ant colony optimization approach. The main idea is that such subgraphs represent data from the same class, therefore it can be seen as a new classification method. Another advantage of this method consists in using it as a feature selection technique, since the construction of the graphs gives us flexibility to consider only some attributes.

Further work will include testing other construction methods and its use for unbalanced classification problems.

Acknowledgments. This work was supported by a grant of the Ministry of Research, Innovation and Digitization, CNCS/CCCDI - UEFISCDI, project number 194/2021 within PNCDI III.

References

1. Abid, F., Izeboudjen, N.: Predicting forest fire in Algeria using data mining techniques: case study of the decision tree algorithm. In: Ezziyyani, M. (ed.) AI2SD 2019. AISC, vol. 1105, pp. 363–370. Springer, Cham (2020). https://doi.org/10.1007/978-3-030-36674-2_37
2. Araújo, B., Zhao, L.: Data heterogeneity consideration in semi-supervised learning. Expert Syst. Appl. **45**, 234–247 (2016). https://doi.org/10.1016/j.eswa.2015.09.026, https://www.sciencedirect.com/science/article/pii/S0957417415006545
3. Bertini, J.R., Zhao, L., Motta, R., de Andrade Lopes, A.: A non-parametric classification method based on k-associated graphs. Inf. Sci. **181**(24), 5435–5456 (2011). https://doi.org/10.1016/j.ins.2011.07.043, https://www.sciencedirect.com/science/article/pii/S0020025511003823

4. Brimberg, J., Mladenović, N., Urošević, D., Ngai, E.: Variable neighborhood search for the heaviest k-subgraph. Comput. Oper. Res. **36**(11), 2885–2891 (2009)
5. Carneiro, M.G., Cheng, R., Zhao, L., Jin, Y.: Particle swarm optimization for network-based data classification. Neural Netw. **110**, 243–255 (2019)
6. Carneiro, M.G., Zhao, L.: Organizational data classification based on the importance concept of complex networks. IEEE Trans. Neural Netw. Learn. Syst. **29**(8), 3361–3373 (2018). https://doi.org/10.1109/TNNLS.2017.2726082
7. Ciresan, D.C., Meier, U., Gambardella, L.M., Schmidhuber, J.: Convolutional neural network committees for handwritten character classification. In: 2011 International Conference on Document Analysis and Recognition, pp. 1135–1139. IEEE (2011)
8. Kesavaraj, G., Sukumaran, S.: A study on classification techniques in data mining. In: 2013 Fourth International Conference on Computing, Communications and Networking Technologies (ICCCNT), pp. 1–7 (2013)
9. Kincaid, R.K.: Good solutions to discrete noxious location problems via metaheuristics. Ann. Oper. Res. **40**(1), 265–281 (1992)
10. Kumar, R., Verma, R.: Classification algorithms for data mining: a survey. Int. J. Innov. Eng. Technol. (IJIET) **1**(2), 7–14 (2012)
11. Lu, D., Weng, Q.: A survey of image classification methods and techniques for improving classification performance. Int. J. Remote Sens. **28**(5), 823–870 (2007)
12. Malick, J., Roupin, F.: Solving k-cluster problems to optimality with semidefinite programming. Math. Program. **136**(2), 279–300 (2012)
13. Phyu, T.N.: Survey of classification techniques in data mining. In: Proceedings of the International Multiconference of Engineers and Computer Scientists, vol. 1, pp. 727–731. Citeseer (2009)
14. Silva, T.C., Zhao, L.: Network-based high level data classification. IEEE Trans. Neural Netw. Learn. Syst. **23**(6), 954–970 (2012)
15. Szummer, M., Jaakkola, T.: Partially labeled classification with Markov random walks. In: Advances in Neural Information Processing Systems, vol. 14 (2001)
16. Tasnádi, Z., Gaskó, N.: An ant colony optimisation approach to the densest k-subgraph problem. In: 2022 24th International Symposium on Symbolic and Numeric Algorithms for Scientific Computing (SYNASC), pp. 208–211. IEEE (2022)
17. Tasnádi, Z., Gaskó, N.: A new type of anomaly detection problem in dynamic graphs: an ant colony optimization approach. In: Mernik, M., Eftimov, T., Črepinšek, M. (eds.) BIOMA 2022. LNCS, vol. 13627, pp. 46–53. Springer, Cham (2022). https://doi.org/10.1007/978-3-031-21094-5_4

SMBOX: A Scalable and Efficient Method for Sequential Model-Based Parameter Optimization

Tarek Salhi⬤ and John Woodward⁽✉⁾⬤

Loughborough University, Loughborough , UK
{t.salhi,j.woodward}@lboro.ac.uk

Abstract. The application of Machine Learning (ML) algorithms continues to grow and shows no signs of slowing down. Each ML algorithm has an associated set of hyperparameter values that need to be set to achieve the best performance for each problem. The task of selecting the best parameter values for the problem at hand is known as Hyperparameter Optimisation (HPO). Traditionally this has been carried out manually or by unguided programmatic approaches such as grid or random search. These approaches can be extremely time-consuming and inefficient, especially when dealing with more than a handful of parameters. More advanced methods involving Evolutionary Heuristics [23] or Bayesian Optimisation [17,28] use a guided search approach and are widely considered as the gold standard approach for hyperparameter optimisation.

In this paper, we introduce SMBOX (https://github.com/smbox/smbox), a novel HPO search strategy developed to rival the state-of-the-art, SMAC [15]. Our benchmarking on public classification datasets, against both SMAC and a Random search baseline, shows that SMBOX not only challenges SMAC in tuning hyperparameters for two prevalent ML algorithms, but it also excels in finding good hyperparameter values quicker than SMAC. This rapid optimisation capability is extremely powerful, particularly in situations where time or computational resources are constrained or costly.

Keywords: Hyperparameter Optimisation · Machine Learning · Classification

1 Introduction

The hyperparameter values of a Machine Learning (ML) algorithm can have a significant impact on how well the algorithm performs at the task at hand. Research into the best approaches to select hyperparameter values has gained momentum recently with the increased adoption of ML and is now seen as a fundamental requirement when using an ML algorithm.

© The Author(s), under exclusive license to Springer Nature Switzerland AG 2024
G. Nicosia et al. (Eds.): LOD 2023, LNCS 14506, pp. 149–162, 2024.
https://doi.org/10.1007/978-3-031-53966-4_12

The problem of selecting optimal values for these hyperparameters is non-trivial and is referred to as Hyperparameter Optimisation (HPO). HPO has several properties that make it a complex task:

- Firstly, ML algorithms can have a large number of interrelated and conditional hyperparameters, thus each parameter cannot be optimised in isolation. The large parameter search space which can include a mixture of variable types, such as integer, continuous, categorical, and/or conditional adds additional complexity to the search. This high dimensional complex space renders techniques such as exhaustive search and large grid searches infeasible.
- Secondly, the evaluation of ML algorithms can be very computationally expensive and extremely time-consuming. Due to the ever-growing sheer size of the datasets being generated and therefore analysed this exacerbates the computational resources required to evaluate ML algorithms. Furthermore, to achieve robust evaluation metrics that generalise well to non-training data a rigorous K-Fold Cross-Validation evaluation strategy should be implemented which in turn increases the required computation resources required.
- Thirdly, every dataset will result in a different HPO objective function (error function). Thus, the optimal set of hyperparameters for one dataset will be different from those of another dataset. Similar to there being no one ML algorithm that outperforms the rest, there is not one set of hyperparameters that works best across all datasets.

With the exponential growth in model complexity [18] the need for efficient HPO techniques only continues to grow.

In order to resolve the above-stated problems we present an efficient, lightweight and intuitive Sequential Model-Based Optimisation (SMBO) routine called SMBOX. SMBOX, which is short for Sequential Model-Based Optimization eXpress introduces two notable adaptions to the traditional SMBO routine and achieves promising performance results on a range of classification datasets. Not only does our approach challenge the current state-of-the-art, but it also finds good parameter values quicker making it ideal for time and computational resource-limited problems, such as near real-time online learning or resource-expensive tasks which are common when working with large datasets.

This paper is structured as followed. Section 2 describes the field of HPO, followed by the most common and current state-of-the-art HPO techniques. Section 3 introduces our search routine, and details the adaptions made to a traditional SMBO routine. Section 4 provides the full experiment overview. Section 5 presents the results and findings of our experiment. Section 6 covers potential future work and advancements. Section 7 concludes this research paper.

2   Background

The precise definition of a hyperparameter is often subject to debate. Some researchers have extended the traditional definition from strictly an ML algorithm's configurable parameters to the actual selection of the ML algorithm itself, such as in [9,29]. This wider definition is also known as the Combined Algorithm Selection and Hyperparameter (CASH) problem.

More recently, the scope has been stretched even further to include a wide range of possible feature engineering options, such as whether to normalise continuous variables and missing value imputation strategies [9,20,32]. This wider definition of HPO is commonly known as AutoML, see [16] or [23] for an in-depth introduction.

In our research, HPO is defined using the conventional definition - that is, tuning the configurable parameters of the chosen ML algorithm. The most widely adopted HPO techniques are grid search and manual search as demonstrated in [14,19,22]. Although both strategies are capable of finding good hyperparameter values, their effectiveness is generally limited to low-dimensional search spaces. These methods are not efficient when dealing with larger, high-dimensional search spaces.

Grid search works by evaluating every possible hyperparameter configuration within a predefined search space. Random search is another similar approach which randomly selects possible hyperparameter configurations from the predefined search space. Random search has been found to be empirically and theoretically more efficient than grid search [3] however this strategy is in no way optimal.

When working with complex ML algorithms that contain a larger number of hyperparameters the unguided nature of both grid search and random search becomes a significant weakness. This flaw becomes even more apparent when working with large datasets due to the increased training time and computational resources required. With the ever-increasing volume of the data being collected and analysed together with the application of deep learning, techniques such as Recurrent Neural Networks [12] and Large Language Models [5] the need for efficient HPO techniques is evident and hence worthwhile pursuing. More advanced HPO techniques generally leverage the learnings from SMBO or Evolutionary algorithms such as genetic algorithms [1,21]. See [2,8] for an overview of Evolutionary algorithms. SMBO and Evolutionary algorithms aim to efficiently balance a trade-off between the exploration and the exploitation of the search space.

SMBO works by successively fitting a cheap-to-compute surrogate model to estimate an expensive objective function. The surrogate model adapts to better fit the true objective function once actual configuration outputs become known.

A Gaussian Process (GP) [26] is the most commonly used surrogate model, and provides a natural way to define distributions. For an in-depth introduction to GPs refer to [31]. GPs are however limited to the optimisation of continuous variables which is problematic for HPO applications which tend to include mixed variable types. To get around the mixed variable type weakness [15] proposed the use of a Random Forest (RF) [4] rather than a GP due to the way an RF elegantly handles categorical variables. A simpler approach could be to use a standard Decision Tree, however, the random ensemble of Decision Trees that make up an RF generally results in more accurate predictions [4]. This new and improved approach developed by [15] is called Sequential Model-based Algorithm Configuration (SMAC) and has become a popular choice when optimising hyperparameters.

The research area of HPO has gained increased attention recently due to the widespread adoption of ML and the development of more complex cutting-edge ML algorithms. Due to the performance gains, HPO can yield when implemented correctly, it is now considered a fundamental requirement when using an ML algorithm.

3   SMBOX Design

We present a new HPO technique called Sequential Model-Based Optimisation eXpress (SMBOX). While SMBOX retains the core principles of conventional SMBO methods, its introduction of two strategic modifications enables SMBOX to identify optimal parameter values more efficiently and rapidly than existing techniques.

1. Firstly we incorporate a low-fidelity random search for a minimum number of iterations. The random search configuration results are used to rapidly train an initial surrogate model. A minimum number of iterations has been set based on experimentation, however, there is scope for this meta parameter to be selected using a more intelligent approach, such as schema designs or meta-rules.

2. Secondly we implement a more sophisticated surrogate model than existing SMBO techniques which enables SMBOX to estimate behaviour of the objective function more accurately. SMBOX iteratively fits a Gradient Boosting (GB) [11] model rather than the RF [4] that is used within SMAC. GB algorithms are known to achieve state-of-the-art performance on a range of supervised ML problems, and thus provide an optimal choice for the task at hand, adding further precision and reliability to our proposed search strategy.

We selected the CatBoost [7] GB implementation due to its strong performance and inbuilt algorithm to handle categorical features with minimal information loss during training rather than at the pre-processing stage. This feature enables our surrogate model to optimally handle any categorical hyperparameter.

SMBOX Design

STEP 1: Initialise SMBOX by evaluating lf_n randomly generated parameter configurations on the lf_ratio training dataset.
STEP 2: Train the surrogate model on all previously evaluated parameter configurations
STEP 3: Predict the $alpha\_n$ best parameter configurations using our trained surrogate model
STEP 4: Generate $alpha\_n$ pseudo-random parameter configurations
STEP 5: Evaluate the selected $2*alpha\_n$ parameter configurations on the training dataset
STEP 6: Repeat **STEPS 2, 3, 4, 5** until the allocated time budget is reached.

SMBOX includes three main configurable parameters that are required to be set at the start of the HPO search.

The first parameter is lf_ratio which controls the sampling rate of the initialization dataset. Setting this to 0 would result in using complete training set for initialization which would increase the time required but may yield more robust results. Based on our experience when developing SMBOX setting this value to 0.5 proved a suitable ratio.

The second parameter is lf_n, which controls the number of iterations to evaluate during the initialization phase. Setting this too big may expend the entire time budget, thus resulting in a completely random search. Setting this too small will result in a poorer understanding of the solution space and thus a poorer-performing surrogate model. We decided to set this to 35 for the time budget and dataset sizes used in our experiment.

The last parameter $alpha\_n$ controls the number of configurations to evaluate at each iteration. Currently, we evaluate an even number of surrogate model-based and pseudo-random parameters to ensure diversity. In our study, we set $alpha\_n$ to 2.

A high-level overview of the SMBOX design is presented below together with a pseudo-code implementation in Algorithm 1.

Algorithm 1. SMBOX

Require: data ($Dtrain$), model (M), parameter space (S), time (T)
Require: lf_n, lf_ratio, $alpha\_n$ ▷ SMBOX parameters
  Sample $Dtrain$ based on lf_ratio –> $Dtrain\_lf$
  while $t < T$ **do**
    Initialisation start
    Randomly generate lf_n parameter cfg's from S –> CFG_0
    for cfg in CFG_0 **do**
      Evaluate using K-fold CV
      $P = M(cfg)$ on $Dtrain\_lf$ ▷ store $[cfg, P]$
    end for
    Train surrogate model F on $[cfg, P]$
    Initialisation end
    while $t < T - T*$ **do**   ▷ T* elapsed time
      Predict $\hat{P}$ best $alpha\_n$ cfg's using F –> CFG_i
      Randomly generate $alpha\_n$ cfg's from S –> CFG_i
      for cfg in CFG_i **do**
        Evaluate using K-fold CV
        $P = M(cfg)$ on $Dtrain$ ▷ append $[cfg, P]$
      end for
      Retrain surrogate model F on $[cfg, P]$
    end while
  end while
  Return best $[cfg, P]$

---

4   Experiment

In our research, we investigated how well our search routine performs against the current state-of-the-art procedure, SMAC using its default configuration parameters. Our specific research questions were:

1. Can SMBOX find better-performing hyperparameter values than SMAC within the same time budget?
2. Can SMBOX find better-performing hyperparameter values in less time than SMAC?

4.1   Datasets

In order to answer these research questions, we evaluated each search strategy and a random search baseline on 8 popular publicly available binary classification datasets from the OpenML [30] repository. OpenML contains one of the largest open-source data repositories, currently with over 2100 data sets.

Table 1. Experiment datasets and dataset attributes from the OpenML [30] repository

OpenML dataset ID	Number of instances	Number of features	Balance ratio	Positive class rate
38	3,772	29	15.33	0.939
179	48,842	14	0.31	0.239
298	9,822	85	15.76	0.940
917	1,000	25	0.83	0.454
1049	1,458	37	7.19	0.878
1111	50,000	230	55.18	0.982
1120	19,020	10	1.84	0.648
1128	1,545	10,935	0.29	0.223

Table 1 presents the datasets selected for this study, along with relevant metadata, such as the number of features and the ratio of the positive class. These specific datasets were chosen based on their use in similar studies, specifically [10] and [32], which also evaluated search strategy techniques.

4.2   Machine Learning Algorithms

In our study, we focus on optimizing the hyperparameters of two prevalent and high-performing supervised ML algorithms. These algorithms were selected based on their widespread usage in the industry, primarily attributed to their robust performance on noisy tabular data and their intuitive design that can be easily communicated to non-technical stakeholders. The specific ML algorithms chosen for this study are:

1. A RF with 6 configurable parameters (2 integers, 3 continuous, 1 categorical) implemented within the popular Sckit-learn [24] ML Python framework.
2. A GB algorithm more specifically the XGBoost [6] implementation with 13 configurable parameters (2 integer, 10 continuous, 1 categorical) using the XGBoost [6] Python package.

The exact hyperparameters optimised are provided in Table 2.

These ML algorithms are both tree-based ensemble methods that require minimal data preprocessing making them ideal for running experiments across a large number of datasets. That being said, the HPO technique presented in this paper can be easily applied to any ML algorithm such as Neural Networks.

Table 2. ML Algorithms and associated hyperparameters optimised

Algorithm	Framework	Hyperparameter	Type
RF	SCIKIT LEARN	max features	Int
		n estimators	Int
		max depth	Int
		min samples leaf	Float
		min samples split	Float
		class weight	Categorical
XGBoost	XGBOOST	n estimators	Int
		max depth	Int
		learning rate	Float
		booster	Categorical
		gamma	Float
		min child weight	Float
		min delta step	Float
		subsample	Float
		column sample by tree	Float
		colsample by level	Float
		colsample by node	Float
		reg alpha	Float

4.3   Incumbent

We benchmark SMBOX against SMAC using the Python implementation, version 1.4 and its default configuration parameters.

4.4   Evaluation

We split each dataset in a training and test subset, and tune the hyperparameters on the training dataset using K-fold cross-validation (with equal to 4) for 10 min to maximise the Area Under the Curve (AUC). The best-performing hyperparameter configurations are later evaluated on the test (hold out) dataset to understand how well the found parameters generalise to unseen data. To increase the robustness of the experiment we repeat each experiment 10 times with a different random seed, and then average the best found solution across the 10 experiments.

4.5   Computation

We run all of our experiments sequentially on a Dell XPS 13 i7 Processor (8M Cache) using the Ubuntu Linux 18.04 operating system.

Table 3. Performance comparison of each HPO search strategy applied to a Random Forest classifier on a set of OpenML Datasets. The best mean performance values per HPO search strategy and dataset are highlighted in bold. Statistical significance, determined through a one-way ANOVA and Tukey's HSD test, is indicated by underlined values.

Dataset	Random	SMAC	SMBOX
38	0.99123	0.99466	**0.99531**
179	0.89693	0.88375	**0.90715**
298	0.75209	0.75003	**0.75475**
917	0.95807	**0.96114**	0.96109
1049	0.93121	0.93515	**0.93745**
1111	**0.81887**	0.80839	0.81519
1120	0.90896	0.89243	**0.92662**
1128	0.96896	0.97031	**0.97192**

(a) After 5 min.

Dataset	Random	SMAC	SMBOX
38	0.99140	**0.99618**	0.99556
179	0.89693	0.89077	**0.90717**
298	0.75209	0.75189	**0.75564**
917	0.95839	**0.96297**	0.96114
1049	0.93227	**0.93848**	0.93811
1111	0.82076	0.81177	**0.82261**
1120	0.90896	0.92199	**0.92783**
1128	0.96896	0.96832	**0.97697**

(b) After 10 min.

5   Results

5.1   Random Forest Results

The dataset-level results for each HPO search strategy are provided in Tables 3a and 3b. The bold highlights represent the best mean performance achieved across the 10 runs for each dataset. Statistical significance among the search strategies is determined through a one-way ANOVA, assessing the overall difference among the three search strategy results. Subsequently, Tukey's HSD test is conducted to identify specific search strategies that exhibit significant differences. Only results that are significantly different from both of the other search strategies are indicated with underlined values.

After half of the allowable run time budget of 5 min, SMBOX performed the strongest on 6 out of 8 datasets, with SMAC and Random search performing the best on the remaining 2 datasets (Datasets 917 and 1111 respectively).

After allowing the search strategies to run for the full-time budget of 10 min, SMBOX remained the strongest HPO search strategy on the majority of datasets, achieving the best performance on 5 out of 8 datasets. The experiment findings are summarized in Table 4.

However, a couple of findings emerged with respect to specific datasets. For Datasets 38 and 1049, SMAC now outperformed SMBOX when allowed to run for the additional 5 min. This suggests that SMBOX may be better suited to more time-limited HPO problems, while SMAC demonstrates its strength with longer time budgets.

This theory is supported when visualising the HPO performance across each run. Figure 1 demonstrates that SMBOX is capable of rapidly converging on strong hyperparameter values, whereas SMAC requires a longer time budget. Further exploration is required to understand why SMBOX plateaued whereas SMAC kept improving. This could be due to the caching procedure implemented

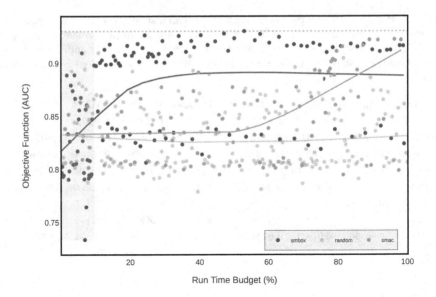

Fig. 1. HPO performance evolution for a Random Forest on Dataset 1120. The x-axis denotes the percentage of a 10-min run time budget. The y-axis displays the objective function value. The shaded green region indicates SMBOX initialization; the dotted grey line represents the best-found solution. (Color figure online)

within SMBOX to avoid reevaluating known hyperparameter configurations in order to improve efficiency.

Another notable finding is that Dataset 1111, the largest dataset in our experiment is better tuned by SMBOX rather than Random search when allowed to run for the full 10 min. This observation was somewhat expected since SMBOX requires a minimum number of iterations (lf_n) before the surrogate model is trained and can start converging on strong hyperparameter values.

An additional finding emerged when studying the HPO performance visualisations of individual runs. Figure 1 shows that SMBOX continues to oscillate between two distinct solution space regions. This behaviour suggests that the diversity mechanism within SMBOX could benefit from further refinement to more efficiently avoid known poor regions of the search space. Enhancing the diversity mechanism has the potential to improve the overall performance of SMBOX.

Table 4. Tabulated summary of the HPO search strategies that yielded best-found Random Forest performance at 50% and 100% of the allocated run time budgets

Search strategy	After 5 min	After 10 min
SMBOX	6	5
SMAC	1	3
Random	1	0

(a) All results

Search strategy	After 5 min	After 10 min
SMBOX	2	3
SMAC	0	0
Random	0	0

(b) Statistically significant results

5.2   XGBoost Results

The dataset-level results from the XGBoost HPO experiment are provided in Tables 5a and 5b. The experiment findings are less conclusive than those for the RF in Sect. 5.1, with no single search strategy clearly outperforming the others.

After running the experiments for 5 min SMAC was the strongest search strategy on half (4 out of 8) of the datasets, with SMBOX performing the strongest on 3 and Random search on the remaining 1 dataset (Dataset 1111).

After the full allowable time budget of 10 min, there were no changes in the findings as summarised in Table 6. XGBoost is a more complex algorithm than RF containing more hyperparameters. These inconclusive results suggest that a longer time budget may be required to tune ML algorithms with more complex search spaces.

Further work is required to understand why SMBOX finds better hyperparameter values than SMAC when optimising the parameters of a RF compared to those of a XGBoost. This could indicate that SMBOX is better suited to less complex search spaces, however, this needs to be studied further.

Table 5. Performance comparison of each HPO search strategy applied to a XGBoost classifier on a set of OpenML Datasets. The best mean performance values per HPO search strategy and dataset are highlighted in bold. Statistical significance, determined through a one-way ANOVA and Tukey's HSD test, is indicated by underlined values.

Dataset	Random	SMAC	SMBOX
38	0.97261	**0.97843**	0.97785
179	0.90886	0.90443	**0.90966**
298	0.72625	**0.73433**	0.72812
917	0.88815	**0.93057**	0.89341
1049	0.91744	**0.92907**	0.91129
1111	**0.77829**	0.77246	0.77677
1120	0.92097	0.87500	**0.92355**
1128	0.97792	0.97712	**0.98098**

(a) After 5 min.

Dataset	Random	SMAC	SMBOX
38	0.97352	**0.98114**	0.97834
179	0.90938	0.90569	**0.91043**
298	0.72984	**0.73490**	0.72843
917	0.89269	**0.93644**	0.89434
1049	0.91945	**0.93287**	0.91158
1111	**0.78097**	0.77246	0.77876
1120	0.92097	0.91397	**0.92375**
1128	0.97818	0.97712	**0.98251**

(b) After 10 min.

Table 6. Tabulated summary of the HPO search strategies that yielded the best-found XGBoost performance at 50% and 100% of the allocated run time budgets.

Search strategy	After 5 min	After 10 min
SMBOX	3	3
SMAC	4	4
Random	1	1

(a) All results

Search strategy	After 5 min	After 10 min
SMBOX	3	2
SMAC	0	2
Random	0	0

(b) Statistically significant results

6   Further Work and Discussion

While SMBOX already demonstrates strong performance, we have identified a couple of areas for improvement that we would like to address in future work, such as:

1. The three configuration parameters (lf_ratio, lf_n, $alpha\_n$) that are required to be set at the start of a SMBOX search could be further explored to provide better default values.
2. In addition to providing better default values, we aim to develop a dynamic initialization routine to intelligently set the aforementioned parameters based on the specific dataset, ML algorithm, and available time budget.
3. We also aim to enhance the initialisation process of the surrogate model. One strategy could be to incorporate Transfer Learning techniques to provide a warm start for SMBOX based on good parameter values identified in previous HPO experiments. Research conducted by [25] and [13] indicates that this approach can lead to HPO performance improvements.

In addition to the aforementioned points, we would like to evaluate SMBOX on a wider range of ML algorithms, such as Neural Networks, and a wider range of ML problem types, such as multi-class classification, regression and natural language processing. We expect SMBOX to perform well on other ML algorithms and domain problems however we would like to validate this through empirical experiments to enhance the community's adoption of SMBOX.

SMBOX is written in Python, leveraging both well-established frameworks and custom functions. We have open-sourced the approach as a lightweight Python module. Moving forward, we plan to continue enhancing its speed, efficiency, and usability.

7   Conclusion

In this paper, we present a novel hyperparameter search strategy called SMBOX and demonstrate its ability to challenge the current state-of-the-art approach, SMAC. SMBOX improves the traditional SMBO search routine by incorporating two performance-enhancing modifications.

First, we introduce a low-fidelity initiation stage to enable the rapid development of a high-performing surrogate model. Secondly, we employ the more advanced ML algorithm, CatBoost within the surrogate model. This algorithm is known to outperform RF [27] and excels with limited training data. This implementation of Gradient Boosting optimally handles categorical parameters, making this approach well-suited to hyperparameter configuration data from any ML algorithm. The native handling of categorical parameters eliminates the need for additional data pre-processing such as one-hot encoding, which reduces the complexity of the search routine whilst increasing its speed.

Secondly, our experiments show that SMBOX is capable of finding good hyperparameter configurations quicker than SMAC on several of the datasets evaluated. Based on our findings, we believe that SMBOX is a promising hyperparameter search routine that should be considered when tuning supervised ML algorithms. Owing to its rapid convergence on good hyperparameter values, we recommend SMBOX as the practitioner's first choice when optimizing hyperparameters under limited time resources. The intuitive modifications that we introduce to the traditional SMBO routine can be easily adopted within other HPO search strategies.

References

1. Banzhaf, W., Nordin, P., Keller, R., Francone, F.: Genetic Programming: an introduction on the automatic evolution of computer programs and its applications (1998)
2. Bäck, T.H.W., et al.: Evolutionary algorithms for parameter optimization-thirty years later. Evol. Comput. **31**(2), 81–122 (2023)
3. Bergstra, J., Bengio, Y.: Random search for hyper-parameter optimization. J. Mach. Learn. Res. **13**(10), 281–305 (2012)
4. Breiman, L.: Random forests (2001)
5. Brown, T.B., et al.: Language models are few-shot learners (2020)
6. Chen, T., Guestrin, C.: XGBoost: a scalable tree boosting system (2016)
7. Dorogush, A.V., Ershov, V., Yandex, A.G.: CatBoost: gradient boosting with categorical features support (2018)
8. Eiben, A.E., Smith, J.E., et al.: Introduction to Evolutionary Computing, vol. 53. Springer, Heidelberg (2003). https://doi.org/10.1007/978-3-662-44874-8
9. Erickson, N., et al.: AutoGluon-tabular: robust and accurate AutoML for structured data (2020)
10. Feurer, M., Klein, A., Eggensperger, K., Springenberg, J., Blum, M., Hutter, F.: Efficient and robust automated machine learning. In: Cortes, C., Lawrence, N., Lee, D., Sugiyama, M., Garnett, R. (eds.) Advances in Neural Information Processing Systems, vol. 28. Curran Associates, Inc. (2015)
11. Friedman, J.H.: Greedy function approximation: a gradient boosting machine. Ann. Stat. 1189–1232 (2001)
12. Graves, A., Mohamed, A.R., Hinton, G.: Speech recognition with deep recurrent neural networks. In: 2013 IEEE International Conference on Acoustics, Speech and Signal Processing, pp. 6645–6649. IEEE (2013)
13. Hellan, S.P., Shen, H., Salinas, D., Klein, A., Aubet, F.X.: Obeying the order: introducing ordered transfer hyperparameter optimisation. In: AutoML Conference 2023 (2023)

14. Hinton, G.: A practical guide to training restricted Boltzmann machines (2010)
15. Hutter, F., Hoos, H.H., Leyton-Brown, K.: Sequential model-based optimization for general algorithm configuration. In: Coello, C.A.C. (ed.) LION 2011. LNCS, vol. 6683, pp. 507–523. Springer, Heidelberg (2011). https://doi.org/10.1007/978-3-642-25566-3_40
16. Hutter, F., Kotthoff, L., Vanschoren, J.: The springer series on challenges in machine learning automated machine learning methods, systems, challenges (2022)
17. Kotthoff, L., Thornton, C., Hoos, H.H., Hutter, F., Leyton-Brown, K.: Auto-WEKA 2.0: automatic model selection and hyperparameter optimization in WEKA. J. Mach. Learn. Res. $18(25)$, 1–5 (2017)
18. Kurshan, E., Shen, H., Chen, J.: Towards self-regulating AI. In: Proceedings of the First ACM International Conference on AI in Finance. ACM (2020)
19. Larochelle, H., Erhan, D., Courville, A., Bergstra, J., Bengio, Y.: An empirical evaluation of deep architectures on problems with many factors of variation, pp. 473–480. Association for Computing Machinery (2007)
20. Le, T.T., Fu, W., Moore, J.H.: Scaling tree-based automated machine learning to biomedical big data with a feature set selector. Bioinformatics $36(1)$, 250–256 (2020)
21. Leung, F.H., Lam, H.K., Ling, S.H., Tam, P.K.: Tuning of the structure and parameters of a neural network using an improved genetic algorithm. IEEE Trans. Neural Netw. 14, 79–88 (2003)
22. LeCun, Y., Bottou, L., Orr, G. B., Müller, K. R.: efficient backprop (2012)
23. Olson, R.S., Bartley, N., Urbanowicz, R.J., Moore, J.H.: Evaluation of a tree-based pipeline optimization tool for automating data science. In: Proceedings of the Genetic and Evolutionary Computation Conference 2016, GECCO 2016, pp. 485–492. ACM, New York (2016)
24. Pedregosa, F., et al.: Scikit-learn: machine learning in Python. J. Mach. Learn. Res. 12, 2825–2830 (2011)
25. Perrone, V., Shen, H., Seeger, M., Archambeau, C., Jenatton, R.: Learning search spaces for Bayesian optimization: another view of hyperparameter transfer learning (2019)
26. Rasmussen, C.E.: Gaussian processes in machine learning. In: Bousquet, O., von Luxburg, U., Rätsch, G. (eds.) ML 2003. LNCS (LNAI), vol. 3176, pp. 63–71. Springer, Heidelberg (2004). https://doi.org/10.1007/978-3-540-28650-9_4
27. Sahin, E.K.: Assessing the predictive capability of ensemble tree methods for landslide susceptibility mapping using XGBoost, gradient boosting machine, and random forest. SN Appl. Sci. $2(7)$, 1–17 (2020)
28. Snoek, J., Larochelle, H., Adams, R.P.: Practical Bayesian optimization of machine learning algorithms. In: Advances in Neural Information Processing Systems, vol. 25 (2012)
29. Thornton, C., Hutter, F., Hoos, H.H., Leyton-Brown, K.: Auto-WEKA: combined selection and hyperparameter optimization of classification algorithms (2012)
30. Vanschoren, J., van Rijn, J.N., Bischl, B., Torgo, L.: OpenML: networked science in machine learning (2014)
31. Williams, C.K.: Prediction with gaussian processes: from linear regression to linear prediction and beyond. In: Jordan, M.I. (ed.) Learning in Graphical Models. NATO ASI Series, vol. 89, pp. 599–621. Springer, Dordrecht (1998). https://doi.org/10.1007/978-94-011-5014-9_23
32. Zimmer, L., Lindauer, M., Hutter, F.: Auto-pytorch tabular: multi-fidelity meta learning for efficient and robust AutoDL (2020)

Accelerated Graph Integration with Approximation of Combining Parameters

Taehwan Yun[1]📵, Myung Jun Kim[3]📵, and Hyunjung Shin[1,2(✉)]📵

[1] Department of Artificial Intelligence, Ajou University, Suwon 16499, South Korea
{youndh0101,shin}@ajou.ac.kr
[2] Department of Industrial Engineering, Ajou University, Suwon 16499, South Korea
[3] Soda, INRIA Saclay, Palaiseau, France
myung.kim@inria.fr

Abstract. Graph-based models offer the advantage of handling data that resides on irregular and complex structures. From various models for graph-structured data, graph-based semi-supervised learning (SSL) with label propagation has shown promising results in numerous applications. Meanwhile, with the rapid growth in the availability of data, there exist multiple relations for the same set of data points. Each relation contains complementary information to one another, and it would be beneficial to integrate all the available information. Such integration can be translated to finding an optimal combination of the graphs, and several studies have been conducted. Previous works, however, incur high computation time with a complex design of the learning process. This leads to a low capacity of applicability in multiple cases. To circumvent the difficulty, we propose an SSL-based fast graph integration method that employs approximation in the maximum likelihood estimation process of finding the combination. The proposed approximation utilizes the connection between the co-variance and its Neumann series, which allows us to avoid explicit matrix inversion. Empirically, the proposed method achieves competitive performance with significant improvements in computational time when compared to other integration methods.

Keywords: Graph-based semi-supervised learning · Graph integration · Maximum likelihood estimation · Neumann series

1 Introduction

Graph-based models have become prevalent machine learning and data mining communities as they offer the flexibility of analyzing datasets with complex and relational structure. Given a graph, semi-supervised learning (SSL) [3,35,36] has become a popular choice with its incorporation of both labeled and unlabeled data in cases where the former is scarcely given compared to the latter. The key idea behind label propagation is the spreading process of labels to neighboring

G. Nicosia et al. (Eds.): LOD 2023, LNCS 14506, pp. 163–176, 2024.
https://doi.org/10.1007/978-3-031-53966-4_13

unlabeled data points through the edges in a graph. Through label propagation and basic kernel of graphs using graph Laplacian [7], we obtain predictive values for unlabeled data points, which we can utilize for various machine learning tasks.

In numerous practical applications, heterogeneous sources of data can exist. For example, in protein function prediction, multiple descriptions of protein relations, such as genetic interactions or co-participation in a protein complex, can be used to construct independent graphs [22,28]. In social networks, an edge may be defined differently based on friend relations, emotional expressions, comments on a posted article, or 'likes' between users [19]. In such cases, each of the graph sources may contain complementary information, and thus combining available graphs is desirable for comprehensive analyses and performance enhancements in machine learning tasks.

The problem of combining multiple graphs boils down to deriving an integrated graph, expressed as a linear combination of the associated graphs [11,14,24]. In previous works, various approaches have been proposed to estimate the optimal combination. In [14–16], semi-definite programming with support vector machine was proposed to predict functional classes of protein using heterogeneous sources of data. In [1,28,34], the Lagrange multiplier for the dual problem of the graph-based regularization was used to obtain the optimal combination. And, works of [17,18,32] proposed a method to minimize the least squares error between the integrated graph and the target graph. In [12,30], expectation-maximization style algorithm was proposed.

Despite the promising results, the main downside of previous works is the required computation time for optimization. The aforementioned works, in essence, require iterative optimization (or computation), which can be time consuming for large-sized datasets. To circumvent the difficulty, we propose a simple and fast algorithm for estimating the optimal combination as a closed-form solution of maximum likelihood estimation (MLE) [4,10]. Starting from the graph-based regularization for multiple graphs [21], we formulate the log-likelihood function of the labels with the regularized graph-based kernel [23]. While there is no closed-form solution for the associated MLE, we employ an approximation scheme on Neumann series expansion [26] of the graph kernel, which endows us with a closed-form solution. Moreover, through experiments on several datasets, we show the strength of the proposed method in terms of both performance and computation time.

2   Graph-Based Semi-supervised Learning

For the problem of graph-based SSL [35,37], a dataset can be represented by a graph $G(V, E)$ composed of the node set (V) and the edge set (E). For a graph, the former denotes data points, and the latter denotes the similarity between the nodes. The similarities are defined by the weight (or adjacency) matrix W, in which the elements represent the strength of connection between nodes. Given n number of data points with the associated label set $y = \{y_l, y_u\}$, we can assign

$y_l \in \{-1, 1\}$ for n_l known data points and $y_u = 0$ for the n_u unknown data points. For SSL, we usually have $n_u \gg n_l$. The goal of SSL is to determine the output vector (same as predicted labels) $f = (f_1, f_2, \cdots, f_n)^T$ by minimizing the following objective function [2,6]

$$\min_{f} \quad (f - y)^\top (f - y) + \alpha f^\top L f, \tag{1}$$

where $L = I - D^{(-1/2)} W D^{(-1/2)}$ is the normalized graph Laplacian [7] with $D = \text{diag}(d_i)$ and $d_i = \sum_j W_{ij}$. The first term in (1) is the loss for consistency with labeled data, and the second term is the *smoothness* that enforces similar data points to have more similar outputs. The parameter α is for the trade-off between the two terms. The solution of (1) is given by

$$f = (I + \alpha L)^{-1} y, \tag{2}$$

where I is the identity matrix. It should be noted that Eq. (2) requires an inverse operation, but the predictive values can be obtained by solving a large sparse linear system $(I + \alpha L) f = y$ [25], once the parameter α is obtained.

To determine the optimal parameter α^*, the cross-validation technique is commonly employed. For cross-validation, however, a portion from already limited amount of labeled data (with $n_u \gg n_l$) must be held out. This may lead to imprecise optimization of α^*. In this paper, we employ the minimization of the negative log-likelihood (NLL). If optimal parameter α^* is determined using this approach, there is no need to construct a separate validation set. For SSL, the term $(I + \alpha L)^{-1}$ can be considered as the graph kernel or the covariance matrix over f with $f \sim \mathcal{N}(0, C)$ where $C = (I + \alpha L)^{-1}$ [13, 29, 37]. With such condition, NLL over α is given as

$$\text{NLL}(\alpha) = \frac{1}{2} y^\top C^{-1} y + \frac{1}{2} \log |C| + \frac{n}{2} \log 2\pi \tag{3}$$

To find α^*, we take the partial derivative, which yields

$$\frac{\partial \text{NLL}}{\partial \alpha} = -\frac{1}{2} y^\top C^{-1} \frac{\partial C}{\partial \alpha} C^{-1} y + \frac{1}{2} \text{tr} \left(C^{-1} \frac{\partial C}{\partial \alpha} \right)$$

$$= \frac{1}{2} y^\top L y - \frac{1}{2} \text{tr}(LC),$$

where we used $\frac{\partial C}{\partial \alpha} = -CLC$ from derivative of an inverse. Since α is embedded in the inverse, there is no closed-form solution form for α and gradient-based optimization must be employed.

3   Graph-Based SSL with Multiple Graphs

3.1   Problem Formulation

Assuming that we have P different sources of information, we are given with a set of graphs $G = \{G_1, G_2, \cdots, G_P\}$, where each G_i represents a graph constructed from a specific source of information. Each graph contains independent

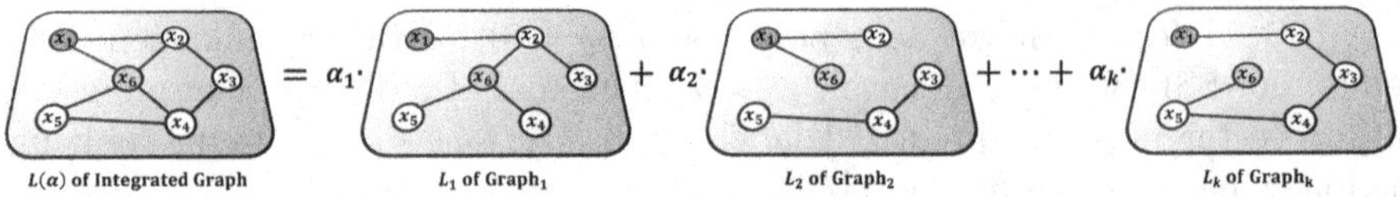

Fig. 1. A schematic description of graph integration. Prediction with L_2 produces zero predictive values for x_3, x_4, and x_5 whereas the combined graph yields non-zero values through additional information from other graphs. (Color figure online)

but complementary information to one another. Thus, it is desirable to formulate a framework that integrates all the available information. The goal of graph integration is to find the optimal linear combination of individual graphs, or technically speaking, individual graph Laplacians. With linear combination of individual graph Laplacians in mind, the formulation is given by [22]

$$\min_f \quad (f - y)^\top (f - y) + \sum_{k=1}^{P} \alpha_k f^\top L_k f, \tag{4}$$

where L_k is the normalized graph Laplacian of G_k. Similar to (2), the solution for (4) is given as

$$f = \left(I + \sum_{k=1}^{P} \alpha_k L_k\right)^{-1} y. \tag{5}$$

The benefit of graph combination is that it allows more comprehensive analysis with performance improvements compared to single networks. In Fig. 1, if we only use L_2 for prediction, the predictive outputs would be zero for nodes x_3, x_4, and x_5 due to disconnection between the labeled nodes (blue and green nodes). On the other hand, with the combined graph Laplacian, it becomes possible to obtain predictive outputs as connections from other Laplacians induces connections between x_3, x_4, and x_5 with labeled nodes in the combined graph.

3.2   Parameter Optimization

The value of the coefficients of the linear combination, namely combining coefficients, determines how much information of associated graphs is reflected. To find the coefficients, we may employ a similar procedure to that with a single graph. The difference between Eqs. (2) and (5) is that the single Laplacian matrix is replaced by a linear combination of Laplacian matrices. By substituting $C = \left(I + \sum_{k=1}^{P} \alpha_k L_k\right)^{-1}$ and taking the partial derivative with respect to α_i (without loss of generality), we obtain

$$\frac{\partial \text{NLL}}{\partial \alpha_i} = \frac{1}{2} y^\top L_i y - \frac{1}{2} \text{tr}(L_i C). \tag{6}$$

Similar to that of single graphs, (6) does not have a closed form solution since the parameters are embedded in C which has the form of an inverse. To avoid

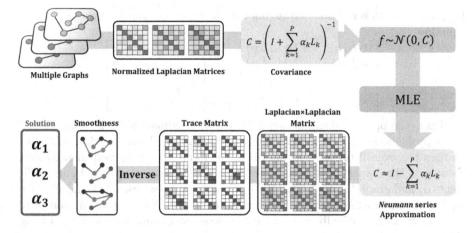

Fig. 2. Overall process of the proposed method. The figure presents the case where there are 3 graphs is taken as an example.

the inversion procedure, we approximate the covariance by the first order of its Neumann expansion. In general, a Neumann expansion of the form $(I + S)^{-1}$ is defined as [20]

$$(I + S)^{-1} = \sum_{k=0}^{\infty} (-1)^k S^k, \tag{7}$$

where $S^n \to 0$ for $n \to \infty$. Since $L_i^n \to 0$ for $n \to \infty$, Eq. (7) holds if we set $0 < \alpha_i < 1$ with $\sum_{i=1}^{P} \alpha_i \leq 1$. By taking the first order approximation ($k = 1$), we approximate the covariance C as the following:

$$C = \left(I + \sum_{j=1}^{P} \alpha_j L_j \right)^{-1} \approx I - \sum_{j=1}^{P} \alpha_j L_j. \tag{8}$$

With the approximation, we replace (8) into (6), which results in

$$\frac{\partial \text{NLL}}{\partial \alpha_i} = \frac{1}{2} y^\top L_i y - \frac{1}{2} \text{tr} \left(L_i \left(I - \sum_{j=1}^{P} \alpha_j L_j \right) \right)$$

$$= \frac{1}{2} y^\top L_i y - \frac{n}{2} - \frac{1}{2} \text{tr} \left(L_i \sum_{j=1}^{P} \alpha_j L_j \right) \tag{9}$$

By setting (9) to zero and rearranging the terms, we get the following system of equation:

$$\begin{bmatrix} \operatorname{tr}(L_1 L_1) & \cdots & \operatorname{tr}(L_1 L_P) \\ \operatorname{tr}(L_2 L_1) & \cdots & \operatorname{tr}(L_2 L_P) \\ \vdots & \ddots & \vdots \\ \operatorname{tr}(L_P L_1) & \cdots & \operatorname{tr}(L_P L_P) \end{bmatrix} \begin{bmatrix} \alpha_1 \\ \alpha_2 \\ \vdots \\ \alpha_P \end{bmatrix} = \begin{bmatrix} n - y^\top L_1 y \\ n - y^\top L_2 y \\ \vdots \\ n - y^\top L_P y \end{bmatrix}$$

Thus, the optimal combination with the approximation is the solution the above problem. For better understanding of the proposed framework, Fig. 2 summarizes the overall process. As shown in the figure, the proposed method first constructs the covariance matrix C as a linear combination of combining coefficients α_i and graph Laplacian L_i. With the covariance, we establish the negative log-likelihood function and use the first term of the Neumann expansion of C for approximation. Finally, we obtain the closed-form solution for α by solving the system of linear equation.

3.3 Remarks on the Proposed Method

When we observe the closed-form solution for α, it is dependent on the smoothness of the labels over the given graphs and the similarity in structure of given graphs. In regards to the smoothness, the coefficient α_i has smaller value with larger smoothness value with the given label over the graph G_i. Meanwhile, the term $\operatorname{tr}(L_i L_j)$ depends on the number of same connections that graphs G_i and G_j have.

This naturally leads us to discuss about the computational complexity of the proposed method. Although the proposed method requires the operation of matrix inversion, its complexity is $O(N_g^3)$, where N_g is the total number of graphs to integrate, most of the cases where graph integration applied are cases in which the number of nodes is large and the number of graphs is relatively very small. Therefore, our proposed method, which is absolutely affected by the number of graphs, can perform graph integration quickly. Also, the major computational component of the proposed method is $\operatorname{tr}(L_i L_j)$, which depends on the number of same connections for G_i and G_j. Since graphs are sparse in general, the proposed method takes the huge benefit in computational perspective.

4 Related Works

4.1 Fast Protein Function Classification

The work of fast protein function classification [28] is the basis for the proposed method. The formulation begins by constructing the following convex problem:

$$\min_{f, \xi, \gamma} \quad (f - y)^\top (f - y) + c\gamma + c_0 \sum_{k=1}^{P} \xi_k$$

$$\text{s.t.} \quad f^\top L_k f \le \gamma + \xi_k, \ \xi_k \ge 0, \ \gamma \ge 0.$$

where c and c_0 are positive constants, $\{\xi_k\}_{k=1}^P$ are slack variables, and γ is the upper bound of the smoothness term $f^T L_k f$. To obtain the optimal coefficients, the work utilizes the dual problem which is

$$\min_{\alpha} \quad y^\top \left(I + \sum_{k=1}^P \alpha_k L_k \right)^{-1} y$$

$$\text{s.t.} \quad 0 \leq \alpha_k \leq c_0, \ \sum_{k=1}^P \alpha_k \leq c.$$

where α_k is the Lagrange multiplier [5] and can be interpreted as combining coefficient of graph k. To estimate the Lagrange multiplier α, the partial derivative with respect to α is used as the gradient.

4.2   Robust Label Propagation on Multiple Networks

Robust label propagation on multiple networks [12] is an Expectation Maximization [8,33] style iterative algorithm that runs the following two equations until convergence:

$$\hat{f} = \left(G + \frac{\beta_{bias}}{\beta_y} I + \frac{\beta_{net}}{\beta_y} \sum_{k=1}^P \alpha_k L_k \right)^{-1} Gy \tag{10}$$

$$\alpha_k = \frac{\nu + n}{\nu + \beta_{net} f^\top L_k f} \tag{11}$$

where v, β_v, β_{bias}, β_{net} are constants, and G is the diagonal indicator matrix for the labeled data. Observing (10) and (11), this approach also penalizes for large values of smoothness but does not consider the similarity in the edge structure of the given graphs.

4.3   Gene Multiple Association Network Integration Algorithm

Gene Multiple Association Network Integration Algorithm [18] is based on the following linear regression problem:

$$\min_{\alpha} \quad \text{tr}((T - W^*)^\top (T - W^*))$$

$$\text{s.t.} \quad W^* = \sum_{k=1}^P \alpha_k W_k, \ \alpha_k \geq 0.$$

where W_k is the similarity matrix of graph k, T_{ij}, elements of the target graph T. The target graph T is a signed graph with edge values dependent on the number of positive and negative labels. By optimizing the above objective function, the integrated graph W^* get closer to the target graph. However, this approach does not take the smoothness into consideration.

Table 1. Summary of datasets and density of each network

Dataset	# Nodes	# Classes	W	Description
Protein	3,588	11	W_{ST}	Similarity of protein structure
			W_{BP}	Co-participation
			W_{PPI}	Protein-protein interactions
			W_{GI}	Genetic interactions
			W_{GE}	Gene expression measurements
Cora	2,708	7	W_{CIT}	Original citation network
			W_{SWV}	Similarity of word vector
IMDB	4,919	4	W_{MDM}	Same director
			W_{MAM}	Same actor
Amazon	11,944	2	W_{USV}	Same star rating within one week
			W_{UVU}	Similarity of text (TF-IDF)
Yelp	45,954	2	W_{RSR}	Same product with same star rating
			W_{RTR}	Same product posted in same month
			W_{RUR}	Same user

5 Experiments

In this section, experimental results of the proposed framework is presented. To validate the proposed graph integration framework, we evaluated the performance and computation time on five datasets with classification tasks. For comparing algorithms, we consider SSL with single graph, fast protein function classification (FPC), robust label propagation on multiple networks (RLPMN), Gene multiple association network integration algorithm (GMANIA), and Gaussian process classification (GPC). For single graphs, we recorded the best performance among the plurality of graphs. For GPC, it should be noted that we used the linear combination of the weight matrices as the kernels, although the combination may not satisfy conditions for a kernel. All experiments were carried out on a AMD Ryzen Threadripper 3960X 24-Core Processor 3.79 GHz PC with 256 GB memory. In the following subsections, we present brief summary of data, experiment settings and detailed discussions of the results.

5.1 Datasets

Table 1 summarizes the five datasets (Protein function, Cora, IMDB, Amazon, and Yelp) used for the experiments. The following are brief descriptions for each.

- **Protein function** is used for predicting functional class of yeast proteins. The function of each protein is labeled according to the MIPS Comprehensive Yeast Genome Database (CYGD-mips.gsf.de/proj/yeast) [28]. In each dataset, each protein may belong to 13 functional classes. We performed experiments on 11 of these classes. This dataset has five types of edge (W_{ST}, W_{BP}, W_{PPI}, W_{GI}, and W_{GE}) derived from five different measures. Table 1 summarizes networks derived from five different measures.

- **Cora** is a citation network for scientific publication classification task. Nodes in the graph represent publications and each publication can belong to only one of the seven categories. The original dataset has only one type of edge (W_{CIT}), but we make an additional type (W_{SWV}) with the node feature vector using the cosine similarity.
- **IMDB** contains information about movies, such as genres, actors, and directors. We created two types of edge from this dataset. 1) W_{MDM}: connect movies made by the same director; 2) W_{MAM}: connect movies featuring at least one same actor. Each movie can belong to one of four genres (*Action, Comedy, Drama, Thriller*) [31].
- **Amazon** is a benchmark graph for fraudulent user detection task. Nodes in the graph represent users and there are two types of edge useful for classification. 1) W_{USV}: connect users having at least one same star rating within one week; 2) W_{UVU}: connect users with top 5% text similarities (measured by TF-IDF) among all users [9].
- **Yelp** is a benchmark graph for spam review detection task. Nodes in the graph represent reviews written by users. There are three types of edge. 1) W_{RSR}: connect reviews under the same product with the same star rating (1–5 stars); 2) W_{RTR}: connect two reviews under the same product posted in the same month; 3) W_{RUR}: connect reviews posted by the same user [9].

5.2   Results and Discussion

For the five datasets, the performance was measured with area under receiver operating characteristic curve (AUC) [27]. For multi-class problems, we performed binary classification for each of the class to measure the AUC. For the label set, we randomly assigned 20% of labels to induce the SSL setting. Experiment for each dataset was repeated 30 times.

An overall comparison of performance across the five datasets in terms of AUC is shown in Fig. 3. The results show two aspects. *First*, an integrated graph outperforms a single graph regardless of which integration method is used. This is because each graph provides independent and complementary information, so from this point of view, the integrated graph contains more information than any single graph. So it leads to better performance. The outperformance of the integrated graph over the individual graphs can be observed explicitly in Fig. 4. For each repeated experiment, the AUC values of the single graph (with the best performance) and the integrated graph (by the proposed method) were indicated as x-coordinates and y-coordinates, respectively, and marked as a dot. For visibility, only one class is used for each dataset. A dot located above the diagonal line indicate higher AUC value of the method on the y-axis. The figure shows that most dots are above the diagonal line, standing for better performance of the integrated graph by the proposed method. Similar results can be

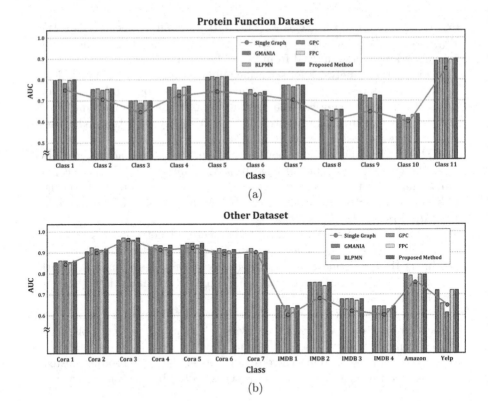

Fig. 3. The overall AUC performance of comparing algorithms on (a) protein function dataset (b) other datasets. The proposed method achieves similar results to that of comparing algorithms, which shows its competitiveness in the problem of graph integration.

expected for other integration method. This implies that incorporating various complementary information indeed leads to improved prediction results.

Second, the most recent representative studies on graph integration–FPC, RLPMN, GMANIA, and GPC, including the proposed method, are comparable to each other in terms of accuracy performance. For instance, we tested the difference in AUC between the proposed method and the winner method for cases where the proposed method did not yield the best performance. And in most cases, the difference was not statistically significant. Most of the p-value of the two-tailed pairwise t-test had high values. This means, those methods have reached to the most achievable accuracies(at least so far) for given task. If so, that suggests the focus of the competition should shift to computing efficiency, theoretical soundness, and simplicity of implementation.

In computational time perspective, Table 2 summarizes the average optimization time (in seconds) on the five datasets for comparing algorithms that reflect the integrated graph. Overall, it can be seen that the proposed integration frame-

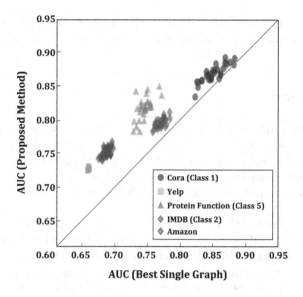

Fig. 4. Performance of integrated graph derived from the proposed method versus best performance of single graph per each repetition.

Table 2. Comparison of computation time (s) with existing methods for five datasets. We repeated the experiment 30 times and reported mean ± standard deviation. It was confirmed that our proposed method showed the fastest speed

	Protein	Cora	IMDB	Amazon	Yelp
GMANIA	1.1548 ± 0.0947	0.1802 ± 0.1646	0.2271 ± 0.0109	1.9630 ± 1.2117	212.8583 ± 251.5614
RLPMN	6.1869 ± 0.2415	2.8716 ± 0.0294	9.6062 ± 0.0914	121.0524 ± 7.1263	4269.9623 ± 149.2684
GPC	31.1244 ± 15.3637	5.9858 ± 0.7860	8.0670 ± 0.0462	1084.2671 ± 220.7073	16220.4962 ± 4058.6283
FPC	4.0471 ± 0.7729	0.7448 ± 0.1157	1.4109 ± 0.0343	20.8813 ± 0.9692	982.6038 ± 69.2546
Proposed	**0.0052 ± 0.0074**	**0.0016 ± 0.0047**	**0.0053 ± 0.0073**	**0.3434 ± 0.0183**	**0.0772 ± 0.0014**

work achieves the fastest time for optimization. Most notably, on the largest dataset, Yelp, the proposed method only required less than 0.1 s for optimization while comparing algorithms required at least 2.5 fold with three order of magnitude from the proposed method. Moreover, the change in computation time with respect to increase in the number of data is the least sensitive for the proposed method. For the proposed method, the critical factor of computation time is the network density of the associated graphs (as shown in the result of Amazon dataset). Overall, the comparison results in performance and computation bolster that the proposed integration framework is competitive against comparing algorithms and has a huge advantage of scalabilty when time becomes a critical factor.

6   Conclusion

With the growing availability of data in our hands, it is becoming more crucial to handle multiple sources of data that contain different but complementary information for a given set of data points. By optimizing the linear combination of multiple graphs with semi-supervised learning framework, existing approaches exhibit performance improvements, but with the hurdle of large computational time required for optimization.

To overcome the scalability issue, we proposed an integration method that obtains the set of coefficients in a closed-form solution. The proposed method utilizes the maximum likelihood estimation with the first-order approximation on the Neumann expansion of the covariance. Empirically, the proposed method exhibited competitiveness in prediction performance against other algorithms on several datasets. In computational perspective, the proposed method showed superiority over comparing algorithms, showing its favoring characteristic of fast optimization.

However, there are issues, including future works, that need to be addressed. First, a closer look on the approximation should be conducted. In general, the justification of approximation on the Neumann expansion depends on the eigenvalue of S (see Eq. (7)). The proposed method, however, neglects this point. Thus, a more rigorous and sophisticated design of approximation should be considered. Furthermore, the proposed integration can be sensitive density, which implies that the proposed method can hurt its performance with presence of noisy edges. Deeper investigations into such problems are left for our future study.

An integrated graph by the proposed method can be utilized for graph convolutional networks (GCNs). The results of GCN are improved if the graph contains rich information, mainly contained in the edges connecting the nodes. When there are many graphs representing different relationships to a set of nodes, each set of edges provides independent and complementary information to the set of nodes. From this point of view, an integrated graph combining individual graphs will contain a better representation than any single graph, and better results can be expected when used as input to a GCN. On the other hand, we can consider a method of integrating (latent) representations of hidden layer of respective GCNs. That is, multiple GCNs are prepared and each GCN is independently trained using each graph. The resulting graphs from the final hidden layer of each GCN are then combined. The latter remark remains as our future work.

Acknowledgements. This research was supported by Institute for Information communications Technology Promotion(IITP) grant funded by the Korea government (MSIT) (No. 2022-0-00653, Voice Phishing Information Collection and Processing and

Development of a Big Data Based Investigation Support System), BK21 FOUR program of the National Research Foundation of Korea funded by the Ministry of Education(NRF5199991014091), the National Research Foundation of Korea (NRF) grant funded by the Korea government (MSIT) (No. 2021R1A2C2003474) and the Ajou University research fund.

References

1. Argyriou, A., Herbster, M., Pontil, M.: Combining graph Laplacians for semi-supervised learning. In: Advances in Neural Information Processing Systems, vol. 18 (2005)
2. Belkin, M., Matveeva, I., Niyogi, P.: Regularization and semi-supervised learning on large graphs. In: Shawe-Taylor, J., Singer, Y. (eds.) COLT 2004. LNCS (LNAI), vol. 3120, pp. 624–638. Springer, Heidelberg (2004). https://doi.org/10.1007/978-3-540-27819-1_43
3. Bengio, Y., Delalleau, O., Le Roux, N.: Label propagation and quadratic criterion, pp. 183–206. MIT Press (2006)
4. Bishop, C.M., Nasrabadi, N.M.: Pattern Recognition and Machine Learning, vol. 4. Springer, Heidelberg (2006)
5. Boyd, S.P., Vandenberghe, L.: Convex Optimization. Cambridge University Press, Cambridge (2004)
6. Chapelle, O., Weston, J., Schölkopf, B.: Cluster kernels for semi-supervised learning. In: Advances in Neural Information Processing Systems, vol. 15 (2002)
7. Chung, F.R., Graham, F.C.: Spectral Graph Theory, vol. 92. American Mathematical Society (1997)
8. Dempster, A.P., Laird, N.M., Rubin, D.B.: Maximum likelihood from incomplete data via the EM algorithm. J. Roy. Stat. Soc.: Ser. B (Methodol.) $39(1)$, 1–22 (1977)
9. Dou, Y., Liu, Z., Sun, L., Deng, Y., Peng, H., Yu, P.S.: Enhancing graph neural network-based fraud detectors against camouflaged fraudsters. In: Proceedings of the 29th ACM International Conference on Information & Knowledge Management, pp. 315–324 (2020)
10. Duda, R.O., Hart, P.E., et al.: Pattern Classification. Wiley, Hoboken (2006)
11. Gönen, M., Alpaydın, E.: Multiple kernel learning algorithms. J. Mach. Learn. Res. 12, 2211–2268 (2011)
12. Kato, T., Kashima, H., Sugiyama, M.: Robust label propagation on multiple networks. IEEE Trans. Neural Netw. $20(1)$, 35–44 (2008)
13. Kim, M., Lee, D.G., Shin, H.: Semi-supervised network regression with gaussian process. In: 2022 IEEE International Conference on Big Data and Smart Computing (BigComp), pp. 27–30. IEEE (2022)
14. Lanckriet, G.R., Cristianini, N., Bartlett, P., Ghaoui, L.E., Jordan, M.I.: Learning the kernel matrix with semidefinite programming. J. Mach. Learn. Res. 5(Jan), 27–72 (2004)
15. Lanckriet, G.R., De Bie, T., Cristianini, N., Jordan, M.I., Noble, W.S.: A statistical framework for genomic data fusion. Bioinformatics $20(16)$, 2626–2635 (2004)
16. Lanckriet, G.R., Deng, M., Cristianini, N., Jordan, M.I., Noble, W.S.: Kernel-based data fusion and its application to protein function prediction in yeast. In: Biocomputing 2004, pp. 300–311. World Scientific (2003)

17. Mostafavi, S., Morris, Q.: Fast integration of heterogeneous data sources for predicting gene function with limited annotation. Bioinformatics **26**(14), 1759–1765 (2010)
18. Mostafavi, S., Ray, D., Warde-Farley, D., Grouios, C., Morris, Q.: GeneMANIA: a real-time multiple association network integration algorithm for predicting gene function. Genome Biol. **9**(1), 1–15 (2008)
19. Musiał, K., Kazienko, P.: Social networks on the internet. World Wide Web **16**(1), 31–72 (2013)
20. Petersen, K.B., Pedersen, M.S., et al.: The matrix cookbook. Tech. Univ. Denmark **7**(15), 510 (2008)
21. Shin, H., Tsuda, K.: Prediction of protein function from networks, pp. 343–356. MIT Press (2006)
22. Shin, H., Tsuda, K., Schölkopf, B.: Protein functional class prediction with a combined graph. Expert Syst. Appl. **36**(2), 3284–3292 (2009)
23. Smola, A.J., Kondor, R.: Kernels and regularization on graphs. In: Schölkopf, B., Warmuth, M.K. (eds.) COLT-Kernel 2003. LNCS (LNAI), vol. 2777, pp. 144–158. Springer, Heidelberg (2003). https://doi.org/10.1007/978-3-540-45167-9_12
24. Sonnenburg, S., Rätsch, G., Schäfer, C., Schölkopf, B.: Large scale multiple kernel learning. J. Mach. Learn. Res. **7**, 1531–1565 (2006)
25. Spielman, D.A., Teng, S.H.: Nearly-linear time algorithms for graph partitioning, graph sparsification, and solving linear systems. In: Proceedings of the Thirty-Sixth Annual ACM Symposium on Theory of Computing, pp. 81–90 (2004)
26. Stewart, G.W.: Matrix Algorithms: Volume 1: Basic Decompositions. SIAM (1998)
27. Swets, J.A.: Signal Detection Theory and ROC Analysis in Psychology and Diagnostics: Collected Papers. Psychology Press (2014)
28. Tsuda, K., Shin, H., Schölkopf, B.: Fast protein classification with multiple networks. Bioinformatics **21**(suppl_2), ii59–ii65 (2005)
29. Verbeek, J.J., Vlassis, N.: Gaussian fields for semi-supervised regression and correspondence learning. Pattern Recogn. **39**(10), 1864–1875 (2006)
30. Wang, M., Hua, X.S., Hong, R., Tang, J., Qi, G.J., Song, Y.: Unified video annotation via multigraph learning. IEEE Trans. Circuits Syst. Video Technol. **19**(5), 733–746 (2009)
31. Wang, X., Ji, H., Shi, C., Wang, B., Ye, Y., Cui, P., Yu, P.S.: Heterogeneous graph attention network. In: The World Wide Web Conference, pp. 2022–2032 (2019)
32. Warde-Farley, D., et al.: The GeneMANIA prediction server: biological network integration for gene prioritization and predicting gene function. Nucl. Acids Res. **38**(suppl_2), W214–W220 (2010)
33. Wu, C.J.: On the convergence properties of the EM algorithm. Ann. Stat. 95–103 (1983)
34. Ye, J., Akoglu, L.: Robust semi-supervised classification for multi-relational graphs. arXiv preprint arXiv:1510.06024 (2015)
35. Zhu, X.: Semi-supervised learning with graphs. Carnegie Mellon University (2005)
36. Zhu, X., Ghahramani, Z., Lafferty, J.D.: Semi-supervised learning using gaussian fields and harmonic functions. In: Proceedings of the 20th International conference on Machine learning (ICML-03), pp. 912–919 (2003)
37. Zhu, X., Lafferty, J., Ghahramani, Z.: Semi-supervised learning: from Gaussian fields to Gaussian processes. School of Computer Science, Carnegie Mellon University (2003)

Improving Reinforcement Learning Efficiency with Auxiliary Tasks in Non-visual Environments: A Comparison

Moritz Lange[1]([✉]) [iD], Noah Krystiniak[1], Raphael C. Engelhardt[2] [iD],
Wolfgang Konen[2] [iD], and Laurenz Wiskott[1] [iD]

[1] Institute for Neural Computation, Faculty of Computer Science,
Ruhr-University Bochum, Bochum, Germany
{moritz.lange,noah.krystiniak,laurenz.wiskott}@ini.rub.de
[2] Cologne Institute of Computer Science, Faculty of Computer Science
and Engineering Science, TH Köln, Gummersbach, Germany
{raphael.engelhardt,wolfgang.konen}@th-koeln.de

Abstract. Real-world reinforcement learning (RL) environments, whether in robotics or industrial settings, often involve non-visual observations and require not only efficient but also reliable and thus interpretable and flexible RL approaches. To improve efficiency, agents that perform state representation learning with auxiliary tasks have been widely studied in visual observation contexts. However, for real-world problems, dedicated representation learning modules that are decoupled from RL agents are more suited to meet requirements. This study compares common auxiliary tasks based on, to the best of our knowledge, the only decoupled representation learning method for low-dimensional non-visual observations. We evaluate potential improvements in sample efficiency and returns for environments ranging from a simple pendulum to a complex simulated robotics task. Our findings show that representation learning with auxiliary tasks only provides performance gains in sufficiently complex environments and that learning environment dynamics is preferable to predicting rewards. These insights can inform future development of interpretable representation learning approaches for non-visual observations and advance the use of RL solutions in real-world scenarios.

Keywords: Representation learning · Auxiliary tasks · Reinforcement learning

1 Introduction

In reinforcement learning (RL), the complex interplay of observations, actions, and rewards means that algorithms are often sample-inefficient or cannot solve

Supported by the research training group "Dataninja" (Trustworthy AI for Seamless Problem Solving: Next Generation Intelligence Joins Robust Data Analysis) funded by the German federal state of North Rhine-Westphalia.

problems altogether. State representation learning tackles this issue by making information encoded in observations, and possibly actions, more accessible. Mnih et al. [17] were the first to introduce deep RL to extract information from the high-dimensional observations provided by Atari games. The neural networks of deep RL agents make it possible to implicitly extract representations of the input and thus enables the agent to find a good policy.

Munk et al. [18] additionally introduced predictive priors, learning targets that differ from the RL task but are also based on data generated by the environment. Other authors such as Legenstein et al. [13], Wahlstrom et al. [27], Anderson et al. [1], and Schelhamer et al. [25] have proposed further learning targets and started to call these auxiliary tasks.

Both Munk et al. [18] and Stooke et al. [26] argue for decoupling representation learning with auxiliary tasks from solving the RL task of maximizing cumulative rewards. Approaches with separate representation learning modules are versatile as representations can replace raw observations and actions as inputs to arbitrary RL algorithms. The individual parts of such agents have distinct purposes; auxiliary tasks and RL task do not interfere with each other and representations are agnostic to the RL task. Segmenting systems into such distinct parts provides flexibility and aids interpretability as representations used as inputs become explicit rather than being hidden within layers inside networks. The distinction of integrated and decoupled representation learning however is not always easy, e.g. when a deep RL agent is simply furnished with additional prediction heads to enhance internal representations within its neural networks.

Few works in RL learn interpretable representations (an exception is e.g. [13]) and methods such as autoencoders, which can supposedly learn semantically meaningful representations (used for RL in [10, 27]), have been shown to be unreliable in this regard [16]. Another shortcoming of the field is that these and most other works study complex, visual environments such as Arcade games [25] or race tracks [4]. Many RL problems, however, provide non-visual observations that are often of lower dimensionality and do not have the same properties as visual data. Observations are thus not necessarily suited for methods designed for visual data, such as CNN-based autoencoders, and are also not as easily interpreted. In real-world applications, e.g. control problems such as system control in factory production lines, RL is underrepresented due to concerns about reliability and sample efficiency [5]. We believe that new methods for interpretable, modular representation learning will be required to change this.

To the best of our knowledge, OFENet by Ota et al. [20] is to date the only available method for decoupled representation learning with auxiliary tasks that works also with low-dimensional observations and has been used on non-visual environments. We use it in this paper to learn representations on different auxiliary tasks and compare them on non-visual environments of different complexity. Despite the fact that OFENet as a method does not produce interpretable representations, we are confident that our findings concerning auxiliary tasks will help researchers in developing new, interpretable methods for representation learning in RL.

We conduct our comparison by investigating returns and sample efficiencies achieved with common auxiliary tasks on five diverse environments. These environments cover a range of observation and action dimensionalities, and varying levels of complexity in the relationship between observations, actions, and rewards. Decoupled representations are computed with OFENet and used as inputs to the off-policy RL algorithms TD3 and SAC [7,8]. Since one environment, FetchSlideDense-v1, cannot be solved with baseline TD3 or SAC, we additionally train agents with hindsight experience replay (HER, [2]).

Our results show that representation learning with auxiliary tasks increases both maximum returns and sample efficiency for environments that are sufficiently complex and high-dimensional, but has little effect on simpler, smaller environments. We find that learning representations based on environment dynamics, for instance by predicting the next observation, is superior to using reward prediction. We discover that decoupled representation learning with an inverse dynamics task does not work with actor critic algorithms because gradients cannot be backpropagated. Interestingly, adding representation learning to TD3 makes it possible to train agents on the FetchSlideDense-v1 environment, even if baseline TD3 does not learn anything at all.

2 Related Work

Many works in recent years have made use of an auxiliary task to learn state representations. We cite multiple of these in Sect. 3. Ota et al. [20] for instance, whose OFENet representation learning network we use here, predict the next observation from current observation and action. There are however few papers which compare auxiliary tasks to each other. Lesort et al. [14] have written a thorough survey of state representation learning, which summarizes different auxiliary tasks and includes a comprehensive list of publications. It is however a purely theoretical discussion of methods without any empirical comparisons or results. There are two empirical comparisons of auxiliary tasks [4,25], which differ in various aspects from ours. Shelhamer et al. [25], like us, compare auxiliary tasks on various environments. In contrast to us, they use Atari games with visual observations. Another difference is that they do not fully decouple representation learning from the RL algorithm. Instead, they merely connect a different prediction head to train the initial, convolutional part of the deep RL algorithm on auxiliary targets. Their results generally vary across environments. An interesting feature of their paper is the comparison of individual auxiliary tasks to a combination of several. Curiously, the combination never clearly outperforms the respective best individual tasks. The second comparison, by de Bruin et al. [4], uses only one car race environment but with several race tracks. It provides multimodal observations which again include visual data. In contrast to the decoupled module we use, loss functions of auxiliary tasks and RL task are linearly combined, and auxiliary tasks are investigated by removing their individual loss terms from the combination. This means representation learning is inherently integrated into the RL agent and happens implicitly, rather than in a decoupled way such as in our work.

3   Auxiliary Tasks

In this section we present five common auxiliary tasks according to Lesort et al. [14]. Of these five, we will empirically compare three while the last two do not work with our setup. An overview of the tasks is presented in Fig. 1. To discuss these tasks, we first need to briefly formalize the reinforcement learning problem: An environment provides reward r_t and observation o_t at time step t. The agent then performs an action a_t, which generates a reward r_{t+1} and leads to the next observation o_{t+1}. This cycle is modeled by a Markov decision process, which means that there can be randomness in the transition from o_t to o_{t+1}, given some a_t. The Markov property implies that o_{t+1} only depends on o_t and a_t which already contain all past information. It does not depend on previous states or actions. The goal of the RL agent is to maximize cumulative expected reward (return). Altogether, these components are the ones available to auxiliary tasks, and various possible combinations are used.

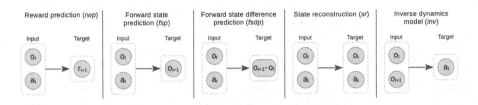

Fig. 1. An overview of inputs and prediction targets of common auxiliary tasks.

Reward prediction (rwp) is the task of predicting r_{t+1} from o_t and a_t. Works that use rwp include [9,11,18,19,25]. With decoupled representation learning, rwp is limited in that it can only be applied to environments that provide non-trivial rewards, i.e. rewards that are not constant or sparse. Representations learned within the model might otherwise become decoupled from o_t and a_t since the model output r_{t+1} is (nearly) independent of model inputs o_t, a_t. A RL algorithm trained on these representations could not learn anything at all. It can be argued that representations based on reward prediction have an advantage over those learned with other auxiliary tasks as they are optimized towards the actual RL task. On the other hand, rwp is somewhat redundant to the RL task of maximizing returns, although it only considers the immediate next reward and is therefore less noisy [25].

Forward state prediction (fsp) is the task of predicting o_{t+1} from o_t and a_t. It is a popular task and used e.g. in [10,18,20,21,27]. In contrast to our work, several of these deal with high-dimensional image observations and therefore try to predict the next internal representation rather than the next observation in time. The fsp task, as opposed to rwp, can be applied to any kind of environment without conditions. Its task amounts to learning environment dynamics similar to e.g. in model-based RL. The fsp task can thus be considered model-based RL, although we combine it with model-free RL algorithms.

Forward state difference prediction (*fsdp*) describes the task of predicting $(o_{t+1} - o_t)$ from o_t and a_t. The papers [1,11] use *fsdp*. Conceptually, it is very similar to *fsp*. While *fsdp* also learns environment dynamics, Anderson et al. [1] claim that successive observations are very similar and predicting only the difference thus gives more explicit insight into environment dynamics. The *fsdp* task requires a notion of difference, though this is not a practical issue as observations are usually encoded as numerical vectors. In comparison with *fsp*, the *fsdp* task should provide an advantage in environments without excessive noise or significant changes between successive observations. That would mean *fsp* is more robust, but *fsdp* is particularly suited for environments simulating real-world physics.

State reconstruction (*sr*) (used e.g. in [11,25]) is the task of reconstructing o_t (and possibly a_t) from o_t and a_t. This is the classical autoencoder task, but does not make sense in our setup where data dimensionality is expanded by concatenation (see Sect. 4.1). Our representations thus always contain the raw o_t and a_t, and reconstruction would amount to simply filtering these out. No useful representations could be learned. On a technical level, *sr* is similar to *fsp* but, crucially, does not learn environment dynamics. It therefore seems reasonable to assume that *fsp* will in most cases be a better choice for learning representations for RL. The results of both [11] and [25] confirm this.

The **inverse dynamics model** (*inv*), framed as a learning task, predicts a_t from o_t and o_{t+1}. Works using this task include [21,25]. While *fsp* and *fsdp* focus on learning transition probabilities of the environment, the *inv* task considers how actions of the agent affect changes in the environment. The *inv* task does not work with actor-critic algorithms; using its representation as input to the critic renders the actor untrainable. Gradients would have to pass through the – usually not differentiable – environment in order to be propagated back from critic to actor. Figure 2 provides a visualization of this problem.

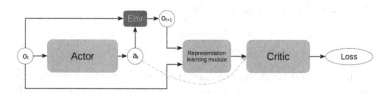

Fig. 2. Diagram of information flow in an actor-critic setup with the *inv* auxiliary task where the critic, or representation learning module, receives o_t and o_{t+1} (and potentially a_t) as input. If o_{t+1} is part of the input to the critic, directly or through the module, the gradient of the critic loss cannot be propagated back to the actor as long as the environment (red) is not differentiable. Even if the action is additionally passed into the critic directly (dashed grey line) the actor will not get the true gradient. (Color figure online)

Various **other priors** have been proposed by different authors (for a list, see [14]). Noteworthy examples include the slowness principle [13] and the robotics prior [12]. However, these are not as commonly used as the tasks above, and many are even problem-specific. We thus exclude them from our comparison.

In addition to the tasks above, there are several works on combining auxiliary tasks. A popular combination is *fsp* or *fsdp* with *rwp* (e.g. [11,18]), various others exist. Lin et al. [15] have even proposed a method to adaptively weigh different auxiliary tasks.

4   Methods

This section explains the neural network we use to learn representations with auxiliary tasks, the RL algorithms we train on these representations and the environments we use for training.

4.1   Representation Learning Network

To train decoupled representations on auxiliary tasks, we use the network architecture of OFENet from [20]. The architecture is composed of two parts. Its first part calculates a representation z_{o_t} of o_t, and the second part calculates a representation z_{o_t,a_t} of z_{o_t} and a_t. Internally, the parts stack MLP-DenseNet blocks which consist of fully connected and concatenation layers. The whole arrangement is visualized in Fig. 3. For our experiments we give both parts of OFENet the same internal structure (apart from input dimensionality), but adjust parameters to different environments as described in Table 1. The auxiliary loss is calculated as the mean squared error between predicted and actual target.

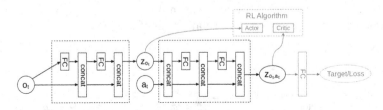

Fig. 3. Sketch of the OFENet architecture, modified from [20]. Observation o_t and action a_t are used to calculate representations z_{o_t} and z_{o_t,a_t}. These are passed into the RL algorithm (light grey). The prediction target necessary to evaluate the auxiliary loss, e.g. o_{t+1}, is calculated with a fully connected layer (FC, light grey) from z_{o_t,a_t}. (Color figure online)

OFENet is a good choice for comparing auxiliary tasks as it is a rather generic architecture for learning representations of expanded dimensionality. Besides OFENet, we are not aware of any other decoupled approaches used in RL that learn representations without dimensionality reduction. Most works use autoencoders (variational or otherwise), which have been shown to be very powerful especially for visual data. OFENet, however, has the advantage that it can be applied to smaller, simpler environments. This allows us to study auxiliary tasks in environments that have far fewer dimensions than visual observations would have and that are less complex than those used in [4,25].

4.2   Reinforcement Learning Algorithms

To solve the RL task of maximizing returns, we use TD3 [7] and SAC [8], two well-known state-of-the-art RL algorithms. They are both model-free off-policy actor-critic methods. Comparing auxiliary tasks against these two presents a trade-off between the computational expense of the runs required for our comparison (hundreds per RL algorithm) and investigating more than one algorithm to avoid results being biased. We chose these two algorithms in particular because they are powerful and also popular, which makes them a testbed that is both non-trivial and particularly relevant to readers.

We study one environment, FetchSlideDense-v1, that is too difficult to solve with baseline TD3 and SAC. It does however become at least partially solvable when adding hindsight experience replay, first proposed in [2]. HER infuses the replay buffer used by off-policy algorithms with additional samples copied from previous episodes. In these copied samples it changes the reward signal to pretend the agent had performed well in order to present it with positive learning signals. Additional supposedly successful episodes provide a stronger incentive for the agent to learn, which makes learning in complex environments easier. Nowadays it is wide-spread practice to use HER for robotics tasks such as FetchSlideDense-v1.

4.3   Environments

We perform our study on five different environments: A simulated pendulum, three MuJoCo control tasks and a simulated robotics arm. They span a large range of size and complexity. Size, here, refers to the dimensionality of observation and action space, while complexity concerns how difficult it is to learn a sufficient mapping between observation space, action space, and rewards. The three MuJoCo control tasks differ in size but are controlled by similar dynamics, which allows for a very direct comparison. All studied environments are depicted in Fig. 4. Sizes of observation and action spaces, and of the corresponding representations learned with OFENet, are listed in Table 1.

In the following, all five environments we use are briefly described. For further details on the first four we refer the reader to OpenAI Gym's documentation [3].

Pendulum-v1 is a simple and small classic control environment in which a pendulum needs to be swung upwards and then balanced in this position by applying torque. Its observations quantify angle and angular velocity of the pendulum. The reward at each time step is an inversely linear function of how much the angle differs from the desired goal, how much the angle changes, and how much torque is applied.

Fig. 4. Sample images rendered to visualize the environments. The image of FetchSlideDense-v1 is taken from [22].

Table 1. Dimensions of observations, actions, representations, and OFENet parameters used to achieve them. Layers per part describes the total amount, and individual width, of fully connected layers per OFENet part.

Environment	dim (o_t)	d (a_t)	dim (z_{o_t})	dim (z_{o_t,a_t})	Layers/part
Pendulum-v1	3	1	23	44	2×10
Hopper-v2	11	3	251	494	6×40
HalfCheetah-v2	17	6	257	503	8×30
Humanoid-v2	292	17	532	789	8×30
FetchSlideDense-v1	31	4	271	515	8×30

Hopper-v2 is one of three MuJoCo control tasks we consider here. It is based on a physical simulation of a two-dimensional single leg with four parts, which can be controlled by applying torque to three connecting joints. This makes it comparatively small and simple. The observation contains certain angles and positions of parts and joints, and their velocities. The reward at a given time step mostly depends on how much the hopper has moved forward, plus a constant term if it has not collapsed.

HalfCheetah-v2 is another two-dimensional MuJoCo control task, similar to Hopper-v2 but larger and more complex. It already consists of 9 links and 8 joints, with action and observation space similar in nature to those of Hopper-v2 but consequently of larger dimensionality. The reward is again based on how much the HalfCheetah-v2 has moved forward since the last time step.

Humanoid-v2 is the third MuJoCo control task we use in our comparison. As opposed to the others, it is three-dimensional. It roughly models a human, which leads to actions and observations similar to those of Hopper-v2 and HalfCheetah-v2, but of far higher dimensionality. Again, the reward is primarily based on forward movement plus a constant term if the robot has not fallen over.

FetchSlideDense-v1 is a simulated robotics task presented in [22]. It is not much larger than HalfCheetah-v2, but much more complex than any of the other tasks. A three-dimensional arm needs to push a puck across a low-friction table such that it slides to a randomly sampled goal position out of reach of the arm. The action controls movement of the tip of the arm, while the observation encodes position and velocities of arm and puck as well as the goal location. The reward is the negative distance between puck and goal, and thus constant until the arm hits the puck. In their technical report, the authors show that this task is very difficult to solve even with state-of-the-art methods, unless additional methods such as HER are deployed. FetchSlideDense-v1 is evaluated by success rate instead of return. Success rate describes in how many cases out of 100 the puck ended up closer than some threshold to its goal.

5   Experiments

To compare the auxiliary tasks, we train agents with baseline TD3 and SAC on raw observations (baseline) and on representations learned with auxiliary tasks. We do this for each of the five environments. Additionally, for FetchSlideDense-v1, we combine TD3 and SAC with HER and train these on raw observations as well as on representations learned with auxiliary tasks. All of the aforementioned experiments are conducted five times with the same set of random seeds. We do regular evaluations over several evaluation episodes throughout training, and their average return/success rate is what we report here.

For our experiments we use the PyTorch implementations of TD3 [6] and SAC [28], together with our own PyTorch implementation of OFENet based on the Tensorflow code provided with [20]. For the experiments with HER, we modified the Stable-Baselines3 code [24] to include OFENet. For the experiments done with Stable-Baselines3, we took hyperparameters from the RL Baselines3 Zoo repository [23]. In all other experiments, hyperparameters are the default ones provided by the respective RL algorithm implementation or the OFENet implementation of [20], configured as indicated in Table 1.

We pretrain OFENet with 1000 steps for Pendulum, and 10,000 steps for the other environments. This pretraining data is sampled using a random policy. After that, the system alternates between training OFENet on its auxiliary task and the RL algorithm on its RL task, while freezing the weights of the respective other. Representations are thus continuously updated during the training process and become optimized on those states and actions relevant to the agent. In each iteration OFENet and agent are trained on the same sampled observations and we count this as one training step of the overall system. In other words, we only count training steps of the RL algorithm for better clarity and comparability.

In terms of computation time, adding OFENet to the RL algorithms roughly doubles to triples the training time of our agents, which appears little given the large increase in dimensionality. We speculate that this factor is caused by a doubling in backward passes for gradient updates plus some additional overhead in handling two separate networks for separate tasks, while additional gradients due to the increased network width can be computed in parallel by PyTorch.

6   Results

The returns or success rates on all different environments are shown in Figs. 5 and 6 for all auxiliary tasks and baseline algorithms. For a direct, normalized comparison Fig. 7 plots the normalized maximum return/success rate against sample efficiency. To measure sample efficiency, we calculate the fraction of training steps (and therefore samples) which are required to reach 80% of the maximum return of the baseline algorithm, calibrated against the untrained baseline algorithm since the initial reward is not always 0. We choose 80% instead of 100% because at this lower threshold we can capture increases in sample efficiency even where maximum return/success rate are similar to that of the baseline. We call our measure SE80.

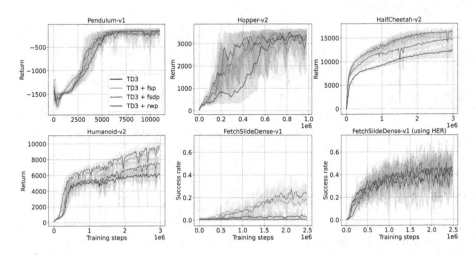

Fig. 5. Returns/success rates achieved with TD3 and different auxiliary tasks on various environments. The shaded areas show minimum and maximum performance achieved across 5 runs, while the lines represent the means. Values have been smoothed slightly for better visualisation.

In the following, the word *performance* shall refer to the combination of maximum return and sample efficiency. If only one of the two is concerned, we will state that explicitly. It is apparent that all three auxiliary tasks lead to a significant increase in performance for complex, higher-dimensional environments. For the low-dimensional and simple Pendulum-v1, a slight increase in sample efficiency but not in best return can be achieved. In fact, improvements in sample efficiency are achieved across almost all environments. Increases in maximum returns follow a certain pattern: They seem to increase with problem complexity rather than strictly dimensionality, although the two go hand in hand. When using HER to solve FetchSlideDense-v1, however, representation learning only leads to minor improvements. This is a special case discussed in Sect. 6.1.

6.1   Representation Learning for Different Types of Environments

Our experiments show different behavior for small and simple environments compared to larger and more complex ones. For the very small and simple Pendulum-

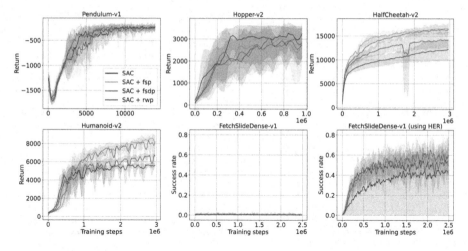

Fig. 6. Returns/success rates achieved with SAC and different auxiliary tasks on various environments. The shaded areas show minimum and maximum performance achieved across 5 runs, while the lines represent the means. Values have been smoothed slightly for better visualisation.

v1 environment, representation learning with auxiliary tasks does not significantly benefit return or sample efficiency. For the slightly larger and less simple Hopper-v2 environment, the picture is ambiguous with an increase in sample efficiency for TD3 but not for SAC. For the remaining larger and more complex environments, however, representation learning with auxiliary tasks provides clear performance gains over baseline TD3 and SAC. These gains seem to scale with complexity rather than size of the environments, as the difference in performance between HalfCheetah-v2, FetchSlideDense-v1 and Humanoid-v2 is not proportionate to their difference in size.

An interesting case is the FetchSlideDense-v1 environment. It is too complex for any learning to occur with baseline TD3 or SAC (without HER). Because of its initially constant rewards, *rwp* is not able to learn anything at all. Adding HER to the RL algorithm, however, seems to speed up learning enough to generate meaningful rather than trivial reward signals very soon and to successfully train *rwp*, as evidenced by the fact that agents using *rwp* are competitive with those trained on other tasks.

The authors proposing HER argue that in cases such as FetchSlideDense-v1 too few learning impulses, in the form of rewards, are provided to meaningfully update network weights in the RL algorithm. When using TD3 with HER, the auxiliary tasks do not seem to offer any benefits. For SAC with HER, the agents trained with auxiliary tasks are on average better than those without. Baseline SAC with HER can in fact perform as well as with auxiliary tasks, but is less reliable; its mean is lowered considerably by two agents which did not learn at all. These results suggest that adding a learning signal through HER in principle enables the RL algorithm itself to learn meaningful patterns from original

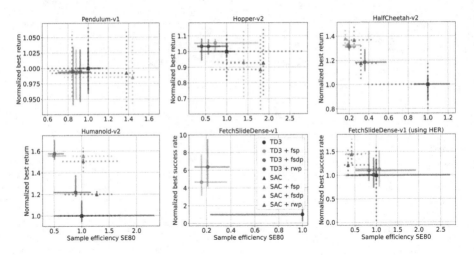

Fig. 7. Sample efficiency on different environments compared to normalized best returns/success rates. Note the different scales on the axes. The markers describe average performance, error bars (solid for TD3 and dotted for SAC) mark best and worst case out of 5 runs. Where markers or error bars are missing, agents in question never reached the return/success rate required to calculate SE80.

observations (i.e. HER significantly reduces the complexity of the problem), but it only does so reliably when adding representations learned on auxiliary tasks.

Furthermore, FetchSlideDense-v1 becomes at least partially solvable for TD3, even without HER, when *fsp* or *fsdp* are used. This interesting result shows that even if an environment is too complex for a RL algorithm, adding representation learning might still make training of agents possible. There is however no equal improvement in the same experiment with SAC, which shows that this strategy has its limits. We hypothesize that the learned representations recast observations, actions and thereby the entire RL problem into a less complex manifold. At least some dimensions of the representation learned with OFENet would then contain more informative features than the original observation. For FetchSlideDense-v1 the representation might for instance contain a feature encoding distance between arm and puck, instead of just the absolute positions from raw observations. When the arm accidentally hits the puck, the RL algorithm could consequently relate observation and reward more easily. Another possible factor, proposed in [20], is that the added depth and width of OFENet enable the agent to learn more complex and therefore more successful solutions. In this case, however, additional expressivity through added weights alone does not reduce problem complexity which is caused by initially constant rewards. It can therefore not explain why *fsp* and *fsdp* make FetchSlideDense-v1 learnable for TD3 without HER. We thus consider simplification of the learning problem to be the dominant factor at least for this setting.

6.2   Comparison of Auxiliary Tasks

This section presents a direct comparison of auxiliary tasks across the different algorithms and environments. Since HER seems to significantly distort the performance of auxiliary tasks compared to just using baseline RL algorithms, the FetchSlideDense-v1 solved with HER will not be considered.

In the remaining cases, the *rwp* task performs worst out of all investigated tasks. For the complex and high-dimensional environments it is quickly outperformed by *fsp* and *fsdp*, even though it appears competitive for environments with less complex dynamics where differences in performance are minimal. The performances of *fsp* and *fsdp* are approximately similar, although one usually outperforms the other by a slight but noteworthy margin. There is however no apparent pattern to this. When used with TD3, there might be a slight tendency for *fsdp* to outperform *fsp*, but results are too inconclusive to confidently make this claim, especially since it cannot be observed with SAC-based agents.

There are three potential causal factors which might explain why *rwp* performs worse. Firstly, due to its dimensionality alone, the prediction target r_{t+1} of *rwp* can not convey the same amount of information as the prediction targets of *fsp* and *fsdp*. Secondly, the nature of the information differs. Learning representations on environment dynamics makes environment information accessible that is much harder to access when using reward signals. Thirdly, the reward signals are provided to both the agent and OFENet, which underlines the redundancy claim regarding *rwp*. However, neither of these factors is easy to investigate without studying the representations. We hope to conduct such an investigation as future work to better understand these explanations.

The absence of a consistent difference in performance between *fsp* and *fsdp* suggests that the theoretical advantages of each (Sect. 3) are either not important or cancel each other out. Our studied environments are well behaved as they all simulate real-world physics. Consequently, they do not confront the algorithm with abrupt state changes or excessive noise. The fact that *fsp* on average still works about as well as *fsdp*, despite those properties, suggests that the advantages proposed for *fsdp* in particular do not play a large role in practice.

7   Conclusion

In this paper we compare auxiliary tasks for decoupled representation learning on non-visual observations in RL. To this end we use five common benchmark environments and two different state-of-the-art off-policy RL algorithms. We find that representation learning with auxiliary tasks can significantly improve both sample efficiency and maximum returns for larger and more complex environments while it makes little difference with simpler environments that are easy to solve for the baseline agent. In those latter cases, we observe a slight increase in sample efficiency at most. Auxiliary tasks that encourage learning environment dynamics generally outperform reward prediction. Particularly encouraging is that the FetchSlideDense-v1 environment, a simulated robotics arm, becomes

partially solvable when adding representation learning to the otherwise unsuc-
cessful TD3 algorithm. We interpret this as an indication that decoupled representation learning with auxiliary tasks can reduce problem complexity in RL.
Across all experiments, we found that results might vary between RL algorithms,
even when using the same representation learning techniques.

Despite this variability we are confident that the patterns we found can contribute to the future development of representation learning algorithms for RL,
in particular for decoupled and interpretable representation learning approaches
for real-world applications.

References

1. Anderson, C.W., Lee, M., Elliott, D.L.: Faster reinforcement learning after pretraining deep networks to predict state dynamics. In: 2015 International Joint
   Conference on Neural Networks (IJCNN), pp. 1–7 (2015). https://doi.org/10.1109/
   IJCNN.2015.7280824
2. Andrychowicz, M., et al.: Hindsight experience replay. In: Advances in Neural
   Information Processing Systems, vol. 30 (2017)
3. Brockman, G., et al.: OpenAI gym. arXiv:1606.01540 [cs] (2016)
4. de Bruin, T., Kober, J., Tuyls, K., Babuska, R.: Integrating state representation
   learning into deep reinforcement learning. IEEE Robot. Autom. Lett. 3(3), 1394–
   1401 (2018). https://doi.org/10.1109/LRA.2018.2800101
5. Dulac-Arnold, G., Mankowitz, D., Hester, T.: Challenges of real-world
   reinforcement learning (2019). https://doi.org/10.48550/arXiv.1904.12901,
   arXiv:1904.12901 [cs, stat]
6. Fujimoto, S.: TD3 implementation in PyTorch (2022). https://github.com/sfujim/
   TD3
7. Fujimoto, S., van Hoof, H., Meger, D.: Addressing function approximation
   error in actor-critic methods (2018). https://doi.org/10.48550/arXiv.1802.09477,
   arXiv:1802.09477 [cs, stat]
8. Haarnoja, T., Zhou, A., Abbeel, P., Levine, S.: Soft actor-critic: off-policy maximum entropy deep reinforcement learning with a stochastic actor (2018). https://
   doi.org/10.48550/arXiv.1801.01290, arXiv:1801.01290 [cs, stat]
9. Hlynsson, H.D., Wiskott, L.: Reward prediction for representation learning and reward shaping (2021). https://doi.org/10.48550/arXiv.2105.03172,
   arXiv:2105.03172 [cs, stat]
10. van Hoof, H., Chen, N., Karl, M., van der Smagt, P., Peters, J.: Stable reinforce-
    ment learning with autoencoders for tactile and visual data. In: 2016 IEEE/RSJ
    International Conference on Intelligent Robots and Systems (IROS), pp. 3928–3934
    (2016). https://doi.org/10.1109/IROS.2016.7759578
11. Jaderberg, M., et al.: Reinforcement learning with unsupervised auxiliary tasks
    (2016). https://doi.org/10.48550/arXiv.1611.05397, arXiv:1611.05397 [cs]
12. Jonschkowski, R., Brock, O.: Learning state representations with robotic priors.
    Auton. Robot. 39(3), 407–428 (2015). https://doi.org/10.1007/s10514-015-9459-7
13. Legenstein, R., Wilbert, N., Wiskott, L.: Reinforcement learning on slow features
    of high-dimensional input streams. PLoS Comput. Biol. 6(8), e1000894 (2010).
    https://doi.org/10.1371/journal.pcbi.1000894

14. Lesort, T., Díaz-Rodríguez, N., Goudou, J.F., Filliat, D.: State representation learning for control: an overview. Neural Netw. **108**, 379–392 (2018). https://doi. org/10.1016/j.neunet.2018.07.006

15. Lin, X., Baweja, H., Kantor, G., Held, D.: Adaptive auxiliary task weighting for reinforcement learning. In: Advances in Neural Information Processing Systems, vol. 32 (2019)

16. Locatello, F., et al.: Challenging common assumptions in the unsupervised learning of disentangled representations. arXiv:1811.12359 [cs, stat] (2019)

17. Mnih, V., et al.: Playing Atari with deep reinforcement learning (2013). https:// doi.org/10.48550/arXiv.1312.5602, arXiv:1312.5602 [cs]

18. Munk, J., Kober, J., Babuska, R.: Learning state representation for deep actor-critic control. In: 2016 IEEE 55th Conference on Decision and Control (CDC), pp. 4667–4673. IEEE, Las Vegas (2016). https://doi.org/10.1109/CDC.2016.7798980

19. Oh, J., Singh, S., Lee, H.: Value prediction network. In: Advances in Neural Information Processing Systems, vol. 30. Curran Associates, Inc. (2017)

20. Ota, K., Oiki, T., Jha, D., Mariyama, T., Nikovski, D.: Can increasing input dimensionality improve deep reinforcement learning? In: International Conference on Machine Learning, pp. 7424–7433. PMLR (2020)

21. Pathak, D., Agrawal, P., Efros, A.A., Darrell, T.: Curiosity-driven exploration by self-supervised prediction. In: Proceedings of the 34th International Conference on Machine Learning, pp. 2778–2787. PMLR (2017)

22. Plappert, M., et al.: Multi-goal reinforcement learning: challenging robotics environments and request for research (2018). https://doi.org/10.48550/arXiv.1802. 09464, arXiv:1802.09464 [cs]

23. Raffin, A.: RL Baselines3 zoo (2020). https://github.com/DLR-RM/rl-baselines3-zoo

24. Raffin, A., Hill, A., Gleave, A., Kanervisto, A., Ernestus, M., Dormann, N.: Stable-baselines3: reliable reinforcement learning implementations. J. Mach. Learn. Res. **22**(268), 1–8 (2021)

25. Shelhamer, E., Mahmoudieh, P., Argus, M., Darrell, T.: Loss is its own reward: self-supervision for reinforcement learning. arXiv:1612.07307 [cs] (2017)

26. Stooke, A., Lee, K., Abbeel, P., Laskin, M.: Decoupling representation learning from reinforcement learning. In: International Conference on Machine Learning, pp. 9870–9879. PMLR (2021)

27. Wahlström, N., Schön, T.B., Deisenroth, M.P.: From pixels to torques: policy learning with deep dynamical Models (2015). https://doi.org/10.48550/arXiv.1502. 02251, arXiv:1502.02251 [cs, stat]

28. Yarats, D., Kostrikov, I.: Soft actor-critic (SAC) implementation in PyTorch (2022). https://github.com/denisyarats/pytorch_sac

A Hybrid Steady-State Genetic Algorithm for the Minimum Conflict Spanning Tree Problem

Punit Kumar Chaubey and Shyam Sundar[(✉)]

Department of Computer Applications, National Institute of Technology Raipur,
Raipur 492010, India
{pkchaubey.phd2019.mca,ssundar.mca}@nitrr.ac.in

Abstract. This paper studies a hybrid approach for the minimum conflict spanning tree (MCST) problem, where the MCST problem deals with finding a spanning tree (T) with the minimum number of conflicting edge-pairs. The problem finds some important real-world applications. In this hybrid approach (hSSGA), a steady-state genetic algorithm generates a child solution with the help of crossover operator and mutation operator which are applied in a mutually exclusive way, and the generated child solution is further improved through a local search based on reduction of conflicting edge-pairs. The proposed crossover operator is problem-specific operator that attempt to create a fitter child solution. All components of SSGA and local search effectively coordinate in finding a conflict-free solution or a solution with a minimal number of conflicting edge-pairs. Experimental results, particularly, on available 12 instances of type 1 benchmark instances whose conflict solutions are not known show that the proposed hybrid approach hSSGA is able to find better solution quality in comparison to state-of-the-art approaches. Also, hSSGA discovers new values on 8 instances out of 12 instances of type 1.

Keywords: Conflict · Spanning Tree · Steady-state Genetic Algorithm · Local search

1 Introduction

The minimum spanning tree problem having conflict on edge-pairs (MSTC) is a variant of well-known classical minimum spanning tree problem. The MSTC is a NP-hard problem which seeks to find a conflict-free spanning tree (T) of minimum cost in a given connected, undirected and weighted graph $G(V, E, w, P)$, where V is the set of vertices, E is the set of edges, a weight ($w(\{u, v\})$) is assigned to each edge ($\{u, v\}$) connecting u and v vertices, P is the set of conflicting edge-pairs. Darmann *et al.* [5,6] were the first one who introduced this problem and proved its $\mathcal{NP}$-hardness. Later, Zhang *et al.* [25] proved that the problem can be solved in polynomial time for a particular class of instances, i.e.,

G. Nicosia et al. (Eds.): LOD 2023, LNCS 14506, pp. 192–205, 2024.
https://doi.org/10.1007/978-3-031-53966-4_15

instances whose pairs in P follow the transitive property, i.e. if $\{\{uv\}, \{rs\}\} \in P$ and $\{\{rs\}, \{gh\}\} \in P$, then even $\{\{uv\}, \{gh\}\} \in P$ also, where $\{uv\}, \{rs\}$ and $\{gh\} \in E$.

There are numerous optimization problems that are variants of spanning tree problem, such as degree-constrained spanning tree [20], tree t-spanner problem [19], bounded degree spanning tree problem [23], bounded diameter spanning tree problem [21], quadratic spanning tree problem [8,22]. There are also other conflict-based optimization problems such as the minimum cost perfect matching with conflict pair constraints [12], the bin packing problem with conflicts [17], the maximum flow problems with disjunctive constraints [14], and the knapsack problem with conflict constraints [13].

The MSTC problem has some important real-world applications such as in designing an offshore wind farm network [11], for the solution of quadratic bottleneck spanning problem [25], in road map where some roads are not permitted to turn left or right [10], in the setting-up of an oil pipeline system that joins several countries [5].

By definition, the MSTC problem deals with finding a conflict-free spanning tree of minimum cost in a given graph; however, it becomes infeasible when the graph does not have a conflict-free spanning tree solution. To continue this, Zhang $et\ al.$ [25] proposed several heuristics based on construction methods, Lagrangian relaxation, local search, tabu search and tabu thresolding for the MSTC problem. These proposed heuristics only give the number of conflict present in the returned solution in case of no conflict-free spanning tree is found. Samer and Urrutia [18] proposed with a branch-and-cut approach. Carrabs $et\ al.$ [2] also proposed a branch-and-cut approach which is superior to the previous one branch-and-cut approach [18].

Carrabs $et\ al.$ [1] observed this issue (i.e. when no conflict-free spanning tree is found) and came up with idea of resolving this issue by proposing a new variant of the MSTC problem. This new variant deals with two objectives in which the first objective is to deal with finding a spanning tree (T) that contains the minimum number of conflicting edge-pairs, and the second objective only deals with finding a conflict-free spanning tree of minimum cost, when, the first objective finds a conflict-free spanning tree on a given input graph. They referred to this new variant as minimum conflict weighted spanning tree (MCWST) problem. Authors presented a multiethnic genetic algorithm (Mega) for this MCWST problem. The Mega framework starts with dividing the starting population into 3 sub-populations and allows to evolve independently in 3 different environments. Once Mega is over, three local search methods are applied on each chromosome of the final population. The first local search method aims to reduce the number of conflicts, and if the resultant spanning tree is found to be a conflict-free, then two local search methods aims to reduce the weight of this conflict-free spanning tree. Authors showed that their Mega performs superior to tabu search (TS) [25]. Cerrone $et\ al.$ [3] proposed a genetic algorithm (GA) in which GA generates the next generation (new population) by preserving the best half of the current population that goes to the next generation, and the remaining half of the next population is generated through genetic operators (crossover and mutation). For crossover, GA selects two parent solutions (p_1 and p_2) respectively from the best half of the current population and from the remaining half of the current population. If p_1 is not conflict-free, then the first variant of crossover operator is applied that uses four random indices (i_1, i_2, i_3, i_4) from 1 to $|V|$-1 to generate a child solution. A subset of edges from i_1 to i_3 in p_1 are copied to empty child soluton (*child*). After that another subset of edges from i_2 to i_4 in p_2 are picked and only those edges from this subset that reduce the number of disconnected component of partial *child*. If *child* is still infeasible, then the spanning tree of *child* is completed through adding random edges. However, if p_1 is conflict-free, another variant of crossover is applied which is similar to the first variant except addition of edges from p_2 and random edges does not introduce a conflict into the child solution (*child*). Once the crossover operator is over, the mutation operator is applied. This mutation operator uses two local search methods, where first local search aims to reduce the number of conflict iff the *child* solution is not conflict free, otherwise, the second local search method aims to reduce the weight of conflict free *child* solution. The performance of GA on available benchmark instances is superior to TS [25] and Mega [1]. Recently, Punit and Sundar [4] proposed two phases of metaheuristic techniques (TPMT) to tackle the two objectives of the MCWST problem, where the role of a hybrid artificial bee colony algorithm (hABC) that acts as the first phase of TPMT is to tackle the first objective of the problem, and as soon as hABC finds a conflict-free solution, the role of an iterated local search (ILS) that starts with this conflict free solution , where ILS that acts as the second phase of TPMT tackles the second objective of the problem starting with this conflict free solution. Results of TPMT shows its superiority over TS [25], Mega [1] and GA [3].

This paper motivates from the work on TPMT that tackles the MCWST problem in phase-wise in which the first phase (hABC) of TPMT tackles the first objective of the MCWST problem, i.e., minimizing the number of conflicts, and iff, the first phase returns a conflict free solution, then the second phase (ILS) of TPMT that tackles the second objective of the MCWST problem, i.e., minimizing the weight of conflict-free solution returned by hABC. In this paper, we only focus on the first objective of the MCWST problem which is referred to as *MCST* problem and present a hybrid approach (hSSGA) combining a steady-state genetic algorithm (SSGA) and a local search based on the reduction of conflicting edge-pairs for this MCST problem. Note that our proposed hSSGA is quite different from the existing GA [3] for this problem. Computational results on available two set of type 1 and type 2 instances indicate the superiority of hSSGA over the existing TS [25], Mega [1], GA [3] and TPMT [4] approaches.

The organization of the remaining of the paper is as follows: Sect. 2 discusses a hybrid steady-state genetic algorithm for the MCST problem; Sect. 3 discusses the computational results; and Sect. 4 presented some concluding remarks.

2 Hybrid Steady-State Genetic Algorithm

Genetic algorithm that comes under the class of evolutionary algorithms is well-known search method based on the principles of natural evolution [9]. GA that starts from a population of individuals searches for optimum value in the solution space of a given optimization problem through applying techniques based on the natural evolution, such as selection, crossover and mutation operators. Genetic algorithm, in general, has two variants, such as generational genetic algorithm (GGA) and steady-state genetic algorithm (SSGA). This paper works on a hybrid approach (hSSGA) combining a SSGA and a local search for the MCST problem, as SSGA [7] has some advantages over GGA. In GGA, usually an equal number of newly generated child solution through crossover and mutation operators replaces the whole current parent population in every generation; whereas in SSGA, one or two child solution is generated through crossover and mutation operators in each generation. If the generated child solution is different from the individuals of the current parent population, then such a child solution usually replaces a worst individual (based on the fitness of the chromosome) of the current parent population. SSGA does not have a duplicate individual in the population as compared to GGA. Having duplicate individuals in the population hampers the search. Besides, SSGA keeps both the best-so-far generated individual and newly included generated child solution in the current population, helping in generating a better child solution in lesser computational time as compared to GGA.

The following subsections describe the main ingredients of hSSGA for the MCST problem in detail.

2.1 Solution Representation

Each chromosome (solution or spanning tree) is encoded as a set of $|V|$-1 edges of $G(V, E, w, P)$ [16]. This encoding provides high heritability, locality, and can be adapted with problem-specific knowledge.

2.2 Initial Solution Generation

A spanning tree of each initial solution of the population is generated randomly using a random variant of Prim's algorithm [15] so that diversity of the population is preserved. The size of the population is Pop which is a parameter that is to be determined empirically.

2.3 Fitness Computation

The fitness $(f(T_r))$ of a solution (T_r) is the number of conflicting edge-pairs that exists in the solution.

2.4   Crossover Operator

Algorithm 1: The pseudo-code of crossover operator

Input: Two parent solutions $Par1$ and $Par2$
Output: A feasible child solution C
$Prob \leftarrow \frac{1/(f(Par1))^2}{(1/(f(Par1))^2)+(1/(f(Par2))^2)}$;
$S \leftarrow \Phi; U \leftarrow V$;
$C \leftarrow \Phi$; flag $\leftarrow 0$;
Select a vertex (say v_i) randomly from U;
Select a random vertex (say v_j) from U and v_j must be adjacent to v_i in G;
$C \leftarrow \{v_i, v_j\}$;
$S \leftarrow S \cup \{v_i, v_j\}; U \leftarrow U \setminus \{v_i, v_j\}$;
while $U \neq \Phi$ **do**
    //$u01$ is a uniform variate
    if $u01 < Prob$ **then**
        $Jump_{Par1}$:
        Search a candidate edge (say $\{v_x, v_y\}$) randomly that joins a selected
        vertex (v_x) in S to an unselected vertex (v_y) in U from $Par1$ and that
        must be conflict-free with the edges of partial child solution (C).
        if *the search is successful* **then**
            $C \leftarrow \{v_x, v_y\}$;
            $S \leftarrow S \cup \{v_y\}; U \leftarrow U \setminus \{v_y\}$;
            flag $\leftarrow 0$;
        else
            flag $\leftarrow$ flag+1;
            if *flag == 1* **then**
                goto $Jump_{Par2}$;
    else
        $Jump_{Par2}$:
        Search a candidate edge (say $\{v_x, v_y\}$) randomly that joins a selected
        vertex (v_x) in S to an unselected vertex (v_y) in U from $Par2$ and that
        must be conflict-free with the edges of partial child solution (C).
        if *the search is successful* **then**
            $C \leftarrow \{v_x, v_y\}$;
            $S \leftarrow S \cup \{v_y\}; U \leftarrow U \setminus \{v_y\}$;
            flag $\leftarrow 0$;
        else
            flag $\leftarrow$ flag+1;
            if *flag == 1* **then**
                goto $Jump_{Par1}$;
    if *flag == 2* **then**
        flag $\leftarrow 0$;
        Select a candidate edge (say $\{v_s, v_u\}$) from $Par1$ and $Par2$ that should
        have minimum conflict with the edges of partial C, where v_s is selected
        from S and v_u is selected from U;
        $C \leftarrow \{v_s, v_u\}$;
        $S \leftarrow S \cup \{v_u\}; U \leftarrow U \setminus \{v_u\}$;
return C;

The crossover operator has a significant impact on the performance of genetic algorithm. The role of this operator is to create a child solution by inheriting the traits of good genes from their parents. With this motivation, we design a problem-specific crossover operator ($Xover$) that attempts to generate a new child solution that should have a better potential of adaptation, i.e. its fitness values should increase.

The proposed $Xover$ first selects two parent solutions (say $Par1$ and $Par2$) from the current parent population using the binary tournament selection procedure, then follows similar to Prim's algorithm [15] to generate a child solution using spanning trees of $Par1$ and $Par2$. Initially, the child solution (say C) is an empty spanning tree. All vertices of G is assigned to a set (say U) and a set (say S) is initially empty. $Xover$ first selects a vertex (say v_i) randomly, then an unselected vertex (say v_j) adjacent to selected vertex v_i is selected randomly. This way leads to an edge ($\{v_i, v_j\}$) which is assigned to C. The vertices v_i and v_j are added to S and removed from U. After this, iteratively, $Xover$, at each step, selects a candidate edge randomly that joins a selected vertex in S to an unselected vertex in U from one (say $Par1$) of two parent solutions ($Par1$ and $Par2$). With probability ($Prob$), $Par1$ is selected; otherwise, $Par2$ is selected, where $Prob$ is determined by $\frac{1/(f(Par1))^2}{(1/(f(Par1))^2)+(1/(f(Par2))^2)}$. Here, a candidate edge means an edge that must be conflict-free with the edges of partial child solution (C). If such an edge is not available, then it moves to another parent solution (say $Par2$) in search of an edge that joins a selected vertex in S to an unselected vertex in U. If such an edge is still not found, then $Xover$ searches a candidate edge from $Par1$ and $Par2$ that should have minimum conflict with the edges of partial C. The selected edge (say $\{v_s, v_u\}$) is added to the partial C, where v_s is selected from S and v_u is selected from U. v_u is added to S and removed from U. This way of selecting an edge for C is an attempt to create a child solution that carries out a minimal number (or may be zero number) of conflicting edges. This iterative procedure is called repeatedly until U becomes empty set. At this point, a spanning tree of C is generated.

Algorithm 1 describes the pseudo-code of the proposed crossover operator.

The role of probability ($Prob$), in case of an available conflict-free candidate edge with the edges of the partial C, is to encourage that parent solution which is better (in terms of fitness) between $Par1$ and $Par2$, which in turn, a fitter child solution is created.

It is to be noted that our proposed crossover operator is different from one in Mega [1] and GA [3].

2.5   Mutation Operator

The role of mutation operator is inject diversity in the population. The mutation operator (say mutation_operator) used here is similar to shaking procedure used in [4]. This operator first selects an individual from the current population using a binary tournament selection procedure, and create a copy (C) of this selected solution. Then, the operator cycles over a certain number (noi) of iterations, where $noi = max(1, (int)(P_m * (|V|\text{-}1))))$. P_m is a parameter. In every iteration, a non-leaf edge (say $\{v_t, v_s\}$) of spanning tree (say T_m) of C is removed randomly, leading to a partition of T_m into two components. To connect these two components, a new candidate edge $\in E \setminus E_{T_i}$ (not $\{v_t, v_s\}$) that can join these two components is selected randomly. This iterative procedure is called repeated for noi times.

One should note that the mutation_operator always removes a different edge of T_m in each iteration, and the operator may stop earlier if no new candidate edge $\in E \setminus E_{T_i}$ (not e_{ij}) is possible to connect these two components of T_m.

The crossover operator and the mutation operator are applied in a mutually exclusive manner. With probability (P_x), the crossover operator is applied; otherwise the mutation operator is applied. P_x is a parameter. Both operators applied in a mutually exclusive manner generate a single child solution (C).

2.6   Conflict Reduction Operator

A local search based conflict reduction operator (say $CRLS$) is applied to reduce the number of conflicting edge pairs in the spanning tree (say T) newly generated child (C) which is taken from [4]. This operator begins with deleting a conflict edge (say $\{uv\}$) in T whose association of conflicts with other edges in C is maximum. This causes T into two components. To make T feasible, a candidate edge in $E \setminus E_T$ that does not only join these two components, but also has the minimum number of conflicts with other edges in T, when it is compared with the deleted edge $\{uv\}$, is selected, resulting T into a spanning tree. If such candidate edge while searching finds conflict free with the remaining edges of T, it is immediately added to T. If no such candidate edge exists, then $\{uv\}$ is added back to T. The operator repeats it another edge that has the next maximum conflict edge in T until improvement is not possible in T. Also, this operator stops to execute when the operator is not able to reduce the number of conflicting edge-pairs in T after a fixed (say $Para_{ls}$) consecutive number of trials, where $Para_{ls}$ is a parameter. Note that if the reduction in the number of conflicting edge-pairs in T causes a conflict-free T at any time, then this operator stops.

2.7   Replacement Operator

Once a child solution (say C) generated by either the crossover operator or the mutation operator is completed, C is examined whether it is a copy of any individual in the current parent population. If it is so, then C is discarded; otherwise, C replaces a worst individual (in terms of fitness) of the current parent population.

Algorithm 2 illustrates the pseudo-code of hSSGA for the MCST problem that returns the best solution obtained by hSSGA.

Algorithm 2: Pseudo-code of hSSGA for the MCST problem

Generate a population of initial feasible solution $(T_1, T_2, \ldots, T_{Pop})$;
Select a best (*best_so_far*) solution from $<T_1, T_2, \ldots, T_{Pop}>$;
while (*stopping criterion is not met*) **do**

   if ($u01 < P_x$) **then**

      Apply binary tournament selection procedure twice to select two
      parents (say *Par1* and *Par2*) from $<T_1, T_2, \ldots, T_{Pop}>$;
      $C \leftarrow Xover(Par1, Par2)$;

   else

      Apply binary tournament selection procedure to select an
      individual (say *Par*) from $<T_1, T_2, \ldots, T_{pop}>$;
      $C \leftarrow mutation\_operator(Par)$;

   $C \leftarrow CRLS(C)$;

   if (*the fitness of C is less than that of best_so_far solution*) **then**
      best_so_far $\leftarrow C$;

   Apply replacement operator;

return *best_so_far*;

---

3 Computational Results

This section presents a performance evaluation of hSSGA for the MCST problem. For this problem, an empirical comparisons of hSSGA with TS [25], Mega [1], GA [3] and TPMT [4] is carried out based on available benchmark instances.

3.1 Benchmark Instances

Our proposed hSSGA uses the same benchmark instances that were used by TS [25], Mega [1], GA [3] and TPMT for the MCST problem. Zhang *et al.* [25] generated two set of instances – type 1 and type 2 using different-2 number of vertices, edges and conflicting edge-pairs. Each instance has Instance ID. The set type 1 consists of 23 instances (from Instance ID 1 to 23), whereas the set type 2 consists of 27 instances (from Instance ID 24 to 50). Each instance in the set type 1 may have no conflict free solutions, whereas each instance in the set type 2 have at least one conflict free solution.

From the literature of TS [25], Mega [1], GA [3] and TPMT, the results of Instance ID from 10 to 23 except 13 and 16 of type 1 instances, i.e. total 12 instances of type 1 are not known for conflict free solutions, whereas, the results of remaining instances including type 2 are known for conflict free solutions.

3.2 Experimental Setup

Our proposed hSSGA for the MCST problem is coded in C language. For the execution of computational experiments, a Linux-based computer with Intel Core i5 CPU 3.4 GHz $\times$ 4 with 4 GB RAM for execution.

Similar to TS [25], Mega [1], GA [3] and TPMT [4], hSSGA is allowed to execute 5 runs for each instance. hSSGA is stopped to execute when the best-so-far solution does not improve over a successive of 2000 generations.

Given its stochastic nature, hSSGA uses 5 parameters (Pop, P_{btms}, P_x, P_m, $Para_{ls}$) whose tuning have impact on the solution quality. Although tuning of parameters is an arduous task, we have taken a range of each parameter (reported in Table 1) which are based on our preliminary observation. Such range does not cover all values, but help in approximating the values. For tuning, we change one parameter in its range, while keeping the other parameters to their default (final) values. Based on our experimental observation, we select $Pop = 100$, $P_{btms} = 0.95$, $P_x = 0.8$, $P_m = 0.1$, and $Para_{ls} = 4$ as the final value of each parameter (highlighted in bold in Table 1) as they produce overall the best results.

Table 1. Range of each parameter used for parameter tuning

Parameter	Range
Pop	<50, **100**, 150>
P_{btms}	<0.85, 0.9, **0.95**, 0.98>
P_x	<0.6, 0.7, **0.8**, 0.9>
P_m	<0.05, **0.01**, 0.15>
$Para_{ls}$	<2, 3, **4**, 5>

3.3   Comparison with State-of-the-Art Approaches

In this section, the performance of our proposed hSSGA is evaluated against four approaches (i.e. TS [25], Mega [1], GA [3] and TPMT [4]). It is to be noted that conflict-free solutions for 12 instances of type 1 (Instance ID from 10 to 23 except 13 and 16) are not known. For this, Table 2 only summarizes the empirical results of these 12 instances of type 1 obtained by hSSGA, TS [25], Mega [1], GA [3], and TPMT [4]. Results reported in Table 2 are in the form of the number of conflicting edge-pairs (*Conf*), average of the number of conflicting edge-pairs over 5 runs (*AvgConf*), average total execution time (*AvgTime*) in seconds over 5 runs. the best value is highlighted in bold. Our proposed approach hSSGA clearly dominates all four existing approaches (TS, Mega, GA and TPMT) in terms of solution quality. Out of these 12 instances of type 1, hSSGA finds new values on 8 instances (highlighted in bold in Table 2). In terms of computational time, hSSGA takes almost same time as compared to GA and TPMT approaches. Although we have not reported their results in this paper for remaining instances of type 1 and type 2, hSSGA finds conflict free solutions for thse instances within seconds.

Table 2. Comparative results of hSSGA with TS, Mega, GA and TPMT on 12 instances of type 1

ID	n	m	p	TS		Mega			GA			TPMT			hSSGA		
				Conf	Time	Conf	AvgConf	AvgTime	Conf	AvgConf	AvgTime	Conf	AvgConf	AvgTime	Conf	AvgConf	AvgTime
10	100	300	1344	13	6.77	10	10.40	2.90	8	9	3.4	7	8.80	2.47	6	7.60	2.29
11	100	500	6237	11	15.17	8	9.40	5.00	7	7.8	3.8	5	6.40	7.59	4	5.20	7.71
12	100	500	12,474	41	14.64	35	37.80	6.90	34	34	8.74	34	34.80	13.60	34	36.20	11.35
14	200	600	3594	67	70.24	57	60.00	24.00	55	56	20.8	54	55.80	18.54	51	53.80	15.25
15	200	600	5391	149	80.12	142	143.20	31.60	136	138.8	28.8	135	138.60	25.95	134	136.80	28.98
17	200	800	6392	39	98.01	23	27.60	28.40	18	18	20.6	17	18.60	29.43	14	17.20	23.25
18	200	800	9588	95	97.10	87	88.60	38.30	83	84.4	31.8	78	80.60	45.73	77	83.00	32.86
19	200	800	15,980	178	104.93	172	177.20	48.00	171	171.2	40.4	171	172.60	46.69	171	173.60	52.91
20	300	800	3196	63	239.63	52	55.00	62.80	47	48.4	41.4	44	46.00	26.95	46	48.60	26.95
21	300	1000	4995	38	303.04	21	23.40	79.30	18	18	20.2	12	15.00	35.86	10	11.80	35.13
22	300	1000	9990	207	345.25	176	180.60	119.10	166	167.6	100.4	167	170.00	108.20	168	171.60	77.01
23	300	1000	14,985	351	381.28	329	330.20	141.30	321	322.8	90.6	320	324.60	93.60	318	322.20	132.04
Avg					138.10			44.45			34.25			37.87			37.14

3.4   Convergence Behaviour of hSSGA

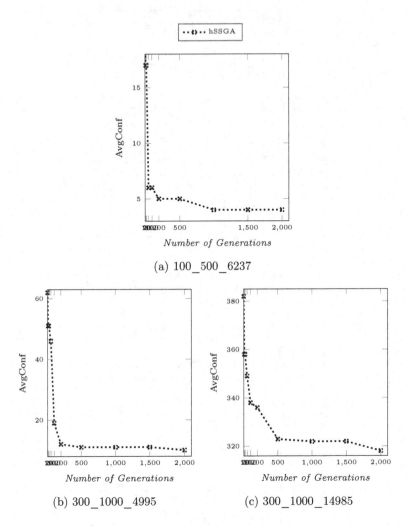

Fig. 1. Evolution of solution quality (*AvgConf*) over *Number of Generations*

In this subsection, we perform experiments on three instances (100_500_6237, 300_1000_4995, 300_1000_14985) to examine the convergence behaviour of hSSGA. Figure 1(a–c) shows the evolution of solution quality (average of the number of conflicting edge-pairs over 5 runs (*AvgConf*)) as the number of generations increases on considered instances. It is clear from Fig. 1(a–c) that hSSGA rapidly converges towards better and better solution quality with the increase of generations.

3.5 Statistical Analysis

In this subsection, we perform a statistical analysis using a two-tailed non-parametric Wilcoxon's signed rank test with significance criterion 0.05 on type 1 instances in order to compare hSSGA with state-of-the-art approaches (TPMT, GA, MEGA and TS) in terms of the best (Conf) value over 5 runs. We use a calculator of this test available at [24]. Results of this test are reported in Table 3 and show that hSSGA has significant difference with GA, MEGA and TS except TPMT.

Table 3. Results of statistical test

Approaches	p-value	Significant
hSSGA vs TPMT	0.07508	No
hSSGA vs GA	0.01242	Yes
hSSGA vs MEGA	0.00222	Yes
hSSGA vs TS	0.00222	Yes

4 Conclusions

We presented a hybrid approach (hSSGA) combining a steady-state genetic algorithm with a local search based on the reduction of conflicting edge-pairs for the minimum conflict spanning tree (MCST) problem. This problem finds some real-world applications. In hSSGA, we proposed a problem-specific crossover operator that inherits higher traits of good genes to the child solution from a better parent of two selected parents. This crossover operator combines with other components of hSSGA helps in finding solutions of high quality. hSSGA finds conflict free solutions within seconds for all instances of type 1 and type 2 except 12 instances of type 1 instances whose conflict free solution are not known. On these 12 instances of type 1, hSSGA finds new values on 8 instances. The performance of hSSGA is much better than the existing approaches in terms of solution quality. In terms of computational time, hSSGA is at par with state-of-the-art approaches.

In future work, the final solution returned by hSSGA can be used as an initial solution for tackling the second objective of the minimum conflict spanning tree (MCWST) problem.

References

1. Carrabs, F., Cerrone, C., Pentangelo, R.: A multi-ethnic genetic approach for the minimum conflict weighted spanning tree problem. Networks **74** (2019). https://doi.org/10.1002/net.21883
2. Carrabs, F., Cerulli, R., Pentangelo, R., Raiconi, A.: Minimum spanning tree with conflicting edge pairs: a branch-and-cut approach. Ann. Oper. Res. **298**(1), 65–78 (2018)

3. Cerrone, C., Di Placido, A., Russo, D.D.: A genetic algorithm for minimum conflict weighted spanning tree problem. In: Paolucci, M., Sciomachen, A., Uberti, P. (eds.) Advances in Optimization and Decision Science for Society, Services and Enterprises. ASS, vol. 3, pp. 445–455. Springer, Cham (2019). https://doi.org/10.1007/978-3-030-34960-8_39

4. Chaubey, P.K., Sundar, S.: Two phases of metaheuristic techniques for the minimum conflict weighted spanning tree problem. Appl. Soft Comput. 110205 (2023)

5. Darmann, A., Pferschy, U., Schauer, J.: Determining a minimum spanning tree with disjunctive constraints. In: Rossi, F., Tsoukias, A. (eds.) ADT 2009. LNCS (LNAI), vol. 5783, pp. 414–423. Springer, Heidelberg (2009). https://doi.org/10.1007/978-3-642-04428-1_36

6. Darmann, A., Pferschy, U., Schauer, J., Woeginger, G.J.: Paths, trees and matchings under disjunctive constraints. Discret. Appl. Math. **159**(16), 1726–1735 (2011)

7. Davis, L.: Handbook of Genetic Algorithms. Van Nostrand Reinhold, New York (1991)

8. Fu, Z.H., Hao, J.K.: A three-phase search approach for the quadratic minimum spanning tree problem. Eng. Appl. Artif. Intell. **46**, 113–130 (2015)

9. Holland, J.H.: Adaptation in Natural and Artificial Systems: An Introductory Analysis with Applications in Biology, Control, and Artificial Intelligence. University of Michigan Press, Ann Arbor (1975)

10. Kanté, M.M., Laforest, C., Momège, B.: Trees in graphs with conflict edges or forbidden transitions. In: Chan, T.-H.H., Lau, L.C., Trevisan, L. (eds.) TAMC 2013. LNCS, vol. 7876, pp. 343–354. Springer, Heidelberg (2013). https://doi.org/10.1007/978-3-642-38236-9_31

11. Klein, A., Haugland, D., Bauer, J., Mommer, M.: An integer programming model for branching cable layouts in offshore wind farms. In: Le Thi, H.A., Pham Dinh, T., Nguyen, N.T. (eds.) Modelling, Computation and Optimization in Information Systems and Management Sciences. AISC, vol. 359, pp. 27–36. Springer, Cham (2015). https://doi.org/10.1007/978-3-319-18161-5_3

12. Öncan, T., Zhang, R., Punnen, A.P.: The minimum cost perfect matching problem with conflict pair constraints. Comput. Oper. Res. **40**(4), 920–930 (2013)

13. Pferschy, U., Schauer, J.: The knapsack problem with conflict graphs. J. Graph Algorithms Appl. **13**(2), 233–249 (2009)

14. Pferschy, U., Schauer, J.: The maximum flow problem with disjunctive constraints. J. Comb. Optim. **26**(1), 109–119 (2013)

15. Prim, R.: Shortest connection networks and some generalizations. Bell Syst. Tech. J. **36**, 1389–1401 (1957)

16. Raidl, G.R., Julstrom, B.A.: Edge sets: an effective evolutionary coding of spanning trees. IEEE Trans. Evol. Comput. **7**(3), 225–239 (2003)

17. Sadykov, R., Vanderbeck, F.: Bin packing with conflicts: a generic branch-and-price algorithm. INFORMS J. Comput. **25**(2), 244–255 (2013)

18. Samer, P., Urrutia, S.: A branch and cut algorithm for minimum spanning trees under conflict constraints. Optim. Lett. **9**(1), 41–55 (2015)

19. Singh, K., Sundar, S.: Artificial bee colony algorithm using problem-specific neighborhood strategies for the tree t-spanner problem. Appl. Soft Comput. **62**, 110–118 (2018)

20. Singh, K., Sundar, S.: A hybrid genetic algorithm for the degree-constrained minimum spanning tree problem. Soft. Comput. **24**(3), 2169–2186 (2020)

21. Singh, K., Sundar, S.: Artificial bee colony algorithm using permutation encoding for the bounded diameter minimum spanning tree problem. Soft. Comput. **25**(16), 11289–11305 (2021)

22. Sundar, S., Singh, A.: A swarm intelligence approach to the quadratic minimum spanning tree problem. Inf. Sci. **180**(17), 3182–3191 (2010)
23. Sundar, S., Singh, A., Rossi, A.: New heuristics for two bounded-degree spanning tree problems. Inf. Sci. **195**, 226–240 (2012)
24. Wilcoxon, F.: Wilcoxon signed-rank test calculator (1945)
25. Zhang, R., Kabadi, S.N., Punnen, A.P.: The minimum spanning tree problem with conflict constraints and its variations. Discret. Optim. **8**(2), 191–205 (2011)

Reinforcement Learning
for Multi-Neighborhood Local Search
in Combinatorial Optimization

Sara Ceschia⬤, Luca Di Gaspero⬤, Roberto Maria Rosati$^{(\boxtimes)}$⬤,
and Andrea Schaerf⬤

DPIA, University of Udine, via delle Scienze 206, 33100 Udine, Italy
{sara.ceschia,luca.digaspero,robertomaria.rosati,
andrea.schaerf}@uniud.it

Abstract. This study investigates the application of reinforcement learning for the adaptive tuning of neighborhood probabilities in stochastic multi-neighborhood search. The aim is to provide a more flexible and robust tuning method for heterogeneous scenarios than traditional offline tuning. We propose a novel mix of learning components for multi-neighborhood Simulated Annealing, which considers both cost- and time-effectiveness of moves. To assess the performance of our approach we employ two real-world case studies in timetabling, namely *examination timetabling* and *sports timetabling*, for which multi-neighborhood Simulated Annealing has already obtained remarkable results using offline tuning techniques. Experimental data show that our approach obtains better results than the analogous algorithm that uses state-of-the-art offline tuning on benchmarking datasets while requiring less tuning effort.

1 Introduction

Machine learning applied to adaptive tuning of parameters of metaheuristics for optimization is a growing research field [1,39]. This is motivated by the fact that offline tuning, in which values are fixed in advance, remains a time-consuming and somewhat ineffective activity, despite the availability of many statistically-principled black-box automatic configuration tools (see [16]).

Indeed, tuning procedures struggle to find a single best configuration when instances differ considerably, either in size or in problem-specific features [34]. Some classical approaches to face scenarios with heterogeneous instances are instance clustering [18], instance space analysis [36] and feature-based tuning [6]. However, they are, in general, time-consuming and may require in-depth problem-specific knowledge, and so the need for best practices is still felt as a priority by the optimization community [4].

On the contrary, online parameter control reduces the time spent for the tuning phase and is suitable for scenarios in which there is no parameter configuration that fits all instances or some validation instances end up differing substantially from the training ones. In addition, it also helps in cases in which

G. Nicosia et al. (Eds.): LOD 2023, LNCS 14506, pp. 206–221, 2024.
https://doi.org/10.1007/978-3-031-53966-4_16

the best parameter configuration changes, within a single run, in different stages of the search procedure and in different areas of the search space.

For these reasons, we investigate the application of reinforcement learning (RL) methods to adjust parameters for metaheuristic search; in particular, the neighborhood probabilities in *multi-neighborhood* local search. Indeed, in many complex combinatorial problems a combination of neighborhoods is necessary to effectively explore the search space. In this work, we use Simulated Annealing, according to which a move is chosen at random at every iteration and different probabilities are assigned to the atomic neighborhoods. Traditionally, a tuning procedure is used to find those rates, in such a way that a higher probability of being chosen at each iteration is given to those neighborhoods that deliver the highest contribution to the improvement of the solution.

Among possible applications of RL, we consider the *exponential recency-weighted average bandit algorithm* [38] for updating neighborhoods rates after every temperature, with a reward function that considers both the relative improvement in objective function and the computational cost of neighborhoods.

We tested our approach on two real-world optimization problems, namely *examination timetabling* [9] and *sports timetabling* [41]. We selected these problems because multi-neighborhood Simulated Annealing has already obtained state-of-the-art results upon the corresponding benchmarking datasets. Therefore, our aim is not only to improve on the version without learning, but also to obtain truly competitive results for these problems.

Our search method reaches better results on benchmarking datasets, without reducing its efficiency. In addition, we found that the tuning of the learning-related hyperparameters is more robust, thus requiring less computational effort, than the direct offline tuning of the parameters.

2 Local Search for Optimization

Local Search is an algorithmic paradigm for search and optimization problems, which has been shown to be very effective in many AI contexts (e.g., SAT [22]). Many metaheuristics based on local search have been proposed successfully to address a variety of both academic and practical problems. Examples of local search techniques are Simulated Annealing, Tabu Search, Variable Neighborhood Search, Scatter Search, Guided Local Search, Iterated Local Search, and Large Neighborhood Search, just to name the most well-known ones [14].

The key element of the whole local search paradigm is indisputably the neighborhood relation, that defines the atomic movements that allow for navigating the search space, looking for better solutions. Indeed, the crucial issue of local search is how to *escape* from local minima of the search space created by the structure defined by the neighborhood relation, by balancing exploration and exploitation. In fact, all the above-mentioned techniques include some "smart" mechanisms to move away from the basin of attraction of a local minimum.

The use of a neighborhood composed by multiple atomic ones is an additional/alternative way to overcome the possibility of getting stuck, given that a local minimum of a neighborhood is not necessarily a local minimum of another one. In a multi-neighborhood approach, the decision of which neighborhood to probe at each iteration is relevant for efficiency and effectiveness of the search.

As the metaheuristic that guides the search, we make use of Simulated Annealing (SA), a relatively old, but still very effective technique (see [13] for a modern introduction and review).

The SA algorithm starts from a random initial solution and then, at each iteration, draws a random move. This is always accepted if it is improving or sideways, whereas worsening moves are accepted based on the time-decreasing exponential distribution (known as *Metropolis Acceptance*) $e^{-\Delta F/t}$, where ΔF is the difference of the value of the objective function (to be minimized), and t is a control parameter called *temperature*.

SA starts with an initial high temperature T_0, which is decreased after a fixed number of samples $\mathcal{N}_s$ are drawn according to the geometric cooling scheme ($t_i = \alpha\, t_{i-1}$), where α (with $0 < \alpha < 1$) is the *cooling rate*. The search is stopped when the final temperature T_f is reached. In order to speed up the early stages of the search, we add the customary *cut-off* mechanism, such that the temperature decreases also if a maximum number of moves has been accepted. We introduce the parameter ρ (with $0 < \rho \leq 1$) which represents the maximum fraction of accepted moves with respect to $\mathcal{N}_s$.

Within the context of SA, which relies on the iterated drawing of random moves, the standard way of using multiple neighborhoods is to select the move in two stages: first a *biased* random selection to establish which atomic neighborhood should be sampled, and then a *uniform* selection within the chosen neighborhood. The probabilities for the first selection, called *rates*, are parameters that are subject to offline tuning, along with the parameters of SA.

3   Learning for Multi-neighborhood Search

SA, like other metaheuristics, relies on a number of parameters to be fixed a priori. In the case of SA, they are the initial temperature (T_0), the final temperature (T_f), the cooling rate (α), the number of iterations at each temperature ($\mathcal{N}_s$), and the cut-off ratio (ρ). In the multi-neighborhood setting, we consider also the rates $\sigma_1, \ldots, \sigma_K$ of the K neighborhoods, where σ_i is the probability of random selection of the $i - th$ neighborhood, with $\sum_{i=1}^{K} \sigma_i = 1$. Thus, we add $K - 1$ rate parameters, given that one of them can be computed from the others.

To select the parameters, the typical option is to use an automatic tuning tool, such as Iterated F-Race (I/F-Race) [25] or SMAC [17], which selects the parameters based on runs on a training dataset. I/F-Race samples a set of parameter configurations according to particular distributions, and iteratively tests them against an increasingly large set of instances, drawn uniformly from the training dataset. At each step, the Friedman's non-parametric two-way analysis of variance by ranks is applied to select the non-inferior configurations by means of racing [28]. Then the sampling distributions are updated to increase the probability of sampling, in the future, parameter values from the best configurations. The procedure is repeated until the computational budget, expressed in terms of total number of experiments or overall running time, is exhausted. The current version of the `irace` package [25] implements also a "soft-restart" mechanism to avoid premature convergence and an elitist variant for preserving the best configurations found so far.

The alternative option to offline tuning is the online control of the parameters during the search. Obviously, not all parameters can be modified online, as for some of them the value is used immediately at the beginning of the search. For example, the initial temperature of SA must necessarily be fixed offline.

In our proposal, the rates σ_* are adapted online based on RL, whereas the parameters of SA are tuned offline. The learning mechanism is applied after a given batch of iterations, and not upon every single move execution. In this context, the natural option is to synchronize the learning action with the annealing step, so that the new rates are computed in the very same iteration in which the temperature is decreased. This choice limits the overhead of the learning procedure upon the search method. On the contrary, given that normally the number of executed moves in SA is extremely large, learning at any move execution would result in an unacceptable overhead.

According to the classification of Karimi-Mamaghan et al. [20], our method implements a low-level integration of RL into metaheuristics, with the purpose of operator selection during the search evolution based on a credit assigned to each operator from its historical performance. We chose an exponential recency-weighted average bandit algorithm [38] to update the rates

$$\sigma_{i,t+1} = (1 - \lambda)\sigma_{it} + \lambda r_{it} \tag{1}$$

where σ_{it} and r_{it} are the rate and the reward of operator N_i at temperature level t, respectively, and $\lambda \in (0, 1]$ is the learning rate. Given that we operate in batch mode, we aggregate the performance of all the operators over an entire temperature level (Average Credit Assignment - ACA).

In order to guarantee that a neighborhood can always be selected, we set a minimum probability σ_{min}, so that it does not disappear from the search (Probability Matching Selection - PMS). We do this by correcting Eq. 1 so that $\sigma_{i,t+1}$ is equal to the maximum between the computed value and σ_{min}. In case of a correction, all other σ_* are rescaled accordingly, so that their sum remains equal to one. To define the reward r_{it} in Eq. 1, we first compute the score φ of neighborhood N_i at temperature level t as

$$\varphi_{it} = \frac{\sum_j^{n_{it}} |\min\{\Delta F_j, 0\}|}{n_{it}} \tag{2}$$

where n_{it} is the total number of drawn moves of neighborhood N_i at temperature level t and the summation term is the contribution to cost minimization. In other words, we consider as score the total improvement generated by the neighborhood upon the previous temperature divided by the number of drawn moves, so as not to bias the reward toward neighborhoods with high rates.

Given that the computational cost for the construction and the evaluation of a move might be different from neighborhood to neighborhood, the computational effort has to be taken into account in the learning phase, by applying a reward formula that penalizes the neighborhoods that take more time (see [29]). To this aim, we insert the average running time of moves of N_i at temperature level t, called τ_{it}, as denominator of the formula that computes the reward (Eq. 3). However, the penalization should be smoothed in order to give value to the moves that find improvements, even if in a computationally expensive way. We perform such smoothing by using the following exponential function

$$R_{it} = \frac{\varphi_{it}}{(\tau_{it})^m} \tag{3}$$

where m (with $0 < m \leq 1$) is a novel real-valued hyperparameter. To ensure that the rates add up to one, the rewards are normalized by dividing each reward by the sum of all rewards as follows:

$$r_{it} = \frac{R_{it}}{\sum_{j=1}^{K} R_{jt}} \tag{4}$$

This normalization guarantees that all the terms used in Eq. 1 add up to one and, consequently, also the updated selection rates will always sum up to one.

In conclusions, the hyperparameters of the learning procedure are the learning rate (λ), the rate threshold (σ_{min}), and the exponent m of Eq. 3. To these three, we add the SA cooling rate α, given that it governs not only the duration of the annealing step, but also the learning batch. The best values for these four hyperparameters are computed by the tuning procedure discussed in Sect. 6.

4   Related Work

The literature in the field of parameter control is extremely vast. We refer to the comprehensive and up-to-date work of Adriaensen *et al.* [1] for a review of the area, and to Doerr and Doerr [11] for theoretical results about its effectiveness.

The interaction between machine learning techniques and optimization methods for solving hard combinatorial optimization problems has been a hot topic for both the operations research and artificial intelligence communities in recent years [7,20,37,39]. In particular Karimi-Mamaghan *et al.* [20] reviewed the recent contributions in the integration of machine learning into metaheuristics and proposed a taxonomy to classify the different types of integration.

To the best of our knowledge, the only previous works that employ RL techniques for the neighborhood selection within SA have been presented by Mosadegh *et al.* [30] and Shahmardan and Sajadieh [35].

Mosadegh *et al.* integrated a *Q-learning* algorithm into a hyper SA framework to learn the most suitable action through the search. They identified 16 admissible actions, each one composed by a triple of move operators to be executed in sequence. The reward is computed as the difference between the value of the objective function before and after applying the three operators.

Shahmardan and Sajadieh experimented several RL-based selection mechanisms, including Q-learning, all updating the credits at each iteration of the search process. A unit reward is assigned to a move if the value of the objective function is equal to or better than the current one. In order to guarantee a good balance between exploitation and exploration, they adopted a ϵ-greedy selection policy, such that any operator is randomly selected with probability ϵ, while the operator with the maximum credit is selected with probability $(1-\epsilon)$. The methods have been evaluated on benchmarks of an assembly line sequencing problem [30] and a truck scheduling problem [35].

Other examples of integration of RL methods into other metaheuristics for the purpose of operator selection can be found in [2,12,15,33,40], with applications to Evolutionary Algorithms, Variable Neighborhood Search, Late Acceptance Hill-Climbing, and Iterated Local Search.

The use of feedback about neighborhood operators' performance for dynamically adjusting the probabilities of operators in the different stages of the search process is also referred to as Adaptive Neighborhood Search [24,26]. It has found wide application in particular in combination with the Large Neighborhood Search paradigm, resulting in the Adaptive Large Neighborhood Search metaheuristic (ALNS) [27,31], where the rates of destroying and repairing operators are dynamically adjusted using the recorded performance of the neighborhoods.

Lastly, the selection of the appropriate operator during the search process was also studied in the context of *hyper-heuristics* [8]; however, in this case the operator represents a low-level heuristic [21,29], instead of a single move. Recently, Kallestad *et al.* [19] proposed a selection hyper-heuristic framework that integrates Deep RL into ALNS.

5 Case Studies

We propose as case studies two real-world problems coming from the field of timetabling: examination timetabling and sports timetabling. Both come along with a challenging dataset that has been used extensively in the literature.

5.1 Examination Timetabling

Examination timetabling (ETT) is one of the problems that every university has to deal with on a regular basis. Many formulations of ETT have been proposed in the literature. We consider here the classical and essential version proposed by Carter *et al.* [9], which is the most studied one (see the recent survey [10]).

The input data in this formulation is just the Boolean-valued *enrollment* matrix, that stores for each pair ⟨student, exam⟩ the information about whether the student has to take the exam or not. Two exams with one or more students in common are in conflict, and so they cannot be scheduled in the same period. The objective function is based on the distance between exams with students in common. Distances are penalized in the following way: the cost of scheduling two exams with k students in common at distance equal to 1, 2, 3, 4, and 5 periods is $16k$, $8k$, $4k$, $2k$, and k, respectively.

Many authors have tackled this problem by using local search in general and multi-neighborhood search in particular. We refer here to the approach by Bellio *et al.* [5] that propose a two-stage SA procedure based on a comprehensive set of neighborhoods. They obtained state-of-the-art results on the Toronto dataset, which is the standard ground for comparison for ETT and has been used in many previous studies. In detail, they collected and implemented five different neighborhood relations from the literature. The first three neighborhoods reschedule one, two, or three exams, respectively; the fourth swaps all the exams of two periods, and the last one performs a Kempe chain. They are called MoveExam (ME), KickExam (KE), DoubleKickExam (DKE), SwapPeriods (SP), and KempeChain (KC), respectively. These neighborhoods exhibit different computational costs for constructing and evaluating the move. In particular, Bellio *et al.* showed experimentally that the computation cost of the construction and evaluation of a DKE or KC move is more than 30 times the cost of a ME move. For this reason, their tuning procedure which uses F-Race, takes into account the average computational cost of each neighborhood. They compare the alternative configurations by fixing the total running time of SA. Coherently, the temperature is decreased after a given allotted time, and not based on the given number of samples.

The first stage of Bellio *et al.* aims at obtaining the first feasible solution and it is rather fast. For this reason, we focused on the second stage. For the second stage, we use the same neighborhoods and also the same configuration for the SA parameters; by contrast, all rates σ_* are uniformly initialized to $1/K$ (i.e., 0.2 in the specific case, since $K = 5$) and are adjusted dynamically by the learning procedure.

5.2 Sports Timetabling

Sports timetabling (STT) is an active research field, due to the commercial interest in the maximization of fan attendance to sport events and to the quest for timetables that are fair toward all teams. In round-robin tournaments, the most frequent format for team sports, as well as the most studied one in optimization for its combinatorial complexity, teams play against all other teams.

We consider the formulation proposed by Van Bulck and Goossens [41] for the fifth International Timetabling Competition (ITC2021). The competition problem consists of building a compact double round-robin tournament on instances. A solution is feasible when all hard constraints are satisfied, and the objective consists in minimizing the penalties from violated soft constraints. A peculiarity of this formulation is that every single specific constraint can appear in either its hard or soft version, and can be repeated several times in the same instance, involving different sets of teams and/or rounds. As for the match sequence, in ITC2021 instances it can be completely free or *phased*, which means that the tournament is split in two legs in which every team plays against all other teams.

Several instances of this problem are remarkably challenging already from a point of view of the feasibility. To the best of our knowledge, only the matheuristic approach proposed by Lamas-Fernandez *et al.* [23], the winner of the competition, was able to find a feasible solution for all 45 instances. However, to give a measure of the grade of hardness of the instances, 100 h of computation on 4 CPU were needed to find a feasible solution for the most challenging instance (M_2). An overview of the current state-of-the-art is provided by Van Bulck and Goossens [41].

We focus here on the solver proposed by Rosati et al. [32], the runner-up of ITC2021, that was able to find feasible solutions for 44 instances out of 45. They proposed a Three-Stage Multi-Neighborhood Simulated Annealing that makes use of a portfolio of six local search neighborhoods. The three stages aim, respectively, at finding a feasible solution that does not violate hard constraints, at navigating both feasible and infeasible regions, and at finding a better local minimum by just exploring the feasible region.

Five of the six neighborhoods employed are classical ones for Sports Timetabling, namely SwapHomes (SH), SwapTeams (ST), SwapRounds (SR), Partial-SwapTeams (PST) and PartialSwapRounds (PSR) [3], plus an original contribution, named PartialSwapTeamsPhased (PSTP).

Rosati et al. tuned offline the probabilities of the neighborhoods, assigning a unique value for the three stages, but differentiated between phased and not phased instances. Like for ETT, we employ their neighborhoods and the same configuration for the SA parameters, but instead of fixed rates we implement the learning algorithm described in Sect. 3, starting every stage with equal probabilities of $1/6$ for every neighborhood (given that $K = 6$).

6   Experimental Results

Our search methods are implemented in C++ using the library EASYLO-CAL++ (bitbucket.org/satt/easylocal-3), building upon the code of Bellio et al. and Rosati et al., available at opthub.uniud.it and github.com/robertomrosati/sa4stt, respectively. The code was compiled on Ubuntu Linux 20.4 using g++ (v. 11.3) and the experiments were run on AMD Ryzen Threadripper PRO 3975WX 32-Cores (3.50 GHz), with one single core dedicated to each experiment.

The tuning procedure was performed using I/F-race separately for the two problems on the learning hyperparameters λ, σ_{min}, m, and α. The total budget assigned for each problem was 5000 experiments, with an average running time for each experiment of 300 s and 2000s for ETT and STT, respectively. Table 1 shows the initial ranges of the hyperparameters and the two best configurations.

For both problems, it turned out that there is statistical significance difference between the "winning" configuration and all the configurations with $\lambda = 0.0$, which are consistently eliminated by the race procedure, confirming that the learning mechanism is necessary. On the contrary, there are many other configurations with $\lambda > 0.0$ that are statistically equivalent, demonstrating that the results are robust with respect of the hyperparameter configuration.

It is worth noting that the selected value for σ_{min} is 0.0 for ETT. This means that it is actually possible that a given neighborhood becomes useless from some point onward in the search.

Table 1. Learning hyperparameter configurations.

Hyperparameter	Initial range	Best configuration	
		ETT	STT
λ	[0.0, 0.5]	0.038	0.093
σ_{min}	[0.0, 0.01]	0.000	0.001
m	[0.16, 1.0]	0.197	0.259
α	[0.98, 0.995]	0.987	0.983

6.1 Results for Examination Timetabling

The results for 30 repetitions on the Toronto dataset using the configuration of Table 1 are shown in Table 2. They are granted the same running time[1] of Bellio et al. [5], which are the best known ones among those with relatively short running time. The gaps reported in Table 2 are computed as the percentage difference with respect to Bellio et al..

We consider for our learning procedure two alternative settings: with uniform initial rates discussed so far (0.2 for all) and with the initial rates reported by Bellio et al. (namely 0.757, 0.144, 0.001, 0.058, and 0.04).

Table 2. Comparison results for ETT.

	time	Bellio et al. (2021)	reinforcement learning			
			uniform initial rates		tuned initial rates	
Inst	sec	avg	avg	% gap	avg	% gap
car91	688.1	**4.44**	**4.44**	0.000	**4.44**	0.000
car92	544.6	3.80	3.80	0.000	**3.79**	−0.263
ear83	249.6	32.89	**32.85**	−0.122	32.91	0.061
hec92	228.5	10.16	10.17	0.098	**10.15**	−0.098
kfu93	217.5	13.06	**13.03**	−0.230	13.06	0.000
lse91	209.3	10.09	10.03	−0.595	**10.02**	−0.694
pur93	1382.0	4.32	**4.29**	−0.694	4.30	−0.463
rye93	281.7	8.10	**8.07**	−0.370	**8.07**	−0.370
sta83	136.6	157.05	**157.04**	−0.006	157.05	0.000
tre92	296.6	7.85	**7.82**	−0.382	7.83	−0.255
uta92	575.8	3.13	**3.11**	−0.639	3.12	−0.319
ute92	130.8	**24.82**	24.83	0.040	24.83	0.040
yor83	429.6	34.93	**34.91**	−0.057	34.96	0.086
avg	413.1	24.20	**24.18**	−0.227	24.19	−0.175

---

[1] The running time was scaled based on the ratio of performance on the different machines.

The outcome is that both configurations perform better than the offline tuning with no learning. Even though the improvement is quite small in absolute terms, it is actually statistically significant. This is confirmed by the Wilcoxon signed-rank test that returns a p-value close to 0 (equal to $7.192 \cdot 10^{-6}$) upon 15 repetitions on each instance, thus strongly rejecting the null hypothesis.

Between the two learning configurations, there is a minimal advantage to the uniform one, which finds the best average results (in boldface) in 9 out of 13 instances. However, using the same test there is no statistical significant difference between these two. Nonetheless, we believe that the reason for the slightly better performance of the uniform initial rates come from the fact that in the beginning of the search all types of moves are useful to "shake" the initial solution. Going on, the most effective moves become prominent. This is confirmed by Fig. 1 that shows the evolution of the rates for one specific run of each configuration (uniform left and tuned right, for instance ear83). We see that indeed the rates tend to stay uniform or become uniform in the initial part of the search, and then diversify in the final part.

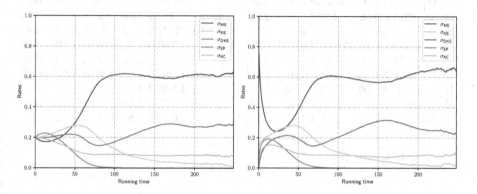

Fig. 1. Evolution of σ_* for ETT (inst. ear83) for uniform and tuned initial rates.

6.2 Results for Sports Timetabling

The results for 15 repetitions on the ITC2021 dataset using the configuration of Table 1 are shown in Table 3, with the same (normalized) running time of Rosati et al. [32]. The column "feas" reports the rate of feasible solutions found.

Our approach yields better results (in boldface) for 27 out of 45 instance, and equal value for 5. The average percentage improvement is 0.54 with a high variance, having differences that go from −16.53 to 45.32. The improvement is statistically supported by the Wilcoxon signed-rank test with a p-value equal to 0.03, even if less significantly than for ETT. It is worth mentioning that the results of Rosati et al. are outperformed only by the matheuristic approach by Lamaz-Fernandez et al. [23], who however grant much longer running times.

The evolution of the rates for two instances of STT is shown in Fig. 2. We notice that the behavior is rather different in the two cases, demonstrating that there is no single configuration that fits all cases.

Table 3. Comparison results for STT.

Inst.	Rosati et al. avg feas	learning avg feas	% gap	Inst.	Rosati et al. avg feas	learning avg feas	% gap
E_1	540.7 1.00	**530.8** 1.00	-1.83	M_9	772.1 1.00	**722.8** 1.00	-6.39
E_2	384.6 1.00	**375.6** 1.00	-2.34	M_10	1687.5 1.00	**1648.9** 1.00	-2.29
E_3	1176.5 1.00	**1175.1** 1.00	-0.12	M_11	2996.5 1.00	**2987.3** 1.00	-0.31
E_4	1007.8 0.56	**943.8** 0.56	-6.35	M_12	**1054.2** 1.00	1061.5 1.00	0.69
E_5	- 0.00	- 0.00		M_13	**479.3** 1.00	486.9 1.00	1.59
E_6	4543 1.00	**4411** 1.00	-2.91	M_14	1304.6 1.00	**1267.6** 1.00	-2.84
E_7	6721.7 1.00	**6263.3** 1.00	-6.82	M_15	1099.7 1.00	**1095.3** 1.00	-0.40
E_8	1151.9 1.00	**1147.5** 1.00	-0.38	L_1	2372.7 1.00	**2327.9** 1.00	-1.89
E_9	228.7 1.00	**190.9** 1.00	-16.53	L_2	6085.5 0.49	**6040** 0.38	-0.75
E_10	- 0.00	- 0.00		L_3	**2718.0** 1.00	2731.3 1.00	0.49
E_11	5784.5 1.00	**5521.8** 1.00	-4.54	L_4	**0.0** 1.00	**0.0** 1.00	0.00
E_12	1200.2 1.00	**1194.4** 1.00	-0.48	L_5	- 0.00	- 0.00	
E_13	233.8 1.00	**232.2** 1.00	-0.68	L_6	**1121.3** 1.00	1138.6 1.00	1.54
E_14	**82.3** 1.00	119.6 1.00	45.32	L_7	**2226.5** 1.00	2252.1 1.00	1.15
E_15	3945.8 1.00	**3933.9** 1.00	-0.30	L_8	**1155.3** 1.00	1158.5 1.00	0.28
M_1	6075.0 0.06	**5936** 0.06	-2.29	L_9	881.2 1.00	**851.8** 1.00	-3.34
M_2	- 0.00	- 0.00		L_10	3527.3 0.05	**3407.0** 0.12	-3.41
M_3	11403.1 0.23	**11379.7** 0.20	-0.21	L_11	**289.3** 1.00	291.4 1.00	0.73
M_4	33.0 1.00	**32.1** 1.00	-2.73	L_12	4830.6 1.00	**4384.7** 1.00	-9.23
M_5	**624.4** 1.00	631.5 1.00	1.14	L_13	**2285.5** 1.00	2299.6 1.00	0.62
M_6	**2186.3** 1.00	2316.9 1.00	5.97	L_14	**1326.3** 1.00	1329.2 1.00	0.22
M_7	**2452.7** 1.00	2474.2 1.00	0.88	L_15	82.8 1.00	**82.4** 1.00	-0.48
M_8	196.6 1.00	**191.1** 1.00	-2.80	avg	2152.9 0.83	**2111.4** 0.83	-0.54

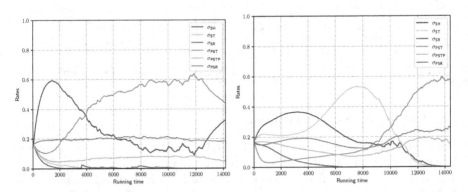

Fig. 2. Evolution of σ_* for STT (instances M_10 and M_12, stage II).

6.3   Discussion

As mentioned earlier, a wide range of hyperparameter configurations can yield statistically similar outcomes. This suggests that the computational effort associated with fine-tuning the learning hyperparameters can be significantly reduced, which is typically not the case when optimizing search parameters offline.

Further validation experiments were conducted using different configurations, which confirmed the findings. For instance, we obtained the same average gap by using the optimal hyperparameter settings of STT for ETT. This indicates that the process of fine-tuning hyperparameters can be performed in a cross-domain manner, leading to significant computational savings.

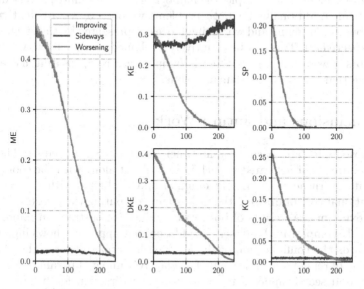

(a) Neighborhood behavior for ETT (instance kfu93)

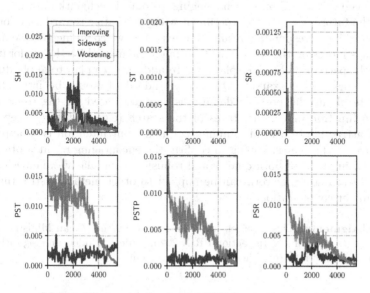

(b) Neighborhood behavior for STT (instance Late_10)

Fig. 3. Examples of neighborhood behavior for the two case studies.

Figures 3a and 3b give some insights on the behavior of the different neighborhoods over time for one instance of ETT and STT, respectively. Specifically, they show the proportion of executed moves that are improving, sideways, and worsening for each neighborhood, relative to the total number of evaluated moves.

The analysis reveals significant disparities in the move quality among the various neighborhoods. For example, the number of sideways moves is very different among them. Additionally, for both problems we observe a consistent trend in the number of improving and worsening moves (represented by the green and red lines, respectively). This can be intuitively explained by the fact that whenever SA accepts a worsening move, there is frequently a compensating move of the same type that recovers the "damage" done.

7   Conclusions and Future Work

We proposed a RL approach for a multi-neighborhood SA procedure. The learning procedure is not activated at each iteration, but along with the cooling step, so that its overhead is absolutely minimal and negligible with respect to the computational cost of drawing the moves and computing their costs.

The experimental results of both case studies demonstrate an improvement over already competitive baselines. In addition, our approach shows high robustness towards the configuration of the learning hyperparameters. Therefore, the application of reinforcement learning for adaptive tuning of parameters in multi-neighborhood search appears to be a promising direction for future research. Notably, the results on ETT improve over a very large body of research that has already been conducted on the challenging Toronto benchmark dataset.

For the future, we plan to experiment with alternative reward functions and to compare the corresponding results. For example, we could include the number of sideways moves on the reward, as they might be helpful for the exploration of the search space in presence of plateaux. In addition, we plan to apply different learning schemes, for example Q-learning. Indeed, it would be interesting to investigate about the possible definitions of state based on the trend of the objective function in various stages of the search (e.g., plateaux, steep slope, basin of attraction of a local minimum). Finally, our intention is to apply this scheme to new problems, hoping to confirm the encouraging results obtained in this study, in terms of both quality and robustness. Similarly, we envision that the same learning framework can be applied to other metaheuristics that rely on parameter configuration.

Acknowledgments. This research has been funded by the Italian Ministry of University and Research under the action PRIN 2020, project "Models and algorithms for the optimization of integrated healthcare management".

References

1. Adriaensen, S., et al.: Automated dynamic algorithm configuration. J. Artif. Intell. Res. **75**, 1633–1699 (2022)
2. Alicastro, M., Ferone, D., Festa, P., Fugaro, S., Pastore, T.: A reinforcement learning iterated local search for makespan minimization in additive manufacturing machine scheduling problems. Comput. Oper. Res. **131**, 105272 (2021)
3. Anagnostopoulos, A., Michel, L., Van Hentenryck, P., Vergados, Y.: A simulated annealing approach to the traveling tournament problem. J. Sched. **9**(2), 177–193 (2006)
4. Bartz-Beielstein, T., et al.: Benchmarking in optimization: best practice and open issues. CoRR, arXiv abs/2007.03488 (2020)
5. Bellio, R., Ceschia, S., Di Gaspero, L., Schaerf, A.: Two-stage multi-neighborhood simulated annealing for uncapacitated examination timetabling. Comput. Oper. Res. **132**, 105300 (2021)
6. Bellio, R., Ceschia, S., Di Gaspero, L., Schaerf, A., Urli, T.: Feature-based tuning of simulated annealing applied to the curriculum-based course timetabling problem. Comput. Oper. Res. **65**, 83–92 (2016)
7. Bengio, Y., Lodi, A., Prouvost, A.: Machine learning for combinatorial optimization: a methodological tour d'horizon. Eur. J. Oper. Res. **290**(2), 405–421 (2021)
8. Burke, E.K., Hyde, M.R., Kendall, G., Ochoa, G., Özcan, E., Woodward, J.: A classification of hyper-heuristic approaches: revisited. In: Gendreau, M., Potvin, J.Y. (eds.) Handbook of Metaheuristics. International Series in Operations Research & Management Science, vol. 272, pp. 453–477. Springer, Cham (2019). https://doi.org/10.1007/978-3-319-91086-4_14
9. Carter, M., Laporte, G., Lee, S.: Examination timetabling: algorithmic strategies and applications. J. Oper. Res. Soc. **74**, 373–383 (1996)
10. Ceschia, S., Di Gaspero, L., Schaerf, A.: Educational timetabling: problems, benchmarks, and state-of-the-art results. Eur. J. Oper. Res. **308**(1), 1–18 (2023)
11. Doerr, B., Doerr, C.: Theory of parameter control for discrete black-box optimization: provable performance gains through dynamic parameter choices. In: Doerr, B., Neumann, F. (eds.) Theory of Evolutionary Computation. NCS, pp. 271–321. Springer, Cham (2020). https://doi.org/10.1007/978-3-030-29414-4_6
12. Fialho, A., Da Costa, L., Schoenauer, M., Sebag, M.: Analyzing bandit-based adaptive operator selection mechanisms. Ann. Math. Artif. Intell. **60**(1–2), 25–64 (2010)
13. Franzin, A., Stützle, T.: Revisiting simulated annealing: a component-based analysis. Comput. Oper. Res. **104**, 191–206 (2019)
14. Glover, F., Kochenberger, G.: Handbook of Metaheuristics, vol. 57. Springer, Berlin, Heidelberg (2006). ISBN 978-1402072635
15. Gunawan, A., Lau, H.C., Lu, K.: Adopt: combining parameter tuning and adaptive operator ordering for solving a class of orienteering problems. Comput. Ind. Eng. **121**, 82–96 (2018)
16. Huang, C., Li, Y., Yao, X.: A survey of automatic parameter tuning methods for metaheuristics. IEEE Trans. Evolut. Comput. **24**(2), 201–216 (2020)
17. Hutter, F., Hoos, H., Leyton-Brown, K.: Sequential model-based optimization for general algorithm configuration. In: Coello, C.A.C. (eds.) Learning and Intelligent Optimization. LION 2011. LNCS, vol. 6683, pp. 507–523. Springer, Berlin, Heidelberg (2011). https://doi.org/10.1007/978-3-642-25566-3_40
18. Kadioglu, S., Malitsky, Y., Sellmann, M., Tierney, K.: ISAC-instance-specific algorithm configuration. In: ECAI 2010, pp. 751–756. IOS Press (2010)

19. Kallestad, J., Hasibi, R., Hemmati, A., Sörensen, K.: A general deep reinforcement learning hyperheuristic framework for solving combinatorial optimization problems. Eur. J. Oper. Res. **309**(1), 446–468 (2023)

20. Karimi-Mamaghan, M., Mohammadi, M., Meyer, P., Karimi-Mamaghan, A., Talbi, E.G.: Machine learning at the service of meta-heuristics for solving combinatorial optimization problems: a state-of-the-art. Eur. J. Oper. Res. **296**(2), 393–422 (2022)

21. Kheiri, A., Gretsista, A., Keedwell, E., Lulli, G., Epitropakis, M., Burke, E.: A hyper-heuristic approach based upon a hidden Markov model for the multi-stage nurse rostering problem. Comput. Oper. Res. **130**, 105221 (2021)

22. KhudaBukhsh, A., Xu, L., Hoos, H., Leyton-Brown, K.: SATenstein: automatically building local search SAT solvers from components. Artif. Intell. **232**, 20–42 (2016)

23. Lamaz-Fernandez, C., Martinez-Sykora, A., Potts, C.: Scheduling double round-robin sports tournaments. In: Proceedings of the 13th International Conference on the Practice and Theory of Automated Timetabling, vol. 2 (2021)

24. Li, J., Pardalos, P.M., Sun, H., Pei, J., Zhang, Y.: Iterated local search embedded adaptive neighborhood selection approach for the multi-depot vehicle routing problem with simultaneous deliveries and pickups. Expert Syst. Appl. **42**(7), 3551–3561 (2015)

25. López-Ibáñez, M., Dubois-Lacoste, J., Cáceres, L.P., Birattari, M., Stützle, T.: The irace package: iterated racing for automatic algorithm configuration. Oper. Res. Perspect. **3**, 43–58 (2016)

26. Lü, Z., Hao, J.K.: Adaptive neighborhood search for nurse rostering. Eur. J. Oper. Res. **218**(3), 865–876 (2012)

27. Mara, S., Norcahyo, R., Jodiawan, P., Lusiantoro, L., Rifai, A.P.: A survey of adaptive large neighborhood search algorithms and applications. Comput. Oper. Res. **146**(105903) (2022)

28. Maron, O., Moore, A.: The racing algorithm: model selection for lazy learners. Artif. Intell. Rev. **11**(1), 193–225 (1997)

29. Mischek, F., Musliu, N.: Reinforcement learning for cross-domain hyper-heuristics. In: Proceedings of the Thirty-First International Joint Conference on Artificial Intelligence IJCAI-22, pp. 4793–4799 (2022)

30. Mosadegh, H., Ghomi, S., Süer, G.: Stochastic mixed-model assembly line sequencing problem: mathematical modeling and Q-learning based simulated annealing hyper-heuristics. Eur. J. Oper. Res. **282**(2), 530–544 (2020)

31. Pisinger, D., Ropke, S.: Large neighborhood search. In: Gendreau, M., Potvin, J.Y. (eds.) Handbook of Metaheuristics. International Series in Operations Research & Management Science, vol. 272, pp. 99–127. Springer, Cham (2019). https://doi.org/10.1007/978-3-319-91086-4_4

32. Rosati, R.M., Petris, M., Di Gaspero, L., Schaerf, A.: Multi-neighborhood simulated annealing for the sports timetabling competition ITC2021. J. Sched. **25**(3), 301–319 (2022)

33. dos Santos, J., de Melo, J., Neto, A., Aloise, D.: Reactive search strategies using reinforcement learning, local search algorithms and variable neighborhood search. Expert Syst. Appl. **41**(10), 4939–4949 (2014)

34. Schneider, M., Hoos, H.: Quantifying homogeneity of instance sets for algorithm configuration. In: Hamadi, Y., Schoenauer, M. (eds.) Learning and Intelligent Optimization. LION 2012. LNCS, vol. 7219, pp. 190–204. Springer, Berlin, Heidelberg (2012). https://doi.org/10.1007/978-3-642-34413-8_14

35. Shahmardan, A., Sajadieh, M.: Truck scheduling in a multi-door cross-docking center with partial unloading-reinforcement learning-based simulated annealing approaches. Comput. Ind. Eng. **139**, 106134 (2020)
36. Smith-Miles, K., Baatar, D., Wreford, B., Lewis, R.: Towards objective measures of algorithm performance across instance space. Comput. Oper. Res. **45**, 12–24 (2014)
37. Song, H., Triguero, I., Özcan, E.: A review on the self and dual interactions between machine learning and optimisation. Progress Artif. Intell. **8**(2), 143–165 (2019)
38. Sutton, R., Barto, A.: Reinforcement Learning: An Introduction. MIT Press, Cambridge (2018)
39. Talbi, E.G.: Machine learning into metaheuristics: a survey and taxonomy. ACM Comput. Surv. (CSUR) **54**(6), 1–32 (2021)
40. Toffolo, T.A., Christiaens, J., Van Malderen, S., Wauters, T., Vanden Berghe, G.: Stochastic local search with learning automaton for the swap-body vehicle routing problem. Comput. Oper. Res. **89**, 68–81 (2018)
41. Van Bulck, D., Goossens, D.: The international timetabling competition on sports timetabling (ITC2021). Eur. J. Oper. Res. **308**(3), 1249–1267 (2023)

Evaluation of Selected Autoencoders in the Context of End-User Experience Management

Sven Beckmann[✉] and Bernhard Bauer

Institute of Computer Science, University of Augsburg, Universitätsstraße 2, 86159 Augsburg, Germany
{sven-ralf.beckmann,bernhard.bauer}@uni-a.de

Abstract. Empirical research shows that a significant portion of employees regularly faces IT-related challenges in their workplace, resulting in lost productivity, customer dissatisfaction, and increased employee turnover [1]. Although the significant impact of these problems, keeping the IT administration informed about ongoing issues is a major challenge. End-User Experience Management (EUEM) aims to help IT administrators address this problem. For example, in the context of EUEM, telemetry data collected from employees' devices can help IT administrators to identify potential issues [2]. Machine learning algorithms can automatically detect anomalies in the collected telemetry data, providing IT administration with essential insights to optimize the end-user experience [2]. This paper examines the advantages and disadvantages of three different autoencoder-based algorithms identified in the literature as well-suited for detecting anomalies applied in this paper to hardware telemetry: Autoencoder (AE), Variational Autoencoder (VAE), and Deep Autoencoding Gaussian Mixture Model (DAEGMM). The results show that all three models provide anomaly detection in hardware telemetry data, though with significant differences. While the AE is the fastest Algorithm, the VAE offers the most stable results. The DAEGMM provides the best separation of endpoints into outliers and normal data points but has the most extended runtime. For all models, data aggregation has a significant potential for data reduction by aggregating the measurements over a longer time interval.

Keywords: Machine Learning · Neural Networks · Deep Learning · Autoencoder · Outlier Detection · Endpoint Monitoring · End-User Experience Management

1 Introduction

1.1 Motivation

Empirical data demonstrates that 40% of employees experience at least one IT problem weekly when using the IT tools and infrastructure their employer provides. Following an unanticipated disruption, it requires approximately 25 min for an employee to

G. Nicosia et al. (Eds.): LOD 2023, LNCS 14506, pp. 222–236, 2024.
https://doi.org/10.1007/978-3-031-53966-4_17

regain complete concentration on their tasks. In addition, 54% of IT managers admit that IT disruptions have often or sometimes led to unpleasant situations with customers and business partners. Indeed, 28% of surveyed employees aged 25 to 34 contemplate resignation due to bad IT user experiences. [1]

Given these findings, one might presume that employees encountering IT problems would promptly seek assistance from IT administration to resolve their issues as swiftly as possible. A survey reveals that merely 15% would seek a solution from IT support. [1]

Thus, a problem arises: on the one hand, there is an enormous need for the employer to provide a good user experience, and on the other hand, the employer does not even know about the actual problems that reduce the productivity of the employees. Hence End-User Experience Management (EUEM) is an increasingly important part of modern IT infrastructure management. A satisfying user experience of the employees is essential to ensure they can work efficiently while experiencing a positive work environment. Organizations focusing on improving user experience can increase employee satisfaction, productivity, and loyalty. A positive user experience is also crucial in employee recruitment and retention. In an increasingly competitive workplace, companies are forced to differentiate themselves from other employers.

Another important role of EUEM is to ensure IT security and compliance. Organizations can prevent data breaches or other security issues by detecting malicious behaviors or unusual user experiences. In addition, monitoring the user experience enables ensuring compliance with required security policies.

Moreover, EUEM facilitates the reduction of IT costs by eliminating unnecessary spending. By monitoring IT resource utilization, companies can determine which resources are being used and which are not. In this way, companies can eliminate outdated or unused applications or devices.

Overall, EUEM is an essential tool that organizations have to consider when shaping their IT strategy. Concentrating on user experience empowers organizations to increase employees' productivity and satisfaction, optimize IT systems and applications, curtail IT costs, and contribute to improved IT security and compliance. Organizations should integrate EUEM into their IT strategy to ensure that their employees can work effectively and that IT systems are up to par.

1.2 Problem Statement

A significant challenge for IT administration is the increasing complexity of IT infrastructures. It is crucial to clarify that the term IT infrastructure in this paper refers specifically to the user devices, also called endpoints, rather than the broader scope of servers, network architectures, and related technologies often associated with the phrase. Accordingly, increasing complexity refers to the growing number and heterogeneity of endpoints associated with different network topologies, configurations and workloads. The high level of complexity makes it more challenging to maintain an overview and identify potential disruptions or errors. In addition, the trend toward mobile working leads to a growing number of devices being used remotely in an unmanaged, insecure environment. These evolutions lead to increasing difficulty for IT administrators to ensure a secure, efficient and productive infrastructure for their employees. Therefore, IT infrastructure monitoring and surveillance are becoming increasingly important to quickly identify and rectify potential problems before they lead to major disruptions.

Monitoring all endpoints generates a considerable amount of data. This data includes, for example, comprehensive hardware, software and network telemetry information for each endpoint. The IT administration is subsequently faced with the task of analyzing this vast amount of information and gaining valuable insights. These insights can help to determine the necessary actions in order to solve existing and prevent potential problems and finally improve the overall user experience.

One way to filter out essential information is to identify anomalies in the monitored data. Beckmann et al. [2] introduced an approach how to detect anomalies in hardware telemetry data by using autoencoders. These insights can be used to conclude the end-user experience of employees in the future. This paper aimed to validate whether it is possible to automatically detect anomalies that can be presented to IT administrators for problem assessment and resolution. The objective was to reduce the volume of data to the most relevant information using machine learning techniques. The resulting anomalies could be presented to IT administrators for assessment and possible solutions.

1.3 Autoencoder-Based Approaches

Having a look at the research reveals there are already several application scenarios with comparable problems in which autoencoder-based approaches were successful, for example:

Meidan et al. developed a network-based approach for IoT environments that can detect IoT bot attacks using autoencoders. For detection, the algorithms were trained with statistical features of normal behavior per device. If the autoencoder fails to reconstruct a snapshot correctly, the behavior is classified as conspicuous, indicating a possible attack. Using autoencoders, Meidan et al. successfully reduced the rate of false alarms compared to other methods. [3]

Borghesi et al. described a way to detect anomalous system states in high-performance computers using autoencoders. For this purpose, a set of autoencoders was trained on the normal behavior of the computer. With the trained models, the authors were able to detect anomalies with high accuracy. [4]

Mirsky et al. have created a network intrusion detection system to detect attacks on the local network. The core algorithm of the developed system is an ensemble of autoencoders, which were used to identify anomalies in the network. [5]

The research of Beckmann et al. in [2] indicates the feasibility of anomaly detection on hardware telemetry data using an autoencoder. Furthermore, as evidenced by the literature, autoencoders have also demonstrated their effectiveness in anomaly detection in similar application domains. Notably, different autoencoder-based approaches have also demonstrated their suitability for different application scenarios. Therefore, in this paper, we continue to focus on autoencoder-based approaches and highlight the differences between various concepts.

1.4 Improving End-User-Experience Using Autoencoder

Having demonstrated the practicability of the concepts from Beckmann et al. in [2], which involves using an autoencoder to detect anomalies to improve the End-User Experience, the next issue at hand is to determine the optimal implementation method for our application. Selecting the optimal algorithm is critical for achieving exceptional and consistent results. A well-suited algorithm should provide optimal results and meet other requirements: Ideally, the algorithm should be ready for production after a short training period. A brief training lets administrators get helpful information about their infrastructure quickly. In addition, it is a great advantage if the algorithms can process data at a higher level of abstraction to significantly reduce the amount of collected and stored data required for learning purposes. Another valuable feature is reliable training. If the training results are too variable, the training must be repeated more frequently, which significantly increases the required computing capacity and time. Runtime is another criterion that should be kept as low as possible. Finally, the most crucial criterion is the correct identification of conspicuous endpoints.

Section 1.2 shows that autoencoders have been used successfully for anomaly detection in similar application domains. Furthermore, autoencoders are able to successfully detect anomalies in hardware telemetry data [2]. Therefore, we continue to focus on autoencoders in this work. In addition, as the dimensionality of input data escalates, two-stage methodologies are prevalently embraced [6]. This entails employing dimensionality reduction where the autoencoder can be used, followed by the execution of density estimation [6]. In this paper, we investigate different complex autoencoder algorithms to determine their advantages and disadvantages and finally find the most suitable approach for anomaly detection on hardware telemetry data. This paper aims not to develop a new method but to explore its application in the novel domain of EUEM.

To gain insight into whether additional complexity helps to detect anomalies in hardware telemetry data, three state-of-the-art autoencoders with different complex architectures were chosen: (1) Autoencoder (AE), (2) Variational Autoencoder (VAE) and (3) Deep Autoencoding Gaussian Mixture Model (DAEGMM). The AE and VAE are single-stage models, whereas the DAEGMM is a two-stage approach. To determine the optimal solution for our purpose, we want to compare the selected algorithms on the following criteria: (1) Effects of different aggregation intervals, (2) Required amount of training data, (3) Stability and robustness, (4) Runtime and (5) Performance.

This paper is organized as follows: Sect. 1 motivates the necessity for End-User Experience Management, emphasizing the challenges arising from managing and analyzing enormous volumes of data. Subsequently, the proposed solution is outlined, and the rationale for comparing selected autoencoder-based machine learning outlier detection models is presented. Section 2 introduces the three algorithms chosen for comparison, and the mathematical concepts employed in the models are explained. Section 3 initially presents the testing environment, followed by a detailed analysis of the algorithms, with individual sections addressing the specific aspects selected for comparison. Finally, Sect. 4 provides a comprehensive overview of the comparison and some outlook for further research.

2 Basics

The following sections briefly introduce the concepts of the three selected models (AE, VAE and DAEGMM). In addition, the process during training and subsequent outlier detection is explained. The presented mathematics should illustrate the algorithms and the complexity of the selected models. For more detailed information on the used mathematical concepts, please refer to the corresponding references, but are out of scope for this paper.

2.1 Autoencoder (AE)

An autoencoder (AE) is an artificial neural network that learns to represent input data in a compressed representation and subsequently reconstructs the input data from this representation. The AE consists of two main components: the encoder and the decoder. The encoder maps the input data into a low-dimensional space, also known as the latent space. The decoder transforms the compressed representation back into the original space. The difference between the original input and the reconstructed output is called the reconstruction error. The AE is trained to minimize this reconstruction error on the training data. The idea for anomaly detection is to let the AE train on normal data without anomalies to learn the structure and limits of normality. When provided with new data points after training, the AE should be able to reconstruct the input data accurately, at least if this data is similar to the trained normal state. [5]

In outlier detection, it is assumed that unusual data points exhibit a higher reconstruction error since the model was trained exclusively on normal data. Therefore, the magnitude of the reconstruction error allows for inference on how conspicuous an evaluated data point is. [9]

This paper employed the Root Mean Square Error (RMSE) between the input and the output vector to calculate the reconstruction error. Equation (1) demonstrates the computation of the RMSE for the AE with N samples based on [5]. The input of the AE is described as x while the reconstructed output is defined as $\hat{x}$.

$$loss_{\text{AE}} = \sqrt{\frac{1}{N} \sum_{i=1}^{N} (x_i - \hat{x}_i)^2} \tag{1}$$

2.2 Variational Autoencoder (VAE)

The autoencoder described in the previous step generates an unstructured representation of the data in the latent space. However, many input data points often exhibit similarities. This information is not taken into account in the representation within the latent space so far. The Variational Autoencoder (VAE) is an extension of the AE described above that considers this property. The VAE adds a probabilistic layer to the autoencoder to model the data distribution in the latent representation. Analogous to the AE, the VAE also consists of an encoder and a decoder. However, the encoder of the VAE does not learn to represent the input data as a single point in the latent space. Instead, it generates a distribution over the latent space. This distribution is given as a mean and covariance matrix. VAEs learn a continuous, stochastic representation of the data within the latent space. Once the encoder has encoded the distribution, a random point is selected from this distribution, which can then be passed to the decoder. The calculation of the reconstruction error in the VAE consists of a reconstruction term and a regularization term. The reconstruction term is the familiar term from the AE and ensures that the error between input and output is minimized. The regularization term is responsible for organizing the generated distributions in the latent space. The regularization term is implemented through the utilization of the Kullback-Leibler divergence. Conceptually, the Kullback-Leibler divergence may be understood as "a measure of the "distance" or "divergence" between statistical populations" [8]. Using the Kullback-Leibler divergence in the VAE ensures that the distributions created by the encoder are as close as possible to a standard normal distribution. [7]

Equation (2) describes the calculation of the loss term of the VAE for N samples based on [7]. The first term calculates the reconstruction error, while the second part calculates the regularization term. The regularization term KL describes the calculation of the Kullback-Leibler divergence as mentioned above. The distribution encoded in the latent space is described as $N(\mu_x, \sigma_x)$, where μ_x is the expectation and σ_x the variance of input data x. For details see [7].

$$loss_{\text{VAE}} = \sqrt{\frac{1}{N} \sum_{i=1}^{N} (x_i - \hat{x}_i)^2} + \text{KL}[N(\mu_x, \sigma_x), N(0, 1)] \qquad (2)$$

2.3 Deep Autoencoding Guassian Mixture Model (DAEGMM)

The Deep Autoencoding Gaussian Mixture Model (DAEGMM) comprises two neural networks: compression and estimation network. The compression network implements the concept of the AE, while the estimation network constructs a Gaussian Mixture Model (GMM). Like the previous models, the compression network consists of an encoder and a decoder. The encoder performs dimensionality reduction on the input data, after which the decoder reconstructs the data. Subsequently, the Relative Euclidean Distance and the cosine similarity between input and output are calculated. These two results are combined with the latent space representation of the data. This combined vector, denoted as z, serves as the input for the estimation network. The estimation network then performs density estimation within the framework of the GMM. During training, the loss function of the DAEGMM is responsible for minimizing the reconstruction error, the energy term, and the singularity term. Minimizing the reconstruction term ensures a low error in the data compression and reconstruction of the autoencoder, analogous to the previous models. The minimization of the energy term ensures the optimization of the probability distribution of the estimation network. A third term ensures that the diagonal entries of the covariance matrices do not take small values, thereby avoiding the singularity problem. [6]

Equation (3) describes the calculation of the value minimized during the training of DAEGMM based on [6]. The calculation is performed for N samples. The first part of the formula calculates the reconstruction error of the compression network. Unlike the previous calculations, the DAEGMM calculates the loss using the L2 norm, which, analogous to the previously used RMSE, calculates the error between the input and output of the reconstruction network. Another goal of DAEGMM is to find the optimal combination of compression and estimation networks that maximizes the probability of observing input samples [6]. This is achieved by minimizing the second term: the energy E. The influence of E can be weighted by the variable λ_1. The third term, which can be weighted via λ_2, ensures that small values on the diagonal of the covariance matrix $\hat{\Sigma}$ are penalized. Thus, the singularity problem is avoided. In the third term, K is defined for the components in the GMM while d describes the number of dimensions of the compressed representation in the latent space. For details see [6].

$$
loss_{\text{DAEGMM}} = \frac{1}{N} \sum_{i=1}^{N} \|x_i - \hat{x}_i\|_2^2 + \frac{\lambda_1}{N} \sum_{i=1}^{N} E(z_i)
$$
$$
+ \lambda_2 \sum_{k=1}^{K} \sum_{j=1}^{d} \frac{1}{\hat{\Sigma}_{kjj}}
\tag{3}
$$

Conceptually, the energy describes the probability of observing the input data [6]. Due to this, once the DAEGMM model is trained, outliers can be identified based on high energy for the input data [6]. Equation (4) describes, based on [6], the calculation of the energy term for the DAEGMM. For each mixture-component, the parameters of the mixture-components are Φ for the mixture-component distribution, μ_k for mixture means and $\hat{\Sigma}$ for mixture covariance. For details see [6].

$$
E(z) = -\log \left(\sum_{k=1}^{K} \Phi_k \frac{\exp\left(-\frac{1}{2}(z - \mu_k)^T \hat{\Sigma}_k^{-1}(z - \mu_k)\right)}{\sqrt{|2\pi \hat{\Sigma}_k|}} \right)
\tag{4}
$$

3   Comparison

As shown in the last section, the selected models are of varying complexity. The following section compares these models to show the advantages and disadvantages for detecting anomalies in hardware telemetry data. For this purpose, the test environment in which the models were used is explained first. Afterward, the results of each comparison are presented, and the differences are discussed.

3.1   Environment

The data for the following analyses were collected on a real-world infrastructure of a medium-sized software manufacturer. An agent was used for the data collection, which collected the telemetry data on the endpoint itself. Collecting data at the endpoint itself gives insight into the actual performance and therefore, the actual user experience.

In total, data were collected from 137 endpoints across all departments. Accordingly, the workload of the endpoints is highly diverse. When selecting the data, we took care to choose measurements whose outliers could be indicators of potential user problems. For example, the network traffic rate is not suitable for analyzing network problems using outlier analysis. The reason is, that as soon as a user starts a download, the transfer rate increases to the maximum, while during normal operation it is usually very low. This leads to the fact that the user's normal behavior, for example, a download, leads to anomalies found in the outlier analysis. This problem can be avoided by selecting the measured values carefully so that actual problem indicators are chosen. In this example, the number of lost packets is a more appropriate measure.

The collection interval of the telemetry data on the endpoints is one minute. Subsequent data pre-processing and aggregation are handled at a central location. This offers the advantage of testing different data pre-processing options on the same data set. In production applications, it is conceivable to perform the pre-processing and aggregation steps on the client side due to the lower data volume. To validate whether the algorithms recognize the endpoints with poor End-User Experience, four computers were equipped with old and slow hardware and used to simulate poor end-user experience.

Those endpoints with poor End-User Experience were labeled as outliers in the used dataset. In addition, users were asked about problems with their devices. The answers were split into five categories: very poor, poor, neutral, good and very good. Endpoints that users rated as very poor were reviewed. If the endpoint exhibited problems, the data were labeled as outliers. The user survey was also used to build the training set. For all users with the categorization very good, the data was manually reviewed and validated to ensure that no outliers were present. In this process, the raw data was meticulously inspected, and various visualization techniques such as histograms, scatter plots, and line graphs were employed to comprehensively analyze the data and identify any potential anomalies. Computers that passed this validation process were subsequently integrated into the training dataset. In total, the training data set consisted of 27 endpoints after a sophisticated selection of the endpoints.

3.2 Aggregation Periods

Ahead of the machine learning, the data passes pre-processing and aggregation. During pre-processing, the data is filtered to eliminate measurement failures and textual values are standardized. During aggregation, the data is aggregated over a fixed time interval. Statistical parameters are calculated over this time to retain as much information as possible about the course of the measured data within the aggregation period. The calculated characteristics are the mean, standard deviation, minimum and maximum, the 25%, 50% and 75% quantiles and the number of measurements made within the aggregated period. The larger the selected aggregation interval, the larger the time periods that must be described by the statistical parameters and the more detailed information is lost. On the other hand, a long aggregation period has the advantage of minimizing the required storage space, the network traffic needed to collect the information and the required computing capacity. In the following scenario, data was aggregated in different aggregation periods and then evaluated using the three selected models. Since this comparison is purely based on the differences in the aggregation intervals, each model was trained multiple times on each aggregation period. Afterwards, the best model for each aggregation was selected for comparison. Figure 1 shows the F1-Score, Accuracy, Precision and Recall for the trained models:

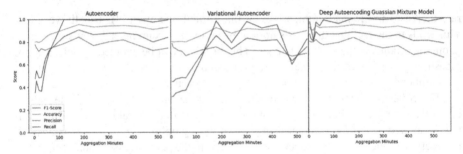

Fig. 1. Comparing different aggregation intervals

In contrast to the assumption that a too-large aggregation interval causes problems, the results in Fig. 1 show that in particular for AE and VAE, a too-short aggregation period and, thus a too-high level of detail has a negative impact on the results. The detected anomalies are mostly correct, but the lower recall indicates that the models can identify less than half of the data labeled as anomalies. Moreover, it can be seen that the VAE is most sensitive to different aggregation intervals. Unlike AE and VAE the DAEGMM can handle even concise aggregation periods. Beyond 180 min, none of the three models offers further advantages. In fact, a slight downward trend in scores can be seen beyond 180 min. Furthermore, the plot shows that all models achieve the best results with an aggregation period of 180 min. The results indicate that proper aggregation offers excellent potential to minimize the required storage capacity and computational power. With a sampling interval of one minute, 180 values are collected over 180 min. These can be reduced to eight characteristics per feature through aggregation. This amounts to a data reduction rate of over 95%.

3.3   Required Training Data

Another important criterion is the training data required for a fully trained model. In practice, a vast array of issues can be identified through hardware telemetry data. Not all problems are immediately discernable. For instance, if a computer's resources are insufficient, this issue can be easily detected. However, identifying an improper load distribution across individual processor cores, as described in [2], proves to be considerably more challenging. This complexity leads to the selection of endpoints used for training algorithms becoming an intricate process. It is essential to ensure that no anomalies are in the training data. Therefore, a model that can operate with a smaller training data set holds significant advantages. To compare the amount of needed training data, multiple training iterations were conducted on variously sized training data sets with all three algorithms. The endpoints selected for the training dataset were shuffled during the examination. Figure 2 illustrates the evolution of the F1-Score, Accuracy, Precision and Recall of the algorithms with different training set sizes:

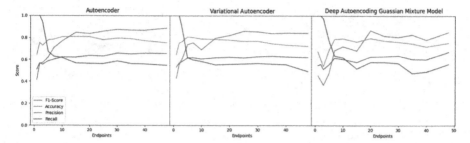

Fig. 2. Model scores for different amounts of Endpoints in the training set

The results demonstrate that the behavior of the three algorithms is similar across training datasets of different sizes. It is noticeable that a very high recall is achieved with a small amount of training data. This can be attributed to the learned normality being very restrictive due to the limited number of endpoints in the training dataset. The concurrently low precision also indicates a relatively high false positive rate. As expected, precision increases with more endpoints in the training dataset, while recall decreases with a larger training dataset. Consequently, the models become more precise yet detect fewer outliers overall.

Since the labeling for the training set was conducted on a per-endpoint basis rather than a per-data point basis, there is an increase in slightly abnormal data points in the training set. This is because no endpoint produces only perfectly normal data. Although the ratio of normal and abnormal data points remains constant, the absolute count of slightly abnormal data increases. Moreover, the variation of normal data expands with a more extensive training dataset. This results in the models learning less restrictive normality, leading to potential misclassification of data from outlier endpoints with low scores, as well as data from normal endpoints exhibiting increased conspicuous behavior.

This behavior emphasizes the importance of selecting a clean training dataset. Additionally, a compromise for the training set size must be found: on the one hand, all outliers should be detected, and on the other hand, false positives should be minimized as much as possible.

In general, it can be observed in Fig. 2 that the AE and VAE yield more stable results from a training size of approximately 15–20 endpoints. The DAEGMM stabilizes at around 20 endpoints but exhibits slightly wider variability than the other two models.

3.4 Stability and Robustness

Factors such as the random initialization of the weights or the random arrangement of the training data provide a non-deterministic course of the training. This leads to different training results if the same model is trained again with the same hyperparameters on the same training data set. Due to this characteristic, choosing a trained model with the best possible recognition rate is essential for the best results. The ideal situation would be to obtain an optimal model after each training session. We consider the trained models from the previous sections to compare the algorithms in this behavior. Figure 3 shows the variance of the scores of all models trained in Sect. 3.2:

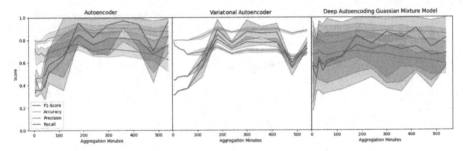

Fig. 3. Model score variation for different aggregation intervals

Although the spread of the model scores is slightly smaller for the AE and VAE with a small aggregation period, based on the results shown in Fig. 3 it can be concluded, that the aggregation period does not have a significant influence on the variation of the models. Similarly, the models compared in Sect. 3.3, which were trained on different amount of training data, achieve similar results. Both results indicates that the VAE provides the most reliable constant results while the DAEGMM shows, by far, the most prominent differences. If this model is to be used for anomaly detection, it is extremely important to choose a model with high scores. One reason for the large spread of the DAEGMM can be the higher complexity of the algorithm. This is partly responsible for the fact that the algorithm often gets stuck in local minima during training.

3.5 Runtime

Another criterion for comparing the algorithms is the execution time of the respective model. Especially using algorithms to detect anomalies in large-scale infrastructures for the improvement of end-user experience, a fast run time is a significant advantage for gaining results quickly and providing real-time analysis to administrators. To compare the execution times of the algorithms, measurements were taken during both the training phase and the data prediction on trained models. The observations show significant differences between the algorithms regarding their execution times. Figure 4 illustrates the measured training times:

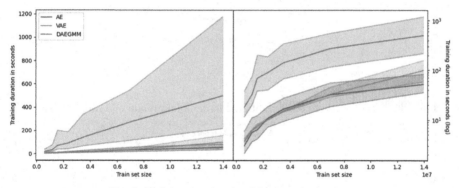

Fig. 4. Training duration for different train set sizes

Figure 4 presents the measured data on the left side using a linear Y-scale and on the right side using a logarithmic Y-scale. The runtime difference between the AE and the DAEGMM increases with the amount of training data increases. Especially the runtime of the DAEGMM increases enormously with an increasingly larger data set. With the highest amount of training points, the DAEGMM takes about ten times longer than the AE. The runtimes of the AE and the VAE are significantly below the DAEGMM from the beginning. Moreover, it can be seen that the VAE develops longer runtimes than the AE on large data sets after they are very similar at the beginning.

3.6 Performance

Another criterion for evaluating the algorithms is identifying anomalous endpoints and correctly distinguishing them from normal behavior. To classify an endpoint as conspicuous, the proportion of data points assessed as anomalies is considered and a threshold is set. Endpoints above this threshold are marked as conspicuous and can be used to assist the administrators in detecting anomalies in the IT infrastructure. Figure 5 shows the percentage of data points marked as anomalies for the individual end devices of the three algorithms.

On the x-axis, the endpoints are ordered by the number of conspicuous data points, while the y-axis shows the percentage of data points classified as conspicuous. The more outlier data points a single endpoint causes, the more conspicuous that endpoint is in the data. The endpoints with the most frequent anomalies are the endpoints that are presented to the administrator for potential issues in production. A threshold can be used to divide the data into normal and conspicuous endpoints.

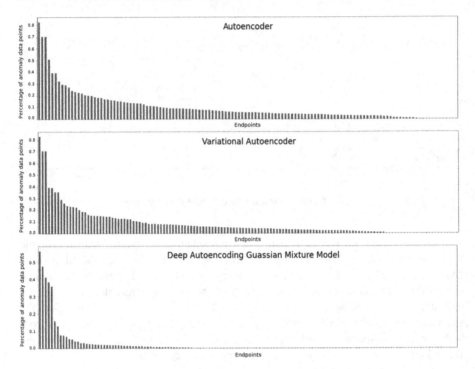

Fig. 5. Percentage of anomalous data points of individual endpoints

Figure 5 shows that the distinction between conspicuous and normal data points improves as the algorithm's complexity increases. This can be seen by the fact that in the left plot of the AE, almost every endpoint contains conspicuous data points, while in the DAEGMM result, more than half of the endpoints have no anomalies. The VAE identifies similar anomalies as in the AE, but the proportion of the respective endpoint is slightly lower. The simulated endpoints with bad end-user experiences described in Sect. 3.1, are among the five endpoints with the most outliers in all three models. Actual anomalies, such as permanently running database tests that affected the work on the endpoint, could also be found in the other devices with high scores. It was impossible to present a visualization of the perfect result because not all the necessary data are available due to the restrictions imposed by data protection regulations.

4   Conclusion

In summary, it can be concluded that all three algorithms could accurately identify anomalies in the endpoints within the data. Furthermore, a higher level of aggregation improves the models' outcomes, thereby serving as an additional means of data reduction. For the required training data set, all three algorithms have similar behavior. The results show that the models can achieve good results even on a small training set. In addition, the results highlight the importance of an accurate training dataset that does not contain any outliers. In the stability tests, the VAE yields the most stable training outcomes. Notably, a high variance in the training results of the DAEGMM is observed. In conjunction with the particularly lengthy training runtime, this represents a substantial drawback compared to the other two algorithms. On the other hand, the strengths of the DAEGMM lie in its ability to separate the data into outliers and normal data points. This can likely be attributed to the algorithm's complexity, which enables it to detect more intricate dependencies within the data. Nevertheless, for the hardware telemetry dataset utilized in this paper, the complexity of all examined models is sufficient. When incorporating additional data points, such as software or network telemetry, the extra resources of the DAEGMM may prove helpful. The performance of the presented concepts on such data needs to be investigated in future work. Furthermore, conceivable approaches could involve combining various models that mutually verify each other, thereby further optimizing the results in the future. In conclusion, this paper has demonstrated the potential of autoencoders to support IT administration in identifying areas of action within their IT environment. By effectively identifying relevant anomalies, autoencoders provide valuable assistance to IT administrators, enabling them to focus on potential issues in their infrastructure. However, it is important to note that the dataset used in this study can be further enriched by incorporating additional data sources such as software and network telemetry. These supplementary data can offer a more comprehensive overview of the IT environment and contribute to a better understanding of potential issues. Exploring software and network data is an important avenue for future research. In addition, research into the usability of the models in different infrastructures of different companies can further improve their applicability and effectiveness. This could provide valuable insights into the generalizability and robustness of the proposed approach.

References

1. Bourne, V.: Digitale Eigensabotage und die Folgen für Unternehmen. https://www.nex think.com/wp-content/uploads/2021/12/Vanson-Bourne-Employee-Centric-Report_DE-1. pdf. Accessed 04 Apr 2023
2. Beckmann, S., Till, J., Bauer, B.: Endpoint-performance-monitoring for a better end-user experience. In: Proceedings of the 2022 European Symposium on Software Engineering, pp. 63–71 (2022). https://doi.org/10.1145/3571697.3571706
3. Meidan, Y., et al.: N-baiot—network-based detection of IoT botnet attacks using deep autoencoders. IEEE Pervasive Comput. **17**(3), 12–22 (2018). https://doi.org/10.1109/MPRV.2018. 03367731
4. Borghesi, A., Bartolini, A., Lombardi, M., Milano, M., Benini, L.: Anomaly detection using autoencoders in high performance computing systems. In: Proceedings of the AAAI Conference on Artificial Intelligence, vol. 33, no. 01, pp. 9428–9433 (2019). https://doi.org/10.1609/ aaai.v33i01.33019428
5. Mirsky, Y., Doitshman, T., Elovici, Y., Shabtai, A.: Kitsune: an ensemble of autoencoders for online network intrusion detection. In: The Network and Distributed System Security Symposium (NDSS) (2018)
6. Zong, B., et al.: Deep autoencoding Gaussian mixture model for unsupervised anomaly detection. In International conference on learning representations (2018)
7. Rocca, J.: Understanding Variational Autoencoders (VAEs). Building, step by step, the reasoning that leads to VAEs (2019). https://towardsdatascience.com/understanding-variational-autoencoders-vaes-f70510919f73. Accessed 04 Apr 2023

8. Kullback, S., Leibler, R.A.: On information and sufficiency. Ann. Math. Stat. **22**(1), 79–86 (1951)
9. Bank, D., Koenigstein, N., Giryes, R.: Autoencoders. arXiv preprint arXiv:2003.05991. (2020). https://doi.org/10.48550/arXiv.2003.05991

Application of Multi-agent Reinforcement Learning to the Dynamic Scheduling Problem in Manufacturing Systems

David Heik$^{(\boxtimes)}$, Fouad Bahrpeyma, and Dirk Reichelt

Faculty of Informatics/Mathematics, University of Applied Sciences Dresden,
01069 Dresden, Germany
david@heik.science, bahrpeyma@ieee.org, dirk.reichelt@htw-dresden.de

Abstract. Most recent research in reinforcement learning (RL) has demonstrated remarkable results on complex strategic planning problems. Especially popular have become approaches which incorporate multiple agents to complete complex tasks in a cooperative manner. However, the application of multi-agent reinforcement learning (MARL) to manufacturing problems, such as the production scheduling problem, has been less frequently addressed and remains a challenge for current research. A major reason is that applications to the manufacturing domain are typically characterized by specific requirements, and impose the research community with major difficulties in terms of implementation. MARL has the capability to solve complex problems with enhanced performance in comparison with traditional methods. The main objective of this paper is to implement feasible MARL algorithms to solve the problem of dynamic scheduling in manufacturing systems using a model factory as an example. We focus on optimizing the performance of the scheduling task, which is mainly reflected in the maskspan. We obtained more stable and enhanced performance in our experiments with algorithms based on the on-policy policy gradient methods. Therefore, this study also investigates the promising and state-of-the-art single-agent reinforcement learning algorithms based on the on-policy method, including Asynchronous Advantage Actor-Critic, Proximal Policy Optimization, and Recurrent Proximal Policy Optimization, and compares the results with those of MARL. The findings illustrate that RL was indeed successful in converging to optimal solutions that are ahead of the traditional heuristic methods for dealing with the complex problem of scheduling under uncertain conditions.

Keywords: Artificial Intelligence · Reinforcement Learning · Manufacturing Systems · Smart Production Systems · Industrial IoT Test Bed

1 Introduction

Automatization of industrial manufacturing systems began with the third industrial revolution, as it made digitalization of production and tools affordable [24]. *Industry 4.0* is primarily driven by technology and cross-linking of these digitalized products and tools. Currently, the scientific community assumes that the

© The Author(s), under exclusive license to Springer Nature Switzerland AG 2024
G. Nicosia et al. (Eds.): LOD 2023, LNCS 14506, pp. 237–254, 2024.
https://doi.org/10.1007/978-3-031-53966-4_18

Table 1. Abbreviation

Abbreviation	Full Form or Explanation	Abbreviation	Full Form or Explanation
A2C	Advantage Actor-Critic	MARL	Multi-Agent Reinforcement Learning
A3C	Asynchronous Advantage Actor-Critic	MARL1C	Multi-Agent Reinforcement Learning
AC	Actor-Critic		with one global critic
ACO	Ant Colony Optimization	MARL4C	Multi-Agent Reinforcement Learning
AI	Artificial Intelligence		with individual critic for each actor
CLDE	Centralized-Learning but	MES	Manufacturing Execution System
	Decentralized-Execution	NP-hard	non-deterministic polynomial-time hardness
CNC	Computerized Numerical Control	OEE	Overall Equipment Effectiveness
CPS	Cyber Physical System	PLC	Programmable Logic Controllers
DDQN	Double Deep-Q Networks	PPO	Proximal Policy Optimization
DQN	Deep-Q Networks		
ESF Plus	European Social Fund Plus	PSO	Particle Swarm Optimization
FIFO	First In First Out	RL	Reinforcement Learning
GA	Genetic Algorithms	RPPO	Recurrent Proximal Policy Optimization
GCN	Graph Convolutional Networks	RR	Round Robin
GMAS	Graph-based Multi-Agent Systems	SARL	Single-Agent Reinforcement Learning
IIoT	Industrial Internet of Things	SPF	Shortest Path First
JSSP	Job-Shop Scheduling Problem	SPT	Shortest Processing Time
LPT	Longest Processing Time	TS	Tabu Search
LSTM	Long Short-Term Memory		

next evolutionary stage will focus on a human-centered approach and will also concentrate on strengthening the resilience of enterprises and the sustainability of processes [31]. Intelligent scheduling and control mechanisms that dynamically allocate resources based on current conditions provide the required technical foundation for a resilient and adaptable manufacturing system. Aside to external fluctuations in the market (customer requirements for product properties, quality and quantity) or within the supply chains, the surrounding production environment, which we focus on in this paper, also exhibits uncertain behavior. The sources of uncertainty are manifold and can influence, among other factors, the operation times, the occurrence of failures, the expected remaining lifetime of components, the quality of the products or the energy consumption. Interdependence and mutual influence in apparent ways and in non-explicit ways add complexity to the challenge of developing a decision-making model for the system [23]. In the literature, reinforcement learning (RL) has been reported to perform well in dealing with such uncertainties, especially when the main objective is to optimize a concrete target parameter such as the overall equipment effectiveness (OEE), tardiness, reliability, profit or makespan. A key pre-requisite for enabling the use of the advantages provided by RL in manufacturing systems is, for the most part, the provision of a mapping from the current state of the system to the data produced continuously in the relevant industrial processes; as the data should be accessed and analyzed in a real-time manner [18]. Currently, many industrial production systems are characterized by a heterogeneous mixture of programmable logic controllers (PLC) and proprietary field devices from different manufacturers. In addition, there are many different approaches to collect the necessary data, often requiring digital retrofitting of the legacy production system [1]. Furthermore, it has to be ensured that the decisions made can be fed back into the production system, and preferably that resource allocations can be changed at runtime if the general conditions are changed.

All these conditions are fulfilled at the Industrial Internet of Things (IIoT) Test Bed at HTW Dresden, which is part of an industrial laboratory for smart

production. The henceforth considered cyber physical system (CPS) is fully automated and has four manufacturing clusters (see Fig. 1, from left to right and top
to bottom: a workstation for manual assembly; a CNC milling machine; a high
rack warehouse; as well as a core out of 13 assembly stations). The stations provide at least one manufacturing operation, with some operations redundantly
existing in the cluster. Because of these redundancies, tasks can be executed
in different manners on different machines. The products that are to be produced are assigned to carriers throughout the entire production process. Driver-
less transport systems ensure that products can be exchanged across different
clusters. Within a single cluster, carriers are transferred from one station to
the next via a conveyor belt. The performance of the CPS is currently limited by the connected static designed manufacturing execution system (MES).
Dynamic events and changed underlying conditions are not taken into account,
so that the existing physical redundancy of the operations cannot be exploited.
This work therefore focuses on exploiting the physical potentials of the system
by applying different heuristic, metaheuristic and RL algorithms in order to
find an optimal and reliable scheduling policy, which is able to minimize the
makespan. Due to scope limitations, this paper focuses on the application of
already known heuristic, metaheuristic, and RL methods to the specific architecture of the IIoT Test Bed, so we do not describe their implementation in
detail. However, in order to make the results of our work comprehensible and to
allow other researchers (without access to the real production system) to continue our work or to design experiments themselves, we have developed a digital
twin that approximates the real behavior. We present this simulation environment after the related work section in the methodology section, and make it
available to the general public, while asking the interested reader to take further
information from the repository [10]. We also provide a list of all abbreviations
used with their full meaning or explanation in Table 1.

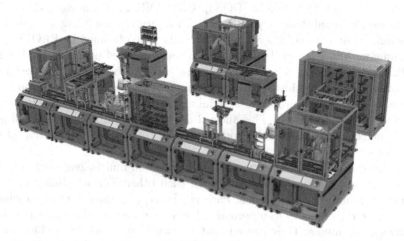

Fig. 1. Industrial Internet of Things Test Bed at HTW Dresden

2   Related Works

Production scheduling is a type of job-shop scheduling problem (JSSP) that has long been a popular combinatorial optimization [27] and is classified as an NP-hard problem [8]. Conventional manufacturing systems are driven by decision making processes that work based on pre-existing data such as available resources or promised delivery dates, and schedule orders in a centralized architecture that is rigid and thus very limited in its ability to respond to changing conditions (due to its overdependence on both centralized controls and limited communication channels). Frequently used and well-known approaches to deal with the problem are heuristic and metaheuristic approaches as suggested by Hussain et al. [14] in a comprehensive survey. Zhang et al. [27] developed a hybrid algorithm that combines two models of Particle Swarm Optimization (PSO) and Tabu Search (TS). They uses PSO to schedule and assign operations on machines, and TS is applied to local search for the scheduling sub-problem originating from each obtained solution.

To achieve self-adaptive production control towards mass customization, RL has emerged as a promising technology for scheduling. In specific, RL is beneficial in handling scheduling tasks under uncertainty, and so has recently received a significant attention from the relevant community. In order to respond better to the dynamics of the system, the rescheduling problem is studied among researcher by Hu et al. [26] for flow shop under rush orders. Their Ant Colony Optimization (ACO) rescheduling method, however, can be stopped at a local optimum. Therefore, the authors suggested the use of the mutation operation to improve the overall performance. Dynamic scheduling using RL has also been studied in [2] and [3], where complex reward functions were adopted as well as resource usage prediction to deal with uncertainty. In a simulation environment that is similar to the one presented in this paper, but characterized by only one operation per station and lower complexity, Heik et al. [11,12] studied the application of Deep-Q Networks (DQN), REINFORCE, Advantage Actor-Critic (A2C), and Proximal Policy Optimization (PPO) to the scheduling parallel operations under uncertainty, where the best results were obtained by PPO. In [25], a dynamic multi-agent scheduling approach was presented using ant intelligence and RL. In their work, the intelligent agent chooses the appropriate behavior in accordance with the historical and an immediate feedback. In addition, the agent also picks the appropriate scheduling algorithm based on the historical incentives of the algorithms. This allowed the local optimization of a small portion of the real-time tasks to be combined with the global optimization of the other tasks.

The most straightforward approach to learning in multi-agent environments is to use agents that learn independent of each other. For scheduling in semiconductor manufacturing systems, Park et al. [21] presented a much advanced variant. Their Double DQN approach allows to minimize the makespan in a decentralized manner. They proved that in such a approach were able to outperform rule-based, metaheuristic, and other RL methods. Zhou et al. [30] presented a distributed architecture with multiple artificial intelligence (AI) sched-

ulers for online scheduling of orders in a smart factory setting. They present
a case study showing their ability to schedule new orders on the shop floor
and take into account that machines can randomly breakdown. Lowe et al. [20]
proposed a multi-agent policy gradient algorithm that considers action policies
of other agents and is able to successfully learn policies that require complex
multi-agent coordination. Therefore, a centralized critic was implemented, which
observes all actions across all agents. A multi-agent RL approach, which integrates self-organization and self-learning mechanisms was proposed by Zhang et
al. [28]. They employed a multi-layer perceptron to establish a decision-making
module, which is periodically trained and updated through a PPO algorithm.
In [4], the authors present a new DRL algorithm, combining policy gradient
algorithms with actor-critic architectures, and interpreting the productions system as a multi-agent system. A method for multi-criteria dynamic planning of
production facilities under both common and sustainable target variables, based
on a PPO algorithm in a multi-agent reinforcement learning (MARL) setting
was proposed by Popper et al. [22]. A centralized-learning but decentralized-
execution (CLDE) multi-agent structure that uses RL is proposed by Jing [15].
Using Graph Convolutional Networks (GCN), they were able to solve the flexible
scheduling problem in a better manner than well-known methods such as First In
First Out (FIFO), Shortest Processing Time (SPT), Longest Processing Time
(LPT), dynamic reinforcement learning, and graph-based multi-agent systems
(GMAS). Lohse et al. [19] combined Monte-Carlo Tree Search and a multi-agent
DQN to solve the Open Shop Scheduling Problem. For the presented use case,
the multi-agent setting provides a scalable solution even for large environments.

3   Problem Statement

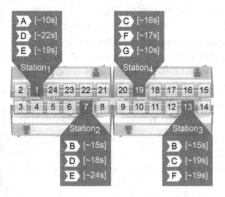

Fig. 2. Structure of the simulation
environment

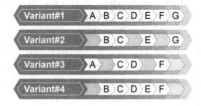

Fig. 3. Product variants

This section describes the methodology presented for accommodating the
scheduling task under uncertainty for varying work plans. In the simulations,

however, the complexity of the real environment has been reduced to be able to focus on the logic and achieve convergence more feasibly. The use case presented in this paper resembles the functionality of the IIoT Test Bed at HTW Dresden. However, we reduced the number of working stations to four as an attempt to account for less complexity. In the simplified simulation, each station offers three operations, illustrated in detail Fig. 2. In total, seven different operations are available, numbered from A to G. The operations B, C, D, E, F are provided by several stations accounting for redundancies. This allows for a significant freedom to deal with bottlenecks, jamming, and additionally an increase in the resilience of the system as a whole. However, operations A and G are only available each only at one specific station and therefore deemed critical. In the experiments, a variety of four different product variants have been considered as illustrated in Fig. 3. In this study, it's only important that a given sequence is specified and remains constant during production. For simplicity, the solution space has been discretized in time and space, leading to the implementation assumption that each step takes only one second. During production, products are transported from one station to the next counterclockwise via a conveyor belt, in an automated manner. The conveyor belt is logically partitioned into 24 slots and arranged in a ring structure so that all the stations (marked in green at the positions: 1, 7, 13 and 19) are indirectly reachable. On the conveyor belt, products are transported by carriers, which may be handled in three different ways. The product can be transported one slot forward if the slot in front of the carrier is free (e.g. from slot 24 to slot 1) and no operations are being performed at the corresponding station. When the following slot be blocked, the transport cannot continue for the successor (the carrier is in a traffic jam). In the event that a carrier is at a station waiting to be processed (or is already being processed), the third option is to wait until the station has completed its process.

The real production environment is affected by uncertainty, which has been taken into account in our experiments. Therefore, under uncertainty, average completion times may vary within a predefined range (for moderate level: $\pm 3\,\mathrm{s}$, for high level: $\pm 5\,\mathrm{s}$). These average completion times were observed from our observations at the Test Bed, shown in Fig. 2. In addition, we have also taken into account a non-ideal behavior in terms of temporal unavailablity of operations. The probability of the occurrence of a fault is independent of other events and exponentially distributed, i.e. it increases with the duration of the fault-free operating time. On average, at what we define as a *moderate* fault rate, the faults occur in an arithmetic mean of ˜840 seconds and last an average of ˜61 seconds. At a *high* disturbance rate, where errors occur on arithmetic average every ˜420 seconds.

In the case that non-redundant operations (A and G) are needed to complete an item, they are always executed directly, assuming the carrier has already reached the corresponding station. For all other operations (B, C, D, E and F), alternatives exist from which the problem solver must choose one. The possible set of treatments for redundant operations are:

- Option#1: Produce now at alternative#1
- Option#2: Leave station, do an additional loop, and then produce on alternative#1
- Option#3: Produce now at alternative#2
- Option#4: Leave station, do an additional loop, and then produce on alternative#2

The option to leave the station for the moment and to make an additional loop allows the system the possibility to transport subsequent carriers as well. This prevents a blocking of the conveyor belt (e.g. due to a slow operation) in the current lap, so that subsequent carriers will have the chance to move to their following stations. In order to make a decision (see response options explained earlier) for each operation for each job in the order, some problem solvers require a representation of the current system state. We represent the state space with a binary representation. For each of the slots on the conveyor, we use 8 bits, with the first indicating whether there is a carrier on the slot ($= 1$) or not ($= 0$). The seven remaining bits each stands for one of the seven operations, indicating whether the corresponding operation is still to be completed ($= 1$) or has already been completed ($= 0$). In addition to the 192 bits, there are another 12 bits (one for each operation of each station) which indicate whether the operation can be performed ($= 0$) or whether there is a defect ($= 1$).

As the number of workpieces or the number of redundant operations increases, the number of possible combinations of decisions increases exponentially, as shown in Eq. (1) and (2).

$$DecisionSpace = (2^{RedundantOperations})^{OrderQuantity} \tag{1}$$

$$DecisionSpace = (2^5)^{50} = 1.8 * 10^{75} \tag{2}$$

If an order with 50 products (all variant#1) is considered, according to Eq. (2), the number of possible decisions will be $1.8 * 10^{75}$. The number of possible initial situations (see Eq. (3) and (4)) depends on:

- the range of variation of the operation times (a margin of ± 3 s results in 7 possible operation times),
- the initial distribution of the carriers on the conveyor belt (the calculation is made using the binomial coefficient, where $n = 24$ (slots) and $k = 4$ (carriers),
- the number of products to be manufactured plus the number of product variants,
- whether operations have failed during the production process.

$$InitialSpace = (Margin^{Operations}) * \binom{Slots}{Carrier}$$
$$* (Producttypes^{Orderscope}) * (Malfunction^{Operations}) \tag{3}$$

$$InitialSpace = (7^{12}) * \binom{24}{4} * (4^{50}) * 2^{12} = 7.6 * 10^{47} \tag{4}$$

Our objective is to obtain a decision policy that minimizes the makespan taking into account the temporal uncertainty, possible disturbances and the dynamics of the underlying system, while regarding the redundant capabilities. To determine the global best solution, it would be necessary to consider all possible initial situations, to each of which all possible decision permutations are applied. Equation (5) and (6) illustrate the corresponding mathematical relationship.

$$SolutionSpace = InitalSpace * DecisionSpace \tag{5}$$

$$SolutionSpace = (7.6 * 10^{47}) * (1.8 * 10^{75}) = 1.37 * 10^{123} \tag{6}$$

4   Methodology

4.1   Reinforcement Learning

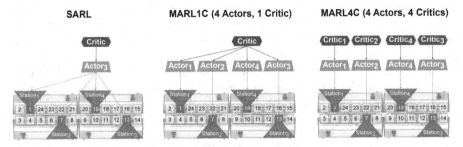

Fig. 4. Architectures

To perform all necessary experiments, even with the utopian assumption that we could compute a quadrillion experiments per second, it would still take ~$4 * 10^{100}$ years to always determine the best global decision. Consequently, it is practically impossible to solve the problem in this manner, so no training datasets with labeled data are available either. Therefore, we resort to problem solving approaches that develop decision strategies in an exploratory way, such as RL. In particular, MARL in combination with an on-policy, policy-gradient approach is cited in the literature as a promising solution strategy when it comes to solving complex problems reliably, with enhanced perfocmace and in reduced training duration [29]. In our previous experiments, we observed the best results in terms of reliability and perfomance with algorithms that use the gradient descent method, following an on-policy strategy [11,12]. Within this contribution, we therefore study in more detail algorithms based on this promising methodology, such as: Asynchronous Advantage Actor-Critic (A3C), PPO and RecurrentPPO (implemented with a Long Short-Term Memory - LSTM) in different configurations of the Actor-Critic architecture (see Fig. 4) to solve the scheduling problem of our CPS. First, the approach where a single actor interacts with the environment and is supervised by a global critic - referred to as SARL. In addition, we consider two multi-agent RL setups. In the first one, only the actor part was swapped, so that each station will henceforth have its own independent actor.

In this configuration, actions are still assessed by only one global critic - refer as MARL1C. In the second MARL setup, we also swap the critic part, so that there is an individual critic for each actor - referred to as MARL4C.

4.2 Training Procedure

In our experiments, the training process is performed episodically. Whenever a new episode starts, a random initial situation is generated. We designed a training scenario in which the amount of products to be manufactured (= 50), the number of carriers on the conveyor (= 4), and the degree of uncertainty (variations in operation times, and occurrence probability for operation failures is moderate) remain constant. However, orders are randomly composed in each episode, carrier distribution is randomized, operation times are initiated randomly and independently, and failures occur randomly at runtime. Subsequently, the step function provided by the simulation environment is used to fast forward until the first decision is required. Then, the simulation is interrupted, the current state of the system is captured as described before, binary coded and provided to the problem solver. The problem solver makes a decision and propagates it back to the simulation environment. There, it will be applied and consequently the step function will be called again. As long as the order was fully processed, the episode is thereby terminated. The history of the operations performed on each product is recorded during the production process and contains information about when and at which station, an operation was performed. In addition, the carrier's route is logged and it is determined how far the product was transported and how long it was held up by a traffic jam (i.e. had to wait for its predecessor).

Makespan is calculated upon completion of an episode. For the RL methods, additionally a reward is calculated. The individual waiting times of each completed product are accumulated. The evaluation is also based on how many seconds of redundant operations were performed simultaneously. It is also counted how many additional loops the carriers made in total, i.e., whether response option #2 or #4 was chosen. Since skipping too often has a negative impact on makespan, only 10 skips are included, anything beyond that is penalized. Based on this information, the reward function is determined as in Eq. (9) from a bonus (Eq. (7)) as well as a penalty (Eq. (8)).

$$Bonus = 3 * AccumulatedParallelProcessingTime \tag{7}$$

$$
\begin{aligned}
Penalty = (2 * AccumulatedTransportTime) \\
+ (3 * AccumulatedWaitingTime) \\
+ (5 * Makespan) \\
+ (100 * QuantityOfSkips)
\end{aligned}
\tag{8}
$$

$$Reward = Bonus - Penalty \tag{9}$$

4.3   Evaluation

For the applied RL methods, a checkpoint is reached after every 50 training episodes, at which the weights of the models are stored. To evaluate the received RL policies and compare them with each other or against other methods, the *Evaluation dataset#1* (same conditions as in the training) containing 1000 records was generated. In order to determine how well the models react to changing conditions and unknown situations without additional adaptation, three additional evaluation datasets were created, also comprising 1000 records. In *Evaluation dataset#2* 8 instead of 4 carriers are used simultaneously, in *Evaluation dataset#3* the total amount of jobs varies randomly in the range between 5 and 500, and in *Evaluation dataset#4* the uncertainty is set from *moderate* to *high*. After the training process is completed, the determined solutions are applied to each problem from the evaluation datasets and the respective makespan is determined.

4.4   Reproducibility

The implementation code for the siumulation environment as well as the evaluation datasets has been made publicly available to enable reproducibility of the results, see [10]. We ask researchers to take further information and details about the implementation from the repository. In addition, the settings and hyperparameters used for the algorithms used are listed in Table 2 and Table 3.

5   Experimental Results

This section presents the results for the application of some state of the arts RL techniques based on actor-critic methods regarding the strategies for single agent and multiple agents deployment mentioned in Sect. 3. To ensure that our results can be applied in practice in a timely manner and with reasonable effort, we limit the horizon of the training procedure to 100,000 episodes. The reward function (see Eq. (9)) was used while training the RL models to achieve the specific objective of this research in the presented setting.

The Fig. 5, 6, 7, 8, 9, 10, 11, 12 and Fig. 13 represent the training curve of ten experiments each for A3C, PPO, and RecurrentPPO, where the experiments were run with the same parameters (environmental conditions, model hyperparameters, and applied reward function) and with the same limitations (maximum 100,000 episodes). In each of these graphs, the Y-axis depicts the accessed reward (+ an offset of 50.000), aiming for the highest possible score. The X-axis, instead, represents the progression over time during training. The graphs were averaged over the last 100 episodes to obtain a smooth representation. We select a representative learning curve for each RL method and architecture, discuss it below, and we end up using the representatives of these series of experiments for evaluation. In doing so, we never selected the best observation, but one that reflects typical behavior.

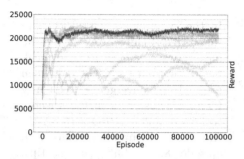

The experiments with A3C report that the SARL case would typically lead to convergence (~80%) of the time after about 20,000 episodes. However, our experiments on A3C have all resulted in failure for the MARL case. The actuators follow different paths during training and thereby confuse each other to such an extent that no knowledge can be derived from the actions.

Fig. 5. A3C SARL

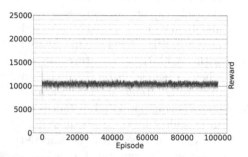

Fig. 6. A3C MARL1C

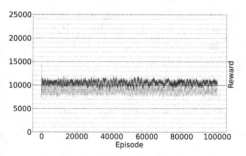

Fig. 7. A3C MARL4C

Table 2. General settings

Parameter	Value	Explanation
Input Dimensions	((1+7)∗24) +(4 ∗ 3) = 204	((Carrier in slot+Remaining operations)*Slots) +(Stations*Operations offered)
Action Dimensions	4	
Max. episodes	100000	
Uncertainty in operation times	±3 s	
Ammount of carriers	4	

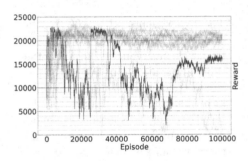

Fig. 8. PPO SARL

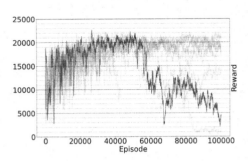

Fig. 9. PPO MARL1C

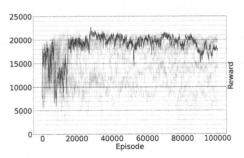

Fig. 10. PPO MARL4C

In the experiments with PPO and the SARL architecture, the results have always shown a local peak before episode 20,000 and then proceed to convergence frequently (~80%). Thus, even in the selected training graph for PPO with SARL, there is an upward trend until episode 8,000, but then it tips. The agent revealed a good decision combination by episode 25,000 through exploration and was thus able to abruptly experience a reward increase. When PPO is implemented with a MARL architecture, we see when a single critic is used, an increasing trend in the learning curve is common in all instances up to episode 50,000. In about 50% of the experiments, with the selected representative run, we observe a significant drop in the learning curve, from which the agent does not recover. When using four independent critics, the results of PPO are much more differentiated and no longer follow a clear scheme. However, in some cases (such as the representative), convergence can be observed from episode ~30,000.

Table 3. Hyperparameter

Hyperparameter	A3C			PPO			RecurrentPPO		
Actors	1	4	4	1	4	4	1	4	4
Critics	1	1	4	1	1	4	1	1	4
Hidden 1 Dimensions	128			256			256		
Hidden 2 Dimensions	–			256			128		
Hidden LSTM Dimensions	–			–			64		
Discount (γ)	0.9			0.99			0.99		
Learning rate	0.0001			0.0003			0.0003		
GAE parameter (λ)	0.95			0.95			0.95		
Betas (β)	0.92, 0.999			–			–		
Minibatch size	–			12			–		
Horizon	–			24			–		
Policy clip	0.2			0.2			0.2		
Num. epochs	–			4			–		

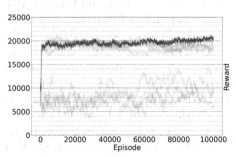

Fig. 11. RPPO SARL

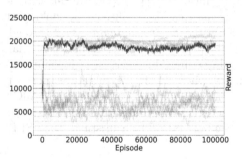

Fig. 12. RPPO MARL1C

RecurrentPPO, however, has shown higher rewards in our experiments when used with the SARL architecture in comparison to the case where MARL was used. In conjunction with SARL, convergence in learning curves can also be detected frequently (~80%) with RecurrentPPO, although in some cases this only occurs after 50,000 episodes (as with the representative curve). When RecurrentPPO is considered in a MARL architecture, the observations fall into two categories. In about half of the models, mutual uncertainty can be seen, analogous to the observations for A3C. The other half quickly learns to work together. Our representative training curve for the MARL1C architecture (see Fig. 12) shows a trend that is continuously increasing and reaches its maximum at the end of the training.

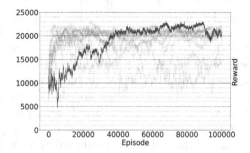

Fig. 13. RPPO MARL4C

For a better interpretation of the results obtained with RL, we also investigated well known heuristic scheduling strategies. Among them is the current job assignment strategy currently used in the IIoT Test Bed at HTW Dresden, the Shortest Processing Time first (SPT) [5,6]. Furthermore, we also consider its counterpart the Longest Processing Time first (LPT) [9], as well as a well-known solver of the Shortest Path First (SPF) problem, Dijkstra's algorithm [7]. For comparison, we also use the Round Robin (RR) [17] strategy commonly used in industry, as well as a pure random resource allocation. Metaheuristics do not incorporate the current state of the system into the solution finding process, as they are designed for static problems (without a moving target). Since the implementation and computational effort is relatively low, they are also occasionally used in practice. Therefore, we furthermore consider Genetic Algorithms (GA) by Holland [13] as well as Particle Swarm Optimization (PSO) by Kennedy and Eberhart in a discrete binary version [16].

For each of the selected nine representatives, those five models were evaluated (with the four previously described evaluation datasets) that achieved the highest reward in the training process. From these evaluation results, the checkpoint with the best result (lowest possible average makespan) was then selected for each method, and henceforth considered for further analysis. Table 4 lists all evaluation results, considering the performance of the problem solvers both individually for each evaluation dataset and collectively. In addition to the average makespan, the minimum as well as maximum observed makespan was also listed for reference. For better interpretation, the results of the average were ranked (four light gray columns). In our study, PPO in combination with a SARL architecture achieved the best overall performance (highlighted in light green), followed by 2nd place RecurrentPPO, also in combination with a SARL architecture. The 3rd and 4th place are again awarded to PPO, but this time in association with a MARL architecture. However, in order to better interpret the observations, the evaluation results are considered in more detail. Figure 14 illustrates in a bar chart for each evaluation dataset how the algorithms perform in relation to the best observed overall performance (PPO SARL). In this context, there is one result in particular which stands out - namely PPO MARL1C in combination the *Evaluation dataset#2*. This algorithm-architecture combination was the only one of the considered experiment series which was able to achieve better results on average and outperform PPO SARL. This observation suggests that under a changing environment (such as the use of 8 instead of 4 carriers) MARL architectures have the potential to outperform other very powerful algorithm-architecture combinations. The above remark can be confirmed by looking at the minimum makespan values for PPO in the *Evaluation datasets#2* through *#4*. Both MARL1C and MARL4C achieved partially better results here than in direct comparison with the SARL architecture. Moreover, enough counterexamples can be found in our experiments, where MARL architectures perform more poorly (e.g., in the case of ReuccrentPPO) or even completely fail (A3C). In the Table 4, for each evaluation dataset, the lowest minimum and maximum makespans have been highlighted by green bold text. The heuristic method SPF,

Table 4. Evaulation results

Evaluation dataset#1 — No. of carriers: 4, Order quantity: 50, Uncertainty: Moderate
Evaluation dataset#2 — No. of carriers: 8, Order quantity: 50, Uncertainty: Moderate
Evaluation dataset#3 — No. of carriers: 4, Order quantity: 5-500, Uncertainty: Moderate
Evaluation dataset#4 — No. of carriers: 4, Order quantity: 50, Uncertainty: High

	Evaluation dataset#1 Makespan				Evaluation dataset#2 Makespan				Evaluation dataset#3 Makespan				Evaluation dataset#4 Makespan				Overall performance	
	Min	Avg	Max	Rank	Min	Avg	Max	Rank	Min	Avg	Max	Rank	Min	Avg	Max	Rank	Avg	Rank
A3C – SARL	2010	2371,6	2855	5	1556	1917,6	2290	8	307	11787,1	24946	5	2006	2487,6	3089	4	4641,0	5
A3C – MARL1C	2136	2491,1	2895	10	1573	1919,3	2349	9	382	12051,3	25778	9	2174	2612,7	3154	9	4768,6	9
A3C – MARL4C	2108	2469,9	2874	9	1501	1882,0	2249	3	336	11914,8	25535	6	2097	2593,3	3175	8	4715,0	7
PPO – SARL	1984	2320,8	2647	1	1549	1880,4	2254	2	311	11555,9	24611	1	2004	2459,3	3023	1	4554,1	1
PPO – MARL1C	2016	2337,5	2737	3	1529	1865,6	2273	1	306	11616,5	24560	3	1985	2492,2	3091	5	4578,0	3
PPO – MARL4C	1984	2338,7	2731	4	1545	1915,9	2356	7	273	11635,3	24898	4	1997	2479,7	3037	3	4592,4	4
Recurrent PPO – SARL	1962	2325,4	2693	2	1581	1883,5	2303	4	283	11567,5	24564	2	1973	2459,4	2970	2	4559,0	2
Recurrent PPO – MARL1C	2057	2403,1	2907	7	1559	1883,7	2380	5	332	11970,6	25490	7	1995	2540,6	3075	6	4699,5	6
Recurrent PPO – MARL4C	2019	2430,8	2857	8	1593	1893,5	2254	6	351	12074,6	25402	10	2071	2551,2	3094	7	4737,5	8
GA	2106	2711,2	3964	12	1494	2096,2	2929	11	328	13239,7	28845	12	2129	3167,4	4429	12	5318,6	12
PSO	2266	2863,4	3999	13	1554	2213,3	3066	14	409	14050,3	30349	13	2351	3359,6	4754	13	5621,8	13
SPT	1987	2524,7	3626	11	1611	2163,9	3220	13	268	12620,2	28232	11	2645	2938,2	3982	11	5061,7	11
LPT	2434	3230,7	4343	16	2116	2837,5	3722	16	416	16004,4	35070	16	2645	3641,4	5093	16	6428,5	16
SPF	1801	2377,1	3357	6	1471	2031,0	3020	10	219	11975,7	27398	8	1812	2785,0	3926	10	4792,2	10
RR	2269	2921,6	4193	15	1470	2150,3	3242	12	421	14864,5	31527	15	2419	3468,7	4763	15	5726,3	15
Random	2266	2894,6	3958	14	1630	2220,8	3148	15	378	14134,8	30663	14	2542	3376,5	5091	14	5657,2	14

Reinforcement Learning: A3C, PPO, Recurrent PPO · Meta-heuristics: GA, PSO · Heuristics: SPT, LPT, SPF, RR, Random

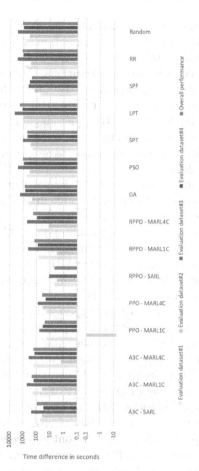

Fig. 14. Procedural comparison to best model (PPO SARL)

Time difference in seconds

which is very primitive compared to RL, delivers in isolated cases - but then with distance - significantly better results than all other considered algorithms. SPF was managed to allocate the resources in such a way that, compared to PPO SARL, it resulted in a partially ~30% shorter makespan (*Evaluation dataset#3* 219 s to 311 s). Thus, we can show that the RL algorithms we investigated provide good and reliable results on average, but in some cases there is still potential for further improvement.

6 Conclusion

In manufacturing systems, the presence of uncertainty can lead to a decline in the performance of the entire system, to the point where required operations and processes can no longer be executed. There are a variety of reasons for this, including interruptions in the supply chain, changes in production requirements, and personnel changes. Production systems in the industrial sector must improve their adaptability and responsiveness in order to fulfill their delivery promises and meet customer expectations. A special type of tasks seen in manufacturing systems is production scheduling where uncertainty appear frequently and need to be addressed in a real-time manner. In this paper, we have therefore proposed the use of RL under single agent and multi-agent settings to address the dynamic scheduling considering possible operation failures and unstable process times within operations as well as fluctuating order compositions. The RL methods used in this paper are A3C, PPO and RecurrentPPO (LSTM). To better evaluate and assess the results, heuristic (SPT, LPT, SPF, RR as well as random assignment) and metaheuristic methods (GA and PSO) were also investigated as potential problem solvers. Contrary to our initial expectations, the results from our experiments indicate that SARL architectures performed better than MARL architectures. The models generated and results obtained with PPO in combination with SARL architecture were able to outperform the other RL methods, heuristics and metaheuristics methods, despite the consideration of strict time constraints during the learning process. The underlying reason that PPO was able to outperform even A3C is mainly the presence of a built-in mechanism in PPO that prevents large gradient updates using a surrogate clipping objective function. A3C, on the other hand, can support the exploration of the problem space through the use of asynchronous updates. In the case of RecurrentPPO, on the other hand, the employed LSTM can help with the memory and the capacity of the algorithm in dealing with complexity in the agents interactions with the environment. Nevertheless, we were able to prove in a concrete scenario that MARL architectures using independent agents are able to outperform other layout arrangements, especially in the case of changing environmental conditions. Our research has identified the reliability of training with MARL architectures as a major issue that needs to be addressed in order to eliminate the observed trade-off in RecurrentPPO between very good learning and fast converging models and models that abstract no knowledge at all. The scaling of the presented application is problematic, since the complexity of the problem increases exponentially with multiple stations, products and operations. In addition, the optimization of hyperparameters is difficult to handle automatically, which complicates the access to the research domain. A remaining problem for future work is the inclusion of additional unforeseen events such as a reject rate that increases over time, or a decreasing quality that is, for example, due to the cause of wear.

Acknowledgements. This research was funded as part of the project "Produktion-ssysteme mit Menschen und Technik als Team" (ProMenTaT, application number: 100649455) with funds from the European Social Fund Plus (ESF Plus) and from tax revenues based on the budget passed by the Saxon State Parliament.

References

1. Alqoud, A., Schaefer, D., Milisavljevic-Syed, J.: Industry 4.0: a systematic review of legacy manufacturing system digital retrofitting. Manuf. Rev. **9**, 32 (2022). https://doi.org/10.1051/mfreview/2022031
2. Bahrpeyma, F., Haghighi, H., Zakerolhosseini, A.: An adaptive rl based approach for dynamic resource provisioning in cloud virtualized data centers. Computing **97**, 1209–1234 (2015)
3. Bahrpeyma, F., Zakerolhoseini, A., Haghighi, H.: Using ids fitted q to develop a real-time adaptive controller for dynamic resource provisioning in cloud's virtualized environment. Appl. Soft Comput. **26**, 285–298 (2015)
4. Burggräf, P., Wagner, J., Saßmannshausen, T., Ohrndorf, D., Subramani, K.: Multi-agent-based deep reinforcement learning for dynamic flexible job shop scheduling. Procedia CIRP **112**, 57–62 (2022). https://doi.org/10.1016/j.procir.2022.09.024
5. Carroll, D.C.: Heuristic sequencing of single and multiple component jobs. Ph.D. thesis, Massachusetts Institute of Technology (1965)
6. Conway, R.W.: Priority dispatching and job lateness in a job shop. J. Ind. Eng. **16**(4), 228–237 (1965)
7. Dijkstra, E.W.: A note on two problems in connexion with graphs. Numer. Math. **1**(1), 269–271 (1959). https://doi.org/10.1007/bf01386390
8. Garey, M.R., Johnson, D.S., Sethi, R.: The complexity of flowshop and jobshop scheduling. Math. Oper. Res. **1**(2), 117–129 (1976). http://www.jstor.org/stable/3689278
9. Graham, R.L.: Bounds on multiprocessing timing anomalies. SIAM J. Appl. Math. **17**(2), 416–429 (1969). http://www.jstor.org/stable/2099572
10. Heik, D.: Discrete-test-bed-environment-with-multiple-operations (v1) (2023). https://doi.org/10.5281/ZENODO.7906613
11. Heik, D., Bahrpeyma, F., Reichelt, D.: An application of reinforcement learning in industrial cyber-physical systems. In: OVERLAY 2022: 4th Workshop on Artificial Intelligence and Formal Verification, Logic, Automata, and Synthesis (2022)
12. Heik, D., Bahrpeyma, F., Reichelt, D.: Dynamic job shop scheduling in an industrial assembly environment using various reinforcement learning techniques. In: 22nd International Conference on Intelligent Systems Design and Applications (ISDA 2022) (2022)
13. Holland, J.H.: Outline for a logical theory of adaptive systems. J. ACM **9**(3), 297–314 (1962). https://doi.org/10.1145/321127.321128
14. Hussain, K., Salleh, M.N.M., Cheng, S., Shi, Y.: Metaheuristic research: a comprehensive survey. Artif. Intell. Rev. **52**(4), 2191–2233 (2018). https://doi.org/10.1007/s10462-017-9605-z
15. Jing, X., Yao, X., Liu, M., Zhou, J.: Multi-agent reinforcement learning based on graph convolutional network for flexible job shop scheduling. J. Intell. Manuf. (2022). https://doi.org/10.1007/s10845-022-02037-5
16. Kennedy, J., Eberhart, R.: A discrete binary version of the particle swarm algorithm. In: 1997 IEEE International Conference on Systems, Man, and Cybernetics. Computational Cybernetics and Simulation. IEEE (1997). https://doi.org/10.1109/icsmc.1997.637339
17. Kleinrock, L.: Analysis of a time-shared processor. Naval Res. Logist. q. **11**(1), 59–73 (1964)

18. Liu, R., Piplani, R., Toro, C.: Deep reinforcement learning for dynamic scheduling of a flexible job shop. Int. J. Prod. Res. **60**(13), 4049–4069 (2022). https://doi.org/10.1080/00207543.2022.2058432

19. Lohse, O., Haag, A., Dagner, T.: Enhancing Monte-Carlo tree search with multi-agent deep q-network in open shop scheduling. In: 2022 5th World Conference on Mechanical Engineering and Intelligent Manufacturing (WCMEIM), pp. 1210–1215 (2022). https://doi.org/10.1109/WCMEIM56910.2022.10021570

20. Lowe, R., Wu, Y., Tamar, A., Harb, J., Abbeel, P., Mordatch, I.: Multi-agent actor-critic for mixed cooperative-competitive environments. In: Proceedings of the 31st International Conference on Neural Information Processing Systems, pp. 6382–6393. NIPS'17, Curran Associates Inc., Red Hook, NY, USA (2017)

21. Park, I.B., Huh, J., Kim, J., Park, J.: A reinforcement learning approach to robust scheduling of semiconductor manufacturing facilities. IEEE Trans. Autom. Sci. Eng. 1–12 (2020). https://doi.org/10.1109/tase.2019.2956762

22. Popper, J., Motsch, W., David, A., Petzsche, T., Ruskowski, M.: Utilizing multi-agent deep reinforcement learning for flexible job shop scheduling under sustainable viewpoints. In: 2021 International Conference on Electrical, Computer, Communications and Mechatronics Engineering (ICECCME), pp. 1–6 (2021). https://doi.org/10.1109/ICECCME52200.2021.9590925

23. de Puiseau, C.W., Meyes, R., Meisen, T.: On reliability of reinforcement learning based production scheduling systems: a comparative survey. J. Intell. Manuf. **33**(4), 911–927 (2022). https://doi.org/10.1007/s10845-022-01915-2

24. Troxler, P.: Making the 3rd Industrial Revolution. Fab Labs: Of Machines, Makers and Inventors. Transcript Publishers, Bielefeld (2013)

25. Xin-li, X., Ping, H., Wan-Liang, W.: Multi-agent dynamic scheduling method and its application to dyeing shops scheduling. Comput. Integr. Manuf. Syst. **16**(03) (2010)

26. Yan-hai, H., Jun-qi, Y., Fei-fan, Y., Jun-he, Y.: Flow shop rescheduling problem under rush orders. J. Zhejiang Univ.-Sci. A **6**(10), 1040–1046 (2005). https://doi.org/10.1631/jzus.2005.a1040

27. Zhang, G., Shao, X., Li, P., Gao, L.: An effective hybrid particle swarm optimization algorithm for multi-objective flexible job-shop scheduling problem. Comput. Ind. Eng. **56**(4), 1309–1318 (2009). https://doi.org/10.1016/j.cie.2008.07.021, https://www.sciencedirect.com/science/article/pii/S0360835208001666

28. Zhang, Y., Zhu, H., Tang, D., Zhou, T., Gui, Y.: Dynamic job shop scheduling based on deep reinforcement learning for multi-agent manufacturing systems. Robot. Comput.-Integr. Manuf. **78**, 102412 (2022). https://doi.org/10.1016/j.rcim.2022.102412

29. Zhang, Z., Ong, Y.S., Wang, D., Xue, B.: A collaborative multiagent reinforcement learning method based on policy gradient potential. IEEE Trans. Cybern. **51**(2), 1015–1027 (2021). https://doi.org/10.1109/TCYB.2019.2932203

30. Zhou, T., Tang, D., Zhu, H., Zhang, Z.: Multi-agent reinforcement learning for online scheduling in smart factories. Robot. Comput.-Integr. Manuf. **72**, 102202 (2021). https://doi.org/10.1016/j.rcim.2021.102202

31. Zizic, M.C., Mladineo, M., Gjeldum, N., Celent, L.: From industry 4.0 towards industry 5.0: a review and analysis of paradigm shift for the people, organization and technology. Energies **15**(14) (2022). https://doi.org/10.3390/en15145221, https://www.mdpi.com/1996-1073/15/14/5221

Solving Mixed Influence Diagrams
by Reinforcement Learning

S. D. Prestwich[(✉)]

School of Computer Science and Information Technology, University College Cork,
Cork, Ireland
s.prestwich@cs.ucc.ie

Abstract. While efficient optimisation methods exist for problems with special
properties (linear, continuous, differentiable, unconstrained), real-world problems often involve inconvenient complications (constrained, discrete, multi-stage,
multi-level, multi-objective). Each of these complications has spawned research
areas in Artificial Intelligence and Operations Research, but few methods are
available for hybrid problems. We describe a reinforcement learning-based solver
for a broad class of discrete problems that we call Mixed Influence Diagrams,
which may have multiple stages, multiple agents, multiple non-linear objectives,
correlated chance variables, exogenous and endogenous uncertainty, constraints
(hard, soft and chance) and partially observed variables. We apply the solver to
problems taken from stochastic programming, chance-constrained programming,
limited-memory influence diagrams, multi-level and multi-objective optimisation. We expect the approach to be useful on new hybrid problems for which
no specialised solution methods exist.

1 Introduction

Optimisation problems are ubiquitous in artificial intelligence, operations research and
a wide range of application areas. They occur in many forms, each with a range of specialised modelling languages and solution methods: mixed integer linear programming
(MILP), constraint programming (CP), stochastic programming, chance-constrained
programming, dynamic programming, bilevel programming and others. Some hybrids
have been investigated — for example multi-objective reinforcement learning with constraints [19] and bilevel multi-objective stochastic integer linear programming [9] —
and CP and MILP have been extended in various directions: stochastic optimisation,
robust optimisation, multiple objectives and multiple decision makers. However, no
method exists for complex hybrids in general. For example there is no obvious way of
solving a bi-level optimisation problem whose leader is a non-linear stochastic program
with chance constraints, and whose followers are a bi-objective constraint satisfaction
problem and a weighted Max-SAT problem. A researcher faced with such a problem
must invent a new approach.

Influence diagrams (IDs) [18] are a particularly expressive formalism that can model
multi-stage decision problems, in which an agent makes decisions by assigning values

G. Nicosia et al. (Eds.): LOD 2023, LNCS 14506, pp. 255–269, 2024.
https://doi.org/10.1007/978-3-031-53966-4_19

to decision variables, observes the (possibly random) results, then makes further decisions, and so on depending on the number of stages. Chance variables model uncertainties and are not controlled by a decision-maker. They may be independent or related by conditional probability tables as in Bayesian networks (which are a special case of IDs without decisions or utilities). Decision variables may appear in these tables and thus affect chance variable distributions, a feature sometimes called *endogenous uncertainty* as opposed to the more common *exogenous uncertainty* where distributions are independent of decisions. Conversely, decisions may depend on chance variable assignments.

IDs have been extended in several ways, and in this paper we combine many of them into a problem class we call Mixed Influence Diagrams (MIXIDs) which allow:

- *Non-linear utilities.* In fact we allow programmable utilities.
- *Multiple agents.* Each decision variable may belong to any one of a set of agents (decision makers) each with its own utility. Agents may be independent, adversarial or cooperative.
- *Multiple objectives.* Each agent may have more than one utility.
- *Constraints.* Many optimisation problems are constrained, and we allow hard, soft and chance constraints.
- *Partially observed variables.* We allow chance and decision variables to be observable or unobservable by subsequent decision variables.

Each of these extensions have been added to IDs in earlier work, at least to some extent (see below), but to the best of our knowledge they have not been combined into a single framework. The contribution of this paper is to show that many problems can be modelled as MIXIDs and solved using a reinforcement learning-based method called MIXID1. In Sect. 2 we describe MIXID1 and provide references to previous ID extensions. In Sect. 3 we apply it to a variety of problems from different optimisation literatures to demonstrate its flexibility and ease of use. In Sect. 4 we summarise the results and discuss future work.

2   Modelling and Solving MIXIDs

We now introduce the MIXID class of problems and the MIXID1 solver. First we introduce standard IDs.

2.1   Influence Diagrams

IDs are popular graphical models in decision analysis, and they can model important relationships between uncertainties, decisions and values. They were initially conceived as tools for formulating problems, but they have also emerged as efficient computational tools for the evaluation stage.

An ID is a directed acyclic graph with three types of node: *decision nodes* correspond to decision variables and are drawn as rectangles; *chance* (or *uncertain*) *nodes* correspond to chance variables and are drawn as ovals; *value nodes* correspond to preferences or objectives and are drawn as rounded rectangles, or polygons such as diamonds.

Each chance variable is associated with a conditional probability table that specifies its distribution for every combination of values for its parent variables in the graph. Each decision variable also has a set of parent variables in the graph, and its value depends only on their values. The decision variables are usually considered to be temporally ordered, and chance variables are observed at different points in the ordering. A standard ID assumption is *non-forgetting* which means that the parents of any decision variable are all its ancestors in the graph: thus a decision may depend on everything that has occurred before. All variables are discrete.

Each value node is associated with a table showing the *utility* (a real number) of each combination of parent variable values. A *decision policy* is a rule for each decision variable indicating how to choose its value from those of its parents. Any policy has a total *expected utility* (it is an expectation because of the chance variables) and *solving* an ID means computing its optimal policy, which has maximum expected utility. An example of an ID is shown in Sect. 3.1.

2.2   The Basic Algorithm

We now describe a reinforcement learning (RL) approach to solving standard IDs, similar to that described in [32]. Suppose we have a single agent with one objective f to be minimised: a standard ID. Each variable v_i ($i = 1, \ldots, n$) is either a decision variable or a chance variable. The variables are strictly ordered $v_1 < \ldots < v_n$ according to their temporal ordering. Denote the current assignment of v by x_v. Each variable v_i has a domain size s_v indicating that it takes values from its domain $D_v = \{0, 1, \ldots, s_v - 1\}$.

> initialise $\epsilon = 1$, $\alpha = 1$ and $q_{v,i} = 0$
> for episode $e = 1 \ldots E$
>   update ϵ and α
>   for $v = v_1 \ldots v_n$
>   if v is chance
>   sample x_v from its distribution
>   else with probability ϵ
>   randomly sample $x_v \in D_v$
>   else
> $x_v = \operatorname{argmin}_{i \in D_v} (q_{v,i})$
>   compute objective $f(\vec{v})$
>   for $v = v_1 \ldots v_n$
> $q_{v,x_v} = \alpha f + (1 - \alpha) q_{v,x_v}$

Fig. 1. The MIXID1 algorithm for standard IDs

The basic algorithm is shown in Fig. 1. It is based on a simple RL algorithm: infinite-step tabular SARSA with ϵ-soft action selection and learning rate α [36]. The RL reward is the objective of the optimisation problem (the ID utility) which is computed at the end of each episode when all variables have been assigned updated values. The user must provide code for this step. Note that in our IDs there is only one value node, which includes all those values that would usually be computed earlier. This simplifies the implementation without loss of generality. *Infinite-step* indicates that the reward is backed up equally to all values in the episode, which is more robust than Q-learning in the presence of unobserved variables [36]. *Tabular* indicates that state-action pairs have values in a table. The discount factor γ is set to 1 as the RL problem is episodic.

The ϵ and α parameters start at 1 and decay to 0. Many decay schemes have been proposed in RL, and we arbitrarily choose $\alpha = \epsilon = ((E - e)/E)^3$. In our use of RL a *state* is an assignment to variables $v_1 \ldots v_i$ for some $i = 1 \ldots n$, an *action* is the assignment of decision variable v_{i+1}, an *episode* assigns all the variables, and the *reward* is the objective value computed at the end of an episode. The q_{v,x_v} are state-action values used to define the policy, which should optimise the expected reward.

Each decision depends on all previous decisions and random events (but see Sect. 2.3) so a policy might involve a huge number of distinct states. To combat this problem, RL algorithms group together states via *state aggregation* methods. For our prototype method we choose a simple form called *random tile coding* [36], specifically *Zobrist hashing* [41] with H hash table entries for some large integer H. This works as follows. To each (decision or chance) variable-value pair $\langle v, x \rangle$ we assign a random integer r_{vx} which remains fixed. At any point during an episode we have some set S of assignments $\langle v, x \rangle$, and we take the exclusive-or of the r_{vx} values (that it, their bit patterns) associated with assignments $X_S = \bigoplus_{\langle v,x \rangle \in S} r_{vx}$. Finally, we use X_S to index an array V with H entries: the value of q_{v,x_v} is stored in $V[X_S \bmod H]$. It might be expected that Zobrist hashing will perform poorly when the number of states approaches or exceeds the size of the hash table, because hash collisions will confuse the values of different state-action pairs. Surprisingly, it can perform well even when hash collisions are frequent, and has been used in chess programming [20].

2.3 Partially Observed Variables

Standard IDs are designed to handle situations involving a single, non-forgetful agent. Limited memory influence diagrams (LIMIDs) [25] are generalizations of IDs that allow decision making with limited information and simultaneous decisions, and can have much smaller policies. They relax the regularity (total variable ordering) and non-forgetting assumptions of IDs. LIMIDs are considered harder to solve optimally than IDs. We handle the limited memory feature in a simple way: during Zobrist hashing, the set S contains only those assignments that are visible to the decision variable (its parent variables, indicated by a link in the ID graph).

2.4 Multiple Objectives

Many real-world applications have multiple objectives and we must find a trade-off. Multi-objective (or multi-criteria) IDs were described in [8, 28]. Objectives can be combined in more than one way in RL, and a simple and popular approach is *linear scalarisation*: take a linear combination of the objectives, giving a single objective that can be used as a RL reward. This has the drawback that it can not generate any solutions in a non-convex region of the Pareto front. It can also be hard to choose appropriate weights, especially when the objectives use different units.

Another approach is to rank the objectives by importance, then search for the lexicographically best result. This has the same problem as linear scalarisation (though recent work addresses this [35]) which can be fixed by thresholding all but the least important objective, a method called *thresholded lexicographic ordering* [11]. However, it has the drawback of using a specific RL algorithm. Moreover, in some of our applications the most important objective is to minimise constraint violations, and this should not be thresholded.

The method we choose is to reduce the multiple objectives to a single objective via *weighted metric scalarisation* [30] in which the distance is minimised between the vector of values f_o for objectives o and a *utopian point* u^* in the multi-objective space:

$$\min_x \left(\sum_o w_o |f_o(x) - u_o^*|^p \right)^{1/p}$$

for some $p \geq 1$ and weights w_o chosen by the user. The need to choose weights is a disadvantage in terms of user-friendliness, but an advantage is that we can tune them to find different points on the Pareto front. The special case of *Chebyshev scalarisation* ($p = \infty$) is theoretically guaranteed to make the entire Pareto front reachable, but the L_p-norms ($p = 1, 2$) may perform better in practice [14] and we found best results with L_1.

The utopian point is often adjusted during search to be just beyond the best point found so far, but MIXID1 uses a fixed u^* provided by the user. Depending on whether each objective is to be maximised or minimised, we choose a value that is optimal or high/low enough to be unattainable.

2.5 Hard and Soft Constraints

Many optimisation problems have constraints that must be met by any solution. Constraints are not modelled in standard IDs but hard constraints were supported in information/relevance IDs [21]. We model soft constraints by minimising their violations as an additional objective. To model hard constraints we use the popular technique of choosing a large weight to prioritise minimising violations to zero.

2.6 Chance Constraints

A *chance* (or *probabilistic*) *constraint* is a constraint that should be satisfied with some probability threshold: $\Pr[C(x, r)] \geq \theta$ for a constraint C on decision variables x and chance variables r. Chance-constrained programming [5] is a method of optimising under uncertainty and has many applications, as it is a natural way of modelling uncertainty. Chance constraints are usually inequalities but we allow any form of constraint. Chance constraints have also been used in the area of Safe (or Constrained) Reinforcement Learning using various approaches [13, 17], and our approach is related to that of [19]. Chance constraints do not seem to have been added to IDs in previous work.

We model chance constraints by adding a new objective for each, with a reward of 1 for satisfaction and 0 for violation, and setting the utopian value for that objective to the desired probability threshold. The weight attached to the objective should be high.

2.7 Multiple Agents and Non-linear Objectives

Many problems in economics, diplomacy, war, politics, industry, gaming and other areas involve multiple agents, which form part of each other's environment. Multi-agent RL can be applied to these problems, as can several forms of ID: bi-Agent IDs [16], multi-agent IDs [23], game theory-based IDs [40], networks of IDs [12] and interactive dynamic IDs [15]. LIMIDs can model multiple agents, but only the cooperative case in which they all have the same objective.

We extend MIXID1 to allow multiple agents numbered $1, 2, \ldots$ Each decision variable v belongs to an agent a_v, and we use 0 as a dummy agent number for chance variables (sometimes called a *chance agent*). Each agent $a \geq 1$ has its own objective f^a, each of which is computed at the end of an episode. The algorithm is modified so that each reward f^a is backed up only to values q_{v,x_v} such that $a_v = a$. This transforms MIXID1 into a multi-agent RL algorithm.

3 Experiments

We now illustrate the flexibility of the method by applying it to a variety of problems taken from diverse literatures. We do not compare it with standard methods in terms of efficiency, as this is not the goal of the paper. For the same reason we do not report runtimes, though they are quite short: typically a few seconds and at most a few minutes. In fact we do not expect MIXID1 to be competitive on any particular class of problem, though as a RL method it should perform well on some applications. Our aim here is only to demonstrate that a single solver can solve a wide variety of optimisation problems, thus filling a gap in optimisation technology. In future work we will explore more efficient RL methods, especially using deep neural networks for state aggregation.

3.1 A Standard ID

As a first example we use the well-known Oil Wildcatter ID shown in Fig. 2. An oil wildcatter must decide either to drill or not to drill for oil at a specific site. Before drilling, they may perform a seismic test that will help determine the geological structure of the site. The test result can be *closed* (indicating significant oil), *open* (indicating some oil) or *diffuse* (probably no oil). The special value *notest* means that test results are unavailable if the seismic test is not done. There are two decision variables Test (T) and Drill (D), and two chance variables Seismic (S) and Oil (O).

The Test decision does not depend on any other variable, but the Drill decision depends on whether a test was made, and if so on its result. The Oil variable is unobservable so no decision depends on it (this is distinct from *forgetting* its value which is addressed in Sect. 2.3). The payoff tables model costs and benefits as utilities: the test payoff is negative (a cost) if the test is performed, otherwise zero; the drill payoff

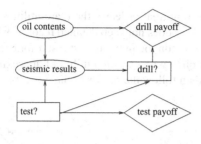

oil contents	$P(O)$		
	dry	wet	soak
	0.5	0.3	0.2

test payoff	$U_1(T)$
Test?	
yes	-10
no	0

Seismic results		$P(S|O,T)$			
Oil cnt.	Test?	closed	open	diffuse	notest
dry	yes	0.1	0.3	0.6	0
dry	no	0	0	0	1
wet	yes	0.3	0.4	0.3	0
wet	no	0	0	0	1
soak	yes	0.5	0.4	0.1	0
soak	no	0	0	0	1

Drill payoff		$U_2(O,D)$
Oil cnt.	Drill?	
dry	yes	-70
dry	no	0
wet	yes	50
wet	no	0
soak	yes	200
soak	no	0

Fig. 2. The oil wildcatter ID

may be positive if oil is found, negative if not, and zero if no drilling occurs. In our implementation the two tables are computed after all variables are assigned values.

We use probabilities and utilities as given in [28]. The known optimal policy for this problem is: apply the seismic test, and drill if the test result is open or closed. MIXID1 finds this solution which has expected utility 22.5.

3.2 Partially Observed Variables

As an example we use a pig breeding problem from [25]. A pig breeder grows pigs for four months then sells them. During this period the pig may or may not develop a disease. If it has the disease when it must be sold, then it must be sold for slaughter and its expected market price is 300. If it is disease-free then its expected market price is 1000. Once a month a veterinary surgeon test the pig for the disease. If it is ill then the test indicates this with probability 0.80, and if it is healthy then the test indicates this with probability 0.90. At each monthly visit the surgeon may or may not treat the pig, and the treatment costs 100. A pig has the disease in month 1 with probability 0.10. A healthy pig develops the disease in the next month with probability 0.20 without treatment and 0.10 with treatment. An unhealthy pig remains unhealthy in the next month with probability 0.90 without treatment, and 0.50 with treatment.

In the LIMID shown in Fig. 3 the h are random variables denoting healthy or unhealthy, the t are random variables denoting positive or negative test results, the d are decision variables denoting treat or leave, the u are utilities, and the numbers 1–4 denote month. The pig breeder does not keep detailed records, which are forgotten, and bases treatment decisions only on the previous test result for each pig. MIXID1 almost always finds the optimal policy with expected utility of approximately 727: ignore the first test and do not treat, then follow the results of the other two tests (treat if positive).

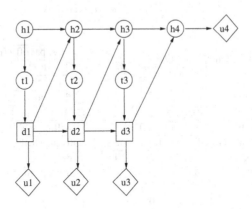

Fig. 3. The pig breeding LIMID.

3.3 Multiple Objectives

We use the bi-objective oil wildcatter ID in [28]. In addition to maximising payoff the aim is to minimise environmental damage. The utility of testing is now $(-10, 10)$ while the utility of drilling is $(-70, 18)$, $(50, 12)$ and $(200, 8)$ for dry, wet and soak respectively: in each case the first value is the original utility while the second value refers to the new utility. Instead of one optimal solution there are four Pareto-optimal solutions, each with two utility values: (1) test, then drill if the result is closed or open, with utility $(22.5, 17.56)$; (2) do not test but drill, with utility $(20, 14.2)$; (3) test, then drill if the result is closed, with utility $(11, 12.78)$; (4) do not test or drill, with utility $(0, 0)$. In multiple runs MIXID1 found policies (1), (2) and (4) using a utopian point $(1000,-1000)$ and weights $(0.55,0.45)$, though it did not find policy (3).

3.4 Hard and Soft Constraints

We take an example from stochastic programming [3] which models and solves problems involving decision and chance variables, with known distributions for the latter. Problems may have one or more stages: in each stage decisions are taken then chance variables are observed. A solution is not a simple assignment of variables, but a policy that tells us how to make decisions given assignments to variables from earlier stages. Stochastic programming dates back to the 1950s and is now a major area of research in mathematical programming and operations research. We use a 2-stage stochastic program from [1]:

$$\max z = 1.5x_1 + 4x_2 + E[Q(x_1, x_2)]$$
where $Q(x_1, x_2)$ is the value of
$$\max 16y_1 + 19y_2 + 23y_3 + 28y_4 \text{ such that}$$
$$2y_1 + 3y_2 + 4y_3 + 5y_4 \leq \xi_1 - x_1$$
$$6y_1 + y_2 + 3y_3 + 2y_4 \leq \xi_2 - x_2$$
where $0 \leq x_1, x_2 \leq 5$, $y + i \in \{0, 1\}$ $(i = 1, 2, 3, 4)$
ξ_1, ξ_2 uniformly distributed on $\{5, 5.5, \ldots, 15\}$

A solution to this problem consists of fixed assignments to x_1, x_2 and assignments to the y_i that depend on the random ξ values. The optimum policy has objective 61.32 with $(x_1, x_2) = (0, 4)$.

We model the problem as a MIXID with two objectives: the first to minimise the number of constraint violations to 0, and the second to maximise z. Decision variables can observe all ancestor variables. We use a utopian point $(-10, 100)$ and weights $(0.99, 0.01)$: note that $0.99 \gg 0.01$, chosen because the constraints are hard. In multiple runs MIXID1 found policies with (x_1, x_2) equal to $(0, 4)$ or $(0, 3)$ and expected reward approximately 60.

3.5 Chance Constraints

We modify the stochastic program of Sect. 3.4 by attaching probabilities to the two hard constraints:
$$\Pr[2y_1 + 3y_2 + 4y_3 + 5y_4 \leq \xi_1 - x_1] = 0.7$$
$$\Pr[6y_1 + y_2 + 3y_3 + 2y_4 \leq \xi_2 - x_2] = 0.9$$

Notice that the chance constraints are of the form $\Pr[\cdot] = p$ instead of the usual $\Pr[\cdot] \geq p$. MIXID1 is not guaranteed to satisfy the probability thresholds because of its multi-objective approach: instead of forcing the satisfaction of a chance constraint to exceed a threshold probability, it tries to match the threshold in a trade-off with other objectives. However, it can be used as an exploratory tool to find a policy with the desired characteristics, and the user can iteratively increase thresholds.

As described in Sect. 2.5 these chance constraints are implemented by adding a new objective for each, but this time with utopian point $(0.7, 0.9, 100)$. To solve this tri-objective MIXID we experimented with different weights and found a variety of policies, with different compromises between the chance constraints and original objective. Not all solutions were useful, but using weights $(0.496, 0.496, 0.008)$ we found a policy with objectives $(0.74, 0.95, 71)$ that satisfies the requirements. Relaxing the hard constraints to chance constraints allowed us to increase z from 60 to 71.

3.6 Multiple Agents and Non-linear Objectives

We use a *tri-level program* studied in at least two publications [2, 29] and shown in Fig. 4. Tri-level programs have been used to model problems in supply chain management, network defence, planning, logistics and economics. They involve three agents:

the first is the *top-level leader*, the second is the *middle-level*) follower, and the third is the (*bottom-level*) follower. They are organised hierarchically: the leader makes decisions first, the middle-level follower reacts, then the bottom-level follower reacts. Many algorithms have been published on tri-level (also bi- and multi-level) programs, but mostly on linear continuous problems and relatively little work has been done on discrete problems.

$$\max_{x_1} z_1 = (x_1 + x_2 + 2x_3 + 4)(-x_1 - x_2 + x_3 + 2x_4 + 1)$$
where x_2 solves
$$\max_{x_2} z_2 = 2x_2 + x_3 + 3x_4$$
where $x_3, x_3, x_3, x_3, x_3, x_3$ solve, for given x_1, x_2
$$\max_{x_3 \ldots x_8} z_3 = \frac{2x_1 + 3x_2 + 2x_3 - 3x_4}{5x_1 + 11x_2 + x_5 + 29}$$
subject to
$$-3x_1 + 7x_2 + x_3 + x_5 = 10 \qquad 14x_1 + 4x_2 + x_6 = 6$$
$$x_1 + x_2 + x_3 - x_4 + x_7 = 5 \qquad 2x_1 + x_2 + 2x_4 + x_8 = 8$$
$$0 \le x_1 \le 5 \qquad 0 \le x_2 \le 2 \qquad 5 \le x_3 \le 20 \qquad 4 \le x_4 \le 30$$
$$1 \le x_5 \le 20 \qquad 0 \le x_6 \le 30 \qquad 0 \le x_7 \le 15 \qquad 0 \le x_8 \le 20$$

Fig. 4. A tri-level non-linear program

This problem is also of interest because its objectives are quadratic for the leader, linear for the first (*middle-level*) follower, and fractional for the second (*bottom-level*) follower. Non-linear objectives were added to IDs in [24] using non-linear optimal control approximations.

We model the problem as a MIXID with 3 agents, each with 2 objectives, and each variable observing the values of all previous variables. The first objective used by all agents is the number of constraint violations, which should be 0: the constraints ensure that any solution is *tri-level feasible* so all agents should be penalised for violations (in this problem no agent is allowed to make a decision that leads to a constraint violation). The secondary objectives are z_1, z_2, z_3. We choose a utopian point for each agent and objective: $((0, 1000), (0, 100), (0, 1))$ and 3 pairs of weights:

$$((0.99999, 0.00001), (0.999, 0.001), (0.99, 0.01))$$

In repeated runs MIXID1 finds the correct policy $(0, 0, 9, 4, 1, 6, 0, 0)$ with objectives $((0, 396), (0, 21), (0, 0.2))$ with approximately 50% success.[1] The suboptimal solutions that it sometimes finds can be recognised by their lower z_1 values:

$$((0, 196), (0, 17), (0, -0.06)), \quad ((0, 240), (0, 18), (0, 0)) \quad \text{and} \quad ((0, 288), (0, 19),$$
$$(0, 0.06)).$$

3.7   Case Study

As a case study to illustrate the system's ease of use, we now show the code required for our most complex example: the tri-level program of Sect. 2.7. MIXID1 is implemented

---

[1] The first objective is given as 612 in [29] but it can be verified that the solution yields $z_1 = 396$.

in C, and the user must provide: (i) a text file specifying the variables, their agent numbers, their domains, their parent variables, how many objectives each agent has, and the conditional probability tables; (ii) a C function that takes as input a 1-dimensional array of (decision and chance) variable values, and computes as output a 2-dimensional array of agent-utility values.

Firstly, the text file describing the problem is shown in Fig. 5 (the format is currently a simple research prototype but it could be adapted to a more user-friendly form). The meaning of the numbers is as follows:

```
8
1 6 8
2 3 0 8
3 16 0 1 8
3 27 0 1 2 8
3 20 0 1 2 3 8
3 31 0 1 2 3 4 8
3 16 0 1 2 3 4 5 8
3 21 0 1 2 3 4 5 6 8
8
8
2 0.0 1000.0 0.99999 0.00001
2 0.0 100.0 0.999 0.001
2 0.0 1.0 0.99 0.01
```

Fig. 5. Text file describing the tri-level MIXID

- The first line simply states that there are 8 variables numbered 0–7. We can therefore use "8" as a list terminator in the rest of the file.
- The next 8 rows describe each variable in turn: which agent it belongs to (0 would indicate a random variable but there are none in this example), its domain size D (values are assumed to be $0, 1, \ldots, D - 1$), then a list of parent variables ending with terminator 8. As the highest agent number was 3, the method now knows that there are 3 agents. In this example each variable has all ancestors as parents.
- Next, each conditional probability table is listed, followed by a terminator. In this example there are no tables.
- Next, the user may provide a list of observed chance variable values, followed by a terminator. This feature is not used in this paper, but we briefly discuss it in Sect. 4.
- Finally, for each agent we input a number ω of objectives (2 in this example) followed by a list of ω weights, and a list of ω utopian values.

Secondly, we show the C utility function that computes the objectives. Its input is a 1-dimensional integer array called values and its output is a 2-dimensional real array called results (we assume that there will be no more than 10 agents and objectives, but these numbers can easily be increased):

```
void utility(int* values, float results[10][10]) {
```

To adapt the notation to that in Fig. 4, and to adjust the variable domains to the desired ranges, we create new integer variables x1–x8:

```
int x1=values[0],    x2=values[1],
    x3=values[2]+5,  x4=values[3]+4,
    x5=values[4]+1,  x6=values[5],
    x7=values[6],    x8=values[7];
```

Compute the number of constraint violations:

```
int violations= !(-3*x1+7*x2+x3+x5==10)+
                !(14*x1+4*x2+x6==6)+
                !(x1+x2+x3-x4+x7==5)+
                !(2*x1+x2+2*x4+x8==8);
```

This becomes the first objective (index 0) for each agent:

```
results[1][0]=violations;
results[2][0]=violations;
results[3][0]=violations;
```

The other objective (index 1) for each agent is its given objective function z:

```
results[1][1]=(x1+x2+2*x3+4)*(-x1-x2+x3+2*x4+1);
results[2][1]=2*x2+x3+3*x4;
results[3][1]=((float) 2*x1+3*x2+2*x3-3*x4)/
             ((float) 5*x1+11*x2+x5+29); }
```

Note that no algorithmic details are required.

4   Conclusion

This paper makes two contributions. Firstly, it combines several known (and some new) ID extensions into a single framework we call a MIXID. Secondly, it proposes a MIXID solution method called MIXID1 based on RL, which has been proposed as a unifying approach to sequential decision making under uncertainty [31]. MIXID1 is a lightweight solver that does not need a graphics processing unit or other specialised hardware. We demonstrated the flexibility of the approach using examples taken from diverse literatures: ID, LIMID, multi-objective ID, stochastic programming, chance-constrained programming and tri-level programming. To the best of our knowledge, no other solver can tackle this range of optimisation problems, and we expect MIXIDs to be useful for hybrid optimisation problems that do not fit well into any particular class. A fruitful source of applications might be Safe RL in which constraints are used to ensure robust and safe policies.

Several methods exist for solving IDs. Some are exact and based on variable elimination [7,22,33,34,37] while others are approximate [4,6,27,38,39]. However, we know of only one other application of RL to solving IDs [32], which seems surprising as both IDs and RL can be used to model and solve sequential decision problems. The main connection usually made between the two is that IDs can model problems that can be tackled by RL, for example Causal IDs have been used to model Artificial General Intelligence safety frameworks which often use RL [10].

An aspect of IDs that we have ignored in this paper is *evidence propagation*. As in Bayesian networks, an ID might be provided with *evidence*: observed values of chance variables. This evidence can be propagated back to earlier chance variables to learn posterior distributions. Rejection sampling was used in [32] and also works with MIXID1, but this becomes impractical when probabilities are very small. Initial experiments with Gibbs sampling are promising, and evidence propagation will be investigated in future work. We are also interested in adding continuous variables, and using more powerful state aggregation than random tile coding: both improvements might be achieved by moving from SARSA to an actor-critic RL algorithm.

Acknowledgments. This material is based upon works supported by the Science Foundation Ireland under Grant No. 12/RC/2289-P2 which is co-funded under the European Regional Development Fund. We would also like to acknowledge the support of the Science Foundation Ireland CONFIRM Centre for Smart Manufacturing, Research Code 16/RC/3918, and the EU H2020 ICT48 project TAILOR under contract #952215.

References

1. Ahmed, S., Tawarmalani, M., Sahinidis, N.V.: A finite branch-and-bound algorithm for two-stage stochastic integer programs. Math. Program. **100**, 355–377 (2004)
2. Arora, R., Arora, S.R.: An algorithm for non-linear multi-level integer programming problems. Int. J. Comput. Sci. Math. **3**(3), 211–225 (2010)
3. Birge, J.R., Louveaux, F.V.: Introduction to Stochastic Programming. Springer, New York (2011). https://doi.org/10.1007/978-1-4614-0237-4
4. Cano, A., Gómez, M., Moral, S.: A forward-backward Monte Carlo method for solving influence diagrams. Int. J. Approx. Reason. **42**, 119–135 (2006)
5. Charnes, A., Cooper, W.W.: Chance-constrained programming. Manag. Sci. **6**(1), 73–79 (1959)
6. Charnes, J.M., Shenoy, P.P.: Multistage Monte Carlo method for solving influence diagrams using local computation. Manag. Sci. **50**(3), 405–418 (2004)
7. Dechter, R.: A new perspective on algorithms for optimizing policies under uncertainty. In: Artificial Intelligence Planning Systems, pp. 72–81 (2000)
8. Diehl, M., Haimes, Y.: Influence diagrams with multiple objectives and tradeoff analysis. IEEE Trans. Syst. Man Cybern. Part A **34**(3), 293–304 (2004)
9. Elshafei, M.M.K., El-Sherberry, M.S.: Interactive Bi-level multiobjective stochastic integer linear programming problem. Trends Appl. Sci. Res. **3**(2), 154–164 (2008)
10. Everitt, T., Kumar, R., Krakovna, V., Legg, S.: Modeling AGI safety frameworks with causal influence diagrams. In: Proceedings of the Workshop on Artificial Intelligence Safety, CEUR Workshop, vol. 2419 (2019)
11. Gábor, Z., Kalmár, Z., Szepesvári, C.: Multi-criteria reinforcement learning. In: Proceedings of the 15th International Conference on Machine Learning, pp. 197–205 (1998)
12. Gal, Y., Pfeffer, A.: Networks of influence diagrams: a formalism for representing agents' beliefs and decision-making processes. J. Artif. Intell. Res. **33**, 109–147 (2008)
13. García, J., Fernández, F.: A comprehensive survey on safe reinforcement learning. J. Mach. Learn. Res. **16**, 1437–1480 (2015)

14. Giagkiozis, I., Fleming, P.J.: Methods for multi-objective optimization: an analysis. Inf. Sci. **293**, 1–16 (2015)
15. Polich, K., Gmytrasiewicz, G.: Interactive dynamic influence diagrams. In: Proceedings of the 6th International Joint Conference on Autonomous Agents and Multiagent Systems, Communications in Computer and Information Science, vol. 288, pp. 623–630 (2007)
16. González-Ortega, J., Insua, D.R., Cano, J.: Adversarial risk analysis for bi-agent influence diagrams: an algorithmic approach. Eur. J. Oper. Res. **273**(3), 1085–1096 (2019)
17. Gu, S., et al.: A Review of Safe Reinforcement Learning: Methods, Theory and Applications. CoRR abs/2205.10330 (2022)
18. Howard, R.A., Matheson, J.E.: Influence Diagrams. Readings in Decision Analysis, Strategic Decisions Group, Menlo Park, CA, chapter 38, pp. 763–771 (1981)
19. Huang, S.H., et al.: A constrained multi-objective reinforcement learning framework. In: CoRL, pp. 883–893 (2021)
20. Hyatt, R.M., Cozzie, A.: The effect of hash signature collisions in a chess program. ICGA J. **28**(3), 131–139 (2005)
21. Jenzarli, A.: Information/relevance influence diagrams. In: Proceedings of the 11th conference on Uncertainty in Artificial Intelligence (UAI), Quebec, Canada, pp. 329–337 (1995)
22. Jensen, F., Jensen, V., Dittmer, S.: From influence diagrams to junction trees. In: Uncertainty in Artificial Intelligence, pp. 367–363 (1994)
23. Koller, D., Milch, B.: Multi-agent influence diagrams for representing and solving games. Games Econ. Behav. **45**(1), 181–221 (2001)
24. Kratochvíl, V., Vomlel, J.: Influence diagrams for speed profile optimization. Int. J. Approx. Reason. **88**, 567–586 (2017)
25. Lauritzen, S.L., Nilsson, D.: Representing and solving decision problems with limited information. Manag. Sci. **47**, 1238–1251 (2001)
26. Lee, J., Marinescu, R., Ihler, A., Dechter, R.: A weighted mini-bucket bound for solving influence diagrams. In: Proceedings of the Conference on Uncertainty in Artificial Intelligence (2019)
27. Marinescu, R., Lee, J., Dechter, R.: A new bounding scheme for influence diagrams. In: Proceedings of the 35th Conference on Artificial Intelligence, pp. 12158–12165 (2021)
28. Marinescu, R., Razak, A., Wilson, N.: Multi-objective influence diagrams. In: Proceedings of the Conference on Uncertainty in Artificial Intelligence (2012)
29. Mishra, S., Verma, A.B.: A non-differential approach for solving tri-level programming problems. Am. Int. J. Res. Sci. Technol. Eng. Math. (2015)
30. van Moffaert, K., Drugan, M.M., Nowé, A.: Scalarized multi-objective reinforcement learning: novel design techniques. In: Proceedings of the IEEE Symposium on Adaptive Dynamic Programming and Reinforcement Learning, pp. 191–199. IEEE (2013)
31. Powell, W.B.: Reinforcement Learning and Stochastic Optimization: A Unified Framework for Sequential Decisions. Wiley, Hoboken (2022)
32. Prestwich, S.D., Toffano, F., Wilson, N.: A probabilistic programming language for influence diagrams. In: Proceedings of the 11th International Conference on Scalable Uncertainty Management (2017)
33. Shachter, R.D.: Evaluating influence diagrams. Oper. Res. **34**(6), 871–882 (1986)
34. Shenoy, P.: Valuation-based systems for Bayesian decision analysis. Oper. Res. **40**(1), 463–484 (1992)
35. Skalse, J., Hammond, L., Griffin, C., Abate, A.: Lexicographic multi-objective reinforcement learning. In: Proceedings of the 31st International Joint Conference on Artificial Intelligence, pp. 3430–3436 (2022)
36. Sutton, R.S., Barto, A.G.: Reinforcement Learning: An Introduction. MIT Press, Cambridge, MA (1998)

37. Dynamic programming and influence diagrams. IEEE Trans. Syst. Man Cybern. **20**(1), 365–379 (1990)
38. Watthayu, W.: Representing and solving influence diagram in multi-criteria decision making: a loopy belief propagation method. In: Proceedings of the International Symposium on Computer Science and Its Applications, pp. 118–125 (2008)
39. Yuan, C., Wu, X.: Solving influence diagrams using heuristic search. In: Proceedings of the International Symposium on Artificial Intelligence and Mathematics (2010)
40. Zhou, L.H., Kevin, L., Liu, W.Y.: Game theory-based influence diagrams. Expert Syst. **30**(4), 341–351 (2013)
41. Zobrist, A.L.: A new hashing method with application for game playing. Technical report 88, Computer Sciences Department, University of Wisconsin, Madison, Wisconsin (1969). Also: International Computer Chess Association Journal 13(2), 69–73, 1990

Multi-scale Heat Kernel Graph Network for Graph Classification

Jong Ho Jhee[1,3]🆔, Jeongheun Yeon[3]🆔, Yoonshin Kwak[3]🆔,
and Hyunjung Shin[2,3(✉)]🆔

[1] Ajou University School of Medicine, Suwon 16499, South Korea
[2] Department of Industrial Engineering, Ajou University, Suwon 16499, South Korea
[3] Department of Artificial Intelligence, Ajou University, Suwon 16499, South Korea
{baical77,yjh970,yoonshin,shin}@ajou.ac.kr

Abstract. Graph neural networks (GNNs) have been shown to be useful in a variety of graph classification tasks, from bioinformatics to social
networks. However, most GNNs represent the graph using local neighbourhood aggregation. This mechanism is inherently difficult to learn
about the global structure of a graph and does not have enough expressive power to distinguish simple non-isomorphic graphs. To overcome
the limitation, here we propose multi-head heat kernel convolution for
graph representation. Unlike the conventional approach of aggregating
local information from neighbours using an adjacency matrix, the proposed method uses multiple heat kernels to learn the local information
and the global structure simultaneously. The proposed algorithm outperforms the competing methods in most benchmark datasets or at least
shows comparable performance.

Keywords: Heat kernel · Graph convolutional networks · Local and
global structure · Graph classification

1 Introduction

Graph neural networks (GNNs) have emerged as a popular application for various graph-related problems. The most pronounced is in the domain of social
networks, chemical/molecular graphs, and bioinformatics [5,23]. GNNs are a
combination of a set of transition functions and a set of transformation functions, representing the node update process by aggregating information from
their neighbouring nodes and by transforming node features through neural networks. This process involves an aggregate function that first takes the features
of each node and their adjacent nodes as inputs and outputs the transformed
features of the nodes.

Many GNN variants have been introduced and are used in tasks such as node
classification, link prediction, graph regression and classification, and more [1]. In
graph convolutional networks (GCN), the algorithm learns node embeddings by
aggregating relevant features from a node's neighbourhood [12,13]. [11] proposed

G. Nicosia et al. (Eds.): LOD 2023, LNCS 14506, pp. 270–282, 2024.
https://doi.org/10.1007/978-3-031-53966-4_20

GraphSAGE, a framework that can generalize the node embedding by changing the inherently transductive settings to inductive settings by sampling a node's neighbours. [16] proposed k-hop GNN, which aggregates k-hop neighbours for graph classification, and [20] proposed graph isomorphism network (GIN), a simple GNN architecture that has discriminative power equal to Weisfeiler-Lehman test.

However, such an approach only aggregates the local information from adjacent nodes and is inherently unable to learn the global structure of a graph. Graphs can have certain types of structures: hierarchies, hubs and peripherals (nodes tend to be preferentially attached to specific nodes), communities (nodes tend to be grouped into several clusters), and so on. Those underlying structures cannot be observed only by looking at local connections alone. One of the ways to mitigate the limitation is to use a heat kernel. The heat kernel has the advantage of capturing local connectivity and global structure by selecting the appropriate diffusion parameters. As the value of the diffusion parameter increases, the influence of the node can propagate over the entire graph, and conversely, as the value decreases, it can only reach its surroundings. Figure 1 shows the extent of influence with changes in diffusion parameters. In this light, the heat kernel can widen the narrow range of the GNN representation for graphs.

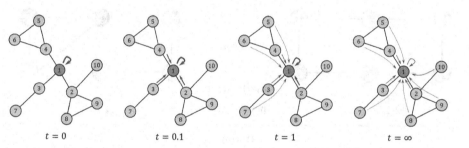

Fig. 1. Node aggregation using heat kernel. Nodes and weights change along with diffusion parameter t.

Another limitation of the GNNs is the inability to distinguish between simple non-isomorphic graphs. [7]. To rephrase this, it is not expressive beyond the Weisfeiler-Lehman test [15,20,21]. Figure 2 shows a typical example, two non-isomorphic graphs. Assuming the values of input features are set to node degrees, the output of GCN convolution and HKGN convolution are shown. The simplest GNN (using only convolution without transformation by learnable weights or activation function) represents both graphs as $[4, 4, 4, 4, 4, 4]$. This is because both are regular graphs, so all nodes in the graph have the same number of adjacent nodes. Therefore, the output of GNN fails to distinguish between the two graphs. On the other hand, the proposed HKGN can identify the two graphs as $[2, 2, 2, 2, 2, 2]$ and $[1, 1, 1, 1, 1, 1]$, respectively. More explanations are provided in Sect. 2.

To solve these problems, we propose a graph classification method based on novel graph embedding using heat kernel convolution. Unlike the previous approach of aggregating local information from neighbours using an adjacency matrix, the proposed method simultaneously learns the local information and the global structure. Multiple heat kernels are incorporated for this purpose. For each of the heat kernels, the extent of diffusion in Fig. 2 is automatically determined by treating it as a trainable parameter. Also, the kernels are combined automatically by employing a multi-head scheme.

The rest of the paper is organized as follows. In Sect. 2, the general background of graph data and the typical GNN model are described. Then, the heat kernel on a graph is derived from the heat equation, following the proposed HKGN, which encompasses graph convolution with heat kernels. Section 3 compares GNN baselines with HKGN on graph classification benchmark datasets. We validate the proposed method on nine graph classification benchmarks. Also, heat kernel analysis is provided to corroborate the effect of the heat kernel on graph representation. The experiments show that our approach outperforms other baseline methods on most benchmarks and learns essential heat kernels for graph representation. Lastly, we conclude with a summarization of the paper and give some directions for future work in the last section.

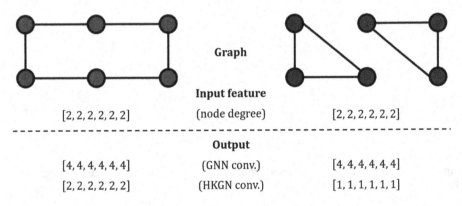

Fig. 2. Example of two non-isomorphic graphs. Input features (top) and Output features (bottom) of simple graph convolutions (GNN conv. And HKGN conv.) are shown.

2   Heat Kernel Graph Networks

The key idea of the proposed method, Heat Kernel Graph Networks (HKGN), is to diversify the graph convolution according to the heat kernels of graphs so that the output of the node aggregation scheme is suitable for the graph representation. This section outlines the GNN model and the heat kernel and describes how the heat kernel on a graph is applied to the GNN model. Then the learning procedure of diffusion parameters and the proposed HKGN with multiple heat kernels are explained.

2.1   Graph Neural Networks

A graph G is represented as (V, E), where $V = \{v_1, v_2, \ldots, v_n\}$ is a set of $n > 0$ nodes and $E \subset V \times V$ is a set of edges. If node v_i and node v_j are connected, then $(v_i, v_j) \in E$. The adjacency matrix of graph G is a matrix $A \in \{0, 1\}^{n \times n}$. The element of A for node v_i and v_j is denoted as A_{ij}, and $A_{ij} = 1$ if $(v_i, v_j) \in E$ and $A_{ij} = 0$ if $(v_i, v_j) \notin E$. Let D is a diagonal matrix with degree element $D_{ii} = \sum_{j=1}^{n} A_{ij}$. Then the graph Laplacian is defined as $L = D - A$. The normalized graph Laplacian can be written as $\hat{L} = D^{-\frac{1}{2}} L D^{-\frac{1}{2}}$. The eigendecomposition of the normalized graph Laplacian is $\hat{L} = \Phi \Lambda \Phi^T$, where $\Lambda = diag(\lambda_1, \lambda_2, \ldots, \lambda_n)$ is the diagonal matrix with eigenvalues in ascending order and $\Phi = (\phi_1, \phi_2, \ldots, \phi_n)$ is the matrix with the corresponding eigenvectors as column vectors. Since $\hat{L}$ is symmetric and positive semi-definite, the eigenvalues are non-negative and bounded by two or less $(0 \leq \{\lambda_i\}_{i=1}^{n} \leq 2)$.

GNNs show outstanding performance in solving tasks with graph-structured data. The essence of GNNs is information retrieval from the node aggregation process. For any node, the neighbour nodes' information is aggregated to update the information of the node. This process is generally referred to as a message-passing neural network (MPNN) [8], which can be formulated as

$$H^{(k)} = MPNN(A, H^{(k-1)}; W^{(k)}).$$

Here, inputs for the MPNN function (in layer k) are the adjacency matrix (or its normalized version) A and the node representation $H^{(k-1)}$. $W^{(k)}$ is the learnable weight matrix. For instance, GCN represents the MPNN function as:

$$H^{(k)} = \sigma(\widetilde{D}^{-\frac{1}{2}} \widetilde{A} \widetilde{D}^{-\frac{1}{2}} H^{(k-1)} W^{(k)}), \tag{1}$$

where $\widetilde{A} = I + A$ is the adjacency matrix with self-loops and $\widetilde{D}$ is the diagonal degree matrix of $\widetilde{A}$. $\widetilde{D}^{-\frac{1}{2}} \widetilde{A} \widetilde{D}^{-\frac{1}{2}}$ is the normalized adjacency matrix. The node representation from layer $k-1$ is transformed by weight matrix $W^k \in \mathbb{R}^{d_{k-1} \times d_k}$. Then neighbour nodes' representations are aggregated using the normalized adjacency matrix. The node representation $H^{(k)}$ of layer k is obtained after applying $\sigma(\cdot)$, the element-wise non-linear activation function such as rectified linear unit (ReLU), to aggregated node representations. Nodes are represented by embedding their adjacent nodes. The entire graph can be represented by applying a readout function to those node embeddings -summarizing representation for all nodes in $H^{(k)}$. Typical readout functions are mean, sum, and max, and differentiable pooling methods also can be used [24].

From Eq. (1), the GCN model only aggregates neighbour nodes in the adjacency matrix, thus, it is difficult to reflect the global structure of the graph. If the graph convolution layer is applied k times, k-hop neighbours can be aggregated to the embeddings. However, as more layers are stacked, the computational complexity increases and the nearest local information may be obscured.

2.2   Heat Kernel Graph Networks (HKGN)

Heat Kernel. Heat kernel is an invariant which is given by the fundamental solution of the heat equation. In graph-structured data, the diffusion time derivative of the heat equation is related to the graph Laplacian [10]. The solution of the heat equation is obtained by exponentiating the Laplacian eigensystem over the diffusion time parameter. Intuitively, the heat (information) flow propagates through the entire graph along the edges as the diffusion parameter increases. More precisely, the fundamental heat equation with the normalized graph Laplacian is defined as:

$$\frac{\partial \Psi_t}{\partial t} = -\hat{L}\Psi_t,$$

where Ψ_t is the heat kernel matrix and t is the diffusion parameter. A solution to the heat equation is:

$$\Psi_t = e^{-t\hat{L}}.$$

Since $\hat{L}$ can be decomposed into its eigenvalues and eigenvectors, the heat kernel can be reformulated as:

$$\psi_t = \sum_{i=1}^{n} e^{-t\lambda_i} \phi_i \phi_i^T = \Phi e^{-t\Lambda} \Phi^T. \tag{2}$$

The diffusion parameter t controls the diffusion scale. The node information diffuses to local neighbours when t is small and reaches the entire graph when t is large. Thus, the heat kernel with varying t enables the model to capture the local and global structure of the graph. Diffusion from node u to node v can be calculated by $\Psi_t(u,v) = \sum_{i=1}^{n} e^{-t\lambda_i} \phi_i(u)\phi_i(v)$. In extreme cases, when t approaches zero, diffusion stays only on the self-node since the heat kernel matrix (2) becomes the identity matrix, $\Psi_t = \Phi\Phi^T = I$. On the other hand, when t gets larger, the heat kernel matrix (2) becomes $\Psi_t = e^{-t\lambda_2}\phi_2\phi_2^T$ where λ_2 is the smallest non-zero eigenvalue and ϕ_2 is Fiedler vector [2]. The Fiedler vector tells the number of disconnected subgraphs (graph components), and the number of zero eigenvalues corresponds to the number of components. Hence, the heat kernel matrix (2) with a large diffusion value captures the structural composition of the graph.

Heat Kernel Convolution. To take those advantages, we employ heat kernel in the proposed method, Heat Kernel Graph Networks (HKGN). The heat kernel substitutes the role of conventional graph convolution based on an adjacency matrix. Moreover, we set the diffusion parameter t as a learnable parameter, which is used to be set as a user parameter in the existing method. More precisely, the following components replace Eq. (1).

$$H^{(k)} = MPNN(\Lambda, \Phi, H^{(k-1)}; t^{(k)}, W^{(k)}),$$

and technically, heat kernel convolution in k^{th} layer is defined as:

$$H^{(k)} = \sigma(\Phi e^{-t^{(k)}\Lambda} \Phi^T H^{(k-1)} W^{(k)}) = \sigma(\Psi_{t^{(k)}} H^{(k-1)} W^{(k)}). \tag{3}$$

As t changes, the range of neighbours changes, and it is implemented by learning the optimal t. Additionally, we employ the following heat trace normalization to prevent the collapsing problem when t goes to infinity [19]. It gives stability and more representation power for discriminating non-isomorphic graphs (e.g., see Fig. 2). Equation (3) is rewritten as

$$H^{(k)} = \sigma(\Psi_t \widetilde{H}^{(k-1)} W^{(k)}),$$

where $\widetilde{H}^{(k-1)} = H^{(k-1)} / \sum_{i=1}^{n} \lambda_i$.

Here, we exemplify the aforementioned advantages of heat kernel convolution using two graphs in Fig. 2. The graph convolution cannot distinguish those non-isomorphic graphs. When t is small, the two graphs are not distinguishable neither in heat kennel convolution (e.g., $t = 0.001, [0.33, 0.33, 0.33, 0.33, 0.33, 0.33]$ for both graphs). But as t increases, the difference between two graphs becomes visible (e.g., $t = 2, [1.57, 1.57, 1.57, 1.57, 1.57, 1.57]$ and $[0.99, 0.99, 0.99, 0.99, 0.99, 0.99]$, respectively). When heat kernel convolution with large diffusion parameter ($t = 100$), the graphs are explicitly distinguishable, i.e. $t = 100, [2, 2, 2, 2, 2, 2]$ and $[1, 1, 1, 1, 1, 1]$, respectively. This shows that non-isomorphic graphs can be discriminated by learning suitable t in the convolution layer.

Graph Embedding. Lastly, a readout function is applied to represent the entire graph Z_G from the respective embeddings of all nodes. For instance, the 'mean' readout function results from the graph representation

$$Z_G = \frac{1}{n} \sum_{v \in V} h_v^{(k)},$$

where $h_v^{(k)}$ denotes the embedding of node v in $H^{(k)}$.

Optimal Diffusion Parameter. Usually, diffusion parameter t is considered a hyperparameter and finds optimal t with trial and error. However, by putting t as a learning parameter, we aim to seek optimal t that fits the data. The optimal t can be obtained by updating with the gradient descent method as $t_{l+1} = t_l - \eta \cdot \partial H / \partial t_l$. The gradient of H with respect to t is calculated as follows. For simplicity, we omit the transformation of input XW and consider the activation function as an identity function.

$$\begin{aligned}\frac{\partial H}{\partial t} &= \frac{\partial}{\partial t}\left(\frac{\sum_{i=1}^{n} exp[-\lambda_i t]\phi_i \phi_i^T}{\sum_{i=1}^{n} exp[-\lambda_i t]}\right) \\ &= -\frac{1}{2}\frac{\sum_{i \neq j} exp[-(\lambda_i + \lambda_j)t](\lambda_i - \lambda_j)(\phi_i \phi_i^T - \phi_j \phi_j^T)}{(\sum_{i=1}^{n} exp[-\lambda_i t])^2}\end{aligned} \quad (4)$$

Thus, t changes at every iteration with the difference between eigenvalues and the difference between eigenvectors.

Multi-head Heat Kernels. Choosing the diffusion parameter t is a difficult problem, but even if t is determined, using one heat kernel reflects only a part of the information in the graph structure. A fine representation of graphs needs to reflect local connectivity to the global structure. If multiple heat kernels are utilized, graphs can be expressed from multiple angles rather than fragments. Furthermore, deciding which heat kernel is more important or less is not easy. The most ideal way is to integrate the information given by each heat kernel according to the characteristics of the graph. HKGN can have more flexibility by varying t with multiple heat kernels. Let $t_p, p = 1, \ldots, m$, denotes each diffusion parameter in m heat kernels. Then, multi-head heat kernel convolution is defined as:

$$MLP\big(Concat(\widetilde{H}_1^{(k)}, ..., \widetilde{H}_p^{(k)}, ..., \widetilde{H}_m^{(k)})\big), \widetilde{H}_p^{(k)} \in \mathbb{R}^{n \times d_k}, \tag{5}$$

where $\widetilde{H}_p^{(k)}$ is the representation of pth heat kernel convolution and $Concat$ denotes the concatenation. After the concatenation of m representations, the MLP function (fully connected layer) is applied before the graph pooling operation. It works as giving weights to more important heat kernels in learning the graph representation.

Figure 3 shows the flow of the HKGN model. From the input graph to graph representation is learned in an end-to-end fashion. The pseudo-code for HKGN graph embedding is described in Algorithm 1.

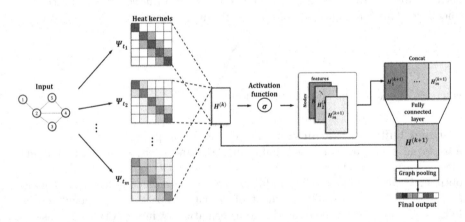

Fig. 3. Architecture of HKGN model. m heat kernels are generated from the input graph. Multi-head heat kernel convolution is performed, and a fully connected layer is followed after concatenation. Graph pooling outputs the final graph representation.

Algorithm 1. Heat Kernel Graph Networks graph embedding

Input: Graph $G = (V, E)$, input features X, eigenvalues and vectors Λ, Φ
Output: Graph representation z_G
1: $H^{(0)} \leftarrow X$
2: **for** $k = 1$ **to** K **do**
3:    **for** $p = 1$ **to** m **do**
4:        $\Psi_{t_p} \leftarrow \Phi exp[-t_p \Lambda] \Phi^T$
5:        $H_p^{(k)} \leftarrow \Psi_{t_p} \widetilde{H}^{(k-1)} W^{(k)}$
6:    **end for**
7:    $H^{(k)} \leftarrow equation(4)$
8: **end for**
9: $z_G \leftarrow readout(H^{(k)})$

---

3 Experiments

To validate the effectiveness of HKGN, a graph classification task is conducted. In particular, we compare HKGN with state-of-the-art methods on nine benchmark datasets. We also compare the two variants of HKGN, averaging heat kernels and concatenating heat kernels, with the varying number of heat kernels and show the difference in their performance and the explanation with the learned heat kernels. All experiments were conducted on a 24-core AMD Ryzen TR 3960X @ 3.8 GHz processor with 256 GB of memory and NVIDIA GeForce RTX 3090 GPUs with 48 GB of memory.

3.1 Datasets

The benchmark datasets from bioinformatics and social networks. Four bioinformatics datasets, including MUTAG, PROTEINS, NCI1, and PTC and three social network datasets, including COLLAB, IMDB-BINARY, and IMDB-MULTI. MUTAG is a dataset of mutagenic aromatic and heteroaromatic nitro compounds with two discrete labels according to their mutagenic activity on Salmonella Typhimurium [14]. PROTEINS is a dataset with amino acid sequences where nodes are secondary structure elements with two classes [3]. NCI1 from the National Cancer Institute (NCI) is a dataset of chemical compounds screened for the ability to suppress or inhibit the growth of a panel of human tumour cell lines with two classes [17]. PTC is a collection of chemical compounds represented as graphs which is the carcinogenicity for rats [14]. COLLAB is a scientific dataset where ego networks of researchers have collaborated. It is derived from three public collaborations: high energy physics, condensed matter physics, and astrophysics [22]. IMDB-BINARY and IMDB-MULTI are relational datasets that consist of a network of 1,000 actors or actresses who played roles in movies in IMDB [22]. A node represents an actor or actress; an edge exists between two nodes if they appear in the same movie. MULTI datasets have more than two classes. The number of graphs in each dataset varies from 188 (MUTAG) to 5,000 (COLLAB) and the average number of nodes ranges from 18 to 509. More details are summarized in Table 1.

Table 1. Statistics of collected datasets.

Name	Graphs	Classes	Nodes (avg.)	Edges (avg.)	Max deg
MUTAG	188	2	17.93	19.79	7
PROTEINS	1,113	2	39.06	72.82	3
NCI1	4,110	2	29.87	32.30	4
COLLAB	5,000	3	74.49	2,457.78	491
IMDB-B	1,000	2	19.77	96.53	135
IMDB-M	1,500	3	13.00	65.94	88
PTC	349	2	14.11	14.48	4

3.2 Graph Classification Results

To evaluate the effectiveness of HKGN, we conducted graph classification. The baseline GNN models for graph classification are GCN [13], Deep Graph CNN (DGCNN) [18], GraphSAGE [11], and GIN [20]. For HKGN, depending on the dataset and graph size, we used from 4 to 32 heat kernels for multi-head heat kernel convolution. A rectified linear unit (Relu) function is applied for activation functions. Batch normalization is applied after the heat kernel convolution layer, and dropout is used after the fully connected layer. The mean readout function is applied for graph pooling, following another fully connected layer to output the graph class with the softmax function. Xavier initialization [9] and uniform initialization from zero to ten are used for the weight matrices and the diffusion parameters, respectively. For each dataset, one hot degree vector was used for input node features and eigenvalues and eigenvectors of the normalized graph Laplacian were used for heat kernel convolution. We followed the experimental procedure as the standard GNN settings on graph classification in benchmarks. The experiment was repeated ten times with 10-fold cross-validation for each dataset. The experiments for baselines were conducted with the help of PyTorch Geometric [6].

A comparison between baselines and HKGN is shown in Fig. 4. The proposed HKGN outperformed other baselines in most datasets and showed competitive performance in the rest of the datasets. HKGN ranked top in MUTAG (91.9%), PROTEINS (76.8%), COLLAB (80.4%) and IMDB-BINARY (76.0%), which is up to 2.8% higher than the best of baselines. Table 2 shows the accuracy on average with standard deviation. The model with the highest accuracy is highlighted in bold. We observed that HKGN works effectively on datasets with relatively small molecules where capturing the global structure is important, as seen in Fig. 2. Also, the model is suitable for social network datasets such as COLLAB or IMDB since these datasets have hub nodes. The diffusion process in the heat kernel reflects the influence of the hub node on its neighbours [4].

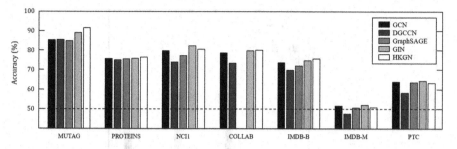

Fig. 4. Graph classification comparison (avg. of 10 cross-validations) between baselines and HKGN. Bars within a group correspond to GCN, DGCCN, GraphSAGE, GIN, and HKGN. For 4 out of 7 datasets, the performance of HKGN with concatenation surpasses other baselines.

Table 2. Graph classification accuracy in percent (avg. $\pm$ std.).

Method	MUTAG	PROTEINS	NCI1	COLLAB	IMDB-B	IMDB-M	PTC
GCN	85.6 ± 5.8	76.0 ± 3.2	80.2 ± 2.0	79.0 ± 1.8	74.0 ± 3.4	51.9 ± 3.8	64.2 ± 4.3
DGCNN	85.8	75.5	74.4	73.7	70.0	47.8	58.6
GraphSAGE	85.1 ± 7.6	75.9 ± 3.2	77.7 ± 1.5	N/A	72.3 ± 5.3	50.9 ± 2.2	63.9 ± 7.7
GIN	89.4 ± 5.6	76.2 ± 2.8	$\mathbf{82.7 \pm 1.7}$	80.2 ± 1.9	75.1 ± 5.1	$\mathbf{52.3 \pm 2.8}$	$\mathbf{64.6 \pm 7.0}$
HKGN	$\mathbf{91.9 \pm 4.0}$	$\mathbf{76.8 \pm 2.3}$	80.9 ± 1.6	$\mathbf{80.4 \pm 1.4}$	$\mathbf{76.0 \pm 5.5}$	52.0 ± 1.2	63.6 ± 8.9

3.3 Effect of Multi-head Heat Kernels

As mentioned in the previous sections, the HKGN model can have multi-head heat kernels. We conducted an ablation study to examine how the performance changes when the number of heat kernels changes. The experiment was conducted by varying the number of heat kernels from one to eight on the PRO-TEINS dataset. Figure 5 shows the accuracy of concatenation (C-HKGN) and mean (M-HKGN) versions of HKGN along with the number of heat kernels. Obviously, a single heat kernel shows the same accuracy for both models. As the number of heat kernels increases, the performance of C-HKGN increases. However, the performance of M-HKGN does not increase and stays on low performance. Thus, it is beneficial to use concatenation rather than averaging (mean) the heat kernels.

Four heat kernel samples from the model learned (with eight heat kernels) on the PROTEINS dataset are shown in Fig. 6. The samples are from the different sizes of the graphs. (a) and (b) have relatively large size graphs compared to (c) and (d). Local structures (i.e., small t) are shown in (a) and (b). (c) and (d) shows global structures (i.e., large t) of the graphs. Thus, not only local information but also global information is necessary for learning graph representation.

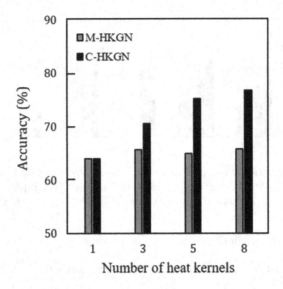

Fig. 5. Accuracy of C-HKGN and M-HKGN. The number of heat kernels varies from 1 to 8.

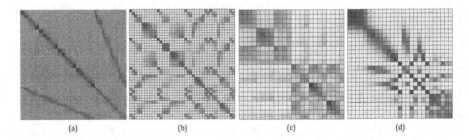

Fig. 6. Heatmap of heat kernel samples from PROTEINS dataset.

4   Conclusion

In this paper, we proposed heat kernel graph networks, a novel method for graph neural networks that focuses on graph representation using heat kernels of the graph. The heat kernel convolution allows the model to learn the local and global structure of the graph according to the diffusion parameter. And the heat trace normalization gives more distinguishability of graphs. We also designed multi-head heat kernels to leverage the flexibility of the model. There are several interesting directions for future work. The computation may be reduced using the approximation methods for heat kernel calculation. Other heat kernel invariants can be utilized for node aggregation or graph pooling methods.

Acknowledgments. This research was supported by Institute for Information communications Technology Promotion (IITP) grant funded by the Korea government (MSIT) (No. 2022-0-00653, Voice Phishing Information Collection and Processing and Development of a Big Data Based Investigation Support System), BK21 FOUR program of the National Research Foundation of Korea funded by the Ministry of Education(NRF5199991014091), the National Research Foundation of Korea (NRF) grant funded by the Korea government (MSIT) (No. 2021R1A2C2003474) and the Ajou University research fund.

References

1. Bacciu, D., Errica, F., Micheli, A., Podda, M.: A gentle introduction to deep learning for graphs. Neural Netw. **129**, 203–221 (2020)
2. Chung, F.R., Graham, F.C.: Spectral graph theory. No. 92, American Mathematical Soc. (1997)
3. Dobson, P.D., Doig, A.J.: Distinguishing enzyme structures from non-enzymes without alignments. J. Mol. Biol. **330**(4), 771–783 (2003)
4. Donnat, C., Zitnik, M., Hallac, D., Leskovec, J.: Learning structural node embeddings via diffusion wavelets. In: Proceedings of the 24th ACM SIGKDD International Conference on Knowledge Discovery & Data Mining, pp. 1320–1329 (2018)
5. Duvenaud, D., et al.: Convolutional networks on graphs for learning molecular fingerprints. arXiv preprint arXiv:1509.09292 (2015)
6. Fey, M., Lenssen, J.E.: Fast graph representation learning with PyTorch Geometric. In: ICLR Workshop on Representation Learning on Graphs and Manifolds (2019)
7. Garg, V., Jegelka, S., Jaakkola, T.: Generalization and representational limits of graph neural networks. In: International Conference on Machine Learning, pp. 3419–3430. PMLR (2020)
8. Gilmer, J., Schoenholz, S.S., Riley, P.F., Vinyals, O., Dahl, G.E.: Neural message passing for quantum chemistry. In: International Conference on Machine Learning, pp. 1263–1272. PMLR (2017)
9. Glorot, X., Bengio, Y.: Understanding the difficulty of training deep feedforward neural networks. In: Proceedings of the Thirteenth International Conference on Artificial Intelligence and Statistics, pp. 249–256. JMLR Workshop and Conference Proceedings (2010)
10. Grigor'Yan, A.: Heat kernels on manifolds, graphs and fractals. In: Casacuberta, C., Miro-Roig, R.M., Verdera, J., Xambo-Descamps, S. (eds.) European Congress of Mathematics. Progress in Mathematics, vol. 201, pp. 393–406. Birkhauser, Basel (2001). https://doi.org/10.1007/978-3-0348-8268-2_22
11. Hamilton, W.L., Ying, R., Leskovec, J.: Inductive representation learning on large graphs. arXiv preprint arXiv:1706.02216 (2017)
12. Kejani, M.T., Dornaika, F., Talebi, H.: Graph convolution networks with manifold regularization for semi-supervised learning. Neural Netw. **127**, 160–167 (2020)
13. Kipf, T.N., Welling, M.: Semi-supervised classification with graph convolutional networks. arXiv preprint arXiv:1609.02907 (2016)
14. Kriege, N., Mutzel, P.: Subgraph matching kernels for attributed graphs. arXiv preprint arXiv:1206.6483 (2012)
15. Morris, C., et al.: Weisfeiler and leman go neural: higher-order graph neural networks. In: Proceedings of the AAAI Conference on Artificial Intelligence, vol. 33, pp. 4602–4609 (2019)

16. Nikolentzos, G., Dasoulas, G., Vazirgiannis, M.: k-hop graph neural networks. Neural Netw. **130**, 195–205 (2020)
17. Wale, N., Watson, I.A., Karypis, G.: Comparison of descriptor spaces for chemical compound retrieval and classification. Knowl. Inf. Syst. **14**(3), 347–375 (2008)
18. Wang, Y., Sun, Y., Liu, Z., Sarma, S.E., Bronstein, M.M., Solomon, J.M.: Dynamic graph CNN for learning on point clouds. ACM Trans. Graph. (tog) **38**(5), 1–12 (2019)
19. Xiao, B., Hancock, E.R., Wilson, R.C.: Graph characteristics from the heat kernel trace. Pattern Recogn. **42**(11), 2589–2606 (2009)
20. Xu, K., Hu, W., Leskovec, J., Jegelka, S.: How powerful are graph neural networks? arXiv preprint arXiv:1810.00826 (2018)
21. Xu, K., Li, C., Tian, Y., Sonobe, T., Kawarabayashi, K.J., Jegelka, S.: Representation learning on graphs with jumping knowledge networks. In: International Conference on Machine Learning, pp. 5453–5462. PMLR (2018)
22. Yanardag, P., Vishwanathan, S.: Deep graph kernels. In: Proceedings of the 21th ACM SIGKDD International Conference on Knowledge Discovery and Data Mining, pp. 1365–1374 (2015)
23. Ying, R., He, R., Chen, K., Eksombatchai, P., Hamilton, W.L., Leskovec, J.: Graph convolutional neural networks for web-scale recommender systems. In: Proceedings of the 24th ACM SIGKDD International Conference on Knowledge Discovery & Data Mining, pp. 974–983 (2018)
24. Ying, R., You, J., Morris, C., Ren, X., Hamilton, W.L., Leskovec, J.: Hierarchical graph representation learning with differentiable pooling. arXiv preprint arXiv:1806.08804 (2018)

PROS-C: Accelerating Random Orthogonal Search for Global Optimization Using Crossover

Bruce Kwong-Bun Tong[1,2]([✉]) [iD], Wing Cheong Lau[2] [iD], Chi Wan Sung[3] [iD], and Wing Shing Wong[2] [iD]

[1] Department of Electronic Engineering and Computer Science, Hong Kong Metropolitan University, Hong Kong, China
`kbtong@hkmu.edu.hk`
[2] Department of Information Engineering, The Chinese University of Hong Kong, Hong Kong, China
[3] Department of Electrical Engineering, City University of Hong Kong, Hong Kong, China

Abstract. Pure Random Orthogonal Search (PROS) is a parameter-less evolutionary algorithm (EA) that has shown superior performance when compared to many existing EAs on well-known benchmark functions with limited search budgets. Its implementation simplicity, computational efficiency, and lack of hyperparameters make it attractive to both researchers and practitioners. However, PROS can be inefficient when the error requirement becomes stringent. In this paper, we propose an extension to PROS, called Pure Random Orthogonal Search with Crossover (PROS-C), which aims to improve the convergence rate of PROS while maintaining its simplicity. We analyze the performance of PROS-C on a class of functions that are monotonically increasing in each single dimension. Our numerical experiments demonstrate that, with the addition of a simple crossover operation, PROS-C consistently and significantly reduces the errors of the obtained solutions on a wide range of benchmark functions. Moreover, PROS-C converges faster than Genetic Algorithms (GA) on benchmark functions when the search budget is tight. The results suggest that PROS-C is a promising algorithm for optimization problems that require high computational efficiency and with a limited search budget.

Keywords: Global Optimization · Pure Random Orthogonal Search · Genetic Algorithm · Blend Crossover

1 Introduction

Metaheuristics and evolutionary algorithms (EA) including Genetic Algorithm (GA) [4], Particle Swarm Optimization (PSO) [5], Differential Evolution (DE) [7] and Cuckoo Search (CS) [10] are widely used in solving practical engineering

© The Author(s), under exclusive license to Springer Nature Switzerland AG 2024
G. Nicosia et al. (Eds.): LOD 2023, LNCS 14506, pp. 283–298, 2024.
https://doi.org/10.1007/978-3-031-53966-4_21

Algorithm 1: Pure Random Orthogonal Search (PROS)

1 $t \leftarrow 0$;
2 Initialize $\mathbf{x}^{(t)} = (x_1, x_2, ..., x_D)$ randomly from
   $\Omega = [a_1, b_1] \times [a_2, b_2] \times \cdots \times [a_D, b_D]$;
3 **repeat**
4 | Sample a random integer $j \in \{1, ..., D\}$;
5 | Sample a random real number $r \sim \mathcal{U}(a_j, b_j)$;
6 | $\mathbf{y} \leftarrow \mathbf{x}^{(t)}$;
7 | $y_j \leftarrow r$;
8 | **if** $f(\mathbf{y}) < f(\mathbf{x}^{(t)})$ **then**
9 | | $\mathbf{x}^{(t+1)} \leftarrow \mathbf{y}$;
10 | **else**
11 | | $\mathbf{x}^{(t+1)} \leftarrow \mathbf{x}^{(t)}$;
12 | **end**
13 | $t \leftarrow t + 1$;
14 **until** *a termination criterion is met*;
15 **return** $\mathbf{x}^{(t)}$;   /* output the best solution vector $\mathbf{x}^{(t)}$ found by PROS */

Algorithm 2: Blend Crossover

   input : two parent vectors $\mathbf{x}^{(p)}$, $\mathbf{x}^{(q)}$, hyperparameter α (typical value = 0.5)
1 **for** $i \leftarrow 1$ *to* D **do**
2 | $d_i \leftarrow |x_i^{(p)} - x_i^{(q)}|$;
3 | $u_i \leftarrow \max(\min(x_i^{(p)}, x_i^{(q)}) - \alpha d_i, a_i)$;
4 | $v_i \leftarrow \min(\max(x_i^{(p)}, x_i^{(q)}) + \alpha d_i, b_i)$;
5 | $x_i^{(new)} \sim \mathcal{U}(u_i, v_i)$;
6 **end**
7 **return** $\mathbf{x}^{(new)}$;   /* output an offspring vector $\mathbf{x}^{(new)}$ */

---

optimization problems. The objective functions of these optimization problems are often non-convex, non-differentiable and in high dimensions. These optimization algorithms often strike a balance between exploration and exploitation well, and treat the functions to be optimized as black-box where prior knowledge (such as mathematical expressions and gradients) and assumption of regularity (such as convexity, continuity, differentiability) are generally not required.

However, these algorithms are in general having various hyperparameters and the performance are often sensitive to the problem dependent hyperparameters [12]. Besides, these algorithms aim to reduce the error of the final solution which often results in high computational and implementation complexity. In view of the above, a computationally efficient algorithm with none or very limited number of hyperparameters would be attractive to practitioners.

Pure Random Orthogonal Search (PROS) is a recently proposed (1+1) Evolutionary Strategy (ES) [6]. PROS is simple, computationally efficient, parameterless and has low implementation complexity. The performance of PROS is promising and it outperforms various well known optimization algorithms on a set of benchmark problems when there is a constraint on the maximum number of objective function evaluations [6]. Two variants, TROS and QROS, have been proposed and shown that theoretically and numerically the convergence rate is faster than that of PROS on the major benchmark problems [9]. But the convergence rates of these simple random orthogonal search algorithms diminish quickly as the number of iterations increases, and they become ineffective in further reducing the errors of solutions because they do not have a balanced exploration and exploitation mechanism.

In this paper, we aim to strengthen PROS by adding a simple crossover operation. The proposed algorithm preserves the simplicity and computational efficiency of PROS and outperforms it significantly. The major contributions of this research work are summarized as follows:

1. An extension to PROS, namely PROS-C, is proposed and the performance on a class of functions that are monotonically increasing in each single dimension is analyzed.
2. A set of numerical experiments on different classes of functions is conducted, and PROS-C converges faster than PROS and GA when the normalized error requirement is from 0.1 to 0.0001 on the benchmark problems.

In this paper, we restrict our attention to global optimization of black-box functions. Black-box functions refer to functions that the explicit forms and gradients are unknown and unavailable to the algorithm. The only information available is obtained by evaluating the function with different inputs. The goal of global optimization is to find $\mathbf{x}^*$ which is a global minimum of f,

$$\mathbf{x}^* = \arg\min_{\mathbf{x} \in \Omega} f(\mathbf{x}) \tag{1}$$

where $f : \mathbb{R}^D \to \mathbb{R}$ is a scalar-valued objective function defined on the decision space $\Omega = [a_1, b_1] \times [a_2, b_2] \times \cdots \times [a_D, b_D] \subseteq \mathbb{R}^D$ and $\mathbf{x} = (x_1, x_2, ..., x_D)$ represents a vector of decision variables with D dimensions.

2 Related Work

In this section, we have an overview on the PROS algorithm, and the blend crossover algorithm that is commonly used by GA for continuous function optimization.

2.1 Pure Random Orthogonal Search (PROS)

The PROS algorithm was originally proposed by Plevris et al in [6]. It has only one parent in the population and generates one child each generation. The better one will be kept in the population and become the parent of the next generation. We describe it with our revised notation as Algorithm 1 as follows:

At the very beginning, the first candidate solution vector $\mathbf{x}^{(0)}$ is generated randomly from Ω (lines 1 and 2). Then, for each iteration t, a decision variable j

is chosen randomly from 1 to D (line 4) and a real number r is drawn uniformly at random within the search space of the j-th decision variable (line 5). A new candidate solution vector $\mathbf{y}$ is obtained by replacing the value of the j-th decision variable of the current best solution vector $\mathbf{x}^{(t)}$ with r (lines 6 and 7). The new candidate solution vector $\mathbf{y}$ is evaluated and compared with the current best solution vector $\mathbf{x}^{(t)}$. The new best solution vector $\mathbf{x}^{(t+1)}$ is updated as $\mathbf{y}$ if $f(\mathbf{y})$ is smaller than $f(\mathbf{x}^{(t)})$ (this is known as *accepted*). $\mathbf{x}^{(t+1)}$ is just $\mathbf{x}^{(t)}$ if $f(\mathbf{y})$ is not smaller than $f(\mathbf{x}^{(t)})$ (this is known as *rejected*) (lines 8 to 12). The iteration counter t is then updated, and the search process is repeated until a termination criterion is met. For example, when a pre-defined maximum number of iterations is reached, or no improvement has been made after a pre-defined number of iterations (lines 13 and 14). Finally, the best solution vector found by the algorithm is returned as the final result (line 15).

2.2   Blend Crossover

In GA, crossover is the major approach to generate offspring solution vectors. For continuous function optimization, blend crossover is commonly used [3]. We describe blend crossover with our revised notation as Algorithm 2 as follows: for each decision variable x_i of the offspring solution vector, it is obtained by sampling uniformly in the interval $[u_i, v_i]$ where $[u_i, v_i]$ are determined by the two parents ($\mathbf{x}^{(p)}$ and $\mathbf{x}^{(q)}$), the hyperparameter (α), and the lower and upper limits of the i-th decision variable (a_i and b_i).

Figure 1 shows examples of performing blend crossover on one-dimensional and two-dimensional vectors. When the two parents are close to each other (on the i-th dimension), or when the α value is small, the interval (of the i-th dimension) will be short. That is, the value of the x_i in the offspring vector will be similar to its parents. This is the major idea to perform exploitation in GA.

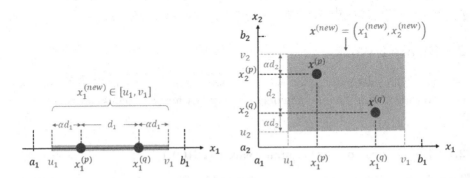

Fig. 1. Examples of Blend Crossover. Left: 1-D blend crossover with $\alpha = 0.5$. The orange interval indicates the possible range of $x_1^{(new)}$. The interval is bounded by $[a_1, b_1]$. Right: 2-D blend crossover with $\alpha = 0.5$. The orange rectangle indicates the possible range of $\mathbf{x}^{(new)} = (x_1^{(new)}, x_2^{(new)})$. The rectangle is bounded by $[a_1, b_1] \times [a_2, b_2]$. (Color figure online)

3   Proposed Algorithm: PROS-C

Inspired by the blend crossover operation of GA, we propose a modified PROS algorithm which combines the advantages of both algorithms and aims to improve the long-term performance of PROS. The proposed algorithm is described as Algorithm 3 as follows:

Initially, two candidate solution vectors, namely $\mathbf{x}^{(p_0)}$ and $\mathbf{x}^{(q_0)}$ are generated randomly from the decision space Ω (lines 1 to 2). In each iteration t (line 3), $\mathbf{x}^{(p_t)}$ and $\mathbf{x}^{(q_t)}$ are updated using different methods.

For $\mathbf{x}^{(p_t)}$, it is updated strictly using the PROS method. That is, to replace the value of one of the decision variable of $\mathbf{x}^{(p_t)}$ by a random value that sampled uniformly from the valid search range of that decision variable. If the new $\mathbf{y}$ is having a smaller evaluated value than $\mathbf{x}^{(p_t)}$, $\mathbf{y}$ will be accepted and become $\mathbf{x}^{(p_{t+1})}$ of the next generation. Otherwise, $\mathbf{y}$ will be rejected and discarded, and $\mathbf{x}^{(p_t)}$ will become $\mathbf{x}^{(p_{t+1})}$ of the next generation (lines 4 to 11).

For $\mathbf{x}^{(q_t)}$, it is updated strictly using one-dimension blend crossover. That is, to replace the value of one of the decision variables of $\mathbf{x}^{(q_t)}$ by a value which is sampled uniformly from u_i to v_i where i is the randomly chosen decision variable for blend crossover, and u_i and v_i are, respectively, the lower limit and upper limit for blend crossover. In this step, the parents for blend crossover are always $\mathbf{x}^{(p_t)}$ and $\mathbf{x}^{(q_t)}$, since they are the only two parents in the population. If the new $\mathbf{z}$ is having a smaller evaluated value than $\mathbf{x}^{(q_t)}$, $\mathbf{z}$ will be accepted and become $\mathbf{x}^{(q_{t+1})}$ of the next generation. Otherwise, $\mathbf{z}$ will be rejected, and $\mathbf{x}^{(q_t)}$ will become $\mathbf{x}^{(q_{t+1})}$ of the next generation (lines 12 to 22).

The best solution vector found by the algorithm $\mathbf{x}^{(t)+}$ is updated and the algorithm proceeds to the next generation until a termination criterion is met (lines 23 to 25). Finally, the best-found solution vector is returned (line 26).

3.1   Motivation

In this subsection, we are going to explain the motivation of the proposed algorithm. PROS-C can be considered as an extension to PROS in the sense that an additional candidate solution $\mathbf{x}^{(q)}$ is maintained in the population. That is, PROS has $\mathbf{x}^{(p)}$ alone while PROS-C has $\mathbf{x}^{(p)}$ and $\mathbf{x}^{(q)}$ in the population. In each iteration, $\mathbf{x}^{(p)}$ is restricted to evolve using the PROS process while $\mathbf{x}^{(q)}$ is restricted to evolve using 1-D blend crossover (using the previous $\mathbf{x}^{(p)}$ and $\mathbf{x}^{(q)}$ pair as parents).

In our proposed algorithm, we use the minimum number of initial points, although a population with 2 points is unusual in population-based algorithms. However, this study shows surprising results that PROS-C demonstrates significant improvements for certain classes of functions. Although it is possible to design algorithms to include more initial solutions, there is a tradeoff between search budget, robustness, and complexity. The benefits of having more initial solutions require a more detailed study, which is out of the scope of this paper.

PROS converges in probability to the region of global optimum for all one-dimensional functions and all totally separable functions [9]. Since $\mathbf{x}^{(p)}$ is always updated using the same method as the PROS algorithm, PROS-C directly inherits the above convergence properties of PROS.

Algorithm 3: PROS with Blend Crossover

---

1 $t \leftarrow 0$;
2 Initialize $\mathbf{x}^{(p_t)}$ and $\mathbf{x}^{(q_t)}$ randomly from Ω;
3 **repeat**
4     $\mathbf{y} \leftarrow \mathbf{x}^{(p_t)}$;   `/* to generate a new `$\mathbf{x}^{(p_t)}$` by PROS */`
5   Sample a random integer $j \in \{1, ..., D\}$;
6   Sample a random real number $y_j \sim \mathcal{U}(a_j, b_j)$;
7   **if** $f(\mathbf{y}) < f(\mathbf{x}^{(p_t)})$ **then**
8 | $\mathbf{x}^{(p_{t+1})} \leftarrow \mathbf{y}$;
9   **else**
10 | $\mathbf{x}^{(p_{t+1})} \leftarrow \mathbf{x}^{(p_t)}$;
11   **end**
12    $\mathbf{z} \leftarrow \mathbf{x}^{(q_t)}$;   `/* to generate a new `$\mathbf{x}^{(q_t)}$` by 1D blend crossover */`
13   Sample a random integer $i \in \{1, ..., D\}$;
14    $d_i \leftarrow |x_i^{(p_t)} - x_i^{(q_t)}|$;
15    $u_i \leftarrow \max(\min(x_i^{(p_t)}, x_i^{(q_t)}) - \alpha d_i, a_i)$;
16    $v_i \leftarrow \min(\max(x_i^{(p_t)}, x_i^{(q_t)}) + \alpha d_i, b_i)$;
17   Sample a random real number $z_i \sim \mathcal{U}(l_i, r_i)$;
18   **if** $f(\mathbf{z}) < f(\mathbf{x}^{(q_t)})$ **then**
19 | $\mathbf{x}^{(q_{t+1})} \leftarrow \mathbf{z}$;
20   **else**
21 | $\mathbf{x}^{(q_{t+1})} \leftarrow \mathbf{x}^{(q_t)}$;
22   **end**
23    $\mathbf{x}^{(t)+} = \arg\min_{\mathbf{x} \in \{\mathbf{x}^{(p_t)}, \mathbf{x}^{(q_t)}\}} (f(\mathbf{x}))$;
24    $t \leftarrow t + 1$;
25 **until** *a termination criterion is met*;
26 **return** $\mathbf{x}^{(t)+}$;   `/* output the best solution vector `$\mathbf{x}^{(t)+}$` */`

In addition, the search procedure of PROS is equivalent to GA's mutation, which aims at exploration, while PROS-C's crossover aims at exploitation. In this sense, PROS only performs exploration on a decision variable but puts no effort on exploitation. PROS may work during the initial stage of exploration to quickly reduce the normalized error down to 10's of percent, but as soon as it enters the stage where the normalized error requirement becomes stringent, say down to a few percent, PROS becomes ineffective in further reducing the objective function because it lacks any exploitation mechanism. In contrast, PROS-C has both exploitation and exploration elements, making it outperform PROS when the normalized error requirement is stringent. The reasons in performance difference are further analyzed and explained in the next subsection using a special class of functions.

3.2 Analysis of PROS-C

In this subsection, we are going to compare the performance between PROS and PROS-C. In particular, we focus on the class of monotonically increasing functions as described below. When an algorithm reaches a domain of attraction, understanding the nature of convergence of this class of functions becomes relevant. For notation simplicity, define $\mathbf{x}_{-i} \triangleq (x_1, ..., x_{i-1}, x_{i+1}, ..., x_D)$.

Definition 1. *A function f is monotonically increasing in each single dimension if it satisfies $f(\mathbf{x}_{-i}, x_i') < f(\mathbf{x}_{-i}, x_i)$, $\forall x_i', x_i \in \Omega_i$, $i \in \{1, ..., D\}$ and $x_i' < x_i$.*

Definition 2 (Expected Movement). *The expected movement by an algorithm is the expected reduction in the error distance between the current best solution vector and the new best solution vector after one-step of search. That is, $\mathbb{E}[\|\mathbf{x}^{(t)} - \mathbf{x}^{(t+1)}\|]$ where $\|\cdot\|$, $\mathbf{x}^{(t)}, \mathbf{x}^{(t+1)}$ denote the Manhattan distance, the current best solution vector and the new best solution vector, respectively.*

We first focus on monotonically increasing functions with $D = 1$ and $\Omega = [a, b]$. Let $\mathbf{x}^{(p_t)} = (p_t)$ and $\mathbf{x}^{(q_t)} = (q_t)$ be the two parents in PROS-C. Let $p = \min(p_t, q_t)$ and $q = \max(p_t, q_t)$, $[u, v]$ be the interval of blend crossover of $\mathbf{x}^{(p)}$ and $\mathbf{x}^{(q)}$, where $u = \max(a, p - \alpha d)$, $v = \min(q + \alpha d, b)$ and $d = q - p$.

Lemma 1. *The conditional expected movement given the current best solution vector, $\mathbf{x} = (p)$, by PROS is $\frac{(p-a)^2}{2(b-a)}$ for all one dimensional monotonically increasing functions.*

Theorem 1. *The conditional expected movement given $\mathbf{x}^{(p_t)}$ and $\mathbf{x}^{(q_t)}$ by PROS-C for all one dimensional monotonically increasing functions is $\frac{(p-u)^2}{2(v-u)}$, and more specifically,*

$$\begin{cases} \frac{(p-a)^2}{2(v-a)} & u = a \text{ and } v < b \\ \frac{\alpha^2 d}{2(1+2\alpha)} & u > a \text{ and } v < b \\ \frac{(\alpha d)^2}{2(b-u)} & u > a \text{ and } v = b \\ \frac{(p-a)^2}{2(b-a)} & u = a \text{ and } v = b. \end{cases}$$

Corollary 1. *The sufficient conditions for PROS-C to have the same as or a larger conditional expected movement over PROS for all one dimensional monotonically increasing functions are $p < a + \beta \alpha d$ when $q < b - \alpha d$, and $p < a + \gamma \alpha d$ when $b - \alpha d \leq q < b$ where $\beta = \sqrt{\frac{b-a}{(1+2\alpha)d}}$ and $\gamma = \sqrt{\frac{b-a}{b-u}}$.*

For monotonically increasing functions with $D > 1$ and $\Omega = [a_1, b_1] \times ... \times [a_D, b_D]$, similar results could be obtained. Let $\mathbf{x}^{(p_t)} = (p_1^{(t)}, ..., p_D^{(t)})$ and $\mathbf{x}^{(q_t)} = (q_1^{(t)}, ..., q_D^{(t)})$ be the two parents in PROS-C. Let $p_i = \min(p_i^{(t)}, q_i^{(t)})$ and $q_i = \max(p_i^{(t)}, q_i^{(t)}) \; \forall i = \{1, ..., D\}$. Then let $[u_i, v_i]$ be the interval of blend crossover of $\mathbf{x}^{(p_t)}$ and $\mathbf{x}^{(q_t)}$ on the i-th dimension where $u_i = \max(a, p_i - \alpha d)$, $v_i = \min(q_i + \alpha d_i, b_i)$ and $d_i = q_i - p_i$.

Corollary 2. *The conditional expected movement given the current best solution vector, $\mathbf{x} = (p_1, ..., p_D)$, by PROS is $\frac{1}{D} \sum_{i=1}^{D} \frac{(p_i - a_i)^2}{2(b_i - a_i)}$. for all D-dimensional monotonically increasing functions.*

Corollary 3. *The conditional expected movement given* $\mathbf{x}^{(p)}$ *and* $\mathbf{x}^{(q)}$ *by PROS-C for all D-dimensional monotonically increasing functions is* $\frac{1}{D}\sum_{i=1}^{D}\frac{(p_i-u_i)^2}{2(v_i-u_i)}$

Due to space limit, we focus mainly on the results and refer the readers to [8] for the proof of the above lemma, theorem and corollaries. Here we would like to elaborate on the results. For one dimensional cases, PROS-C has a larger or at least the same expected movement compared with PROS when p is less than a certain threshold value that depends on q. In other words, crossover becomes effective when the solutions in PROS-C approach the optimum a. These results are inline with the results of the numerical experiments (to be described in the next section) that PROS may be more superior than PROS-C during the initial stage, but after that, its performance may become inferior than PROS-C.

3.3 Sphere Function

In this and the next subsections, we illustrate the performance difference between PROS-C, PROS and GA by using two examples. In the first example, we apply these algorithms to minimize the Sphere function $f(\mathbf{x}) = \sum_{i=1}^{D} x_i^2$. The optimum $\mathbf{x}^* = (x_1, ..., x_D) = (0, ..., 0)$.

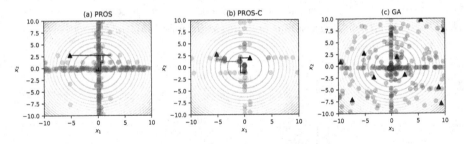

Fig. 2. The search trajectory of a particular instance of various algorithms on the 2D Sphere function with $\Omega = [-10, 10]^2$. The algorithms terminate when a solution with an error less than 10^{-2} is found. (a) PROS carries out 308 function evaluations; (b) PROS-C carries out 94 function evaluations; (c) GA carries out 2320 function evaluations.

Figure 2 shows the search trajectory of PROS, PROS-C, and GA on the Sphere function where $D = 2$, $\Omega = [-10, 10]^2$ and the stopping criterion is that the error of the best solution is less than 10^{-2}. In this instance, PROS takes 308 iterations (308 function evaluations) to find a solution with an error less than 10^{-2}. The black triangle is the location of the starting point (generated randomly). The green dots are the points sampled by the PROS algorithm. The black lines show the trajectory of the sequence of the best-found solutions. The color dots are in the same color, but after superposition, some look darker especially when they are close to each other. It can be seen clearly that the sampled points form a cross shape, meaning the algorithm is able to identify a partial solution (either $x_1 \approx 0$ or $x_2 \approx 0$) quickly. However, the black lines reveal that PROS samples lots of points before finding an accepted solution. This is due to the fact that it blindly picks a point from one of the axes uniformly in each iteration. That makes the algorithm converge relatively slowly.

In same figure, (b) shows the search trajectory of PROS-C on the same function with the same settings. In this instance, it takes 47 generations (94 function evaluations) to find a solution with an error less than 10^{-2}. The black triangles are the locations of the two starting points of PROS-C (one of them is the same as that of PROS since we use the same random seed for these two examples). The red dots are the points sampled by $\mathbf{x}^{(q)}$ (the point that is always evolved by crossover). The red lines show the trajectory of the sequence of the best solutions found by $\mathbf{x}^{(q)}$. The green dots are the points sampled by $\mathbf{x}^{(p)}$ (the point that is always generated by uniform random orthogonal search). The black lines show the trajectory of the sequence of the best solutions found by $\mathbf{x}^{(p)}$. Obviously, the red points are clustered around the global optimum but the green points are not, meaning that the points generated by crossover are often of high quality. This is because it does not sample a point along one of the axes uniformly but with a reduced search range. That makes the algorithm converge relatively quickly.

In the same figure, (c) shows the search trajectory of GA (population size = 10) on the same function with the same settings. In this instance, it takes 232 generations (2320 function evaluations) to find a solution with an error less than 10^{-2}. The black triangles are the location of the 10 starting points (the first generation). The purple dots are the points sampled by GA. Similar to PROS, the points form a cross shape meaning that it can identify a partial solution quickly. But compared with PROS, it explores some more points outside the cross shape. That shows a tradeoff between exploration and exploitation.

3.4   Rastrigin Function

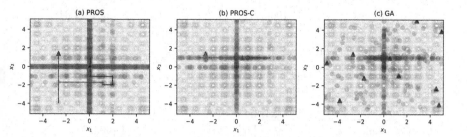

Fig. 3. The search trajectory of a particular instance of various algorithms on the 2D Rastrigin function with $\Omega = [-5.12, 5.12]^2$. The algorithms termination when a solution with an error less than 10^{-2} is found. (a) PROS carries out 2755 function evaluations; (b) PROS-C carries out 436 function evaluations; (c) GA carries out 2340 function evaluations.

In the second example, we apply the algorithms to minimize a classic multimodal benchmark function: Rastrigin function $f(\mathbf{x}) = 10D + \sum_{i=1}^{D}[x_i^2 - 10\cos(2\pi x_i)]$.

The Rastrigin function contains many local minima where the global optimum $(x_1, ..., x_D) = (0, ..., 0)$. Figure 3 shows the search trajectory of a particular instance of PROS, PROS-C, and GA on the Rastrigin function where $D = 2$, $\Omega = [-5.12, 5.12]^2$ and the stopping criterion is the same as the previous example. Similar to the results on the Sphere function, PROS-C converges to the global optimum quickly. PROS, PROS-C, GA carry out 2755, 436 and 2340 function evaluations, respectively, to find a solution with an error less than 10^{-2}. It demonstrates that PROS-C performs well on both unimodal and multimodal functions, despite its simplicity.

4 Numerical Experiments

In order to test the effectiveness of the proposed algorithm, a set of benchmark problems has been selected from the literature [1,9,11] and is shown in Table 1. The benchmark problems include unimodal and highly multimodal functions, convex and non-convex functions, and separable and non-separable functions. Through a wide range of functions, we aim to provide a better understanding of the behaviour of PROS-C. Since PROS-C is an extension of PROS and blend crossover is inspired by GA, we compare PROS-C with PROS and GA.

Table 1. Benchmark Test Problems. The global optimum $\mathbf{x}^* = (0, ..., 0)$ for all problems except for the Rosenbrock function $\mathbf{x}^* = (1, ..., 1)$, and the HGBat and HappyCat functions $\mathbf{x}^* = (-1, ..., -1)$. The global minimum value $f(\mathbf{x}^*) = 0$ for all problems.

No	Function	Formulation	Search Space
f_1	Sphere	$\sum_{i=1}^{D} x_i^2$	$[-10, 10]^D$
f_2	Ellipsoid	$\sum_{i=1}^{D} i x_i^2$	$[-10, 10]^D$
f_3	Bent Cigar	$x_1^2 + 10^6 \sum_{i=2}^{D} x_i^2$	$[-100, 100]^D$
f_4	Schwefel 2.22	$\sum_{i=1}^{D} |x_i| + \prod_{i=1}^{D} |x_i|$	$[-5.12, 5.12]^D$
f_5	Rosenbrock	$\sum_{i=1}^{D-1} [100(x_{i+1} - x_i^2)^2 + (x_i - 1)^2]$	$[-2.048, 2.048]^D$
f_6	Cosine Mixture	$0.1D + \sum_{i=1}^{D} x_i^2 - 0.1 \sum_{i=1}^{D} \cos(5\pi x_i)$	$[-5.12, 5.12]^D$
f_7	Rastrigin	$10D + \sum_{i=1}^{D} [x_i^2 - 10\cos(2\pi x_i)]$	$[-5.12, 5.12]^D$
f_8	Alpine 1	$\sum_{i=1}^{D} |x_i \sin(x_i) + 0.1 x_i|$	$[-10, 10]^D$
f_9	Ackley	$20 - 20\exp(-0.2\sqrt{\frac{1}{D}\sum_{i=1}^{D} x_i^2})$ $- \exp(\frac{1}{D}\sum_{i=1}^{D} \cos(2\pi x_i)) + \exp(1)$	$[-32.768, 32.768]^D$
f_{10}	Griewank	$\sum_{i=1}^{D} \frac{x_i^2}{4000} - \prod_{i=1}^{D} \cos\left(\frac{x_i}{\sqrt{i}}\right) + 1$	$[-600, 600]^D$
f_{11}	HGBat	$|(\sum_{i=1}^{D} x_i^2)^2 - (\sum_{i=1}^{D} x_i)^2|^{1/2} +$ $(0.5\sum_{i=1}^{D} x_i^2 + \sum_{i=1}^{D} x_i)/D + 0.5$	$[-15, 15]^D$
f_{12}	HappyCat	$|\sum_{i=1}^{D} x_i^2 - D|^{1/4} +$ $(0.5\sum_{i=1}^{D} x_i^2 + \sum_{i=1}^{D} x_i)/D + 0.5$	$[-20, 20]^D$

PROS and PROS-C were implemented in Java by the authors. GA was run using pymoo (version 0.6.0), a widely adopted open-source framework in the literature [2]. It is known that GA has a number of hyperparameters and the optimal choice of its hyperparameters is problem dependent. We aim to study the performance of the algorithms on various classes of problems, rather than to find the best possible outcome for each problem by each algorithm. Therefore, the hyperparameters of GA are selected using the default values from the pymoo package, and these are typical values suggested in the literature.

The experiments were carried out on a 3.20 GHz computer with 16GB RAM under a Windows 10 platform. Multiple runs were conducted for each problem by each algorithm with a different seed for the generation of random numbers for each run. For a fair comparison, all algorithms end with the same maximum number of objective function evaluations as function evaluation is assumed to be the most costly part for black-box optimization. The settings are summarized in Table 2.

Table 2. Settings of the numerical experiments.

	Settings	Value
Experiments	Dimension D	10
	Max. No. of Objective Function Evaluations	100,000
	No. of Runs	100
GA	Population Size	$5 \times D = 50$
PROS-C	1-Dimensional Blend Crossover	$\alpha = 0.5$

4.1 Convergence Results of Benchmark Test Problems

Figure 4 shows the convergence curves of the normalized errors (averaged results of 100 independent runs) of PROS, PROS-C, and GA on all benchmark problems with $D = 10$ over 100,000 objective function evaluations (or equivalently 2,000 generations for GA). The x-axes are the number of objective function evaluations and the y-axes are the mean normalized error of the 100 runs of each algorithm. The normalized error of solution vector $\mathbf{x}$ is defined as

$$e(\mathbf{x}) = \frac{f(\mathbf{x}) - f(\mathbf{x}^*)}{f(\mathbf{x}_{max}) - f(\mathbf{x}^*)} \tag{2}$$

where $\mathbf{x}_{max} = \arg\max_{\mathbf{x} \in \Omega} f(\mathbf{x})$ is the global maximum of f.

As shown in Fig. 4, initially PROS converges quickly but soon is superseded by PROS-C after several hundreds of function evaluations. When compared with GA, PROS-C converges faster in the first 10 benchmark problems (f_1 to f_{10}). If more search budget is allowed, GA would supersede PROS-C to obtain a small error on the last 2 benchmark problems (f_{11} and f_{12}).

Table 3 shows the mean of the normalized errors of the solution vectors returned by PROS, PROS-C, and GA for 100 independent runs on each benchmark problem for dimension $D = 10$ after 100,000 objective function evaluations. The corresponding standard deviations are indicated inside the corresponding

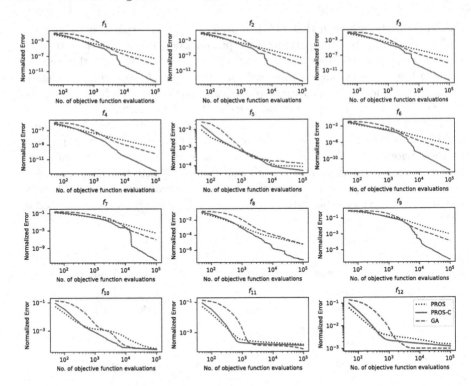

Fig. 4. The convergence curves (averaged results of 100 independent runs) of PROS, PROS-C and GA on the 10-D benchmark problems.

parenthesis. The smallest errors for each benchmark problem are highlighted in bold font. As seen from the table, PROS-C gives the best results on most of the benchmark problems in terms of the mean normalized errors. Compared with PROS, PROS-C is consistently more efficient in reducing the mean normalized errors on the benchmark problems.

4.2   Quality of Solutions

In this subsection, we study how often and how fast an algorithm obtains a solution at an acceptable error level. In practice, we may want an algorithm to run continuously until a certain error threshold has been reached. Unlike the previous experiment in which only a maximum number of objective function evaluations is set as the only stopping criterion, in this experiment the same set of benchmark problems is used but we add an additional stopping criterion: the normalized error of the solution vector is less than a pre-defined threshold ϵ.

Figure 5 shows the box-and-whisker plot of the number of objective function evaluations spent by each algorithm to find a solution with normalized error less than 0.1. The number of evaluations for PROS-C and GA are always multiples of 2, and multiples of 50, respectively, because the population sizes of PROS-C and GA are 2 and 50, respectively. It can be seen clearly that PROS-C spent the

Table 3. The mean normalized errors (with standard deviation) of the solution vectors returned by each of the algorithms for 100 independent runs on the 10-D benchmark functions after 100,000 objective function evaluations.

f	Mean Normalized Error (Standard Deviation)		
	PROS	PROS-C	GA
f_1	2.15E−08 (1.53E−08)	**1.35E−14 (2.06E−14)**	4.82E−10 (4.33E−10)
f_2	2.16E−08 (1.66E−08)	**1.24E−14 (1.96E−14)**	5.22E−10 (5.12E−10)
f_3	2.16E−08 (1.67E−08)	**1.42E−14 (2.28E−14)**	5.87E−10 (5.32E−10)
f_4	4.27E−10 (1.37E−10)	**2.53E−13 (1.35E−13)**	5.48E−11 (2.00E−11)
f_5	8.56E−05 (8.40E−05)	**5.18E−05 (5.78E−05)**	1.26E−04 (7.12E−05)
f_6	2.86E−07 (2.03E−07)	**1.84E−13 (2.39E−13)**	8.15E−09 (5.93E−09)
f_7	2.77E−06 (1.97E−06)	**1.68E−12 (1.93E−12)**	8.24E−08 (6.88E−08)
f_8	5.43E−06 (1.73E−06)	**5.00E−08 (9.35E−08)**	6.06E−06 (9.29E−06)
f_9	8.70E−04 (3.16E−04)	**6.07E−07 (3.78E−07)**	1.35E−04 (4.77E−05)
f_{10}	5.35E−05 (2.74E−05)	**4.58E−05 (2.58E−05)**	4.95E−05 (2.97E−05)
f_{11}	1.63E−04 (8.87E−05)	1.38E−04 (6.58E−05)	**7.87E−05 (4.21E−05)**
f_{12}	1.50E−03 (4.73E−04)	1.21E−03 (5.00E−04)	**9.59E−04 (3.94E−04)**

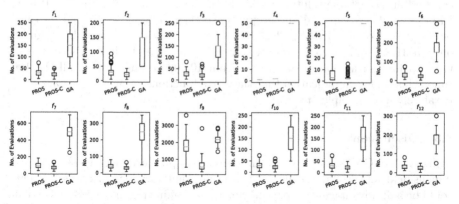

Fig. 5. The box-and-whisker plot of the number of function evaluations required to find a solution with a normalized error less than 0.1 on the 10-D benchmark problems.

smallest number of evaluations to reach the required error level on all benchmark problems. For f_4, PROS-C finds an acceptable solution in its first iteration. For f_4 and f_5, GA finds an acceptable solution in the first generation already.

Figure 6 shows the box-and-whisker plot of the number of objective function evaluations spent by each algorithm to find a solution with normalized error less than 0.0001. Similar to Fig. 5, PROS-C spent the smallest number of evaluations to reach the required error level on all benchmark problems. It has to be noted that some algorithms are unable to find an acceptable solution within 100,000 evaluations on five problems. For f_5. PROS, PROS-C and GA can find

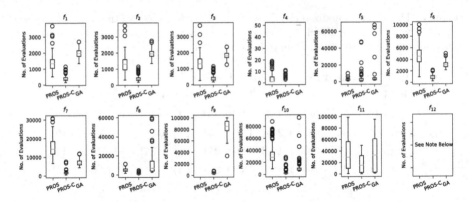

Fig. 6. The box-and-whisker plot of the number of evaluations required to find a solution with a normalized error less than 0.0001 on the 10-D benchmark problems. *Note: For f_9, PROS cannot find a solution to reach the required error level within 100,000 evaluations in all the 100 runs. For f_{12}, PROS, PROS-C and GA cannot find a solution to reach the required error level within 100,000 evaluations in all the 100 runs.*

an acceptable solution in 53, 70 and 28 runs out of the 100 runs, respectively. For f_9, PROS cannot find an acceptable solution in all the 100 runs, while GA can find an acceptable solution in 23 out of the 100 runs only. For f_{10}, PROS, PROS-C and GA can find an acceptable solution only in 94, 98 and 94 runs, respectively. For f_{11}, PROS, PROS-C and GA can find an acceptable solution only in 31, 31 and 74 runs, respectively. For f_{12}, none of PROS, PROS-C and GA can find an acceptable solution within 100,000 evaluations in all the 100 runs. This is also an indication that these five functions, especially f_{12}, are comparatively more complex to solve.

4.3   Discussions

Unlike PROS-C's 1-dimensional blend crossover approach, GA's crossover is D-dimensional and would update all the D decision variables in one step. This could be a disadvantage for separable functions. Taking minimizing the Rastrigin function as an example, given that $D-1$ values in the current candidate solution vector are already equal to the optimal solution 0 and only the i-th decision variable is not, implementing 1-dimensional blend crossover would allow it to keep the $D-1$ already optimized values unchanged.

In addition, GA is not adopting accept-or-reject mechanism while PROS and PROS-C would reject the generated offspring if it is poorer than its parent. That means GA accepts all newly generated offspring no matter how good or bad they are in its new generation. In this setting, it is possible that those partially optimized solutions will be disappeared in the new generation. Even with the elitism feature, only the best solution (or the first few best solutions, for some GA variants) will be kept in the population.

Moreover, GA has to maintain a relatively large pool of population in order to maintain the diversity of the population. A diversified population would help avoiding pre-mature convergence. But this is a tradeoff between convergence rate and diversity. From the experimental results, GA spends more function evaluations at the beginning. After it has identified the promising region, it then converges quickly and surpasses PROS-C (on f_{11} and f_{12}). However, depending on the requirement of the error threshold, PROS-C would be a more favourable and sensible choice when the error threshold requirement is not too strict.

5 Conclusion

In this paper, an effective novel algorithm based on PROS, namely PROS-C, is proposed and it outperforms PROS significantly. It preserves both the advantage of PROS and GA: algorithm simplicity and high convergence rate. We have conducted a numerical experiment on different classes of functions to evaluate the performance of PROS-C. The experimental results show that PROS-C consistently improves the convergence rates and significantly reduces the errors of the obtained solution. PROS-C also outperforms GA in finding more accurate solutions on the benchmark problems when the search budget is tight. The results demonstrate that PROS-C is a good candidate for solving a wide range of problems that require both high computational efficiency and with a limited search budget.

References

1. Ali, M.M., Khompatraporn, C., Zabinsky, Z.B.: A numerical evaluation of several stochastic algorithms on selected continuous global optimization test problems. J. Global Optim. **31**(4), 635–672 (2005). https://doi.org/10.1007/s10898-004-9972-2
2. Blank, J., Deb, K.: Pymoo: multi-objective optimization in python. IEEE Access **8**, 89497–89509 (2020). https://doi.org/10.1109/ACCESS.2020.2990567
3. Eshelman, L.J., Schaffer, J.D.: Real-coded genetic algorithms and interval-schemata. In: Whitley, L.D. (ed.) Foundations of Genetic Algorithms, Foundations of Genetic Algorithms, vol. 2, pp. 187–202. Elsevier (1993). https://doi.org/10.1016/B978-0-08-094832-4.50018-0. https://www.sciencedirect.com/science/article/pii/B9780080948324500180
4. Holland, J.H.: Adaptation in Natural and Artificial Systems. The University of Michigan Press, Ann Arbor (1975)
5. Kennedy, J., Eberhart, R.: Particle swarm optimization. In: International Conference on Neural Networks, ICNN 1995, vol. 4, pp. 1942–1948 (1995). https://doi.org/10.1109/ICNN.1995.488968
6. Plevris, V., Bakas, N.P., Solorzano, G.: Pure random orthogonal search (PROS): a plain and elegant parameterless algorithm for global optimization. Appl. Sci. **11**(11), 5053 (2021). https://doi.org/10.3390/app11115053. https://www.mdpi.com/2076-3417/11/11/5053
7. Storn, R., Price, K.: Differential evolution - a simple and efficient heuristic for global optimization over continuous spaces. J. Glob. Optim. **11**(4), 341–359 (1997). https://doi.org/10.1023/A:1008202821328
8. Tong, B.K.B., Lau, W.C., Sung, C.W., Wong, W.S.: Analysis of pros-c on monotonically increasing functions. https://drive.google.com/file/d/1k0LOJpXBt_98G6Ehvc_JC1Axs_Zt5hQ_/view. Accessed 16 July 2023

9. Tong, B.K.B., Sung, C.W., Wong, W.S.: Random orthogonal search with triangular and quadratic distributions (TROS and QROS): parameterless algorithms for global optimization. Appl. Sci. **13**(3) (2023). https://doi.org/10.3390/app13031391. https://www.mdpi.com/2076-3417/13/3/1391

10. Yang, X.S., Deb, S.: Cuckoo search via lévy flights. In: 2009 World Congress on Nature & Biologically Inspired Computing (NaBIC), pp. 210–214 (2009). https://doi.org/10.1109/NABIC.2009.5393690

11. Yao, X., Liu, Y., Lin, G.: Evolutionary programming made faster. IEEE Trans. Evol. Comput. **3**(2), 82–102 (1999). https://doi.org/10.1109/4235.771163

12. Zamani, S., Hemmati, H.: A cost-effective approach for hyper-parameter tuning in search-based test case generation. In: 2020 IEEE International Conference on Software Maintenance and Evolution (ICSME), pp. 418–429 (2020). https://doi.org/10.1109/ICSME46990.2020.00047

A Multiclass Robust Twin Parametric Margin Support Vector Machine with an Application to Vehicles Emissions

Renato De Leone[1][iD], Francesca Maggioni[2][(✉)][iD], and Andrea Spinelli[2][iD]

[1] School of Science and Technology, University of Camerino,
Via Madonna delle Carceri 9, 62032 Camerino, Italy
renato.deleone@unicam.it

[2] Department of Management, Information and Production Engineering,
University of Bergamo, Viale G. Marconi 5, 24044 Dalmine, Italy
{francesca.maggioni,andrea.spinelli}@unibg.it

Abstract. This paper considers the problem of predicting vehicles smog rating by applying a novel Support Vector Machine (SVM) technique. Classical SVM-type models perform a binary classification of the training observations. However, in many real-world applications only two classifying categories may not be enough. For this reason, a new multiclass Twin Parametric Margin Support Vector Machine (TPMSVM) is designed. On the basis of different characteristics, such as engine size and fuel consumption, the model aims to assign each vehicle to a specific smog rating class. To protect the model against uncertainty arising in the measurement procedure, a robust optimization extension of the multiclass TPMSVM model is formulated. Spherical uncertainty sets are considered and a tractable robust counterpart of the model is derived. Experimental results on a real-world dataset show the good performance of the robust formulation.

Keywords: Multiclass Classification · Support Vector Machine · Robust Optimization

1 Introduction

Support Vector Machine (SVM) is one of the main Machine Learning (ML) techniques commonly used to tackle the problem of binary classification (see [5]). Given two disjoint sets of labelled training data, SVM consists in finding the best classifier capable to predict the category of new unlabelled data. Thanks to its simplicity and predictiveness, SVM has received strong attention in the ML literature, with applications in several fields: chemistry [28], nutrition [33], energy [6,8], bioinformatics [26], finance [7] and green economy [10,29].

Nowadays, sustainability is at the core of policy agendas and debates worldwide. According to the Sustainable Developments Goals 7 and 11 of the UN

G. Nicosia et al. (Eds.): LOD 2023, LNCS 14506, pp. 299–310, 2024.
https://doi.org/10.1007/978-3-031-53966-4_22

2030 Agenda (see [36]) the global greenhouse gas emissions coming from transportations should be effectively reduced. This action can be performed through the promotion of electric and zero emission vehicles by means of new standards, fiscal incentives and improved consumer information.

To raise consumer awareness and promote green attitudes, in 2014 the US introduced the *air pollution score* to measure the amount of pollutants emitted by vehicles (see [12]). Similarly, in Canada the problem of classifying vehicles in terms of their emissions has been addressed by fuel consumption tests, assigning each car to a specific class. Unfortunately, from a practical standpoint, it is not affordable to test all new vehicles to measure fuel consumptions. For this reason, Operational Research and Machine Learning techniques should help policymakers providing new and sustainable solutions.

The aim of this study is to provide a new multiclass classification method under data uncertainty. The novel methodology is applied to address the problem of classifying vehicles in terms of polluting agents under perturbations. In fact, the measurements of pollution parameters are subject to errors due to limited precision of collecting instruments. Ignoring data perturbations in the classification process in general disrupts the reliability of the solution. Robust Optimization (RO) is one of the main paradigms to deal with problems affected with uncertain data (see [2] for an introduction). RO techniques assume that uncertain parameters belong to predefined uncertainty sets. With a min-max approach, a guarantee on the feasibility and optimality of the solution against all possible realizations of the parameters in the uncertainty sets is achieved (see [13]).

Among all the possible variants of the classical SVM, in this work we consider the Twin Parametric Margin Support Vector Machine (TPMSVM) framework, introduced in [30]. This approach achieves better predictive accuracy than other SVM-type techniques, with a reduced computational complexity. Given that real-world observations are corrupted by uncertainty and the classifying categories might be more than two, we specifically focus our attention on the robust and multiclass TPMSVM formulation recently proposed in [9].

In this paper, we consider the problem of predicting vehicles smog rating by proposing a novel multiclass and robust TPMSVM model with linear classifiers. Data uncertainty is explicitly handled within the model by means of spherical sets centered around training observations. The main contributions of the paper are three-fold and can be summarized as follows:

- To formulate a novel TPMSVM model for linear multiclass classification.
- To derive a spherical robust counterpart of the deterministic formulation to protect the model against uncertainty.
- To provide numerical experiments on a real-world dataset with the aim of understanding the advantage of explicitly considering the uncertainty versus deterministic approaches.

The remainder of the paper is organized as follows. Section 2 reviews the existing literature on the problem. In Sect. 3, deterministic binary and robust multiclass formulations for TPMSVM are presented. Section 4 describes data and reports the numerical experiments. Finally, Sect. 5 concludes the paper.

Throughout the paper all vectors will be column vectors, unless transposed by the superscript "$^\top$". For $p \in [1, \infty]$ and $a \in \mathbb{R}^n$, $\|a\|_p$ is the ℓ_p-norm of a. If S is a set, $|S|$ denotes its cardinality.

2 Literature Review

SVM is introduced in [37] with the aim of addressing a binary classification task. The goal is to determine a hyperplane that geometrically separates input obser-vations into two categories, such that the margin between the two sample classes is maximized. The underlying hypothesis of this approach is that input data are linearly separable. To overcome this limitation and capture the misclassification error, in [1,5] a vector of slack variables is introduced in the model. The corre-sponding optimal solution balances the trade-off between the maximization of the margin and the minimization of the misclassification error.

Several alternative formulations of the classical SVM approach have been devised in the literature. In [27] a separating surface induced by a kernel matrix is derived by considering either a quadratic or piecewise-linear objective func-tion. In [18] a parametric margin model (the par-ν-SVM) is introduced to deal with the case of heteroscedastic noise. The positive parameter ν is included in the objective function to bound the fractions of supporting vectors and misclas-sification errors (see [32]). In [22] the two classes are firstly separated by means of two parallel hyperplanes such that the intersection of the convex hulls of the two categories lies in the region between them. Then, the optimal separating hyperplane is searched in that region such that the total number of misclassified observations is minimized. Rather than dealing with parallel hyperplanes, the TWin Support Vector Machine (TWSVM) (see [20]) considers two smaller sized SVM-problems, one for each class. Due to its favourable performances, espe-cially when handling large datasets, many variants of the TWSVM approach have been proposed. In the TPMSVM (see [30]) two nonparallel classifiers are designed, one for each class, such that training observations of the other class are as far as possible from the opposite classifier. This method integrates the fast learning speed of the TWSVM and the flexible parametric margin of the par-ν-SVM. For a comprehensive review on recent developments on TWSVM the reader is referred to [34].

All the approaches discussed so far consider the case of binary classification. However, in real-world applications the classifying categories might be more than two. To tackle the problem of multiclass classification, *all-together* methods and *decomposition-reconstruction* methods have been devised. Since the latter pro-cedures are nowadays considered to be the most effective to achieve multiclass separation (see [11]), we focus our attention on them. Specifically, in this contri-bution we deal with the *one-versus-all* approach (see [19]). Within this technique, one classifier for each class is generated and the final decision function aggre-gates all the results. A review on multiclass models designed for TWSVM can be found in [11].

RO techniques applied to classification problems can be found in [3]. In [13] the robust counterpart of the linear model of [22] is formulated under box and ellipsoidal uncertainty sets. In addition, a moment-based distributionally robust formulation is provided. A recent application of this methodology to a COVID-19 patients classification problem is delivered in [23]. The generalization of the approach of [22] with robust nonlinear classifiers has been derived in [24]. In the spirit of robust TWSVM, in [31] two nonparallel classifiers are proposed in the case of ellipsoidal uncertainty set. The corresponding model is then reformulated as a Second Order Cone Programming (SOCP) problem. In [9] a robust multiclass TPMSVM approach is provided.

In the ML literature, the problem of classifying vehicles have been addressed in several works. In all of these studies, cars and trucks are classified on the basis of images properties or camera detections (see [14]). Recently, in [25] a variant of the classical SVM has been devised to address the problem of classifying vehicles in terms of their emissions. The approach is based on the robust nonlinear classifier presented in [24] and deals with a binary classification task. According to this procedure, each vehicle is assigned to one of the two classes "good smog rating"-"bad smog rating" on the basis of different characteristics, such as engine size, fuel consumption and CO_2 tailpipe emissions. To the best of our knowledge, the current paper is the second work that considers the problem of classifying vehicles in terms of their emissions with a ML perspective.

3   Methods

In this section, we first recall the deterministic formulation presented in [30] for a binary classification task. Then, we consider the robust model proposed in [9] as an extension to the case of multiclass classification.

3.1   Deterministic Binary Formulation

Let $\mathcal{A}$ and $\mathcal{B}$ two disjoint sets of training observations. Each input data is described by a n-dimensional vector x^i of features and characterized by a label y_i denoting the class to which it belongs ("+ 1" for class $\mathcal{A}$ and "− 1" for class $\mathcal{B}$). Let $m_+ = |\mathcal{A}|$ and $m_- = |\mathcal{B}|$ be the cardinalities of the two classes, with $m = m_+ + m_-$ the total number of samples. We denote by $\mathcal{X}_+$ and $\mathcal{X}_-$ the indices of data points in class $\mathcal{A}$ and $\mathcal{B}$, respectively.

The classification task within the linear TPMSVM approach is addressed by means of two nonparallel hyperplanes H_+ and H_- defined by the following equations:

$$H_+ : w_+^\top x + \theta_+ = 0 \qquad H_- : w_-^\top x + \theta_- = 0.$$

Specifically, the normal vectors $w_+, w_- \in \mathbb{R}^n$ and the intercepts $\theta_+, \theta_- \in \mathbb{R}$ are the solutions of the following optimization models:

$$
\min_{w_+, \theta_+, \xi_+} \quad \frac{1}{2} \|w_+\|_2^2 + \frac{\nu_+}{m_-} \sum_{i \in \mathcal{X}_-} \left(x^{i^\top} w_+ + \theta_+ \right) + \frac{\alpha_+}{m_+} \sum_{i \in \mathcal{X}_+} \xi_{+,i}
$$

$$
\text{s.t.} \quad x^{i^\top} w_+ + \theta_+ \geqslant - \xi_{+,i} \qquad i \in \mathcal{X}_+
$$

$$
\xi_{+,i} \geqslant 0 \qquad i \in \mathcal{X}_+
$$

(1)

and

$$
\min_{w_-, \theta_-, \xi_-} \quad \frac{1}{2} \|w_-\|_2^2 - \frac{\nu_-}{m_+} \sum_{i \in \mathcal{X}_+} \left(x^{i^\top} w_- + \theta_- \right) + \frac{\alpha_-}{m_-} \sum_{i \in \mathcal{X}_-} \xi_{-,i}
$$

$$
\text{s.t.} \quad x^{i^\top} w_- + \theta_- \leqslant \xi_{-,i} \qquad i \in \mathcal{X}_-
$$

$$
\xi_{-,i} \geqslant 0 \qquad i \in \mathcal{X}_-.
$$

(2)

The constant values $\nu_+, \nu_- > 0$ and $\alpha_+, \alpha_- > 0$ are regularization parameters controlling the fractions of supporting vectors for each class (see [32]). Components of the slack vectors $\xi_+ \in \mathbb{R}^{m_+}$ and $\xi_- \in \mathbb{R}^{m_-}$ are associated with the misclassified samples in each class (see [5]).

Once models (1)–(2) are solved, each new observation $x \in \mathbb{R}^n$ is classified according to the following decision function:

$$
f(x) = \text{sign} \left(\frac{w_+^\top x + \theta_+}{\|w_+\|_2} + \frac{w_-^\top x + \theta_-}{\|w_-\|_2} \right).
$$

(3)

3.2   Robust Multiclass Formulation

Given that real-world observations may not be always separable into just two categories, for each input data we assume that the corresponding label y_i belongs to the set $\{1, \ldots, C\}$, with C the total number of classes. In addition, to handle uncertainty in data features, we construct around each sample x^i a spherical uncertainty set $\mathcal{U}(x^i)$ defined as:

$$
\mathcal{U}(x^i) = \left\{ x \in \mathbb{R}^n \mid x = x^i + \sigma^i, \left\| \sigma^i \right\|_2 \leqslant \eta_i \right\}.
$$

(4)

The vector of perturbation $\sigma^i \in \mathbb{R}^n$ is unknown but its ℓ_2-norm is bounded by a nonnegative constant η_i (see [35]).

Following the *one-versus-all* approach (see [9]), for each class $c = 1, \ldots, C$ we look for the best separating hyperplane $H_c : w_c^\top x + \theta_c = 0$ as solution of a TPMSVM-type model. Specifically, let m_c be the total number of observations belonging to class c and $\mathcal{X}_c$ the corresponding index set. Conversely, the index set of all points not in class c is denoted as $\mathcal{X}_{-c}$.

For each class $c = 1, \ldots, C$, the normal vector $w_c \in \mathbb{R}^n$ and the intercept $\theta_c \in \mathbb{R}$ of the hyperplane H_c are the solutions of the following min-max model:

$$\min_{w_c, \theta_c, \xi_c} \quad \frac{1}{2} \|w_c\|_2^2 + \frac{\nu_c}{m - m_c} \sum_{i \in \mathcal{X}_{-c}} \max_{\|\sigma^i\|_2 \leq \eta_i} \left[(x^i + \sigma^i)^\top w_c + \theta_c \right] + \frac{\alpha_c}{m_c} \sum_{i \in \mathcal{X}_c} \xi_{c,i}$$

$$\text{s.t.} \quad x^\top w_c + \theta_c \geq -\xi_{c,i} \quad i \in \mathcal{X}_c, \forall x \in \mathcal{U}(x^i)$$

$$\xi_{c,i} \geq 0 \quad i \in \mathcal{X}_c,$$

$$(5)$$

where ξ_c is the slack vector for class $c = 1, \ldots, C$.

Unfortunately, model (5) is intractable due to the infinite possibilities for choosing $x \in \mathcal{U}(x^i)$ in the first set of constraints. Nevertheless, in [9] model (5) has been reformulated in a tractable robust form as:

$$\min_{w_c, \theta_c, \xi_c} \quad \frac{1}{2} \|w_c\|_2^2 + \frac{\nu_c}{m - m_c} \sum_{i \in \mathcal{X}_{-c}} \left(x^{i^\top} w_c + \eta_i \|w_c\|_2 \right) + \nu_c \theta_c + \frac{\alpha_c}{m_c} \sum_{i \in \mathcal{X}_c} \xi_{c,i}$$

$$\text{s.t.} \quad x^{i^\top} w_c + \theta_c - \eta_i \|w_c\|_2 \geq -\xi_{c,i} \quad i \in \mathcal{X}_c$$

$$\xi_{c,i} \geq 0 \quad i \in \mathcal{X}_c.$$

$$(6)$$

By introducing auxiliary variables, the robust model (6) can be expressed as a SOCP problem and easily solved by interior point methods (see [35]).

Once all the C hyperplanes have been determined, one for each class, the final decision function is:

$$g(x) = \arg\min_{c=1,\ldots,C} \frac{|w_c^\top x + \theta_c|}{\|w_c\|_2}.$$

4 Computational Experiments

This section discusses our numerical experiments. In particular in Subsect. 4.1 we describe the features of the real-world dataset regarding vehicles emissions. Then, in Subsect. 4.2 the performance of the model on the basis of classical statistical indicators are discussed.

All computational experiments are obtained using CVX (see [16]) in MAT-LAB (v. 2021b) and solver MOSEK (v. 9.1.9) on a MacBookPro17.1 with a chip Apple M1 and 16 GB of RAM.

4.1 Dataset Description

Real-world data on the fuel consumption ratings on 374 different vehicles in the first months of 2023 are taken from [15]. Among all, vehicles with fuel type "regular gasoline" are considered in this work due to their high pollution rates (see [25]). Each vehicle is described by 7 attributes: engine size (range $1.2 - 8$), number of cylinders (range $3 - 16$), fuel consumptions rating in city (range 4.4

— 30.3), fuel consumptions rating in highway (range $4.4 - 20.9$), combined city-highway fuel consumptions rating (range $4.4 - 26.1$), tailpipe emissions of CO_2 for combined city-highway driving (range $104 - 608$), and tailpipe emissions of CO_2 rating (range $1 - 9$). Each vehicle is labelled according to the tailpipe emissions of smog-forming pollutants rate and assigned to one of the following categories: 3 (worst emissions), 5, 6, 7, 8 (best emissions).

The distribution of vehicles among the different classes is reported in Table 1.

Table 1. Distribution of vehicles among classes in the considered dataset.

Smog rating score	Class	Number of vehicles	Class distribution
3	1	15	4.01%
5	2	127	33.96%
6	3	83	22.19%
7	4	145	38.77%
8	5	4	1.07%

4.2   Model Validation

The dataset has been divided into training set and testing set through a k-fold cross-validation technique. For each fold $s = 1, \ldots, k$ and for each class $c = 1, \ldots, C$, the quality of the solution has been measured by means of an in-class accuracy, defined as:

$$A_c^s := \frac{\text{number of testing points correctly classified in class } c}{\text{number of testing points in class } c}.$$

In addition, for each fold s the overall performance of the classification process has been validated through an aggregate accuracy measure, computed as:

$$A^s := \frac{\text{number of testing points correctly classified}}{\text{number of testing points}}.$$

Finally, the results are averaged:

$$\text{Accuracy for class } c := \frac{\sum_{s=1}^{k} A_c^s}{k}, \qquad \text{Accuracy} := \frac{\sum_{s=1}^{k} A^s}{k}.$$

Since class 5 contains only 4 samples (see Table 1), we set $k = 4$. As in [30], data are normalized such that the features locate in $[0, 1]$. We validate our model by comparing the results with the One-Versus-Rest multiclass TWSVM (OVR TWSVM) presented in [11]. The choice is motivated by the fact that both our model and OVR TWSVM are one-versus-all approaches, considering each class one at a time in the training process. For brevity's sake, regularization parameters are set to $\nu_c = \nu$ and $\alpha_c = \alpha$ for all $c = 1,\ldots,C$ and a grid search procedure is applied to tune their values. Specifically, α is selected from the set $\{2^j | j = -5, -4,\ldots,4,5\}$, and the value of ν/α is chosen from the set $\{0.1, 0.3, 0.5, 0.7, 0.9\}$. The degree of perturbation η_i in the uncertainty set (4) is assumed to be equal to η for all $i = 1,\ldots,m$. In particular, we consider $\eta = 0$, i.e. no perturbation, and three increasing levels of uncertainty, namely 0.001, 0.01 and 0.1.

The results of the simulations are shown in Table 2.

Table 2. Performance of model (6) on real-world data on fuel consumption (see [15]). For each indicator, the best result is highlighted in bold.

	OVR TWSVM [11]	Deterministic model (6)	Robust model (6)		
η	0	0	0.001	0.01	0.1
Accuracy for class 1	0.00%	0.00%	0.00%	0.00%	0.00%
Accuracy for class 2	40.42%	51.99%	45.61%	**67.06%**	61.82%
Accuracy for class 3	7.50%	5.95%	**13.39%**	0.00%	0.00%
Accuracy for class 4	73.84%	73.03%	69.63%	75.11%	**77.46%**
Accuracy for class 5	50.00%	50.00%	50.00%	0.00%	0.00%
Accuracy	44.39%	47.85%	45.98%	**51.89%**	51.05%
Time (s)	0.513	0.717	0.719	0.697	0.662

First of all, we notice that our deterministic formulation (6) with $\eta = 0$ outperforms OVR TWSVM from [11] in terms of overall accuracy (47.85% vs 44.39%). Secondly, the majority of the indicators benefit from including uncertainties in the proposed formulations. The overall accuracy across the different levels of perturbation is higher when compared to the deterministic case (47.85%), with the best result (51.89%) attained at $\eta = 0.01$. Nevertheless, due to the limited number of observations in classes 1 and 5 (see Table 1), both models are no longer able to predict the correct smog rating score in these classes. Specifically, class 5 does not gain additional accuracy improvement adding noise to the data. This is to be expected since in these cases specific techniques for handling rare events should be implemented (see [4,17]). However, this is out of scope of the paper. When considering the confusion matrices, both deterministic formulations predict 60.0% of the observations in class 1 to be in classes 3, 4 or 5, whereas the value decreases to 33.3% with the robust model (6) with $\eta = 0.01$. From a practical perspective, since class 1 is the most polluting, a robust approach may be suitable in order to protect the model against this type of misclassification error.

With the aim of evaluating the performance of our model, we aggregate data with bad smog rating (scores 3–5), and with good smog rating (scores

7–8). With this choice, the dataset is partitioned into three main categories: class $\tilde{1}$ (*bad emissions*), class $\tilde{2}$ (*medium emissions*), class $\tilde{3}$ (*good emissions*). The corresponding distribution of observations in each class is 37.97%, 22.19%, 39.84%, respectively.

The results of the computational experiments on the reduced dataset with 3 classes are shown in Table 3.

Table 3. Performance of model (6) in the case of 3 classes. For each indicator, the best result is highlighted in bold.

	OVR TWSVM [11]	Deterministic model (6)	Robust model (6)		
η	0	0	0.001	0.01	0.1
Accuracy for class $\tilde{1}$	61.81%	54.43%	50.81%	**64.90%**	61.81%
Accuracy for class $\tilde{2}$	10.77%	26.73%	**41.07%**	9.58%	4.76%
Accuracy for class $\tilde{3}$	70.63%	76.49%	69.81%	73.83%	**81.88%**
Accuracy	54.01%	56.67%	56.17%	56.15%	**57.21%**
Time (s)	0.321	0.424	0.411	0.436	0.424

The comparison between the results of OVR TWSVM from [11] and our formulation leads to the same conclusion as in the case of 5 classes, having an overall accuracy of 56.67% vs 54.01% in the deterministic case. Besides, the robust formulation allows to increase the overall accuracy level to 57.21% when $\eta = 0.1$. In addition, the predictive power of the models with 3 classes is always higher than with 5 classes, meaning that unbalanced data may disrupt the reliability of the solution. When including uncertainty, each class benefits at different extents: classes $\tilde{1}$ and $\tilde{3}$ withstand strong degrees of uncertainty ($\eta = 0.01$ or 0.1), whereas class $\tilde{2}$ takes advantages only with low perturbations ($\eta = 0.001$). It is worth noticing that, from a practical perspective, it is worse to misclassify a vehicle with "bad emission"(class $\tilde{1}$) than the other cases. Therefore, it is reasonable to consider a strong robust model with $\eta = 0.01$ or 0.1 which attains a good accuracy for class $\tilde{1}$. Consequently, a trade-off between the performance of the model and its ability to protect against uncertainty needs to be taken into account.

As far as it concerns timing, the OVR TWSVM has a faster learning speed when compared to our formulation, but with a lower predictive power. Moreover, since both OVR TWSVM and model (6) are one-versus-all approaches, requiring to solve C optimization subproblems, the CPU time depends on the number of classes. Within our examples, when passing from 3 to 5 classes, an average increases of 62% of the learning time occurs.

5   Conclusions

This article presents a new multiclass approach to classify vehicles in terms of smog rating emissions under uncertainty as a Twin Parametric Margin Support

Vector Machine task. This technique helps decision makers to rank passenger cars in terms of their pollution emissions. Given the uncertain nature of real-world data features, we formulated a robust optimization model with spherical uncertainty sets around samples. The numerical results show the good performance of the proposed model, especially when including uncertainty. Similar results are obtained in other real-world datasets taken from UCI repository (see [21]) and the corresponding numerical results are reported in [9].

In future works, in order to increase the predictive power of the model, kernel-induced decision boundaries and asymmetric uncertainty sets will be explored.

Acknowledgements. This work has been supported by "ULTRA OPTYMAL - Urban Logistics and sustainable TRAnsportation: OPtimization under uncertainTY and MAchine Learning", a PRIN2020 project funded by the Italian University and Research Ministry (grant number 20207C8T9M).

This study was also carried out within the MOST - Sustainable Mobility National Research Center and received funding from the European Union Next-GenerationEU (PIANO NAZIONALE DI RIPRESA E RESILIENZA (PNRR) - MISSIONE 4 COMPONENTE 2, INVESTIMENTO 1.4 - D.D. 1033 17/06/2022, CN00000023), Spoke 5 "Light Vehicle and Active Mobility" and PNRR MUR project ECS_00000041-VITALITY - CUP J13C22000430001, Spoke 6. This manuscript reflects only the authors' views and opinions, neither the European Union nor the European Commission can be considered responsible for them.

References

1. Bennett, K.P., Mangasarian, O.L.: Robust linear programming discrimination of two linearly inseparable sets. Optim. Methods Softw. **1**(1), 23–34 (1992)
2. Ben-Tal, A., El Ghaoui, L., Nemirovski, A.: Robust Optimization, vol. 28. Princeton University Press (2009)
3. Bertsimas, D., Dunn, J., Pawlowski, C., Zhuo, Y.D.: Robust classification. INFORMS J. Optim. **1**(1), 2–34 (2019)
4. Carreño, A., Inza, I., Lozano, J.A.: Analyzing rare event, anomaly, novelty and outlier detection terms under the supervised classification framework. Artif. Intell. Rev. **53**, 3575–3594 (2020)
5. Cortes, C., Vapnik, V.N.: Support-vector networks. Mach. Learn. **20**, 273–297 (1995)
6. De Cosmis, S., De Leone, R., Kropat, E., Meyer-Nieberg, S., Pickl, S.: Electric load forecasting using support vector machines for robust regression. In: Proceedings of the Emerging M&S Applications in Industry & Academia - Modeling and Humanities Symposium, EAIA and MatH 2013. Society for Computer Simulation International, San Diego (2013)
7. De Leone, R.: Support vector regression for time series analysis. In: Hu, B., Morasch, K., Pickl, S., Siegle, M. (eds.) Operations Research Proceedings 2010. ORP, pp. 33–38. Springer, Heidelberg (2011). https://doi.org/10.1007/978-3-642-20009-0_6
8. De Leone, R., Giovannelli, A., Pietrini, M.: Optimization of power production and costs in microgrids. Optim. Lett. **11**, 497–520 (2017)

9. De Leone, R., Maggioni, F., Spinelli, A.: A robust twin parametric margin support vector machine for multiclass classification (2023). https://arxiv.org/abs/2306.06213

10. De Leone, R., Pietrini, M., Giovannelli, A.: Photovoltaic energy production forecast using support vector regression. Neural Comput. Appl. **26**, 1955–1962 (2015)

11. Ding, S., Zhao, X., Zhang, J., Zhang, X., Xue, Y.: A review on multi-class TWSVM. Artif. Intell. Rev. **52**, 775–801 (2019)

12. Environmental Protection Agency: Fed. Reg. **79**(81), 23414–23886 (2014)

13. Faccini, D., Maggioni, F., Potra, F.A.: Robust and distributionally robust optimization models for linear support vector machine. Comput. Oper. Res. **147**, 105930 (2022)

14. Farid, A., Hussain, F., Khan, K., Shahzad, M., Khan, U., Mahmood, Z.: A fast and accurate real-time vehicle detection method using deep learning for unconstrained environments. Appl. Sci. **13**(5), 30–59 (2023)

15. Fuel Consumption Ratings - Government of Canada. https://open.canada.ca/data/en/dataset/98f1a129-f628-4ce4-b24d-6f16bf24dd64#wb-auto-6. Accessed 05 Mar 2023

16. Grant, M., Boyd, S: CVX: Matlab software for disciplined convex programming, version 2.0 beta (2013). http://cvxr.com/cvx

17. Haixiang, G., Yijing, L., Shang, J., Mingyun, G., Yuanyue, H., Bing, G.: Learning from class-imbalanced data: review of methods and applications. Expert Syst. Appl. **73**(1), 220–239 (2017)

18. Hao, P.Y.: New support vector algorithms with parametric insensitive/margin model. Neural Netw.: Official J. Int. Neural Netw. Soc. **23**(1), 60–73 (2010)

19. Hsu, C.W., Lin, C.J.: A comparison of methods for multiclass support vector machines. IEEE Trans. Neural Netw. **13**(2), 415–425 (2002)

20. Jayadeva, Khemchandani, R., Chandra, S.: Twin support vector machines for pattern classification. IEEE Trans. Pattern Anal. Mach. Intell. **29**(5), 905–910 (2007)

21. Kelly, M., Longjohn, R., Nottingham, K: The UCI Machine Learning Repository (2023). https://archive.ics.uci.edu

22. Liu, X., Potra, F.A.: Pattern separation and prediction via linear and semidefinite programming. Stud. Inform. Control **18**(1), 71–82 (2009)

23. Maggioni, F., Faccini, D., Gheza, F., Manelli, F., Bonetti, G.: Machine learning based classification models for COVID-19 Patients. In: ORAHS 2022 Proceedings (2023, to appear)

24. Maggioni, F., Spinelli, A.: A robust optimization model for nonlinear support vector machine (2023). https://arxiv.org/abs/2306.06223

25. Maggioni, F., Spinelli, A.: A robust nonlinear support vector machine approach for vehicles smog rating classification. Accepted for publication in AIRO Springer Series Optimization and Decision Science (2023)

26. Mancini, A., et al.: Machine learning models predicting multidrug resistant urinary tract infections using "Dsaas". BMC Bioinform. **21**(10) (2020)

27. Mangasarian, O.L.: Generalized support vector machines. In: Advances in Large Margin Classifiers. MIT Press (1998)

28. Marcelli, E., De Leone, R.: Multi-kernel covariance terms in multi-output support vector machines. In: Nicosia, G., et al. (eds.) LOD 2020. LNCS, vol. 12566, pp. 1–11. Springer, Cham (2020). https://doi.org/10.1007/978-3-030-64580-9_1

29. Pellegrini, M., De Leone, R., Maponi, P.: Reducing power consuption in hydrometric level sensor network using support vector machines. In: Proceedings of the PECCS 2013 International Conference on Pervasive and Embedded Computing and Communication Systems. PECCS (2013)

30. Peng, X.: TPMSVM: a novel twin parametric-margin support vector machine for pattern recognition. Pattern Recogn. **44**(10), 2678–2692 (2011)
31. Qi, Z., Tian, Y., Shi, Y.: Robust twin support vector machine for pattern classification. Pattern Recogn. **46**(1), 305–316 (2013)
32. Schölkopf, B., Smola, A., Williamson, R.C., Bartlett, P.L.: New support vector algorithms. Neural Comput. **12**(5), 1207–1245 (2000)
33. Silvi, S., et al.: Probiotic-enriched foods and dietary supplement containing SYN-BIO positively affects bowel habits in healthy adults: an assessment using standard statistical analysis and Support Vector Machines. Int. J. Food Sci. Nutr. **65**(8), 994–1002 (2014)
34. Tanveer, M., Rajani, T., Rastogi, R., Shao, Y.: Comprehensive review on twin support vector machines. Ann. Oper. Res. (2022)
35. Trafalis, T.B., Gilbert, R.C.: Robust classification and regression using support vector machines. Eur. J. Oper. Res. **173**(3), 893–909 (2006)
36. United Nations: Transforming our world: the 2030 Agenda for Sustainable Development (2015). https://wedocs.unep.org/20.500.11822/9814
37. Vapnik, V.N.: Estimation of Dependences Based on Empirical Data. Springer, Heidelberg (1982)

LSTM Noise Robustness: A Case Study for Heavy Vehicles

Maria Elena Bruni[2,3] , Guido Perboli[1], and Filippo Velardocchia[1,3]([✉])

[1] DIGEP, Politecnico di Torino, Turin, Italy
{guido.perboli,filippo.velardocchia}@polito.it
[2] DIMEG, Università della Calabria, Arcavacata, Italy
mariaelena.bruni@unical.it
[3] CIRRELT, Montreal, Canada

Abstract. Artificial intelligence (AI) techniques are becoming more and more widespread. This is directly related to technology progress and aspects as the flexibility and adaptability of the algorithms considered, key characteristics that allow their use in the most variegated fields. Precisely the increasing diffusion of these techniques leads to the necessity of evaluating their robustness and reliability. This field is still quite unexplored, especially considering the automotive sector, where the algorithms need to be prepared to answer noise problems in data acquisition. For this reason, a methodology directly linked to previous works in the heavy vehicles field is presented. In particular, the same is focused on the estimation of rollover indexes, one of the main issues in road safety scenarios. The purpose is to expand the cited works, addressing the LSTM networks performance in case of strongly disturbed signals.

Keywords: transportation systems · heavy vehicles · LSTM networks · noise · rollover risk indicators

1 Introduction

While Artificial intelligence (AI) techniques are becoming more and more popular, the majority of papers deal with their accuracy. Only a few also consider the robustness of the algorithms themselves [8,9,11]. In particular, it has to be noted that LSTM networks represent the most adapt AI technique in case of time-depending data [6], other than being of interest in numerous fields [4] and, for the reason above, especially used in the context of heavy vehicles rollover [14]. This is one of the central themes in road safety scenarios [10,13,15], where prominent results were attained using Long Short-Term Memory (LSTM) networks to define rollover indicators [2,12]. Unfortunately, this was done without specifically addressing the algorithm's robustness to noise-varying situations. Therefore, in this paper, we provide an analysis focused on the noise robustness of an existing LSTM network, developed for heavy vehicle rollover estimation [12]. The main contributions can be synthesized in:

G. Nicosia et al. (Eds.): LOD 2023, LNCS 14506, pp. 311–323, 2024.
https://doi.org/10.1007/978-3-031-53966-4_23

- The validation of the AI algorithm previously developed in [12], taking into account its performance in the analysis of data afflicted by strong disturbance. This has been simulated by introducing white noise on the synthetic data composing the original testing datasets while maintaining noiseless the training of the networks.
- The definition of a specific LSTM network training to further investigate the noise effects and their impact on the final results. This has been done by introducing the disturbance, other than in the datasets used to test the AI algorithm, even in the ones involved in the training. In addition, further datasets have been defined and analyzed to propose a more complete overview of the algorithm's robustness.

Following the introduction, the article is organized in three distinguished sections, the first one being an overall overview of the core elements of the work, Sect. 2. Section 3 provides a detailed explanation of the methodology developed, firstly in terms of data generation and manipulation (Subsect. 3.1) and then focusing on the results provided by the different LSTM networks (Subsect. 3.2). Finally, Sect. 4 concludes the paper, discussing considerations on possible future developments.

2   System Overview

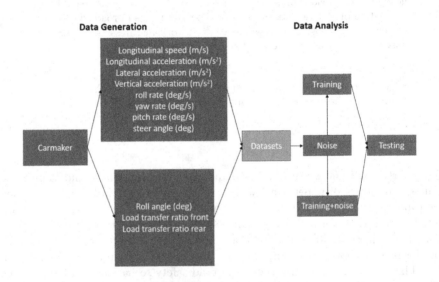

Fig. 1. General system overview

Figure 1 provides a general overview of the developed system. The methodology presented is based on the analysis of the same heavy vehicle (with the exception of two datasets), utilizing synthetic data both for generation and analysis purposes, with the goal of estimating rollover indexes. Concerning the data generation aspects, maneuvers and surfaces are programmed and defined in the Carmaker environment, leading to datasets characterized by inputs and outputs as in [12] for the LSTM networks. Consequently, the novelty relies in the second part of the methodology, the data analysis. In particular, to test the neural networks robustness and validity, white Gaussian noise has been added. This specific kind of noise is present in every environment and, therefore, sensor measurement. Moreover, in simulation, it is possible to program its Signal to Noise Ratio (SNR), highlighting its effects on the affected signals for the desired intensity level. In our specific case, the decision was to add it to the inputs (longitudinal speed etc., Fig. 1) and outputs (roll angle etc.) composing each of the 15 datasets of training, and each of the 8 used for testing. This led to two different types of results in the latter. The first one is obtained by simply using the neural networks trained with noiseless signals (Subsect. 3.2) to estimate rollover indexes from unknown maneuvers (testing phase) characterized by having strongly disturbed signals. This has been done to assess their robustness and by measuring and comparing the outcomes in terms of Mean Squared Error (MSE) and Mean Absolute Error (MAE). The second typology of results has been obtained using a training with inputs and outputs strongly disturbed. In this case, the networks, as can be expected, were performing better than in the previous one, but it has to be noted that the indexes attained were comparable with the first ones (similar order of magnitude).

3   Methodology

3.1   Data Generation: Carmaker

To help a better understanding of the methodology, in this Section a brief recall of the data generation aspects is presented. The Carmaker software has been used to define the heavy vehicle characteristics, focusing then on the programming of the maneuvers (and associated surfaces) of interest. As deeply investigated in [12], the Carmaker software allows the specific definition of several vehicle typologies. A specific interface allows to efficiently interact with the related codes. These permitted, in our case of study, the implementation of 23 new maneuvers directly associated with the heavy vehicle presented in the aforementioned article, with 15 of them used for the training of the LSTM networks and 8 for the testing, as it will be discussed in Subsect. 3.2. It is important to highlight, as anticipated, that the datasets are mostly the same as the ones previously developed in [12], since the goal of the paper is to evaluate the performance that the same LSTM networks provide in case of noise on the datasets of interest. Hence, as can be seen in Fig. 1, the variables collected during the simulation and consequently characterizing the inputs and outputs of the neural networks (Table 1) remained the same.

It is worthwhile noticing that we defined, in output, typical parameters that characterize rollover eventualities [3] while, in input, typical trajectory parameters that can be obtained using an ECU on board. Each of the 15 maneuvers

Table 1. LSTM networks: inputs and outputs

Type	Description	Units
Inputs	Longitudinal speed	m/s
	Longitudinal acceleration	m/s^2
	Lateral acceleration	m/s^2
	Vertical acceleration	m/s^2
	roll rate	deg/s
	yaw rate	deg/s
	pitch rate	deg/s
	steer angle	deg
Outputs	Roll angle	deg
	Load transfer ratio front	–
	Load transfer ratio rear	–

composing the training has been programmed to cover rollover risk situations, other than being of the same length (500 s of duration with every variable sampled every 0.01 s) and, therefore, dataset dimensions. This is particularly important to obtain correctly trained networks, especially in terms of weight. It is highlighted that the methodology is based on synthetic data for different reasons, among which the difficulties in obtaining varied and large enough datasets with on-field acquisitions. These can often lead to excessive expenses in terms of times and costs [1], requesting different solutions. A methodology based on realistic synthetic data allows the experimentation of several techniques to find the aforementioned solutions, enabling a better exploration and understanding of the problem and leading to a first evaluation of the chosen approach in a variety of different situations. This can be extremely relevant especially when the defined AI architecture is implemented on board and, particularly, to evaluate to what extent synthetic data can be used as an effective substitute for on-field data in real applications. This further motivates the investigation of AI noise robustness since this type of disturbance is an inevitable element in on-field testing.

3.2 AI: Data Analysis and Results

Defined the data generation aspects of the methodology, it is important to analyze its core, the AI algorithm. This is based, as anticipated, on LSTM networks, with the inputs and outputs reported in Table 1 and the associated architecture in Table 2.

In order to test the noise robustness of the developed AI algorithm, different tests were conducted. Firstly, white Gaussian noise was introduced in the 8 datasets of testing, unknown to the networks, both for inputs and outputs. This type of disturbance is very common, especially considering road surface vibrations [5,7]. White noise is extremely versatile since it can be added to any

Table 2. LSTM Architecture.

Type	Description	Characteristics
Layers	Sequence input	Number of features (8)
	LSTM	Number of hidden units (100)
	Fully connected	Number of Responses (3)
	Regression	
Main Hyperparameters	Adam	Adaptive moment estimation
	MaxEpochs	850
	GradientThreshold	1
	InitialLearnRate	0.01
	LearnRateDropPeriod	425
	LearnRateDropFactor	0.2

signal defining the considered system. Moreover, it has uniform power across the frequency band, with a normal distribution in the time domain and, adequately regulating its Signal to Noise Ratio (SNR), can replicate the effect of many random noise phenomenons. In particular, in our case, the SNR was set to a value of 20 dB, representing a condition of a strongly disturbed signal (the higher the SNR, the stronger the desired signal is in comparison to background noise). Introduced the disturbance in the 5 datasets of testing, the performance of the neural networks in terms of Mean Absolute Error (MAE) and Mean Squared Error (MSE) was measured. This led, as expected, to good, but less accurate results considering the ones obtained in absence of noise in output, since the 15 datasets of training remained the same. Consequently, we introduced the same noise in the data provided for the training (while maintaining the architecture presented in Table 2). This was done to study the effect of noise in the training of the LSTM network and its effect on the performance. For a better understanding, we propose the outcomes for the same maneuvers presented in [12], adding to the same a resuming table containing the results of the other five datasets developed to expand and further testing the potential applications of the algorithm.

As can be easily noted, the neural network keeps following the real trend (known since obtained by a simulation) of its outputs. This is also confirmed analyzing the correlated MAE and MSE (Table 3) and confronting the same with the ones attained in the case of noiseless signals. It is notable that the considered maneuver, concerning the traversing of the Bernina pass by the programmed heavy vehicle, presents a relevant difference (in relation to the other values) just for the roll angle which is, considering the outputs, the only one that could be attained with an empirical acquisition (while the LTR have to be necessarily calculated trough structured models).

Analyzing the performance of the LSTM network following a training with signals subjected to noise (SNR equal to 20 dB), it is possible to observe (Fig. 3)

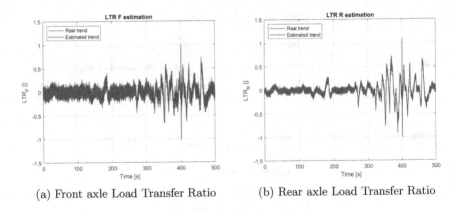

(a) Front axle Load Transfer Ratio   (b) Rear axle Load Transfer Ratio

Fig. 2. Strongly disturbed signals, Bernina Pass

Table 3. Results: Mean Absolute Error and Mean Square Error, Bernina maneuver, networks trained without noise.

Maneuver	Error	Values without noise	Values with noise
Bernina	MAE LTR Front	0.0026	0.0498
	MAE LTR Rear	0.0030	0.0261
	MAE Roll	0.0500 deg	0.5255 deg
	MSE LTR Front	9.7767e−05	0.0040
	MSE LTR Rear	1.1046e−04	0.0012
	MSE Roll	0.0048 deg^2	0.4599 deg^2

the improvement in the results, especially comparing the same with the ones previously presented (Fig. 2b). This is what could be expected, since it shows that the neural network learns to recognize noise casuistry (Table 4). The MAE and the MSE values associated to each outputs results are reduced and nearer to the ideal situation of noiseless signals in exit (and associated networks trained without taking the noise into account).

Table 4. Results: Mean Absolute Error and Mean Square Error, Bernina maneuver, networks trained with noise.

Maneuver	Error	Values without noise	Values with noise
Bernina	MAE LTR Front	0.0026	0.0192
	MAE LTR Rear	0.0030	0.0176
	MAE Roll	0.0500 deg	0.1497 deg
	MSE LTR Front	9.7767e−05	6.6876e−04
	MSE LTR Rear	1.1046e−04	5.7183e−04
	MSE Roll	0.0048 deg^2	0.0565 deg^2

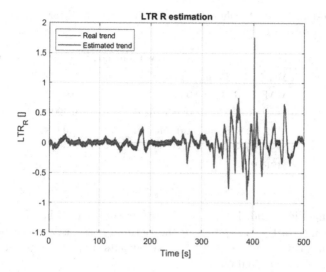

Fig. 3. Estimation of the vehicle rear axle Load Transfer Ratio, in case of strongly disturbed signals and networks trained with noise, Bernina Pass

Considering the second maneuver of interest (Fig. 4a and 4b), a simulation of the heavy vehicle traversing the Stelvio pass, it is possible to notice the same trend observed in Fig. 2a and 2b. In particular, the neural network confirms its robustness, performing clearly not as good as if it was requested an estimate of noiseless signals, but showing an acceptable performance (especially noticing the associated values) in terms of MAE and MSE (Table 5).

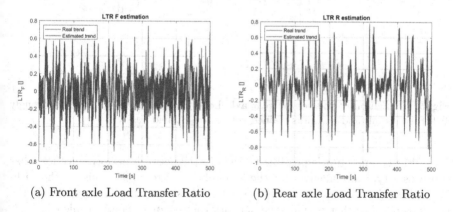

(a) Front axle Load Transfer Ratio   (b) Rear axle Load Transfer Ratio

Fig. 4. Strongly disturbed signals, Stelvio Pass

Table 5. Results: Mean Absolute Error and Mean Square Error, Stelvio maneuver, networks trained without noise.

Maneuver	Error	Values without noise	Values with noise
Stelvio	MAE LTR Front	0.0066	0.0606
	MAE LTR Rear	0.0063	0.0339
	MAE Roll	0.0802 deg	0.5204 deg
	MSE LTR Front	1.0434e−04	0.0058
	MSE LTR Rear	8.8145e−05	0.0018
	MSE Roll	0.0141 deg^2	0.4707 deg^2

Observing Fig. 5 and Table 6, it is further confirmed what is reported for the performance of the AI system in the Bernina pass case. The new network, trained with noisy signals to predict noisy outputs, improves again the already good results of the original one.

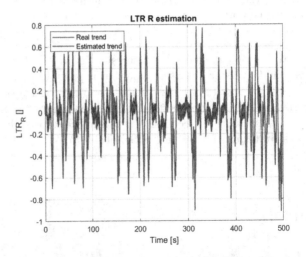

Fig. 5. Estimation of the vehicle rear axle Load Transfer Ratio, in case of strongly disturbed signals and networks trained with noise, Stelvio Pass

In Fig. 6a and 6b it is possible to observe the network performance for what concerns a typical maneuver for the highlighting of rollover casuistry, the Fish-Hook. Even in this case, it is confirmed what has been already observed for the previous ones, further proving the network robustness to noise (Table 7).

As can be noted in Fig. 7 and Table 8, it is confirmed the fact that, as expected, the neural network trained with noisy signals is better suited to estimate the latter. It has to be noted, in this case, the fact that the MSE associated to the LTR of the rear axle results even lower in value in comparison with the one related to an ideal condition, proving again the quality of the training.

Table 6. Results: Mean Absolute Error and Mean Square Error, Stelvio maneuver, networks trained with noise.

Maneuver	Error	Values without noise	Values with noise
Stelvio	MAE LTR Front	0.0066	0.0280
	MAE LTR Rear	0.0063	0.0259
	MAE Roll	0.0802 deg	0.1878 deg
	MSE LTR Front	1.0434e−04	0.0013
	MSE LTR Rear	8.8145e−05	0.0011
	MSE Roll	0.0141 deg^2	0.0686 deg^2

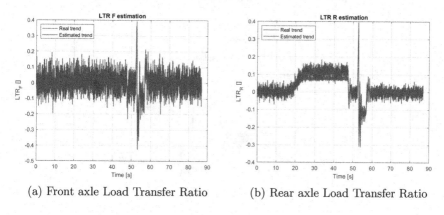

(a) Front axle Load Transfer Ratio   (b) Rear axle Load Transfer Ratio

Fig. 6. Strongly disturbed signals, FishHook maneuver

Finally, Table 9, for brevity reasons, resumes the results attained with the last 5 datasets of testing. These have been composed and defined to assure the actual potential and limits of the AI algorithm. In particular, the first three represents maneuvers conducted with the same vehicle in different conditions, such as different speed, different track, and both of these elements combined, while the last two shows the performance of the LSTM networks applied on a different type of vehicle. It has to be noted that the outcomes contained in Table 9 are, for simplicity, always characterized by noise afflicted parameters. The difference between the two columns containing the MAE and MSE values relies in the training of the networks, characterized in one case by being noiseless, and vice versa for the other.

Entering more in details in what is displayed on Table 9, the first result (Stelvio new speed) reports the performance of the algorithm with the Stelvio track, but changing the heavy vehicle original cruising speed (50 km/h) to a new one (30 km/h). As it is possible to observe, the LSTM system is able to offer comparable results with Table 5 and 6, with even an improvement in terms of LTR accuracy. The second outcome, attained from the Pikes Peak circuit, confirms the same order of magnitude as the ones observed in the examples

Table 7. Results: Mean Absolute Error and Mean Square Error, FishHook maneuver, networks trained without noise.

Maneuver	Error	Values without noise	Values with noise
FishHook	MAE LTR Front	0.0146	0.0453
	MAE LTR Rear	0.0138	0.0243
	MAE Roll	0.1066 deg	0.2224 deg
	MSE LTR Front	6.1685e−04	0.0033
	MSE LTR Rear	5.3661e−04	9.9196e−04
	MSE Roll	0.0291 deg^2	0.0793 deg^2

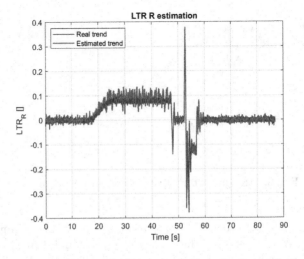

Fig. 7. Estimation of the vehicle rear axle Load Transfer Ratio, in case of strongly disturbed signals and networks trained with noise, FishHook maneuver

Table 8. Results: Mean Absolute Error and Mean Square Error, FishHook maneuver, networks trained with noise.

Maneuver	Error	Values without noise	Values with noise
FishHook	MAE LTR Front	0.0146	0.0207
	MAE LTR Rear	0.0138	0.0138
	MAE Roll	0.1066 deg	0.1163 deg
	MSE LTR Front	6.1685e−04	8.6415e−04
	MSE LTR Rear	5.3661e−04	3.9744e−04
	MSE Roll	0.0291 deg^2	0.0628 deg^2

Table 9. Results: Mean Absolute Error and Mean Square Error, comparison.

Maneuver	Error	Values without noise	Values with noise
Stelvio new speed	MAE LTR Front	0.0477	0.0196
	MAE LTR Rear	0.0235	0.0168
	MAE Roll	0.2980 deg	0.0977 deg
	MSE LTR Front	0.0036	6.3627e−04
	MSE LTR Rear	9.2192e−04	4.5720e−04
	MSE Roll	0.3223 $\deg^2$	0.0206 $\deg^2$
Pikes Peak	MAE LTR Front	0.0750	0.0451
	MAE LTR Rear	0.0487	0.0405
	MAE Roll	1.3757 deg	1.4899 deg
	MSE LTR Front	0.0092	0.0041
	MSE LTR Rear	0.0039	0.0031
	MSE Roll	3.3847 $\deg^2$	3.9642 $\deg^2$
Nascar	MAE LTR Front	0.2949	0.2096
	MAE LTR Rear	0.3050	0.2924
	MAE Roll	15.9339 deg	19.7788 deg
	MSE LTR Front	0.1100	0.0540
	MSE LTR Rear	0.1106	0.1007
	MSE Roll	283.4410 $\deg^2$	461.4968 $\deg^2$
BerninaPass new vehicle	MAE LTR Front	0.1271	0.0628
	MAE LTR Rear	0.0598	0.0456
	MAE Roll	1.6180 deg	0.3408 deg
	MSE LTR Front	0.0219	0.0075
	MSE LTR Rear	0.0091	0.0059
	MSE Roll	3.4861 $\deg^2$	0.1763 $\deg^2$
BerninaPass new vehicle new speed	MAE LTR Front	0.2949	0.0994
	MAE LTR Rear	0.1010	0.0863
	MAE Roll	2.7747 deg	0.4131 deg
	MSE LTR Front	0.0491	0.0184
	MSE LTR Rear	0.0182	0.0138
	MSE Roll	8.7458 $\deg^2$	0.3143 $\deg^2$

previously reported in this paper, with a decreasing in the estimate of the Roll trend. This is particularly pronounced in the third result, precisely developed to investigate the algorithm limits. Indeed, the heavy vehicle has been asked to complete a completely different maneuver from the ones characterizing the training database, both from trajectory and cinematic aspects, simulating a Nascar circuit. As it is possible to see taking in account the MAE and MSE Roll values, there is strong presence of outliers that indicates a great decrease in accuracy. However, this is much less pronounced observing the LTR related parameters, suggesting that there could be room for improvement especially in the treating of roll values during the training of the networks.

The last two results reported in Table 9 are completely different from the others. These have been attained by testing the algorithm on another vehicle, an Audi RTT, characterized, as logical, by completely different mechanics and characteristics if compared with an heavy vehicle. Nonetheless, the LSTM networks performed better than expected, maintaining when requested the analysis of the Bernina Pass an order of magnitude similar to the ones attained for the Stelvio Pass and Pikes Peak. These information, combined with the aforementioned considerations, suggests that the body properties of the vehicle influence in a minor way the accuracy of the algorithm in comparison of a maneuver radically different from the ones designed for the training. In any case, the latter has always shown better outcomes and robustness to noise when considering the same in the training process (Table 9).

4   Conclusions and Possible Future Developments

In light of the obtained results, it is possible to conclude that the methodology previously developed to generate an LSTM network to estimate rollover indicators proved to be sufficiently robust and reliable to noise disturbance. At the same time, the analysis conducted on the same type of neural networks suggests that, in the case of noisy signals, it is better, as expected, to train the AI algorithm to signals with the same type of disturbance (SNR). This can be extremely important to generate realistic tests before the on board vehicle experimentation, other than to develop a robust methodology. It has to be noted, moreover, that the reliability of the defined LSTM network architecture is, in all likelihood, directly correlated with the large and variegated amount of data generated and used in the training of the algorithm. This suggests, for future developments, the necessity to investigate the possibility, with a sufficiently complex heavy vehicle model, of using (at least partially) synthetic data to predict empirical ones.

Acknowledgements. While working on this article, Guido Perboli was the Head of the Urban Mobility and Logistics Systems (UMLS) initiative of the interdepartmental Center for Automotive Research and Sustainable mobility (CARS) at the Politecnico di Torino. Partial funds for the project were given under the Italian "PNRR project, DM 1061". Prof. Maria Elena Bruni acknowledges financial support from: PNRR MUR project PE0000013-FAIR.

References

1. Baldi, M.M., Perboli, G., Tadei, R.: Driver maneuvers inference through machine learning. In: Pardalos, P.M., Conca, P., Giuffrida, G., Nicosia, G. (eds.) MOD 2016. LNCS (LNAI and LNB), vol. 10122, pp. 182–192. Springer, Cham (2016). https://doi.org/10.1007/978-3-319-51469-7_15
2. Chen, X., Chen, W., Hou, L., Hu, H., Bu, X., Zhu, Q.: A novel data-driven rollover risk assessment for articulated steering vehicles using RNN. J. Mech. Sci. Technol. **34**(5), 2161–2170 (2020). https://doi.org/10.1007/s12206-020-0437-4
3. Imine, H., Benallegue, A., Madani, T., Srairi, S.: Rollover risk prediction of heavy vehicle using high-order sliding-mode observer: experimental results. IEEE Trans. Veh. Technol. **63**(6), 2533–2543 (2014). https://doi.org/10.1109/TVT.2013.2292998

4. Le, X.H., Ho, H.V., Lee, G., Jung, S.: Application of long short-term memory (LSTM) neural network for flood forecasting. Water **11**(7) (2019). https://doi.org/10.3390/w11071387

5. Lenkutis, T., Čerškus, A., Šešok, N., Dzedzickis, A., Bučinskas, V.: Road surface profile synthesis: assessment of suitability for simulation. Symmetry **13**(1), 1–14 (2021). https://doi.org/10.3390/sym13010068

6. Lindemann, B., Müller, T., Vietz, H., Jazdi, N., Weyrich, M.: A survey on long short-term memory networks for time series prediction. Procedia CIRP **99**, 650–655 (2021). https://doi.org/10.1016/j.procir.2021.03.088

7. Liu, Y., Cui, D.: Collaborative model analysis on ride comfort and handling stability. J. Vibroeng. **21**(6), 1724–1737 (2019). https://doi.org/10.21595/jve.2019.20454 https://doi.org/10.21595/jve.2019.20454

8. Perboli, G., Arabnezhad, E.: A Machine Learning-based DSS for mid and long-term company crisis prediction. Expert Syst. Appl. 114758 (2021). https://doi.org/10.1016/j.eswa.2021.114758

9. Perboli, G., Tronzano, A., Rosano, M., Tarantino, L., Velardocchia, F.: Using machine learning to assess public policies: a real case study for supporting SMEs development in Italy. In: 2021 IEEE Technology & Engineering Management Conference - Europe (TEMSCON-EUR), pp. 1–6. IEEE (2021). https://doi.org/10.1109/TEMSCON-EUR52034.2021.9488581

10. Sellami, Y., Imine, H., Boubezoul, A., Cadiou, J.C.: Rollover risk prediction of heavy vehicles by reliability index and empirical modelling. Veh. Syst. Dyn. **56**(3), 385–405 (2018). https://doi.org/10.1080/00423114.2017.1381980

11. Sharma, S., Henderson, J., Ghosh, J.: CERTIFAI: a common framework to provide explanations and analyse the fairness and robustness of black-box models. In: Proceedings of the AAAI/ACM Conference on AI, Ethics, and Society, AIES 2020, pp. 166–172. Association for Computing Machinery, New York (2020). https://doi.org/10.1145/3375627.3375812

12. Tota, A., Dimauro, L., Velardocchia, F., Paciullo, G., Velardocchia, M.: An intelligent predictive algorithm for the anti-rollover prevention of heavy vehicles for off-road applications. Machines **10**, 835 (2022). https://doi.org/10.3390/machines10100835

13. Us Department of Transportation: Traffic safety facts 2016: a compilation of motor vehicle crash data from the fatality analysis reporting system and the general estimates system. Technical report, NHTSA (2017)

14. Velardocchia, F., Perboli, G., Vigliani, A.: Analysis of heavy vehicles rollover with artificial intelligence techniques. In: Nicosia, G., et al. (eds.) LOD 2022. LNCS (LNAI and LNB), vol. 13810, pp. 294–308. Springer, Cham (2023). https://doi.org/10.1007/978-3-031-25599-1_22

15. Zhu, T., Yin, X., Li, B., Ma, W.: A reliability approach to development of rollover prediction for heavy vehicles based on SVM empirical model with multiple observed variables. IEEE Access **8**, 89367–89380 (2020). https://doi.org/10.1109/ACCESS.2020.2994026

Ensemble Clustering for Boundary Detection in High-Dimensional Data

Panagiotis Anagnostou[1]([✉]), Nicos G. Pavlidis[2], and Sotiris Tasoulis[1]

[1] Department of Computer Science and Biomedical Informatics,
University of Thessaly, Volos, Greece
panagno@uth.gr
[2] Lancaster University, Lancaster, UK

Abstract. The emergence of novel data collection methods has led to the accumulation of vast amounts of unlabelled data. Discovering well separated groups of data samples through clustering is a critical but challenging task. In recent years various techniques to detect isolated and boundary points have been developed. In this work, we propose a clustering methodology that enables us to discover boundary data effectively, discriminating them from outliers. The proposed methodology utilizes a well established density based clustering method designed for high dimensional data, to develop a new ensemble scheme. The experimental results demonstrate very good performance, indicating that the approach has the potential to be used in diverse domains.

Keywords: Ensemble Clustering · Boundary Data · Minimum Density Hyperplanes

1 Introduction

Clustering boundaries [7] correspond to regions where two or more clusters "meet". These can be considered of special interest since we are frequently interested in understanding which are the small changes that lead to an object (point) being assigned to a different cluster, as well as potential differences between interior and boundary objects. In many applications, boundary objects can correspond to samples that need to be investigated closely. For example, in semi-supervised classification, labeling boundary data improves models when unlabeled samples occur near class boundaries [16]. Selectively sampling boundary points reduces labeling costs and avoids overfitting. For social network analysis, members bridging disparate communities form crucial boundaries for information diffusion [15]. Analyzing boundary roles sheds light on influence and contagion. In summary, boundary data occurs are frequently of interest in cluster analysis tasks, as they can represent transitional, ambiguous, or bridging examples between groupings. Boundary detection techniques provide valuable signals for classification, anomaly detection, segmentation, and other applications.

G. Nicosia et al. (Eds.): LOD 2023, LNCS 14506, pp. 324–333, 2024.
https://doi.org/10.1007/978-3-031-53966-4_24

The presence of noise in the data makes the problem of identifying boundary objects far more difficult. In general, a low level of confidence for the cluster assignment of a given object can be indicative of a boundary point, but such objects can also be outliers. In contrast to approaches in the recent literature [14], we specifically define boundary points in this work as data that lie at the boundary between two or more clusters, discriminating them from outliers (that belong to a cluster with high confidence), and from boundary points that lie at the outer edge of clusters (border data). Figure 1 illustrates the aforementioned categorization.

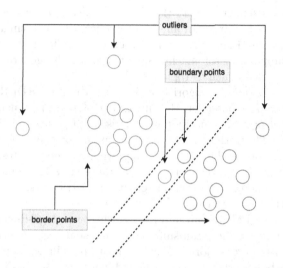

Fig. 1. Visually illustrated discrimination between outliers, boundary points and border points.

One way to discriminate between outliers and boundary points is to use clustering algorithms that assign each object to a specific cluster. Outliers are then objects that do not belong to any cluster, while boundary points are objects that can be assigned to more than one cluster. By examining the cluster assignments, it thus becomes possible to distinguish between outliers and boundary points. At first glance, density-based algorithms that are most successful for the task of boundary data detection are expected to be highly problematic because although boundary points have a lower density than points in the interior of the cluster, outliers (by definition) have even lower density. Hence a characterisation of boundary points based on the density of points within a cluster runs the risk of identifying outliers as boundary points.

In this work, we utilize ensemble clustering as a methodology to identify boundary points. The basis for the proposed ensemble is a hierarchical density-based algorithm that can overcome the aforementioned challenge [11]. We exploit the convergence to different local optima of the clustering objective function of the algorithm in [11] to achieve the necessary diversity required to design a successful ensemble.

The remaining paper is organised as follows: Sect. 2 reviews the existing literature on ensemble clustering and cluster boundary detection. We introduce our proposed ensemble clustering approach in Sect. 3. Section 4 is devoted to a comprehensive evaluation of the proposed approach on various datasets. The paper ends with conclusions and potential future directions in Sect. 5.

2   Related Work

Cluster boundaries refer to points that lie in the decision regions between two or more clusters. Such points can influence significantly the estimation of supervised and unsupervised learning methods. This is true for instance in the case of training of support vector machines, but also for the generation of cluster representatives in tasks such as topic modelling.

One of the most popular algorithms for this task is BORDER, proposed in [17]. The algorithm is based on the utilization of k nearest neighbor search, and thus it can incorrectly identify noisy and isolated points as belonging to the cluster boundary. The authors in [13] propose an improvement of BORDER that smooths the noisy biases, but algorithm performance deteriorates noticeably in high dimensions. Similarly, the methodology proposed in [12] cannot handle high dimensional data due to hyper-parameter tuning restrictions.

More recently, the authors in [3] presented a cluster boundary detection scheme that exploited MeanShift and Parzen window in high-dimensional space. However, the update of the mean-shift vector in a sparse distribution will prefer the subspaces with large ℓ_2 norms. Finally, the authors in [2] propose a methodology for cluster boundary detection in high dimensions using a directed Markov tree, but do not discriminate boundary points from "border points" found at the outer regions of a cluster.

3   Proposed Methodology

Our methodology relies on the minimum density hyperplane (MDH) [11], and the associated divisive hierarchical clustering algorithm minimum density divisive clustering (MDDC) [9]. The underlying assumption of MDH is that clusters are contiguous regions of high probability density, which are separated by contiguous regions of low probability density. For $a \in \mathcal{S}^m = \{x \in \mathbb{R}^m : \|x\|_2 = 1\}$ and $b \in \mathbb{R}$, the hyperplane $H(a, b)$ is defined as $H(a, b) = \{x \in \mathbb{R}^m : a^\top x = b\}$. Assuming that the observed data, $\{x_i\}_{i=1}^n \subset \mathbb{R}^m$, is a sample from an unknown distribution with density p, the density on $H(a, b)$ is defined as the integral,

$$I(a, b) = \int_{H(a,b)} p(x)\mathrm{d}x.$$

In practice p is unknown and computing line integrals is non-trivial. However, if p is approximated with a density estimator using isotropic Gaussian kernels with bandwidth h, then the integral of the estimated density on $H(a, b)$, can be computed exactly by a simple procedure. First, project the data onto a, and then estimate the value of the one-dimensional kernel density estimator at b.

MDH exploits this property to create a projection pursuit algorithm to solve the following optimisation problem,

$$a^* = \underset{a \in \mathcal{S}^m}{\arg \min} \, \phi(a), \tag{1}$$

$$\phi(a) = \min_{b \in [\mu_a - c\sigma_a, \mu_a + c\sigma_a]} \frac{1}{n\sqrt{2\pi h^2}} \sum_{i=1}^{n} \exp\left\{-\frac{(b - a^\top x_i)^2}{2h^2}\right\}, \tag{2}$$

where μ_a and σ_a are the mean and standard deviation (respectively) of the data after being projected onto a, and $c > 0$ is a user-specified constant. To obtain a complete clustering MDDC uses MDHs to recursively bi-partition (subsets of) the data.

The projection pursuit algorithm for computing MDHs is computationally expensive because each evaluation of $\phi(a)$ involves the solution of the one-dimensional optimisation problem in Eq. (2). Furthermore, the objective function in Eq. (1) is not guaranteed to be everywhere differentiable (in particular, $\phi(a)$ can be non-differentiable when the solution to the problem in Eq. (2) is not unique). The justification for the above formulation has been that it allows the algorithm to avoid getting trapped in poor local minima which can occur when one optimises simultaneously over both (a, b).

In this work we take a different approach to computing MDHs. Since we aim at an ensemble clustering the diversity that arises by different locally optimal solutions to the MDH problem is an advantage rather than a disadvantage. We thus optimise directly over both (a, b),

$$(a^*, b^*) = \min_{a, b} \frac{1}{n\sqrt{2\pi h^2}} \sum_{i=1}^{n} \exp\left\{-\frac{(b - a^\top x_i)^2}{2h^2}\right\} \tag{3}$$

$$\text{s.t. } \|a\|_2 = 1, b \in [\mu_a - c\sigma_a, \mu_a + c\sigma_a]. \tag{4}$$

The above objective function is everywhere differentiable, and orders of magnitude cheaper to evaluate compared to $\phi(a)$.

We now describe the ensemble approach we use. Starting with K distinct MDDC clusterings, each denoted as $C^k \in \mathbb{N}^n$, we create a binary indicator matrix $B \in \{0, 1\}^{n \times KM}$ where M is the maximum number of clusters in any C^k. This process is illustrated in Table 1. From B we can obtain a similarity matrix S as,

$$S = \frac{1}{K} BB^\top,$$

where s_{ij} is the proportion of the times points i and j were assigned to the same cluster [18]. The final, ensemble cluster assignment is obtained by applying spectral clustering algorithm on S as discussed in [1].

Our approach to detect boundary points relies on the structure of the similarity matrix S. As discussed before, boundary points occur in regions where

Table 1. Clustering depiction

...	C_k	C_{k+1}	...		...	B_k	B_{k+1}	...
	1	2			1 0 0	0 1 0	
	2	1			0 1 0	1 0 0	
	1	2		$\Longrightarrow$	1 0 0	0 1 0	
...	1	1	...		... 1 0 0	1 0 0	...
	3	3			0 0 1	0 0 1	
	3	3			0 0 1	0 0 1	
	1	2			1 0 0	0 1 0	

clusters "meet", and hence such points can be assigned to different clusters in different executions of the clustering algorithm. Assuming i is a boundary point then we expect the ith row $S_{i,:}$ (or equivalently ith column $S_{:,i}$) of S to contain fewer non-zero values compared to those of a point in the interior of a cluster. Based on this observation we propose two simple statistics to distinguish boundary points: the variance and the entropy of the entries in $S_{i,:}$ for each point. We argue that the smaller the value of the variance of $S_{i,:}$ the more likely is x_i to be a boundary point. Similarly the larger the entropy of $S_{i,:}$ the more likely is x_i to be a boundary point.

4   Experimental Results

To evaluate the effectiveness our methodology in detecting boundary points we design the following experimental procedure. We consider three different soft clustering methods for comparison. The first is fuzzy-c-means (FCM) as implemented in [5]. FCM iteratively assigns points to clusters and updates the cluster centers based on the degree of membership of each point to each cluster. The degree of membership is calculated using fuzzy logic, which allows a data point to belong partially to multiple clusters. FCM terminates when the cluster assignments and centers converge, or a maximum number of iterations is reached. The second method is HDBSCAN [10], a state-of-the-art hierarchical density-based clustering algorithm that builds clusters based on the density of data points. It starts with high density seed regions and expands clusters outward by including lower density areas that are density-reachable from a seed region to form clusters. The algorithm calculates the local density of each point and forms clusters by connecting points that meet a minimum density threshold. A relevant feature HDBSCAN is that does not perform hard assignment but rather returns a cluster membership vector for each point. The third method is the ensemble k-means (ekMeans) [1]. ekMeans generates multiple k-means clusterings, and then we apply the procedure described above to obtain the final clustering to identify boundary points. We use the subscripts (_v) and (_e) to denote when variance and entropy (respectively) is used as a metric to identify boundary points. Finally, we considered the BORDER algorithm [17], mentioned in Sect. 2. BORDER is designed to detect boundary point, not to cluster the data, and thus we combined it with k-Means and MDH clustering algorithms as BORDER_kMeans and BORDER_MDH, respectively.

To visually illustrate the algorithmic results, we initially consider a well known two-dimensional simulated dataset, called S3 [8]. The data is a sample from a mixture of Gaussians and there is considerable overlap between the clusters. Figure 2 illustrates the boundary points identified by the proposed algorithm for this dataset. More specifically, in the top row of the figure boundary points are identified based on the variance of the entries in the rows of S, while in the bottom row of the figure the entropy of the rows of S is used. The middle and right columns correspond illustrate in black colour the boundary points when the user-specified proportion of boundary points is set to 10 and 20 percent respectively. As seen in the figure both the variance and entropy produce very sensible results. The leftmost plot in each row is a kernel density estimator for the values of the variance and entropy across the rows of S. This is information which we are currently not using but it could potentially be informative to automatically identify how many boundary points exist in a given dataset.

Subsequently, we consider the following real-world datasets for empirical evaluation: the Zheng [19] a dataset (3994 samples and 15568 dimensions) of Human cells from transplant patients, and the GSE45719 [4] a dataset of Mice pre-

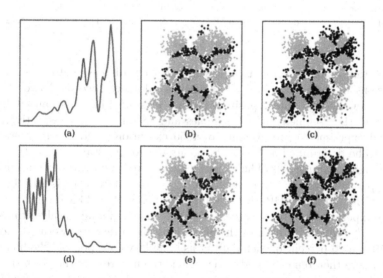

Fig. 2. Visualization of boundary points for the S3 dataset, through variance and entropy (first and second row respectively). Figures (a) and (d) are kernel density estimators of the variance and entropy of the entries each row of the similarity matrix, S. Black points in Figures (b), (c), (e) and (f) indicate the identified boundary points in S3. In Figs. (b) and (e) 10% of the points are considered boundary points, while in Figs. (c) and (e) 20% of points are considered boundary points.

Table 2. Total results.

Method	Purity				NMI			
	Zheng	glass	GSE45719	segment	Zheng	glass	GSE45719	segment
kMeans	0.541	0.589	0.795	0.53	0.345	0.428	0.736	0.471
BORDER_kMeans_0.1	0.52	0.597	0.79	0.545	0.319	0.468	0.744	0.496
BORDER_kMeans_0.2	0.52	0.581	0.793	0.547	0.319	0.462	0.754	0.503
ekMeans	0.473	0.621	0.792	0.608	0.266	0.366	0.651	0.531
ekMeans_0.1_v	0.489	0.648	0.762	0.598	0.334	0.376	0.639	0.55
ekMeans_0.2_v	0.358	0.62	0.732	0.55	0.0	0.135	0.661	0.506
ekMeans_0.1_e	0.586	0.638	0.73	0.715	0.28	0.419	0.664	0.638
ekMeans_0.2_e	0.586	0.638	0.73	0.715	0.28	0.419	0.664	0.638
FCM	0.514	0.607	0.757	0.557	0.222	0.359	0.647	0.478
FCM_0.1	0.551	0.651	0.811	0.599	0.287	0.409	0.666	0.51
FCM_0.2	0.572	0.684	0.806	0.646	0.306	0.441	0.667	0.551
HDBSCAN	0.25	0.453	0.42	0.286	0.0	0.301	0.0	0.316
HDBSCAN_0.1	0.258	0.453	0.438	0.276	0.0	0.301	0.0	0.323
HDBSCAN_0.2	0.262	0.453	0.438	0.276	0.0	0.301	0.0	0.323
MDH	0.683	0.551	0.574	0.597	0.536	0.307	0.397	0.615
BORDER_MDH_0.1	0.668	0.576	0.58	0.612	0.52	0.352	0.438	0.631
BORDER_MDH_0.2	0.668	0.611	0.581	0.619	0.52	0.363	0.441	0.643
eMDH	0.715	0.645	0.839	0.648	0.688	0.371	**0.766**	0.623
eMDH_0.1_v	0.74	0.693	**0.856**	0.675	0.803	0.435	0.731	0.672
eMDH_0.2_v	**0.729**	**0.716**	0.838	**0.716**	**0.832**	**0.483**	0.679	0.712
eMDH_0.1_e	0.692	0.693	0.839	0.684	0.686	0.436	0.746	0.685
eMDH_0.2_e	0.676	**0.716**	0.822	0.709	0.66	0.442	0.738	**0.723**

implantation embryos cells, as representative high dimensional datasets, with dimensions in the scale of ten of thousands. Moreover, we used two popular datasets for the UCI data repository [6] the "glass" (214 samples and 9 dimensions) and the "segment" (2310 samples and 18 dimensions) respectively. All reported experiments are averages over 50 executions, and the percentage of points to be determined as boundary points are 10 (_0.1) and 20 (_0.2) percent of the original sample size. The experiments where conducted on a system with Linux operating system, kernel version 5.11.0, with an Intel Core i7-10700K CPU @ 3.80 GHz and four DDR4 RAM dims of 32 GB with 2133 MHz frequency.

The experiments were executed in the same manner for all the algorithms. We created a baseline with each of the algorithms and afterward, we used them to label 10 and 20 percent of each dataset as boundary data points. The boundary points where then removed and to determine the new results the clustering was re-calculated. In this way we assess the impact of removing these points on the quality of the identified clusters. Based on the boundary point definition we would hope that their removal would lead to an improved clustering result (since the remaining points are more clearly separable).

To assess performance, we utilize are purity and normalised mutual information (NMI) clustering metrics, reported in Tables 2. Purity measures the extent to which the elements within each cluster belong to the same class. Mutual

Table 3. Execution time of the methodologies used, in seconds.

Metric	glass	segment
kMeans	0.034	0.005
BORDER	0.039	0.272
ekMeans	0.032	0.289
ekMeans_v	0.032	0.306
ekMeans_e	0.032	0.333
FCM	0.015	0.202
HDBSCAN	0.005	0.154
MDH	17.061	62.198
BORDER_MDH	17.065	62.475
eMDH	1421.569	2840.011
eMDH_v	1421.57	2840.028
eMDH_e	1421.57	2840.053

information is an information theoretic measure that quantifies how much information we obtain about the actual class labels from the cluster labels. Both purity and NMI range between zero and one with higher values indicating better performance.

The results reported in Tables 2 are quite consistent. First, the proposed ensemble clustering algorithm (eMDH) is competitive with the fuzzy c-means and ensemble k-means on all datasets. Notable, the ensemble methodology we used improves consistently the MDH algorithm but in the case of k-means there are only marginal improvements if any. One possible reason behind this behaviour is the "curse of dimensionality" which plagues the k-means algorithm but it does not affect the MDH due to the projection based data splits. Moreover, removing boundary points improves eMDH performance in all cases with the exception of the GSE45719 dataset, where still performance improved marginally in terms of purity. We also interestingly observe, that using variance as the statistic to identify boundary points overall produces superior results. Finally, we should highlight that MDH is by definition a significantly more computationally demanding method for clustering compared to the rest of methods utilized here since it necessitates the solution of an optimization task in high dimensional spaces. For completeness we also report in Table 3 computational times for each method.

5 Conclusions

We presented a methodology to identify boundary points in unlabelled data. This constitutes an interesting problem which has not been extensively studied. Our literature review suggests that existing methods for this task can confound outliers with boundary points, which is undesirable. We present an approach that is based on an ensemble of divisive hierarchical clustering models obtained by recursively bi-partitioning the data through low density hyperplanes. As we discuss promoting diversity, which is necessary to construct an effective ensemble, is actually beneficial in terms of computational cost. To identify boundary points we consider two very simple metrics on the similarity matrix; the variance and the entropy. Both appear to be effective and the proposed approach compares favorably with alternatives on both artificial and real-world data. For future work we intend to perform a more thorough performance analysis and consider the automatic determination of the number of boundary points in a dataset.

Acknowledgment. We acknowledge support of this work by the project "Par-ICT CENG: Enhancing ICT research infrastructure in Central Greece to enable processing of Big data from sensor stream, multimedia content, and complex mathematical modeling and simulations" (MIS 5047244), which is implemented under the Action "Reinforcement of the Research and Innovation Infrastructure", funded by the Operational Programme "Competitiveness, Entrepreneurship and Innovation" (NSRF 2014–2020) and co-financed by Greece and the European Union (European Regional Development Fund).

References

1. Boongoen, T., Iam-On, N.: Cluster ensembles: a survey of approaches with recent extensions and applications. Comput. Sci. Rev. **28**, 1–25 (2018)
2. Cao, X.: High-dimensional cluster boundary detection using directed Markov tree. Pattern Anal. Appl. **24**(1), 35–47 (2021)
3. Cao, X., Qiu, B., Xu, G.: BorderShift: toward optimal MeanShift vector for cluster boundary detection in high-dimensional data. Pattern Anal. Appl. **22**, 1015–1027 (2019)
4. Deng, Q., Ramsköld, D., Reinius, B., Sandberg, R.: Single-cell RNA-seq reveals dynamic, random monoallelic gene expression in mammalian cells. Science **343**(6167), 193–196 (2014)
5. Dias, M.L.D.: Fuzzy-c-means: an implementation of fuzzy *c*-means clustering algorithm (2019). https://git.io/fuzzy-c-means
6. Dua, D., Graff, C.: UCI machine learning repository (2017). http://archive.ics.uci.edu/ml
7. Ester, M., Kriegel, H.P., Sander, J., Xu, X.: A density-based algorithm for discovering clusters in large spatial databases with noise. In: Proceedings of the Second International Conference on Knowledge Discovery and Data Mining, KDD 1996, pp. 226–231. AAAI Press (1996)
8. Fränti, P., Virmajoki, O.: Iterative shrinking method for clustering problems. Pattern Recogn. **39**(5), 761–765 (2006)
9. Hofmeyr, D., Pavlidis, N.G.: PPCI: an R package for cluster identification using projection pursuit. R J. (2019)
10. McInnes, L., Healy, J., Astels, S.: HDBSCAN: hierarchical density based clustering. J. Open Sour. Softw. **2**(11), 205 (2017)
11. Pavlidis, N.G., Hofmeyr, D.P., Tasoulis, S.K.: Minimum density hyperplanes. J. Mach. Learn. Res. **17**(156), 1–33 (2016)

12. Qiu, B.Z., Yang, Y., Du, X.W.: BRINK: an algorithm of boundary points of clusters detecton based on local qualitative factors. J. Zhengzhou Univ. (Eng. Sci.) **33**(3), 117–120 (2012)

13. Qiu, B.-Z., Yue, F., Shen, J.-Y.: BRIM: an efficient boundary points detecting algorithm. In: Zhou, Z.-H., Li, H., Yang, Q. (eds.) PAKDD 2007. LNCS (LNAI), vol. 4426, pp. 761–768. Springer, Heidelberg (2007). https://doi.org/10.1007/978-3-540-71701-0_83

14. Qiu, B., Cao, X.: Clustering boundary detection for high dimensional space based on space inversion and Hopkins statistics. Knowl.-Based Syst. **98**, 216–225 (2016)

15. Tang, L., Wang, X., Liu, H.: Uncoverning groups via heterogeneous interaction analysis. In: 2009 Ninth IEEE International Conference on Data Mining, pp. 503–512. IEEE (2009)

16. Ting, K.M.: An instance-weighting method to induce cost-sensitive trees. IEEE Trans. Knowl. Data Eng. **14**(3), 659–665 (2002)

17. Xia, C., Hsu, W., Lee, M., Ooi, B.: BORDER: efficient computation of boundary points. IEEE Trans. Knowl. Data Eng. **18**(3), 289–303 (2006)

18. Zhang, M.: Weighted clustering ensemble: a review. Pattern Recogn. **124**, 108428 (2022)

19. Zheng, G.X., et al.: Massively parallel digital transcriptional profiling of single cells. Nat. Commun. **8**(1), 14049 (2017)

Learning Graph Configuration Spaces with Graph Embedding in Engineering Domains

Michael Mittermaier[1,2]($\boxtimes$), Takfarinas Saber[1,3], and Goetz Botterweck[1,2]

[1] Lero - the Science Foundation Ireland Research Centre for Software, Limerick, Ireland
[2] School of Computer Science and Statistics, Trinity College Dublin, Dublin, Ireland
{mittermm,goetz.botterweck}@tcd.ie
[3] School of Computer Science, University of Galway, Galway, Ireland
takfarinas.saber@universityofgalway.ie

Abstract. In various domains, engineers face the challenge of optimising system configurations while considering numerous constraints. A common goal is not to identify the best configuration as fast as possible, but rather to find a useful set of very good configurations in a given time for further elaboration by human engineers. Existing techniques for exploring large configuration spaces work well on Euclidean configuration spaces (e.g., with Boolean and numerical configuration decisions). However, it is unclear to what extent they are applicable to configuration problems where solutions are represented as graphs – a common representation in many engineering disciplines. To investigate this problem, we propose an adaptation of existing techniques for Euclidean configurations, to graph configuration spaces by applying graph embedding. We demonstrate the feasibility of this adapted pipeline and conduct a controlled experiment to estimate its efficiency. We apply our approach to a sample case of HVAC (Heating, Ventilation, and Air-Conditioning) systems in 40,000 simulated houses. By first learning the configuration space from a small number of simulations, we can identify 75% of the best configurations within 7,508 simulations compared to 29,725 simulations without our approach. That is a speed-up of 4.0× and saves more than 15 days if one simulation takes about one minute, as in our experimental set-up.

Keywords: Graph Learning · Graph Space Exploration

1 Introduction

In various engineering fields, we encounter the problem of configuring a system. Configurations define systems and indirectly determine their properties. Engineers explore large spaces of possible configurations while simultaneously considering various constraints and numerous optimisation objectives to find a suitable configuration. Configuring a Linux kernel [4], optimising a building ventilation system [1], planning a public transport network [12]; all these problems

© The Author(s), under exclusive license to Springer Nature Switzerland AG 2024
G. Nicosia et al. (Eds.): LOD 2023, LNCS 14506, pp. 334–348, 2024.
https://doi.org/10.1007/978-3-031-53966-4_25

involve the configuration of a system with an incomprehensibly large configuration space. Choosing the wrong configurations can lead to products with undesirable properties or have much more serious consequences. A common goal is not to identify the best configuration as fast as possible, but rather to find a useful set of very good configurations in a given time for further analysis and elaboration by human engineers. Thus, there is a need for tool support that helps users handle the complexity of the configuration process and narrow it down to humanly comprehensible sets of solutions to explore in more detail.

Pereira et al. conducted a systematic literature review on learning configuration spaces and provide a taxonomy of state-of-the-art strategies and methods applied to software configuration spaces [25]. The considered techniques interpret configurations as a set of Boolean decisions and numerical parameters, which can be represented as a Euclidean vector.

In domains such as civil engineering, applications require graph configurations to accurately reflect relationships between various parts of the system. In industrial practice and when dealing with problems of realistic scale, the optimisation of graph configurations is often done in an iterative and heuristic fashion, thus only exploring a small part of the configuration space that is only as good as the knowledge of the expert conducting the search. This is due to the complexity and arbitrary structure of graphs that raise a variety of challenges when it comes to exploration, including scalability [26], feature engineering [21], heterogeneity [30], and interpretability [22]. Hence, there is a need for advanced exploration tools to efficiently explore large graph configuration spaces.

Furthermore, solely focusing on finding the optimal solution is often shortsighted as (i) not all data/simulations are error-free, (ii) many models are often simplified versions of the real problem which leave out details that are difficult to express, and (iii) some preferences might not have been elucidated by the decision-maker – thus an optimal solution might prove inferior or impractical [9]. Therefore, we do not only want to find the best solution. Instead, we need to search for a selection of k best (i.e., optimal or near-optimal) solutions. These k best solutions create a smaller configuration space that is humanly assessable for further constraints, requirements, or optimisation goals.

In this research, we will investigate how to efficiently explore graph configurations using strategies categorised by Pereira et al. [25] by answering the following research questions:

- *RQ1 (Feasibility): How can we apply configuration space learning methods established for Euclidean configuration spaces to graph configuration spaces?*
- *RQ2 (Efficiency): How does tool-supported learning of graph configuration spaces compare to a random search approach with respect to time consumption and quality of determined graph configurations?*

Contributions. Our paper makes the following contributions:

1. We propose a pipeline of adapted techniques to demonstrate the feasibility/ applicability of these techniques for learning configuration spaces – even when using graphs as configurations (answering *RQ1*).

2. We report on a controlled experiment that explores a dataset of graph configurations describing house ventilation systems by AC consumption with as few MATLAB Simulink simulations as possible (answering *RQ2*) and providing further evidence of the feasibility of the approach. Our results show that the overall approach of learning software configuration spaces is adaptable for graph configurations and offers an alternative to iterative approaches by considering a larger space of graph configuration candidates. In terms of efficiency, first results show that our learning graph configuration space (LEGCS) approach reliably identifies graph configurations with good performances (by AC usage) prior to simulations, thus, we can explore the graph configuration space much faster.

3. We make all generated and measured data (graph configurations, Simulink models, measurements) available in an online appendix [3] to encourage follow-up work and replicability.

The remainder of the paper is structured as follows: following this introduction, we provide a background on learning software configuration spaces (Sect. 2). We then present the research design (Sect. 3), and results on feasibility (Sect. 4.1), and efficiency (Sect. 4.2). The paper closes with an overview of related work (Sect. 5) and conclusion (Sect. 6).

2 Background

In this section, we introduce current practices of learning software configuration spaces (that we will adapt to graph configurations) and graph embedding, which will facilitate the adaptations.

2.1 Learning Software Configuration Spaces

There are established practices in the process of learning software configuration spaces [25]. These practices consist of four phases to explore the software configuration space: sampling, measuring, learning, and evaluating (sometimes also referred to as validation phase) as shown in Fig. 1. Below is a short description and Table 1 presents strategies for each phase.

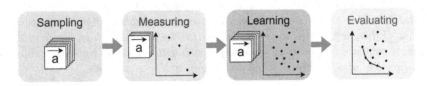

Fig. 1. Phases of learning software configuration spaces.

Table 1. Strategies used in the four phases of learning software configuration spaces, according to Pereira et al. [25].

Phase	Strategy	Description
Sampling		
	Random sampling	Using a random sample to train the machine learning model. There are various strategies and rules of thumb to determine the number of configurations depending on the number of features
	Heuristics	The general motivation to use heuristics is to cover features and their interactions, thus capturing the essence of the configuration space with a smaller sampling size
	Transfer learning	This strategy progressively learns interesting regions in the configuration space by exploiting common similarities
	Arbitrarily chosen sampling	In some instances, domain experts choose a sample from currently available resources
	No sampling	For smaller configuration spaces, it is a valid strategy to explore the whole space and not select a sample in the space
Measuring		
	Execution	Configure the system accordingly, execute the system, and measure the NFPs
	Simulation	Configure the system accordingly, and measure the NFPs during a simulation
	Static analysis	Determine the NFPs by examining code, models, or documentation
	User feedback	Gather NFPs by collecting user feedback
	Synthetic analysis	Combine multiple metrics and measuring methods to assess various features
Learning		
	CART regression	The classification and regression technique partitions a sample into smaller clusters and uses regression techniques on these clusters
	Performance-influence models	This combination of sampling heuristics and step-wise linear regression selects relevant features and determines their influence on the configuration performance
	Other learning algorithms for regression	These strategies use a set of samples to train a regression model until the accuracy is satisfactory
	Learning to rank	In some instances, it is useful to rank configurations instead of predicting exact numerical values
	Transfer learning	Instead of training a machine learning model from scratch, this technique reuses already available knowledge from relevant sources
Evaluating		
	Accuracy metrics	There are plenty of accuracy metrics (e.g., the mean absolute error) which compare the predicted values to exact measurements
	Interpretability metrics	When we use transfer learning techniques, interpretability metrics are a tool to better understand the configuration space
	Sampling cost	One optimisation goal is to keep the sample size low because reducing measuring efforts saves resources

1. *Sampling* – The sampling phase includes multiple challenges: (1) selecting a number of valid configurations (satisfiability problem/constraint satisfaction problem), and (2) labelling a representative sample for the configuration space, so that a learning system in the third phase can generalise.
2. *Measuring* – In the measuring phase, we determine non-functional properties (NFP) of the labelled samples. There are different definitions of what a NFP actually is [13,29]. In this work, we refer to attributes that describe the performance of a configuration (some of them are not directly derivable from the configuration). These attributes are domain dependent, for instance, in the controlled experiment in Sect. 3, we assess the time of AC usage per month in a house ventilation system.
3. *Learning* – Measuring all software configurations sampled in the first phase is technically possible, but cost-intensive. Hence, a machine learning system is often trained in this phase with the results of the measuring phase to predict measuring results for the unassessed configurations in the set.
4. *Evaluating/Validation* – Validating the results of the learning phase means evaluating the quality of the predictions. Ideally, the prediction error and measurement effort are low. The results of this final phase depend on the application of the exploration process. These applications may be pure prediction, interpretability, optimisation, mining constraints, or evolution.

2.2   Graph Embedding

Graph embedding techniques map graph structures into vectors (or matrices) while preserving structural information [14]. The main challenge of efficient graph embedding is keeping relevant information in arbitrary and complex graph structures while storing them in low-dimensional vectors (or matrices). Generally, we find graph embedding applied in graph analysis for node classification, link prediction, and clustering. Different applications require different embeddings. Node and edge embeddings are especially useful for operations within large graphs. For instance, node classification techniques use a node embedding that maps nodes with similar neighbours to vectors with close Euclidean distance. Here, however, we mainly use graph embedding as a similarity measurement between graphs and as regression analysis input for NFP prediction. Thus, we will focus in this section on whole-graph embedding rather than node embedding, edge embedding, or substructure embedding.

Cai et al. [6] point out the key challenge of whole-graph embedding: choosing between the expressiveness of the embedding (preserve information) and the efficiency (embedding time/low dimensionality). In Table 2, we briefly describe the taxonomy of whole-graph embedding techniques that were classified in the same work.

3   Research Design

To answer the research questions raised in the introduction, we propose a pipeline of techniques for learning configuration spaces adapted to graph configurations

Table 2. Families of graph embedding techniques for whole-graph embedding [6].

Family	Description
Matrix factorisation	A structure-preserving dimensionality reduction that factorises a high-dimension matrix representing graph properties such as pairwise node similarity into node embeddings that places nodes in Euclidean space at a distance according to their pairwise similarity. Examples include graph Laplacian eigenmaps and node proximity matrix factorisation
Deep learning	There are two types of deep learning based embedding techniques. In the first one, the deep learning model is fed with random walks over the graph, and in the second type, the deep learning model is trained by the whole graph, with various strategies of breaking down the graph structure (autoencoders, mixture model networks, graph neural networks, etc.)
Edge reconstruction	This embedding technique aims to optimise one of the following three objective functions: (1) maximise the edge reconstruction probability, (2) minimise the edge reconstruction distance-based loss, or (3) minimise the edge reconstruction margin-based ranking loss
Graph kernels	These calculate the inner product of two graphs to measure their similarity. First applied in chemistry, graph kernels now find applications in biology, neuroscience, and natural language processing. Kriege et al. [18] classified graph kernel techniques into approaches based on: neighbourhood aggregation, assignment and matching, subgraph pattern, walks and paths, or others
Generative model	This approach embeds nodes as vectors of latent variables, thus viewing the observed graph as model-generated. Generally, this method is used for node and/or edge embedding for heterogeneous graphs

(*RQ1*) and conduct controlled experiments using this pipeline (*RQ2*). We first give details on the research methodology and then report on the implementation.

3.1   Answering *RQ1* Proposal of a Pipeline of Adapted Techniques

A research question like *RQ1* asks for a "method or means of development" (see the taxonomy suggested by Shaw [28]). We identify a pipeline of exploration techniques and suggest how they can be adapted to graph configurations. Note that the experiments undertaken later provide further evidence that the approach is feasible.

As the starting point of our work and to provide subjects for our controlled experiments, we chose an example from MATLAB Simulink [2], a simulation tool used commonly in various engineering disciplines. Our example is from civil engineering dealing with building ventilation and HVAC (Heating, Ventilation, and Air-Conditioning) systems [1]. Our goal is to explore and filter the space of possible floor plans and air ventilation systems with regard to their monthly AC usage. We interpret and represent these floor plans as graphs configuring a building, making this an exploration problem on graph configurations. Section 3.3 will give more details on the technical aspects.

3.2 Answering *RQ2* Using a Controlled Experiment

A research question like *RQ2* asks for a "design, evaluation, or analysis of a particular instance" (Shaw [28]). We conduct a controlled experiment (Easterbrook et al. [10]) and compare our approach (i.e., the pipeline suggested in *RQ1*) to a naïve simulation approach – in terms of the quality of identified k best graph configurations and the time needed to identify them. While reporting the controlled experiment, we will follow the guidelines of Jedlitschka et al. [15].

3.3 Experimental Set-Up

In this section, we provide an overview of our implementation, i.e., the suggested combination of adapted techniques. We describe (1) how we generated graph configurations that span up the explored configuration space, (2) the domains-specific constraints and objectives affecting the configurations, (3) how we measured the properties resulting from the configurations, i.e., the fitness of a solution, and (4) what frameworks we used for the implementation.

Floor Plan Generation: First, we built a simple floor plan generator based on the work of Lopes et al. [20] that creates JSON files, each containing one configuration. This generator divides the entire floor into a grid, places initial room positions, and then grows rooms to their maximal feasible rectangular size. Finally, some rooms grow into an L-shape to fill the remaining space. Note that in this work, we do not focus on the coverage of graph generation and sampling to the entire valid graph configuration space. Instead, our primary objective is to enhance the efficiency of the subsequent processes (i.e., learning and selection).

Graph Representation of Floor Plans: From the 2D floor plans, we derived the corresponding graphs representing the same building in terms of rooms (with attributes like room area), connections between rooms, and other variation points like an outside door or an AC. Nodes represent rooms, and edges represent the airflow between neighbouring rooms. One room A can be connected to another room B by (1) air exchange, (2) sucking hot air out of room B into room A's AC, or both. These graphs contain between three and ten nodes. Simulink's building ventilation problem [1] sets the constraints for the remaining graph properties. These constraints are:

- The node properties are the room size and if this room is connected to AC or an outside door.
- Every connected subgraph has at least one room with AC.
- Each graph has one room with an outside door.
- There are two edge types. One edge type connects two rooms for regular airflow; each node can have up to three of these edges. The other edge type connects rooms without AC or connection to the outside door with AC. This edge pulls hot air into the AC to cool down; every AC needs one edge of this type.

Graph Floor Plan Measurement: After creating graph representations of floor plans, we built a tool to transform these JSON files into MATLAB models that the MATLAB engine can execute as simulations to calculate AC costs (the number of hours the AC runs within a month). Figure 2 gives an overview of this process.

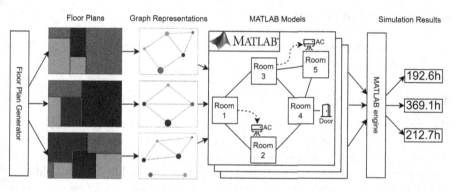

Fig. 2. Overview of the Implementation.

This experiment (i.e., the simulations) was executed on a Linux Mint 21 Cinnamon machine with 16GB of RAM, and an Intel Core i7-6700 CPU. With this set-up, each simulation takes about a minute, which means a brute force simulation approach to explore a graph configuration space of 40,000 graphs takes about 28 days to complete.

Learning Graph Floor Plan Measurements: The goal from here is to accurately predict the outcomes of the simulations for these graphs within seconds (while the simulation of one configuration takes about a minute). This way, we can explore the graph configuration space more efficiently by predicting the simulation results of most configurations and only simulating promising configurations.

At this stage of the research, we chose a combination of embedding and regressor techniques that showed promising results in first experiments. For the embedding, we used the LDP embedding [5] implemented by the Karateclub framework [27] and used it to train a CART regressor. This regressor uses a combination of random forest classification and regression [19]; implementations were provided by Scikit-learn [23]. In terms of training, we have to trade-off between investing more time to run *more* simulations (model available later, but gives better predictions) or *fewer* simulations (model available faster, but predictions less precise). A prestudy with different numbers of simulations indicated that for the considered case, there is a sweet spot of about 200 simulations where the model learns just enough to provide a precise prediction, but adding further simulations does not lead to a increase in precision that would be big enough considering the time required for achieving it.

This combination of state-of-the-art techniques for embedding, the regressor, and the number of simulations provided reliable predictions of system properties in the controlled experiment and is sufficient to demonstrate feasibility. We plan a more systematic evaluation of combinations of techniques and time invested in simulations in future work.

4   Results and Discussion

In this section, we present and discuss the results of our suggested pipeline (*RQ1*) and the controlled experiment (*RQ2*).

4.1   *RQ1:* Feasibility

We chose one path of strategies from learning software configuration spaces to apply to graph configuration spaces. During the process, it was necessary to add the subphase of graph embedding to facilitate the learning phase. Figure 3 gives an overview of the strategies we chose in each phase (highlighted in boldface) and the respective goals of these strategies. We give a rationale for each selection below.

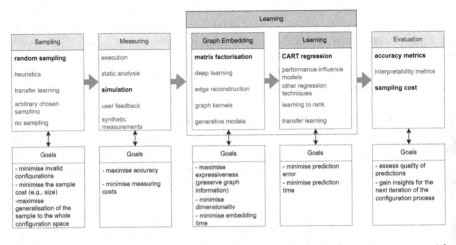

Fig. 3. Strategies we chose in each phase of learning graph configuration spaces with their respective goals.

We used random sampling as it is a widely used baseline for sampling that needed no further adaptations for graphs. Future work will be able to answer if heuristics or transfer learning in software configurations are adaptable for graph spaces. We did not have the resources (domain experts) for arbitrarily chosen sampling, and "no sampling" is not useful due to the large size of the configuration space.

We relied on MATLAB Simulink simulations for measuring and consider the simulation results as a "truth" regarding the properties/fitness of the solution defined by the configuration. If even more fidelity of the performance evaluation is desired one could actually start building houses corresponding to the simulated models; however, due to the associated costs and complexity of the analysis, that is beyond the scope of our work. "User feedback" could theoretically be applied (e.g., aiming to measure satisfaction with a provided floor layout) but is unrealistic with the chosen simulation approach.

We added the phase of graph embedding to the four phases of learning software configuration spaces to facilitate the learning phase. We chose matrix factorisation techniques as a baseline. However, in future work, we plan to compare how well matrix factorisation captures graph structures in vectors with deep learning strategies, such as neural networks, or even apply graph neural networks as a learning strategy.

In our example, the non-functional property (NFP) to optimise is the duration of AC usage in hours per month to maintain a temperature below 26°C. We use CART regression to predict this property as that is a powerful strategy for predictions in learning software configuration spaces. In the following section, we will discuss how well this predictor performs, and in future work, we will compare it to other regression techniques.

We evaluated the predictions by accuracy and sampling size. Interpreting the results is challenging since it involves interpreting the learning model as well as the graph embedding, future work will discuss how interpreting could improve the learning process for graph configuration spaces. In the remainder of this paper, we will refer to the approach described in this section as the learning graph configuration space (LEGCS) approach.

4.2 *RQ2:* Efficiency

We now evaluate how efficiently we can learn the graph configuration space with respect to the accuracy and sampling costs to explore the configuration space. We implemented LEGCS according to Sect. 4.1 and set up the experiment according to Sect. 3.3. In terms of reporting the controlled experiment, we follow the guidelines of Jedlitschka et al. [15].

Goal of the Experiment: The overall goal of the controlled experiment is to analyse the quality of the simulated graph configurations suggested by LEGCS compared to a random selection. Instead of finding the one optimal configuration as fast as possible, we aim to narrow down the graph configuration space to provide the k best solutions (that human experts will further explore) within a limited time. As experimental units, we operate on 40,000 graph configurations generated by the process described earlier (Sect. 3.3).

Figure 4 shows the distribution of the 40,000 graph configurations in the dataset by AC usage per month. The peak on the right end of the spectrum shows configurations where the AC never turns off. We are interested in quickly (with respect to the number of needed simulations) identifying the 1,344 configurations (3.36% of the dataset) in the bin with minimal AC usage (at the left end of the spectrum).

Procedure and Results: We trained, sampled, and measured 200 randomly selected graph configurations, trained a CART regressor with these measurements, and simulated the remaining graph configurations starting the configuration with the lowest AC usage prediction to the highest. The CART regressor's normalised mean absolute error is 0.19 [17]. Figure 5 shows how early we find the minimal AC usage configurations compared to randomly selecting configurations to simulate. After the first 200 simulations (that we use for training the regressor), our LEGCS approach finds minimal AC usage configurations much faster than random selection. We already find the first 100 of those configurations within 534 simulations (including the 200 simulations for training) instead of 3,132 simulations we would need with random selection, that is a speed-up of 5.9×. We identify half of the configurations in the bin with minimal AC usage within the first 11.4% of simulations, 75% of configurations in the first 18.8% of simulations, and 90% in the first 52.1% of simulations. The time spent on embedding the graphs and training the regression model is negligible since it takes less time than one simulation.

Putting this in perspective: engineers, who need to explore a graph configuration space that would take about a month of simulations, can use the LEGCS approach to narrow down the space of configurations to simulate only 18.8% of the space (thus taking less than a week) and still obtain 75% of configurations with minimal AC usage to consider in further exploration. Figure 6 shows the performance of the first 200 graph configurations suggested by LEGCS (in red) as opposed to random selection (in blue). The graph configuration space we explore by using LEGCS shifts left, i.e., towards lower AC usage.

5 Related Work

In this section, we give an overview of related work. First, we introduce work on approaches to find the k best graphs as state-of-the-art for optimising graphs configuring a system. Second, we give a brief overview of using regression analysis on whole graphs.

5.1 Search-Based Graph Optimisation

Finding the k best solutions is a problem that arises in many applications in engineering and several approaches have been proposed. For instance, Dechter

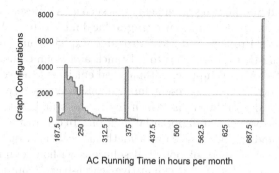

Fig. 4. Histogram of graph configurations in the dataset organised by AC usage in hours per month. The bin size is 25,000 s.eps

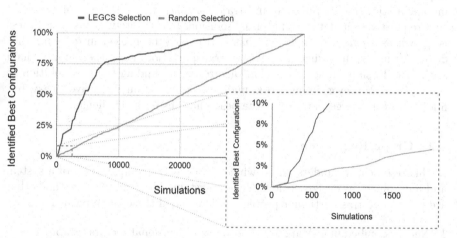

Fig. 5. Number of graph configurations with minimal AC usage in % suggested by LEGCS per completed simulations.

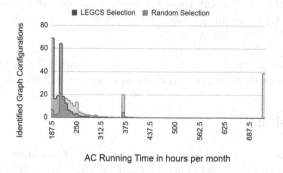

Fig. 6. The graph configuration space for 200 graphs suggested by LEGCS compared to a random selection.

et al. [11] adapted well-known search heuristics such as A*, Best-First, and Depth First to identify the k best solutions over graphical models (e.g., Weighted Constraint Satisfaction Problem and Most Probable Explanation) in Bayesian networks. Comparing their approaches to a Branch and Bound, the authors prove that their adapted A* is complete, optimal, and efficient. However, their graphical models are designed to represent finite domains of discrete variables with a set of real-valued cost functions which limits the search space and streamlines/quickens the assessment of potential solutions. In contrast, our search space is not discrete, and in the absence of a mathematical function that is precise enough to assess the fitness of potential solutions without employing a time-consuming simulation the use of exhaustive search-based approaches becomes impractical.

When the underlying optimisation problem is NP-hard, metaheuristics are required to solve large-scale instances in reasonable computing times. Devising metaheuristics (i.e., population and trajectory-based approaches) for problems that require the simulation of potential solutions to accurately assess their fitness is known as *simheuristics* [8] and has particularly been used in domains that deal with uncertainty such as logistics, transportation, and other supply chain areas [16]. In simheuristics, the challenge resides in finding the best solution to the problem. However, in our work, finding a best solution is relatively easy, but finding all (or a large portion) of the best solutions is the challenge.

5.2   Graph Regression

In this research, we understand a whole graph as the configuration of a system and a graph set as design options for this system. The regression methods that are relevant to this particular problem follow one of these three paths [7] [24]:

1. Network embedding: Map graphs into low-dimensional vectors (graph embedding) and use regression analysis techniques.
2. Graph regularised neural networks: Feed graphs into a neural network with a regularised objective.
3. Graph neural networks: Interpret the graph itself as a neural network, let the nodes learn their neighbourhoods for a couple of iterations, and use global pooling to determine the result [31].

So far, we have implemented the first of these three paths. We will compare these paths with respect to accuracy and training efficiency in future work.

6   Conclusion

In this paper, we investigated how we can adapt current practices of learning software configuration spaces originally devised for Euclidean configurations (consisting of Boolean and numerical variables) to graph configurations (containing structural information). We proposed and implemented a pipeline of strategies (that were already established for software configurations) for graphs as a proof of concept. Subsequently, we evaluated its efficiency for the exploration process. The presented approach can support engineers in exploring configurations of complex graph structures – while considering a large and diverse set of possible candidates in an efficient manner. The proposed LEGCS approach finds 75% of the k best graph configurations while simulating only 18.8% of the space.

Acknowledgement. This work was supported by Science Foundation Ireland grant 13/RC/2094_P2 to Lero - the Science Foundation Ireland Research Centre for Software (www.lero.ie).

References

1. Building Ventilation - MATLAB Simulink. https://www.mathworks.com/help/simscape/ug/building-ventilation.html. Accessed 10 May 2023
2. MATLAB Simulink - Simulation and Model-Based Design. https://uk.mathworks.com/products/simulink.html. Accessed 10 May 2023
3. https://github.com/mittermm/LEGCS . Accessed 10 May 2023
4. Acher, M., et al.: Learning very large configuration spaces: what matters for linux kernel sizes. Inria Rennes-Bretagne Atlantique. hal-02314830 (2019)
5. Cai, C., Wang, Y.: A simple yet effective baseline for non-attributed graph classification. arXiv:1811.03508 (2022)
6. Cai, H., Zheng, V.W., Chang, K.C.C.: A comprehensive survey of graph embedding: problems, techniques, and applications. IEEE Trans. Knowl. Data Eng. **30**(9), 1616–1637 (2018)
7. Chami, I., Abu-El-Haija, S., Perozzi, B., Ré, C., Murphy, K.: Machine learning on graphs: a model and comprehensive taxonomy. J. Mach. Learn. Res. **23**, 89:1–89:64 (2022)
8. Chica, M., Juan, A.A., Bayliss, C., Cordon, O., Kelton, D.: Why simheuristics? benefits, limitations, and best practices when combining metaheuristics with simulation. In: Statistics and Operations Research Transactions (2020)
9. Church, R.L., Baez, C.A.: Generating optimal and near-optimal solutions to facility location problems. Environ. Plan. B: Urban Anal. City Sci. **47**(6), 1014–1030 (2020)
10. Easterbrook, S., Singer, J., Storey, M.A.D., Damian, D.E.: Selecting empirical methods for software engineering research. In: Guide to Advanced Empirical Software Engineering, pp. 285–311. Springer, Heidelberg (2008). https://doi.org/10.1007/978-1-84800-044-5_11
11. Eppstein, D., Kurz, D.: k-best solutions of MSO problems on tree-decomposable graphs. In: 12th International Symposium on Parameterized and Exact Computation (IPEC) (2017)
12. Farahani, R.Z., Miandoabchi, E., Szeto, W.Y., Rashidi-Bajgan, H.: A review of urban transportation network design problems. Eur. J. Oper. Res. **229**(2), 281–302 (2013)
13. Glinz, M.: On non-functional requirements. In: International Requirements Engineering Conference (RE), pp. 21–26 (2007)
14. Goyal, P., Ferrara, E.: Graph embedding techniques, applications, and performance: a survey. Knowl. Based Syst. **151**, 78–94 (2018)

15. Jedlitschka, A., Ciolkowski, M., Pfahl, D.: Reporting experiments in software engineering. In: Guide to Advanced Empirical Software Engineering, pp. 201–228. Springer, Heidelberg (2008)
16. Juan, A.A., Kelton, W.D., Currie, C.S.M., Faulin, J.: Simheuristics applications: dealing with uncertainty in logistics, transportation, and other supply chain areas. In: WSC, pp. 3048–3059. IEEE (2018)
17. Karunasingha, D.S.K.: Root mean square error or mean absolute error? use their ratio as well. Inf. Sci. **585**, 609–629 (2022)
18. Kriege, N.M., Johansson, F.D., Morris, C.: A survey on graph kernels. Appl. Netw. Sci. **5**(1), 6 (2020)
19. Liaw, A., Wiener, M.: Classification and Regression by randomForest. R News **2**(3), 18–22 (2002)
20. Lopes, R., Tutenel, T., Smelik, R.M., De Kraker, K.J., Bidarra, R.: A constrained growth method for procedural floor plan generation. In: Proceedings of International Conference on Intelligent Games Simulation, pp. 13–20 (2010)
21. Makarov, I., Kiselev, D., Nikitinsky, N., Subelj, L.: Survey on graph embeddings and their applications to machine learning problems on graphs. PeerJ Comput. Sci. **7**, e357 (2021)
22. Miao, S., Liu, M., Li, P.: Interpretable and generalizable graph learning via stochastic attention mechanism. In: International Conference on Machine Learning, pp. 15524–15543. PMLR (2022)
23. Pedregosa, F., et al.: Scikit-learn: machine learning in python. J. Mach. Learn. Res. **12**, 2825–2830 (2011)
24. Peng, Y., Choi, B., Jianliang, X.: Graph learning for combinatorial optimization: a survey of state-of-the-art. Data Sci. Eng. **6**(2), 119–141 (2021)
25. Pereira, J.A., Acher, M., Martin, H., Jézéquel, J.M., Botterweck, G., Ventresque, A.: Learning software configuration spaces: a systematic literature review. J. Syst. Softw. **182**, 111044 (2021)
26. Pienta, R.S., Abello, J., Kahng, M., Chau, D.H.: Scalable graph exploration and visualization: sensemaking challenges and opportunities. In: International Conference on Big Data and Smart Computing (BIGCOMP), pp. 271–278. IEEE Computer Society (2015)
27. Rozemberczki, B., Kiss, O., Sarkar, R.: Karate club: an API oriented open-source python framework for unsupervised learning on graphs. In: International Conference on Information and Knowledge Management (CIKM), pp. 3125–3132. ACM (2020)
28. Shaw, M.: Writing good software engineering research paper. In Proceedings of the 25th International Conference on Software Engineering, pp. 726–737. IEEE Computer Society (2003)
29. Siegmund, N., et al.: Scalable prediction of non-functional properties in software product lines. In: SPLC (2011)
30. Wang, X., Bo, D., Shi, C., Fan, S., Ye, Y., Philip, S.Y.: A survey on heterogeneous graph embedding: methods, techniques, applications and sources. IEEE Trans. Big Data **9**(2), 415–436 (2022)
31. Zhou, J., et al.: Graph neural networks: a review of methods and applications. AI Open **1**, 57–81 (2020)

Artificial Intelligence and Neuroscience (ACAIN 2023)

Towards an Interpretable Functional Image-Based Classifier: Dimensionality Reduction of High-Density Diffuse Optical Tomography Data

Sruthi Srinivasan[1]([✉]) [ID], Emilia Butters[1,2] [ID], Flavia Mancini[1] [ID], and Gemma Bale[1,3] [ID]

[1] Department of Engineering, University of Cambridge, Cambridge, UK
ss2814@cam.ac.uk
[2] Department of Psychiatry, University of Cambridge, Cambridge, UK
[3] Department of Physics, University of Cambridge, Cambridge, UK

Abstract. High-density diffuse optical tomography (HD-DOT) is a wearable neuroimaging method that demonstrates high temporal and spatial resolution. While this data contains far richer information as a result, the high dimensionality and presence of complicated interconnections between data points requires the use of dimensionality reduction techniques to simplify the predictive modelling task without eliminating meaningful data features. To interrogate the possibility of designing a physiologically relevant HD-DOT feature set, cortical parcellations were applied to reconstructed images of brain activity to reduce the data dimensionality. A preliminary assessment of the predictive power of these parcel features was conducted on two binary tasks, with reasonable accuracies being achieved using standard classification models. Our results also demonstrated high spatial signal reproducibility across participants, which is promising for the application of image-based classification models that rely on spatial similarities to define separable class boundaries. These results provide insight into how the increased spatial resolution of HD-DOT can be leveraged to perform more accurate classification of neural data.

Keywords: Dimensionality Reduction · Diffuse Optical Tomography · Near Infrared Spectroscopy · Image Reconstruction · Explainable Artificial Intelligence (XAI)

1 Introduction

High-density diffuse optical tomography (HD-DOT) is a noninvasive optical neuroimaging technique which has demonstrated spatial resolutions on the order of functional magnetic resonance imaging (fMRI), vastly improving on the spatial resolutions traditionally offered by functional near-infrared spectroscopy (fNIRS) [1, 2]. However, higher spatial resolution images of neural activity necessitate the use of dimensionality

G. Nicosia et al. (Eds.): LOD 2023, LNCS 14506, pp. 351–357, 2024.
https://doi.org/10.1007/978-3-031-53966-4_26

reduction techniques to both consolidate high-dimensional imaging data into lower-dimensional feature space and account for data interdependencies without significant loss of information [3].

HD-DOT uses a high-density array of light sources and detectors, which provides a larger number of NIRS measurement channels and greater cortical coverage, in conjunction with anatomical priors to produce three-dimensional images of changes in oxygenated (HbO) and deoxygenated hemoglobin (HbR). Despite the recent advances in HD-DOT image quality [4], there are no HD-DOT datasets available at the scale needed for machine learning. This makes it difficult to develop feature engineering methods and train sufficiently robust models that operate solely on HD-DOT data.

Previous work applying machine learning methods to fNIRS data largely focuses on using either raw time series data or statistical features as model inputs [5]. However, this presents a major disadvantage: the features themselves are not inherently interpretable, thus limiting the post-hoc explainability of these models. By contrast, models trained on feature sets that are physiologically relevant can demonstrate greater explainability [6]. As described in fMRI literature, parcellation atlases can serve to reduce data dimensionality [7], with each parcel representing an anatomical or functional brain region. These parcels can then be used to engineer an inherently interpretable feature set.

In this study, we describe a parcellation-based dimensionality reduction technique for HD-DOT data and perform a preliminary assessment of the predictive power of features derived from this parcellation using two binary classification tasks. To achieve this, HD-DOT data were collected from the bilateral auditory cortices and surrounding brain regions across a large population, such that image reconstructions of brain activity could be produced. Here, we aim to demonstrate that anatomical parcellation of HD-DOT-derived cortical activity maps reduces data dimensionality without compromising the benefits of higher spatial resolution imaging. This work serves as a preliminary step in the development of image-based classification models with post-hoc model explainability.

2   Materials and Methods

The experimental protocol for this study was designed such that two separate classification tasks could be investigated. Each experimental run featured two participants concurrently undergoing HD-DOT recording. In task 1, both participants were seated opposite each other, separated by a screen such that they could not see each other, and were each instructed to listen to film music [8] with two affective conditions: positive and negative. In task 2, the screen was removed such that participants could see and interact with each other as the same music was played, allowing for two social setting conditions: solo (from task 1) and duo. Data was recorded from 160 participants (aged 16+) using a wearable HD-DOT device (LUMO; Gowerlabs Ltd., UK) placed over the bilateral auditory cortices, allowing for ~200 possible optical channels per hemisphere, sampled at a rate of 5 Hz.

The optical data were pre-processed by bandpass filtering (cut-off: 0.01–0.2 Hz) and regressing out short channels (0-12 mm). Dataset quality was assessed using measures of optical coupling: sets demonstrating > 50% of useable channels in the short separation range (0−5 mm) were included in the final analysis. Thus, 99 datasets (~62%) were included. Data were block-averaged over two 30s trials per condition, with pre-stimulus baseline defined as the last 15s of the 30s rest period preceding each stimulus block. Normalized spatial image reconstructions were created using the DOT-HUB Toolbox [9] and Toast++ [10].

Following this, the Schaefer parcellation atlas was used to parcellate individual-level data for each condition into 1000 unique regions [11]. The voxel-based parcellation atlas was first transformed to the space of the node-based HD-DOT surface image representation, and nodes were matched to their nearest parcel. Grey matter mesh nodes were classed as sensitive if any channels demonstrated a normalized Jacobian value above 5% of the maximum Jacobian value at that node, across both HbO and HbR (meaning that the HD-DOT measurement device demonstrated sensitivity to absorption changes at that node) [12]. Parcels were then included in the analysis if $\geq$50% of the nodes comprising that parcel were sensitive. HbO and HbR values for each parcel were calculated from the average time series of all sensitive nodes within that parcel.

Two binary classification problems were derived: positive/negative and solo/duo. Separate HbO and HbR feature sets were generated as matrices of parcellated values. The impact of further dimensionality reduction was assessed via feature selection, which was performed in MATLAB R2022b using individual chi-square (X^2) tests to select predictors (parcel values) for which a statistically significant relationship to the class label ($p < 0.001$ for solo/duo, $p < 0.1$ for positive/negative) exists. Multiple standard prediction model classes (decision tree, linear/quadratic discriminant, logistic regression, naïve Bayes, support vector machine, K-nearest neighbor) were used to assess the predictive value of HD-DOT parcellations. Model validation was performed using a 5-fold cross-validation.

To compare similarity in active regions across participants, the 5–10 s period post-onset of auditory stimulus was selected, as this region typically contains the hemodynamic response peak [13], and the data were averaged across these time points per-parcel. If a particular subject's parcel demonstrated an absolute HbO (HbR) concentration change greater than 25% of their absolute maximum HbO (HbR) concentration change, that parcel was included in the active region for a given condition.

3 Results

As expected, models trained on non-parcellated data demonstrated far lower classification accuracies for both tasks (<50%), due to the larger feature space (>30,000 grey matter surface mesh nodes, compared to ~300 parcels following dimensionality reduction), and lack of inter-participant comparability. However, for parcellated feature sets, we found a nearly 12% difference in both the average and maximum classification accuracies between the two binary tasks. Standard classification models performed worse on the positive/negative task, as compared to the solo/duo task. This is in line with results from our feature importance ranking using X^2-tests, where the range of test statistics for the solo/duo task ($X^2 = 6.95 - 10.55$, $p < 0.001$) was higher than that of the positive/negative task ($X^2 = 2.30 - 3.03$, $p < 0.1$), indicating greater response dependence on parcel predictor values for the solo/duo task. Prediction results for each binary task are shown in Table 1.

Table 1. Confusion matrix showing prediction results on the validation set of the *highest performance model* for each binary classification task, following feature selection. The highest performance binary classification models used were logistic regression (for positive/negative classification using HbO parcels, 14 features with $p < 0.1$) and quadratic discrimination (for solo/duo classification using HbR parcels, 10 features with $p < 0.001$).

True Class	Predicted Class				Total
	Positive	Negative	Solo	Duo	
Positive	91	95	–	–	186
Negative	66	120	–	–	186
Solo	–	–	145	41	186
Duo	–	–	77	109	186
Total	157	215	222	150	
True Positive Rate (%)	48.9	64.5	78.0	58.6	
Precision (%)	58.0	55.8	65.3	72.7	
Accuracy (%)	**56.7**		**68.3**		

To assess whether variations in spatial activation patterns across participants may account for the preliminary model performance differences seen between the two binary tasks, participant overlap maps of active regions for both HbO and HbR were derived for each of the four conditions seen in Fig. 1. Higher consistency of spatial response was found for HbR across all conditions as compared to HbO, meaning that more participants had overlapping active regions, and thus a more reproducible response, in the HbR maps. This higher consistency in HbR response was observed in both the left (LH) and right hemispheres (RH). Furthermore, our feature ranking results found that out of the top 20 ranked features for both HbO and HbR feature sets, the most important features were approximately evenly distributed across hemispheres (LH: 12, RH: 8 for HbO; LH: 10, RH: 10 for HbR).

4   Discussion and Conclusion

Cortical parcellation of HD-DOT data was found to be a highly suitable dimensionality reduction method that allows for direct physiological interpretability of the feature set. Models trained on the full set of grey matter surface mesh nodes demonstrated lower validation accuracies (below chance) compared to training accuracies, likely a result of model overfitting due to the relative size of the feature set compared to the number of examples. As seen in Fig. 1, differences in the active areas for the solo/duo data were the most pronounced, likely leading to the higher classification accuracies seen for this binary task. In particular, we observed the highest true positive rate (78.0%) when classifying the solo data from the duo data, suggesting that there are significant regional hemodynamic changes between these two conditions.

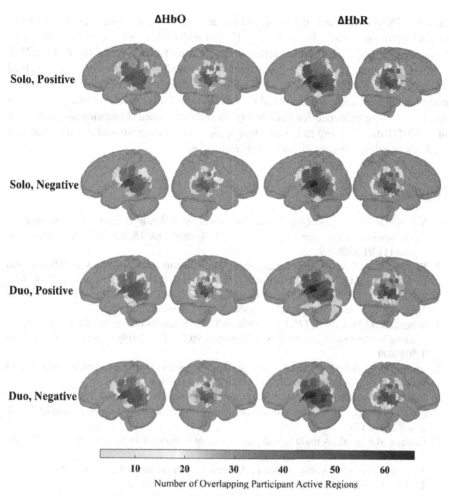

Fig. 1. Number of overlapping active parcels across participants ($N = 93$) for each of the four conditions when using HbO (left two columns) and HbR concentration values (right two columns). Active regions are shown for the period 5–10 s post-onset of stimulus.

The relatively low classification accuracies observed, particularly for the positive/negative classification task, suggest that the classes are not easily separable solely from parcellated hemoglobin data. However, we have achieved ~70% classification accuracy on a single binary task using a physiologically relevant feature set, with further improvements to be made. To realize the full advantage of HD-DOT data, reconstructed and parcellated images should be used as input to a convolutional neural network (CNN) to take advantage of both the spatial representation of activity and the aforementioned dimensionality reduction technique to yield higher accuracy predictions. Our findings indicate that parcels from both hemispheres are important model predictors and support the conclusion that spatial (image-based) representations will more thoroughly capture the data. CNNs with image-based inputs have demonstrated high binary prediction accuracies (>90%) on fNIRS data (e.g., [14, 15]), though to date, no studies have used HD-DOT-derived images as inputs. Furthermore, the use of image-based inputs for CNNs facilitates the application of post-hoc explainability methods, such as Gradient-weighted Class Activation Mapping [16], to identify class-discriminative image regions, with direct anatomical comparisons aided by parcellation. Thus, we view these preliminary results as strong evidence for performing parcellation-based dimensionality reduction of HD-DOT data, leading to the eventual application of image-based machine learning methods to achieve higher classification accuracies.

References

1. White, B.R., Culver, J.P.: Quantitative evaluation of high-density diffuse optical tomography: in vivo resolution and mapping performance. J. Biomed. Opt. **15**, 026006 (2010). https://doi.org/10.1117/1.3368999
2. Eggebrecht, A.T., et al.: Mapping distributed brain function and networks with diffuse optical tomography. Nat. Photonics **8**, 448–454 (2014). https://doi.org/10.1038/nphoton.2014.107
3. Cunningham, J.P., Yu, B.M.: Dimensionality reduction for large-scale neural recordings. Nat. Neurosci. **17**, 1500–1509 (2014). https://doi.org/10.1038/nn.3776
4. Wheelock, M.D., Culver, J.P., Eggebrecht, A.T.: High-density diffuse optical tomography for imaging human brain function. Rev. Sci. Instrum. **90**, 051101 (2019). https://doi.org/10.1063/1.5086809
5. Eastmond, C., Subedi, A., De, S., Intes, X.: Deep learning in fNIRS: a review. Neurophotonics **9**, 041411 (2022). https://doi.org/10.1117/1.NPh.9.4.041411
6. Zytek, A., Arnaldo, I., Liu, D., Berti-Equille, L., Veeramachaneni, K.: The Need for Interpretable Features: Motivation and Taxonomy. arXiv preprint (2022). https://doi.org/10.48550/arXiv.2202.11748
7. Glasser, M.F., et al.: A multi-modal parcellation of human cerebral cortex. Nature **536**, 171–178 (2016). https://doi.org/10.1038/nature18933
8. Putkinen, V., et al.: Decoding music-evoked emotions in the auditory and motor cortex. Cereb. Cortex **31**, 2549–2560 (2021). https://doi.org/10.1093/cercor/bhaa373
9. Frijia, E.M., et al.: Functional imaging of the developing brain with wearable high-density diffuse optical tomography: a new benchmark for infant neuroimaging outside the scanner environment. Neuroimage. **225**, 117490 (2021). https://doi.org/10.1016/j.neuroimage.2020.117490
10. Schweiger, M., Arridge, S.R.: The Toast++ software suite for forward and inverse modeling in optical tomography. J. Biomed. Opt. **19**, 040801 (2014). https://doi.org/10.1117/1.JBO.19.4.040801
11. Schaefer, A., et al.: Local-global parcellation of the human cerebral cortex from intrinsic functional connectivity MRI. Cereb. Cortex **28**, 3095–3114 (2018). https://doi.org/10.1093/cercor/bhx179
12. Uchitel, J., Blanco, B., Vidal-Rosas, E., Collins-Jones, L., Cooper, R.J.: Reliability and similarity of resting state functional connectivity networks imaged using wearable, high-density diffuse optical tomography in the home setting. Neuroimage **263**, 119663 (2022). https://doi.org/10.1016/j.neuroimage.2022.119663
13. Abdalmalak, A., et al.: Assessing time-resolved fNIRS for brain-computer interface applications of mental communication. Front Neurosci. **14**, 105 (2020). https://doi.org/10.3389/fnins.2020.00105

14. Takagi, S., et al.: Application of deep learning in the identification of cerebral hemodynamics data obtained from functional near-infrared spectroscopy: a preliminary study of pre- and post-tooth clenching assessment. J. Clin. Med. **9**, 3475 (2020). https://doi.org/10.3390/jcm 9113475

15. Yang, D., et al.: Detection of mild cognitive impairment using convolutional neural network: temporal-feature maps of functional near-infrared spectroscopy. Front Aging Neurosci. **12**, 141 (2020). https://doi.org/10.3389/fnagi.2020.00141

16. Selvaraju, R.R., Cogswell, M., Das, A., Vedantam, R., Parikh, D., Batra, D.: Grad-CAM: visual explanations from deep networks via gradient-based localization. Int. J. Comput. Vis. **128**, 336–359 (2020). https://doi.org/10.1007/s11263-019-01228-7

On Ensemble Learning for Mental Workload Classification

Niall McGuire[✉] and Yashar Moshfeghi

NeuraSearch Laboratory, Department of Computer and Information Sciences,
University of Strathclyde, Glasgow, Scotland
{niall.mcguire,yashar.moshfeghi}@strath.ac.uk

Abstract. The ability to determine a subject's Mental Work Load (MWL) has a wide range of significant applications within modern working environments. In recent years, techniques such as Electroencephalography (EEG) have come to the forefront of MWL monitoring by extracting signals from the brain that correlate strongly to the workload of a subject. To effectively classify the MWL of a subject via their EEG data, prior works have employed machine and deep learning models. These studies have primarily utilised single-learner models to perform MWL classification. However, given the significance of accurately detecting a subject's MWL for use in practical applications, steps should be taken to assess how we can increase the accuracy of these systems so that they are robust enough for use in real-world scenarios. Therefore, in this study, we investigate if the use of state-of-the-art ensemble learning strategies can improve performance over individual models. As such, we apply Bagging and Stacking ensemble techniques to the STEW dataset to classify "low", "medium", and "high" workload levels using EEG data. We also explore how different model compositions impact performance by modifying the type and quantity of models within each ensemble. The results from this study highlight that ensemble networks are capable of improving upon the accuracy of all their individual learner counterparts whilst reducing the variance of predictions, with our highest scoring model being a stacking BLSTM consisting of 8 learners, which achieved a classification accuracy of 97%.

Keywords: EEG · Mental Workload · Classification · Deep Learning · Ensembles

1 Introduction

The ability to monitor and predict a person's mental workload levels (MWL) has a vast range of impactful applications within modern working environments, such as teaching/training to monitor the performance of a subject and how well they are handling a problem [29], safety systems to detect when a subjects mental workload levels increase to dangers levels that may increase the likelihood of accidents occurring [9,10,21], as well as improving performance by increasing or

G. Nicosia et al. (Eds.): LOD 2023, LNCS 14506, pp. 358–372, 2024.
https://doi.org/10.1007/978-3-031-53966-4_27

decreasing the workload of a subject based on their current MWL [18]. A variety of avenues have already been explored on how to best determine the MWL of a topic, from self-review questionnaires [31] and performance measures [4,17] as well as more direct techniques such as Functional Magnetic Resonance Imaging (fMRI) [23,27,28], Functional Near-Infrared Spectroscopy (fNIRS) [26], and Electroencephalography (EEG) [21,44] which all monitor the neurophysiological features of a subject's brain.

In recent years Electroencephalography (EEG) data has been widely adopted for MWL detection. This is because EEG data is well suited for MWL detection as the signals produced by the subject's brain are strongly linked to their mental workload levels [34,37]. EEG also allows for capturing these signals in real-time due to its high temporal resolution (millisecond scale), allowing for developing systems with lower latency and more responsiveness to a subject's mental workload levels. EEG data has already been used to detect, monitor and predict the subject's mental workload [1,25,35]. Recently there have been significant developments of machine/deep learning techniques designed to perform mental workload classification, which has been shown to produce high accuracy classification [34,44].

Many of these previous studies have employed a variety of machine and deep learning models which make use of single-learner algorithms to achieve results to support their research questions, but often need to look into ways in which to obtain the optimal performance of their models. One such way is the utilisation of ensemble learning techniques. Ensemble learning consists of combining multiple individual machine/deep learning models to improve the system's overall performance [32]. The advantages of ensemble learning can include reduced variance in the models' predictions and an overall increase in accuracy score. Ensemble learning has already demonstrated its effectiveness for improving performance in various applications [46] as well as some tasks that use EEG data [48]. However, the capabilities of different ensemble learning techniques for EEG-based mental workload classification have yet to be thoroughly examined. Thus, in this study, we aim to answer the following research questions:

- **RQ1**: *"Can ensemble learning techniques improve the performance of MWL classification over that of single-learner models?"*
- **RQ2**: *"How do different ensemble techniques compare against each other?"*
- **RQ3**: *"What are the effects of differing ensemble compositions on performance?"*

To answer these research questions, we employ two popular ensemble learning techniques, these being Bagging and Stacking [32]. Each of these techniques is made up of a combination of the following state-of-the-art deep learning models used for MWL classification: Gated Recurrent neural Network (GRU), Short-Term Long Memory (LSTM), Bidirectional Gated Recurrent Neural Network (BGRU), and Bidirectional Long Short-Term Memory (BLSTM) [8,21]. These systems are then trained and evaluated on the open-source mental workload data set STEW [24].

2   Background

The ability to detect, monitor and predict a subject's mental workload level/cognitive load has many potential significant applications in a variety of domains, from safety systems that consider a subject's cognitive load to performance optimisation systems [5]. As such, in recent years, it has become a rapidly evolving area of research. Mental workload (MWL) can be defined as the cognitive resources required by a subject whilst engaging in a task. In such cases, the load on the human subject can directly affect the outcome of the given study [5,30]. Prior research has shown that when an operator's cognitive load increases, their performance may significantly suffer, resulting in lower performance and potential safety risks depending on the task [10,15]. Thus, it is essential to understand the MWL of the subject to ensure that the job is handled appropriately. Early techniques developed to infer MWL employ subjective strategies that utilise responses gathered from the issue. Examples of these emotional techniques include the National Aeronautics and Space Administration's Task Load Index (NASA TLX) [12] and the Subjective Workload Assessment Technique (SWAT) [33]. Alternatives to subjective responses are performance measures; these estimate the MWL of a subject by evaluating their performance on various main and sub-tasks during multi-tasking scenarios [42]. Another potential avenue to infer the workload level of an issue is through monitoring their physiological features such as; heart rate, respiratory activity, saccadic eye movements, and brain activity [7,13,38]. These techniques have moved to the forefront of MWL classification as the other strategies, such as self-rating questionnaires and performance metrics, can only be feasibly measured in a post-hoc manner requiring the user's cooperation, unlike physiological features that can be monitored continuously during the on-goings of the subject's task [45].

2.1   Neuroscience and MWL Detection

Recent works have demonstrated the effectiveness of utilising specific brain activity features to infer a subject's cognitive load directly. These studies have made use of techniques such as magnetoencephalography (MEG) [20], fMRI [28], fNIRS [26], and EEG [2,19,39] to extract neurophysiological features from the brain. However, EEG data is optimal for the specific task of MWL classification due to its strong correlation to the cognitive load of a subject, its portability, its unobtrusiveness, and its ability to capture data in real-time (millisecond scale).

To more accurately determine the cognitive load of a subject using EEG data, Machine Learning (ML) and Deep Learning (DL) algorithms have been developed to perform MWL classification. Works such as [35] extract the EEG from twenty subjects performing four cognitive and motor tasks consisting of arithmetic operations, finger tapping, mental rotation and lexical decisions, ranging from low to medium to high difficulty levels. The researchers used the EEG data recorded from the subjects to train a single learner ML support vector machine (SVM) to predict the current workload level of a subject from their

EEG data. The results showed that they achieved classification accuracy varying from 65%–75% across subjects. Deep learning models were utilised in the works by [6], where in which the researchers employed a deep hybrid model based on a BLSTM and LSTM to perform mental workload classification when trained on the STEW EEG dataset. The results of this study outlined that the model was capable of achieving 82.57% accuracy when predicting whether the mental workload levels of a subject were "low", "medium", or "high" from their EEG data.

2.2 Ensemble Learning for MWL Detection

Prior works that have employed ML and DL algorithms to perform their classification have primarily limited themselves to a single learner model to fulfil the requirements of their study. However, suppose these systems are to be applied to real-world applications, where accurately determining the MWL of a subject may have significant impacts on the task that the operator is engaging with. In that case, it is vital that these systems must be high performing and consistent in their MWL predictions. A potential solution to this would be using an ensemble network, which combines more than one ML and DL model into a single learner to reduce the variance of the predictions and overall improve the model's accuracy. Some recent works have made use of ensemble learning techniques during their MWL classification tasks, such as [43], where the researchers employed a bagging ensemble of stacked denoising autoencoders to perform the binary classification of two mental workload levels of "low" and "high" from the EEG data gathered across eight subjects. Similarly, in the study [36], the researchers examine the performance of several heterogeneous ML and DL algorithms to perform binary classification "low", and "high" mental workload conditions using the EEG data gathered from eight subjects. Although these studies have examined and demonstrated the capabilities of ensemble learning for MWL classification, they have yet to extensively compare the effects of different ensemble variables, such as bagging vs stacking, mixed vs heterogeneous models, and how the number of ensemble members affects overall performance.

3 Methodology

The main steps of this study are EEG signal acquisition, artefact removal, feature engineering, model evaluation and EEG music classification. Details of each step will be described in the following sections.

3.1 Data Set

For mental workload classification, we utilised the open-source Simultaneous Task EEG Workload (STEW) dataset [16]. The STEW data set creates a multi-tasking mental workload environment by employing the study's single-session simultaneous capacity (SIMKAP) experiment [3]. The dataset contains the EEG data recorded from 45 subjects using a 14-electrode system sampled at 128 Hz. The EEG signals were recorded from the subjects while they took part in two experimental conditions, the resting and testing conditions. During the resting state, the participants had to sit in a chair and not perform any tasks or challenges for 3 min whilst their EEG data was recorded. Within the testing conditions, the subjects were tasked to participate in the SIMKAP activity whilst their EEG data was recorded. For the EEG data recorded during the SIMKAP test, the last 3 min of the recording is considered to be the workload condition. The first and last 15 s of the data from each recording were excluded to reduce the effects of any between-task activities; this resulted in each recording being 2.5 min long. After every segment of the experiment conditions, each subject was asked to rate their perceived MWL from a 9-point rating scale of 1 to 9. This was done as a subjective validation that the subject did experience an increase in workload whilst taking part in the SIMKAP test when compared to the resting condition. The rating ranges denote that 1–3 can be considered as a low-level workload, 4–6 as a moderate workload, and 7–9 as a high-level workload.

3.2 EEG Preprocessing

The final data produced when recording of EEG signals can be strongly affected by the environment and electrical activities of the participant. EEG signals are commonly polluted by noise and artefacts, which are unwanted signals that can affect the raw EEG data and alter signals of interest for our task [40]. The removal of noise and artefacts is a critical step in cleaning/processing EEG data. Prior studies have demonstrated the benefits of these procedures on the overall performance of classification, specifically for ML and DL models [16]. Early techniques employed by neuroscientists to perform pre-processing involved the manual or visual analysis of the recorded EEG signals. Although these procedures were very effective in identifying and removing the majority of artefacts from EEG data, they are particularly time-consuming and require expert knowledge of the field. These procedures may also suffer from the possibility of bias when having human operators examine and remove artefacts [41].

As such, within this study, we employ automatic systems to perform these procedures, speeding up the time to process and removing the possibility of human bias interfering with the processed EEG data. To fulfil our pre-processing, we use the MNE-python library [11]. Firstly, we apply a bandpass filter to the data within a threshold of 0.5–40 Hz, EEG data within this frequency band has been shown to strongly correlate with activation frequencies of the brain, and thus we can consider anything out with this threshold to be noise and artefacts caused by electrical interference or subject muscle activity.

3.3 EEG Feature Extraction

After processing and cleaning the EEG signals, we extracted features from the data. Feature extraction of EEG data is important as it can provide the ML and DL models with more meaningful data and gives the models the best chance of understanding the relationships between the data and their associated MWL. The Features extracted are detailed below:

1. **Morphological features:** the number of peaks, average non-linear energy and the curve length were extracted. Features and three morphological features were extracted in two different ways. Firstly, they were calculated at four frequency bands theta (4–8 Hz), alpha (8–13 Hz), beta (13–25 Hz), and low gamma (25–40 Hz).
2. **Statistical features:** skewness, mean, and kurtosis were computed to identify the distribution of the signal. [21] was calculated to describe time-varying processes
3. **non-linear features:** The approximate entropy (ApEn) and Hurst exponent (H) were used to quantify the unpredictability of fluctuations over a time series as well as measure the self-similarity of the time series, respectively.

All of the features above were extracted from each EEG channel using a time window of 1 s over the recorded EEG data. This resulted in 148 samples for each feature generated per trial a subject took part in.

3.4 Models

When examining previous studies that attempt to infer MWLs from EEG data, it can be seen that ML and DL models are commonly employed to perform this classification. Some of the models made use of by these prior studies include the Gated Recurrent Unit (GRU), Long Short-Term Memory (LSTM), Bidirectional Gated Recurrent Unit (BGRU), and the Bidirectional Long Short-Term Memory (BLSTM). These models have seen extensive use in both general EEG and MWL classification due to their effectiveness when learning sequential data with long-term dependencies [47]. Within this study, we will use each of these individual learner models to compose each of our ensemble learners.

Table 1. Deep learning model architectures

Model	Layers/Nodes
GRU	G128-G64-G40-D32(D3)
BGRU	BG128-G64-G32-D16(D3)
LSTM	L128-L64-L40-D32(D3)
BLSTM	BL128-L64-L32-D16(D3)

The architectures for each model can be seen in Table 1. G, BG, L, BL, and D correspond to GRU, BGRU, LSTM, BLSTM, and Dense layer, respectively. For example, BL128-L64-L32-D16(D3) details that there is a BLSTM layer with 128 units, an LSTM layer with 64 units, followed by a Dense-layer with 16 units, and lastly, another Dense-layer with three units (corresponding to the three levels of mental workload we are attempting to predict). SoftMax activation was implemented in the last layer. This work implemented a dropout rate of 0.2, and the Adam optimiser was used within each of the deep-learning models. Additionally, early stopping was utilised to alleviate the potential issue of overfitting. Furthermore, we limited the number of epochs for training to 200.

3.5   Ensemble Strategies

As this study focuses on using ensemble learning techniques to perform MWL classification, we employed two popular ensemble strategies, Bagging and Stacking.

Bagging (short for Bootstrap Aggregating) has two main components in creating an ensemble. Firstly bootstrapping; this is where for each learner (in this study, GRU, LSTM, BGRU, BLSTM) within the ensemble, the training data set is randomly re-sampled into subsets that can contain multiple instances of the same data point. These subsets are then assigned to and used to train an individual model, with the idea being that since each of the models has been trained on its own random subset of the data, the ensemble will be able to better generalise for unseen data. The second function of bagging is aggregating. When the ensemble is presented with a testing dataset, it averages the class predictions across each learner and selects the class with the highest number of "votes". This is done to improve the overall accuracy of the model's predictions and reduce its variance.

Stacking (short for Stacked generalisation), unlike bagging's aggregation of the individual learner's predictions, uses the prediction outputs of the unique models to train a meta-learner that can better weight the output predictions of each model in the ensemble. The first step is to split the training data into two different groups, one that is used to train each model and one that is used to test the models to obtain their prediction outputs. These prediction outputs are then passed to the meta-learner (for this study, a logistic regressor model) to be trained. Once presented with unseen data, the ensemble passes the data to each learner to obtain their prediction outputs which are then given to the meta-learner to make the final prediction.

3.6   Metrics

In this study, the performance of each of the models is evaluated using precision, recall, f1-score and Accuracy. For a classification task such as ours, true positive (TP) is a situation that the model correctly predicted to be positive. A false negative (FN) is when the model incorrectly predicted a case to be negative. A true negative (TN) is a case where the model correctly predicted it to be negative, and a false positive (FP) is a situation where the model incorrectly predicted it be positive. Therefore, each metric is calculated as follows:

Precision refers to the ratio of correct positive examples to the number of actual positive examples.

$$Precision = \frac{TP}{(TP + FP)} \tag{1}$$

Recall is calculated as the sum of true positives across all classes divided by the sum of true positives and false negatives across all categories.

$$Recall = \frac{TP}{(TP + FN)} \tag{2}$$

F1-score is the harmonic mean of the precision and recall.

$$f1score = \frac{2TP}{(2TP + FP + FN)} \tag{3}$$

Accuracy is the sum of the number of true positives and true negatives divided by the total number of examples.

$$Accuracy = \frac{TP + TN}{(TP + TN + FP + FN)} \tag{4}$$

3.7 Experiment Conditions

To effectively address each of the research questions outlined for this paper, we formulated the following experimental conditions, with the goal of predicting the low, medium, and high mental workload of subjects from EEG data.

Experiment Scenario 1. Within this study, we make use of four single-learner baseline models to compose each of our ensembles GRU, LSTM, BGRU, and BLSTM. These models act a baseline for us to compare the performance of our ensembles against (Investigating RQ1). We employ two popular ensemble learning strategies; Bagging and Stacking (Investigating RQ2). To begin with, each of these ensemble strategies (Bagging and Stacking) has 5 variations where they are composed of different single learner models: 4-GRU-Learners, 4-LSTM-Learners, 4-BGRU-Learners, 4-BLSTM-Learners, and a mixed model containing one of each of the models: GRU, LSTM, BGRU, and BLSTM (Investigating RQ3). This means that, in total, we have 10 separate ensemble models for training and evaluation.

Experiment Scenario 2. To further investigate RQ3 we created the following experiment condition. This scenario follows the same training/evaluation process as the first experiment scenario. However, the only difference is the number of individual learners in each of the ensembles. Within this scenario, we increase the number of individual learners in each of the ensembles from 4 learners to 6, to 8, and lastly, 10. This was done to assess how many learners can be added to an ensemble before the performance begins to stagnate, allowing us to determine the optimal number of learners to compose an ensemble network.

3.8   Training Procedure

For each of the experiment scenarios, training and testing data was handled by using a 5-fold time series cross-validation method [22], where 70% of the fold was utilised for training, 10% for validation, and the remaining 20% for testing.

4   Results

In this section, we use a box plot to display the results for the variations of the ensemble models compared against the baseline individual learners that the ensembles are made up of. Each box plot presents five crucial pieces of information: the minimum, first, second (median), third, and maximum quarterlies [14]. Where the red dot in the box represents the mean accuracy for each model.

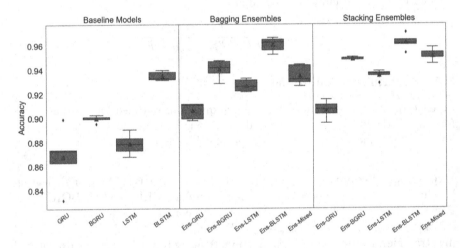

Fig. 1. Baseline and Ensemble Models Accuracy

4.1   Scenario 1

The results for experiment scenario one are displayed in Fig. 1. This figure contains box plots for the 5-fold cross-validation results of each model's accuracy. Within each block, we have the baseline models, the bagging models, and the stacking models. We first compare the ensemble models' results to their respective learner counterparts. It can be observed that every ensemble model for both bagging and stacking strategies has improved accuracy over each of its component learner models. On average, Bagging and Stacking improved the accuracy across each heterogeneous model by 3.7% and 4.2%, respectively. The model that saw the most significant improvement above the baseline was the Stacking LSTM, which achieved an accuracy of 93.5%, a 5.6% increase in performance from the baseline LSTM score of 87.9%. This helps to demonstrate ensemble networks' effectiveness in improving individual learners' performance (RQ1).

Following this, we examine how the two ensemble strategies used compare against each other. On average, the Stacking models had an improved performance of 0.7% over the bagging models. When looking at the highest-performing models for both strategies, we see that the BLSTM for bagging and stacking achieves the highest accuracy scores of 96% and 96.2%, respectively, highlighting that the stacking models slightly outperform their bagging counterparts. This shows that, although very close in performance, the stacking ensemble strategy, on average, offers improved performance over that of bagging (RQ2).

Lastly, we explore how the different model compositions compare against each other. Within the bagging ensemble plots, we can see that the order of lowest to highest-performing models is as such: GRU (90.6%), LSTM (92.7%), Mixed (93.5%), BGRU (94%), and BLSTM (96%). For the stacking ensemble, the order follows GRU (90.6%), LSTM (93.5%), BGRU (94.9%), Mixed (95.1%), and BLSTM (96.2%). Both strategies follow a similar expected performance pattern that mirrors the baseline models; the only exception is the Mixed model, which reliably outperforms GRU and LSTM but differs in its ability to outperform BGRU depending on the ensemble strategy used. This is likely due to the Mixed model containing two weaker learners (GRU, LSTM) than the homogeneous BGRU and BLSTM ensembles. This implies that when given the option of composing a model of varying performance models versus a homogeneous model of strong learners, the latter of the two options will obtain greater performance (RQ3).

Table 2. Stacking Ensembles with Varying Learner Count

Model	Member Count	Precision (S.D)	Recall (S.D)	F1-Score (S.D)	Accuracy (S.D)
GRU	6	0.91 (0.8)	0.91 (0.8)	0.91 (0.8)	0.91 (0.8)
GRU	8	0.92 (0.3)	0.92 (0.3)	0.92 (0.3)	0.92 (0.3)
GRU	10	0.94 (0.1)	0.94 (0.1)	0.94 (0.1)	0.94 (0.1)
BGRU	6	0.95 (0.4)	0.95 (0.4)	0.95 (0.4)	0.95 (0.4)
BGRU	8	0.94 (0.8)	0.94 (0.8)	0.94 (0.8)	0.94 (0.8)
BGRU	10	0.94 (0.5)	0.94 (0.5)	0.94 (0.5)	0.94 (0.5)
LSTM	6	0.93 (0.6)	0.93 (0.6)	0.93 (0.6)	0.93 (0.6)
LSTM	8	0.93 (0.2)	0.93 (0.2)	0.93 (0.2)	0.93 (0.2)
LSTM	10	0.94 (0.3)	0.94 (0.3)	0.94 (0.3)	0.94 (0.3)
BLSTM	6	0.96 (0.3)	0.96 (0.3)	0.96 (0.3)	0.96 (0.3)
BLSTM	8	0.97 (0.5)	0.97 (0.5)	0.97 (0.5)	**0.97 (0.5)**
BLSTM	10	0.96 (0.7)	0.96 (0.7)	0.96 (0.7)	0.96 (0.7)

Table 3. Bagging Ensembles with Varying Learner Count

model	member count	Precision (S.D)	Recall (S.D)	F1-Score (S.D)	Accuracy (S.D)
GRU	6	0.91 (0.2)	0.91 (0.2)	0.91 (0.2)	0.91 (0.2)
GRU	8	0.92 (0.4)	0.91 (0.4)	0.91 (0.4)	0.91 (0.4)
GRU	10	0.91 (0.3)	0.91 (0.3)	0.91 (0.3)	0.91 (0.3)
BGRU	6	0.94 (0.4)	0.94 (0.4)	0.94 (0.4)	0.94 (0.4)
BGRU	8	0.94 (0.5)	0.94 (0.5)	0.94 (0.5)	0.94 (0.5)
BGRU	10	0.94 (0.3)	0.94 (0.3)	0.94 (0.3)	0.94 (0.3)
LSTM	6	0.92 (0.5)	0.92 (0.5)	0.92 (0.5)	0.92 (0.5)
LSTM	8	0.93 (0.4)	0.93 (0.4)	0.93 (0.4)	0.93 (0.4)
LSTM	10	0.93 (0.6)	0.93 (0.6)	0.93 (0.6)	0.93 (0.6)
BLSTM	6	0.94 (0.7)	0.94 (0.7)	0.94 (0.7)	0.94 (0.7)
BLSTM	8	0.94 (0.3)	0.94 (0.3)	0.94 (0.3)	0.94 (0.3)
BLSTM	10	0.94 (0.5)	0.94 (0.5)	0.94 (0.5)	0.94 (0.5)

4.2 Scenario 2

The results for experiment scenario 2 are presented in Table 2, which contains the results for the stacking ensembles, and Table 3, which contains the results for the bagging ensembles. Firstly, when comparing the accuracy scores for each model across both ensemble strategies, we can observe that the bagging ensemble accuracies remain stagnant as the number of member increase from 6 to 8. Whereas it can be observed that a number of the stacking models' accuracy appears to improve, with the highest performing model being the stacking BLSTM achieving an accuracy of 97% using eight learners. These results imply that the stacking strategy appears the benefit from an increased number of learners above the base count of 4. In contrast, to the bagging strategy shows no benefits from increasing the learner count, adding further considerations when selecting optimal ensemble composition (RQ3).

5   Conclusion

To accurately determine a subject's Mental Work Load (MWL) level opens the door to an extensive range of beneficial applications within modern working environments. Areas such as health and safety, teaching, and resource management can benefit directly through monitoring and classifying a person's MWL. In recent years, significant progress has been made in the field of MWL detection thanks to the inclusion of direction brain monitoring techniques such as Electroencephalography (EEG). Prior studies have demonstrated the ability to train single machine/deep learning models on this data to classify MWL levels automatically. However, if these systems are to be applied in real-world scenarios, steps should be taken to explore how these models can be made as accurate and reliable as possible.

Therefore, in this study, we explored the viability of ensemble learning models to provide improved performance over single-learner models when applied to the STEW [24] MWL data set. Furthermore, we examined how different ensemble learning techniques (Bagging and Stacking) compare in performance and explored how ensemble composition (model type and quantity) affects performance. To do this, we composed several Bagging and Stacking ensembles from the following individual learners - GRU, BGRU, LSTM, and BLSTM. Our findings demonstrated the substantial improvement of all ensemble networks over their learner counterparts, with a bagging and stacking ensemble consisting of four learners on average increasing accuracy by 3.7% and 4.2%, respectively, whilst reducing the variation in the prediction output. Moreover, our results also show that performance benefits can be seen by including more members in the ensemble when using the stacking strategy, with the highest achieving ensemble being the BLSTM with eight learners that obtained an accuracy score of 97%.

Future works that aim to perform MWL classification within the context of a real-world scenario should consider the usage of ensemble networks to improve the system's overall performance. Furthermore, the ensemble networks presented in this study can also be applied to other research areas that wish to improve their performance when performing EEG data classification.

References

1. Aghajani, H., Garbey, M., Omurtag, A.: Measuring mental workload with EEG+fNIRS. Front. Hum. Neurosci. **11**, 359 (2017)
2. Allegretti, M., Moshfeghi, Y., Hadjigeorgieva, M., Pollick, F.E., Jose, J.M., Pasi, G.: When relevance judgement is happening? An EEG-based study. In: Proceedings of the 38th International ACM SIGIR Conference on Research and Development in Information Retrieval, pp. 719–722 (2015)
3. Bratfisch, O., Hagman, E.: Simkap-simultankapazität/multi-tasking. Schuhfried GmbH, Mödling (2008)
4. Butmee, T., Lansdown, T.C., Walker, G.H.: Mental workload and performance measurements in driving task: a review literature. In: Bagnara, S., Tartaglia, R., Albolino, S., Alexander, T., Fujita, Y. (eds.) IEA 2018. AISC, vol. 823, pp. 286–294. Springer, Cham (2019). https://doi.org/10.1007/978-3-319-96074-6_31
5. Cain, B.: A review of the mental workload literature (2007)
6. Chakladar, D.D., Dey, S., Roy, P.P., Dogra, D.P.: EEG-based mental workload estimation using deep BLSTM-LSTM network and evolutionary algorithm. Biomed. Signal Process. Control **60**, 101989 (2020)
7. Charles, R.L., Nixon, J.: Measuring mental workload using physiological measures: a systematic review. Appl. Ergon. **74**, 221–232 (2019)
8. Chen, J., Jiang, D., Zhang, Y.: A hierarchical bidirectional GRU model with attention for EEG-based emotion classification. IEEE Access **7**, 118530–118540 (2019)
9. Dehais, F., Somon, B., Mullen, T., Callan, D.E.: A neuroergonomics approach to measure pilot's cognitive incapacitation in the real world with EEG. In: Ayaz, H., Asgher, U. (eds.) AHFE 2020. AISC, vol. 1201, pp. 111–117. Springer, Cham (2021). https://doi.org/10.1007/978-3-030-51041-1_16

10. Deng, P.Y., et al.: Detecting fatigue status of pilots based on deep learning network using EEG signals. IEEE Trans. Cogn. Dev. Syst. **13**(3), 575–585 (2020). https://doi.org/10.1109/TCDS.2019.2963476

11. Gramfort, A., et al.: MEG and EEG data analysis with MNE-Python. Front. Neurosci. 267 (2013)

12. Hart, S.G.: NASA-task load index (NASA-TLX); 20 years later. In: Proceedings of the Human Factors and Ergonomics Society Annual Meeting, vol. 50, pp. 904–908. Sage, Los Angeles (2006)

13. Henelius, A., Hirvonen, K., Holm, A., Korpela, J., Muller, K.: Mental workload classification using heart rate metrics. In: 2009 Annual International Conference of the IEEE Engineering in Medicine and Biology Society, pp. 1836–1839. IEEE (2009)

14. Hofmann, T.: Collaborative filtering via Gaussian probabilistic latent semantic analysis. In: Proceedings of the 26th Annual International ACM SIGIR Conference on Research and Development in Information Retrieval, pp. 259–266 (2003)

15. Hu, X., Lodewijks, G.: Detecting fatigue in car drivers and aircraft pilots by using non-invasive measures: the value of differentiation of sleepiness and mental fatigue. J. Safety Res. **72**, 173–187 (2020). https://doi.org/10.1016/j.jsr.2019.12.015

16. Islam, M.K., Rastegarnia, A., Yang, Z.: Methods for artifact detection and removal from scalp EEG: a review. Neurophysiologie Clinique/Clin. Neurophysiol. **46**(4–5), 287–305 (2016)

17. Jafari, M., Zaeri, F., Jafari, A., Najafabadi, A., Al-Qaisi, S., Hassanzadeh Rangi, N.: Assessment and monitoring of mental workload in subway train operations using physiological, subjective, and performance measures. Hum. Factors Ergon. Manuf. Serv. Ind. **30**(3), 165–175 (2020). https://doi.org/10.1002/hfm.20831

18. Kandemir, C., Handley, H.A.: Work process improvement through simulation optimization of task assignment and mental workload. Comput. Math. Organ. Theory **25**, 389–427 (2019)

19. Karameh, F.N., Dahleh, M.A.: Automated classification of EEG signals in brain tumor diagnostics. In: Proceedings of the 2000 American Control Conference. ACC (IEEE Cat. No. 00CH36334), vol. 6, pp. 4169–4173. IEEE (2000)

20. Kauppi, J.P., et al.: Towards brain-activity-controlled information retrieval: decoding image relevance from MEG signals. NeuroImage **112**, 288–298 (2015)

21. Kingphai, K., Moshfeghi, Y.: On EEG preprocessing role in deep learning effectiveness for mental workload classification. In: Longo, L., Leva, M.C. (eds.) H-WORKLOAD 2021. CCIS, vol. 1493, pp. 81–98. Springer, Cham (2021). https://doi.org/10.1007/978-3-030-91408-0_6

22. Kingphai, K., Moshfeghi, Y.: On time series cross-validation for deep learning classification model of mental workload levels based on EEG signals. In: Nicosia, G., et al. (eds.) LOD 2022. LNCS, vol. 13811, pp. 402–416. Springer, Cham (2022). https://doi.org/10.1007/978-3-031-25891-6_30

23. Lim, J., Wu, W.C., Wang, J., Detre, J.A., Dinges, D.F., Rao, H.: Imaging brain fatigue from sustained mental workload: an ASL perfusion study of the time-on-task effect. NeuroImage **49**(4), 3426–3435 (2010)

24. Lim, W.L., Sourina, O., Wang, L.P.: STEW: simultaneous task EEG workload data set. IEEE Trans. Neural Syst. Rehabil. Eng. **26**(11), 2106–2114 (2018)

25. Lim, W.L., Sourina, O., Liu, Y., Wang, L.: EEG-based mental workload recognition related to multitasking. In: 2015 10th International Conference on Information, Communications and Signal Processing (ICICS), pp. 1–4 (2015). https://doi.org/10.1109/ICICS.2015.7459834

26. Midha, S., Maior, H.A., Wilson, M.L., Sharples, S.: Measuring mental workload variations in office work tasks using fNIRS. Int. J. Hum. Comput. Stud. **147**, 102580 (2021)
27. Moshfeghi, Y., Pinto, L.R., Pollick, F.E., Jose, J.M.: Understanding relevance: an fMRI study. In: Serdyukov, P., et al. (eds.) ECIR 2013. LNCS, vol. 7814, pp. 14–25. Springer, Heidelberg (2013). https://doi.org/10.1007/978-3-642-36973-5_2
28. Moshfeghi, Y., Triantafillou, P., Pollick, F.E.: Understanding information need: an fMRI study. In: Proceedings of the 39th International ACM SIGIR conference on Research and Development in Information Retrieval, SIGIR 2016, New York, NY, USA, pp. 335–344. Association for Computing Machinery (2016)
29. Orru, G., Gobbo, F., O'Sullivan, D., Longo, L.: An investigation of the impact of a social constructivist teaching approach, based on trigger questions, through measures of mental workload and efficiency. In: McLaren, B.M., Reilly, R., Zvacek, S., Uhomoibhi, J. (eds.) Proceedings of the 10th International Conference on Computer Supported Education, pp. 292–302. SciTePress - Science and Technology Publications (2018)
30. Pandey, V., Choudhary, D.K., Verma, V., Sharma, G., Singh, R., Chandra, S.: Mental workload estimation using EEG. In: 2020 Fifth International Conference on Research in Computational Intelligence and Communication Networks (ICRCICN), pp. 83–86. IEEE (2020)
31. Paxion, J., Galy, E., Berthelon, C.: Mental workload and driving. Front. Psychol. **5** (2014). https://doi.org/10.3389/fpsyg.2014.01344
32. Polikar, R.: Ensemble learning. In: Zhang, C., Ma, Y. (eds.) Ensemble Machine Learning: Methods and Applications, pp. 1–34. Springer, New York (2012). https://doi.org/10.1007/978-1-4419-9326-7_1
33. Reid, G.B., Nygren, T.E.: The subjective workload assessment technique: a scaling procedure for measuring mental workload. In: Advances in Psychology, vol. 52, pp. 185–218. Elsevier (1988)
34. Singh, U., Ahirwal, M.K.: Mental workload classification for multitasking test using electroencephalogram signal. In: 2021 IEEE International Conference on Technology, Research, and Innovation for Betterment of Society (TRIBES), pp. 1–6 (2021). https://doi.org/10.1109/TRIBES52498.2021.9751676
35. So, W.K., Wong, S.W., Mak, J.N., Chan, R.H.: An evaluation of mental workload with frontal EEG. PLoS ONE **12**(4), e0174949 (2017)
36. Tao, J., Yin, Z., Liu, L., Tian, Y., Sun, Z., Zhang, J.: Individual-specific classification of mental workload levels via an ensemble heterogeneous extreme learning machine for EEG modeling. Symmetry **11**(7), 944 (2019)
37. Teplan, M.: Fundamental of EEG measurement. Meas. Sci. Rev. **2**, 1–11 (2002)
38. Tokuda, S., Obinata, G., Palmer, E., Chaparro, A.: Estimation of mental workload using saccadic eye movements in a free-viewing task. In: 2011 Annual International Conference of the IEEE Engineering in Medicine and Biology Society, pp. 4523–4529. IEEE (2011)
39. Tzallas, A.T., Tsipouras, M.G., Fotiadis, D.I.: Epileptic seizure detection in EEGs using time-frequency analysis. IEEE Trans. Inf. Technol. Biomed. **13**(5), 703–710 (2009)
40. Urigüen, J.A., Garcia-Zapirain, B.: EEG artifact removal-state-of-the-art and guidelines. J. Neural Eng. **12**(3), 031001 (2015)
41. Vaid, S., Singh, P., Kaur, C.: EEG signal analysis for BCI interface: a review. In: 2015 Fifth International Conference on Advanced Computing & Communication Technologies, pp. 143–147. IEEE (2015)

42. Xie, B., Salvendy, G.: Prediction of mental workload in single and multiple tasks environments. Int. J. Cogn. Ergon. **4**(3), 213–242 (2000)
43. Yang, S., Yin, Z., Wang, Y., Zhang, W., Wang, Y., Zhang, J.: Assessing cognitive mental workload via EEG signals and an ensemble deep learning classifier based on denoising autoencoders. Comput. Biol. Med. **109**, 159–170 (2019)
44. Yin, Z., Zhang, J.: Cross-session classification of mental workload levels using EEG and an adaptive deep learning model. Biomed. Signal Process. Control **33**, 30–47 (2017). https://doi.org/10.1016/j.bspc.2016.11.013
45. Zarjam, P., Epps, J., Lovell, N.H.: Beyond subjective self-rating: EEG signal classification of cognitive workload. IEEE Trans. Auton. Ment. Dev. **7**(4), 301–310 (2015)
46. Zhang, C., Ma, Y.: Ensemble Machine Learning: Methods and Applications. Springer, New York (2012)
47. Zhao, R., Yan, R., Wang, J., Mao, K.: Learning to monitor machine health with convolutional bi-directional LSTM networks. Sensors **17**(2), 273 (2017)
48. Zheng, X., Chen, W., You, Y., Jiang, Y., Li, M., Zhang, T.: Ensemble deep learning for automated visual classification using EEG signals. Pattern Recogn. **102**, 107147 (2020). https://doi.org/10.1016/j.patcog.2019.107147

Decision-Making over Compact Preference Structures

Andrea Martin[1] and Kristen Brent Venable[1,2(✉)]

[1] University of West Florida, Pensacola, FL 32514, USA
bvenable@uwf.edu
[2] Institute for Human and Machine Cognition, Pensacola, FL 32502, USA

Abstract. We consider a scenario where a user must make a set of correlated decisions and we propose a computational cognitive model of the deliberation process. We assume the user compactly expresses her preferences via soft constraints and we study how a psychology-based model of human decision-making, namely Multi-Alternative Decision Field Theory (MDFT), can be applied in this context. We design and study sequential and synchronous procedures which combine local decision-making on each variable, with constraint propagation, as well as a one-shot approach. Our experimental results, which focus on tree-shaped Fuzzy Constraint Satisfaction Problems, suggest that decomposing the decision process along the preference structure allows to find solutions of high quality in terms of preferences, maintains MDFT's ability to replicate behavioral effects and is more efficient in terms of computational cost.

Keywords: Cognitive modeling · Decision-making · Soft Constraints

1 Introduction

Preferences play a crucial role in decision-making. As such they have been studied in multiple disciplines, such as psychology, philosophy, business and marketing. Preference reasoning has grown into an important topic in Artificial Intelligence [15] where preference models are currently used in many applications, such as scheduling [1] and recommendation engines [7]. On the other hand, in psychology, Multi-Alternative Decision Field Theory (MDFT) [14] formalizes the evolution of preferences during the process of deliberation. While MDFT has mainly tackled the problem of making a single decision, both in real life and in artificial intelligence applications, scenarios are more complex and it can be helpful to organize them in a combinatorial structure over which decisions can be applied sequentially or synchronously. To this end, we propose here an approach which brings together compact preference models and MDFTs.

In particular, we use soft constraints to support a deliberation process performed through MDFT. Preferences over the set of alternatives according to each attribute are modeled by soft constraint satisfaction problems defined over the same set of variables. We compare the results obtained applying the deliberation process to each variable in a fixed sequential order and in a synchronous fashion to those obtained by a single deliberation step over the entire combinatorial domain. As expected the sequential approaches outperform the one-shot

G. Nicosia et al. (Eds.): LOD 2023, LNCS 14506, pp. 373–387, 2024.
https://doi.org/10.1007/978-3-031-53966-4_28

procedure in terms of time. Moreover, they focus on a relatively small set of alternatives compared to the size of the choice space and the deliberated assignments are, on average, of high quality with respect to the preferences in the soft constraint problems representing the attributes.

This work achieves two objectives: the first one is to provide a computational model which can help understand human decision-making over complex domains; the second one is to investigate MDFT as means of incorporating a form of uncertainty into the soft constraint formalism. To the first end, we note that our fixed-order sequential approach and the synchronous approach appear to be cognitively more plausible as it is far more likely for humans to break complex decisions into interconnected sub-problems, rather than try to deliberate directly on a large set of complex objects. We further support this claim by showing that the fixed-order sequential procedure is capable of capturing the three well known behavioral effects which have been observed in human decision-making, namely, the similarity, attraction and compromise effect.

Our approach can be used to provide a high-fidelity model of how humans make complex decisions as well as a predictive and simulation tool that enables more personalized decision support. This will play a crucial role in the design of AI-based decision-support systems [12] which increasingly gained traction is several domains, such as health-care [19] and policy-making [16]. In terms, of using MDFT as a paradigm to extend soft constraints, we note that the MDFT simulation of deliberation, which is modeled as an oscillation between evaluation criteria for different scenarios, can be used to model uncertainty about preferences via the attributes and their weights, which cannot otherwise be captured by a fixed soft constraint problem. While a large literature has been dedicated to studying the human deliberation process in settings with few, unstructured alternatives, understanding how humans make decisions over complex or combinatorial structures is for the most part an unexplored topic. In a recent paper, [17] the authors developed a theory of decision-making on combinatorial domains. They adopt compositionally structured utility functions as a means to represent preferences and they use probabilistic reasoning to predict preferences over new unseen alternatives. Our goal is to model the variability which is observed in human decision-making when preferences come from different criteria as opposed to predicting preferences over unseen options from known ones.

Extensions of MDFT to more complex domains have been proposed in the literature, for example addressing dynamic environments [10]. In [5] an extension is presented to model trust and reliance on automation. Similarly to what we consider here, the authors study an iterated decision process where previous decisions may influence later ones. However, while we represent the correlation between decisions via soft constraints they link sequential decision processes by a dynamic updating of beliefs.

2   Background

2.1   Multi-alternative Decision Field Theory (MDFT)

MDFT [2] models preferential choice as an accumulative process in which the decision maker attends to a specific attribute at each time point in order to derive comparisons among options, and updates the estimate of the decision maker's preferences accordingly. Ultimately the accumulation of those preferences forms the decision maker's choice. In MDFT an agent is confronted with multiple options and equipped with an initial personal evaluation for them according to different criteria, called attributes. For example, a student who needs to choose a main course among those offered by the cafeteria will have in mind an initial evaluation of the options in terms of how tasty and healthy they look. More formally, MDFT, in its basic formulation [14], is composed of the following elements.

Personal Evaluation: Given a set of options $O = \{o_1, \ldots, o_k\}$ and set of attributes $A = \{a_1, \ldots, a_l\}$, the subjective value of option o_i on attribute a_j is denoted by m_{ij} and stored in matrix $\mathbf{M}$. In our example, let us assume that the cafeteria options are *Salad (S)*, *Burrito (B)* and *Vegetable pasta (V)*. Matrix $\mathbf{M}$, containing the student's preferences, could be defined as shown in Fig. 1 (left), where rows correspond to the options (S, B, V) and the columns to the attributes *Taste* and *Health*.

$$\mathbf{M} = \begin{vmatrix} 1 & 5 \\ 5 & 1 \\ 2 & 3 \end{vmatrix} \quad C = \begin{vmatrix} 1 & -1/2 & -1/2 \\ -1/2 & 1 & -1/2 \\ -1/2 & -1/2 & 1 \end{vmatrix} \quad S = \begin{vmatrix} +0.9000 & 0.0000 & -0.0405 \\ 0.0000 & +0.9000 & -0.0047 \\ -0.0405 & -0.0047 & +0.9000 \end{vmatrix}$$

Fig. 1. Example of Evaluation ($\mathbf{M}$), Contrast ($\mathbf{C}$), and Feedback ($\mathbf{S}$) matrices.

Attention Weights: Attention weights express the attention allocated to each attribute at a particular time t during deliberation. We denote them by vector $\mathbf{W}(t)$, where $W_j(t)$ represents the attention to attribute a_j at time t. We adopt the common simplifying assumption that, at each point in time, the decision maker attends to only one attribute [14]. Thus, $W_j(t) \in \{0, 1\}$ and $\sum_j W_j(t) = 1$, $\forall t, j$. In our example, where we have two attributes, at any point in time t, we will have $\mathbf{W}(t) = [1, 0]$, or $\mathbf{W}(t) = [0, 1]$, representing that the student is attending to, respectively, *Taste* or *Health*. The attention weights change across time according to a stationary stochastic process with probability distribution $\mathbf{p}$, where p_j is the probability of attending to attribute a_j. In our example, defining $p_1 = 0.55$ and $p_2 = 0.45$ means that at each point in time, the student will be attending *Taste* with probability 0.55 and *Health* with probability 0.45; i.e., *Taste* matters slightly more than *Health* to this student.

Contrast Matrix: Contrast matrix $\mathbf{C}$ is used to compute the advantage (or disadvantage) of an option with respect to the other options. In the MDFT literature [3,4,14], $\mathbf{C}$ is defined by contrasting the initial evaluation of one alternative against the average of the evaluations of the others, as shown for the case with three options in Fig. 1 (center).

At any moment in time, each alternative in the choice set is associated with a **valence** value. The valence for option o_i at time t, denoted $v_i(t)$, represents its momentary advantage (or disadvantage) when compared with other options on some attribute under consideration. The valence vector for k options $o_1, \ldots, o_k$ at time t, denoted by column vector $\mathbf{V}(t) = [v_1(t), \ldots, v_k(t)]^T$, is formed by $\mathbf{V}(t) = \mathbf{C} \times \mathbf{M} \times \mathbf{W}(t)$. In our example, the valence vector at any time point in which $\mathbf{W}(t) = [1, 0]$, is $\mathbf{V}(t) = [1 - 7/2, 5 - 3/2, 2 - 6/2]^T$.

In MDFT, preferences for each option are accumulated across iterations of the deliberation process until a decision is made. This is done by using **Feedback Matrix S**, which defines how the accumulated preferences affect the preferences computed at the next iteration. This interaction depends on how similar the options are in terms of their initial evaluation expressed in $\mathbf{M}$. Intuitively, the new preference of an option is affected positively and strongly by the preference it had accumulated so far, while it is inhibited by the preference of other options which are similar. This lateral inhibition decreases as the dissimilarity between options increases. Figure 1 (right) shows $\mathbf{S}$ computed for our running example following the MDFT standard method described in [6] details of which we omit here.

At each time-step, the preference of each alternative is calculated by $\mathbf{P}(t + 1) = \mathbf{S} \times \mathbf{P}(t) + \mathbf{V}(t+1)$, where $\mathbf{S} \times \mathbf{P}(t)$ is the contribution of the past preferences and $\mathbf{V}(t + 1)$ is the valence computed at that iteration. Starting with $\mathbf{P}(0) = 0$, preferences are then accumulated for either a fixed number of iterations, and the option with the highest preference is selected, or until the preference of an option reaches a given threshold. In the first case, MDFT models decision-making with a *specified* deliberation time, while, in the latter, it models cases where deliberation time is *unspecified* and choice is dictated by the accumulated preference magnitude.

MDFT Model: Given set of options $O = \{o_1, \ldots, o_k\}$ and set of attributes $A = \{a_1, \ldots, a_l\}$, an MDFT Model is defined by the n-tuple $Q = \langle \mathbf{M}, \mathbf{C}, \mathbf{p}, \mathbf{S} \rangle$, where: $\mathbf{M}$ is the $k \times l$ personal evaluation matrix; $\mathbf{C}$ is the $k \times k$ contrast matrix; $\mathbf{p}$ is a probability distribution over attention weights vectors; and $\mathbf{S}$ is the $k \times k$ feedback matrix.

In this paper we make the following assumption. The contrast and feedback matrices $\mathbf{C}$ and $\mathbf{S}$ are defined in the standard way according to the MDFT literature in [3] and [6]. We will, thus, omit them in the specification of the MDFT models that will be denoted just as the pair $\langle \mathbf{M}, \mathbf{p} \rangle$ consisting of the evaluation matrix and the attention weights probability distribution.

Different runs of the same MDFT model may return different choices due to the uncertainty on the attention weights distribution. If we run the model a sufficient number of times on the same set, we obtain a proxy of its choice probability distribution. MDFT can effectively replicate bounded-rational behaviors observed in humans [3], such as, the *similarity effect*, where adding a new similar candidate decreases the probability of an option to be chosen, the *attraction effect*, where adding a similar but slightly worst option increases the chances of an option to be selected and, the *compromise effect*, where including a diametrically opposed option may increase the choice probability of a compromising one [14].

2.2   Soft Constraint Satisfaction Problems

To represent preferences over large combinatorial domains we consider soft constraints [11]. A soft constraint requires a set of variables and associates each instantiation of its variables to a value from a partially ordered set. More precisely, the underlying algebra is a c-semiring which consist of the following, $\langle A, +, \times, 0, 1 \rangle$, where A is the set of preference values, binary operator $+$ induces an ordering over A (where $a \leq b$ iff $a + b = b$), binary operator $\times$ is used to combine preference values, and 0 and 1 are respectively the worst and best element of A. A Soft Constraint Satisfaction Problem (SCSP) is a tuple $\langle V, D, C, A \rangle$ where V is a set of variables, D is the domain of the variables and C is a set of soft constraints (each one involving a subset of V) associating values from A to tuples of values of the constrained variables. An instance of the SCSP framework is obtained by choosing a specific preference structure. Choosing $S_{FCSP} = \langle [0, 1], max, min, 0, 1 \rangle$, where FCSP stands for Fuzzy Constraint Satisfaction Problem [11,18], means that $A = [0, 1]$, that is, preference values are in $[0, 1]$, and we want to maximize the minimum preference value, the worst preference value is 0, and the best preference value is 1. We use FCSPs because they have been extensively studied as a way to model degrees of preference satisfaction. Most importantly FCSPs have computationally tractable sub-classes for which finding and optimal solution and preference propagation can be done in polynomial time [11].

Figure 2 shows the constraint graph of an FCSP where $V = \{X, Y, Z\}$, $D = \{a, b\}$ and $C = \{c_X, c_Y, c_Z, c_{XY}, c_{YZ}\}$. Each node models a variable and each arc models a binary constraint, while unary constraints define preferences over variables' domains. For example, c_Y is defined by the preference function f_Y that associates preference value 0.4 to $Y = a$ and 0.7 to $Y = b$. Default constraints such as c_X and c_Z, where all variable assignments get value 1, will often be omitted in the following examples.

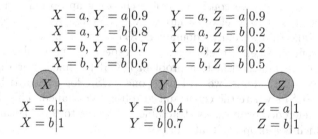

$$
\begin{array}{ll}
X = a, Y = a \mid 0.9 & Y = a, Z = a \mid 0.9 \\
X = a, Y = b \mid 0.8 & Y = a, Z = b \mid 0.2 \\
X = b, Y = a \mid 0.7 & Y = b, Z = a \mid 0.2 \\
X = b, Y = b \mid 0.6 & Y = b, Z = b \mid 0.5
\end{array}
$$

$$
\begin{array}{lll}
X = a \mid 1 & Y = a \mid 0.4 & Z = a \mid 1 \\
X = b \mid 1 & Y = b \mid 0.7 & Z = b \mid 1
\end{array}
$$

Fig. 2. Example of an FCSP.

Given an SCSP, a complete assignment to all of its variables is associated with a preference value obtained combining via the $\times$ operator the preferences associated to its projection on each of the constraints. In our example where $\times$ is min, the preferences of $(X = a, Y = b, Z = b)$ is $0.5 = min(1, 0.9, 0.7, 0.5, 1)$.

These global preferences induce, in general, a partial order with ties over the set of complete assignments, which is total for FCSPs. Often, solving an

SCSP is interpreted as finding an optimal solution, that is, a complete assignment with an undominated preference value (e.g. $(X = a, Y = b, Z = b)$ in this example). Unless certain restrictions are imposed, such as a tree-shaped constraint graph, finding an optimal solution is NP-hard [11]. Constraint propagation may facilitate the search for an optimal solution. In particular, given a variable ordering o, a FCSP is directional arc-consistent (DAC) if, for any two variables X and Y linked by a constraint, such that X precedes Y in ordering o, we have that, for each a in the domain of X, $f_X(a) = max_{b \in D(Y)}(min(f_X(a), f_{XY}(a, b), f_Y(b)))$, where f_X, f_Y, and f_{XY} are the preference functions of c_X, c_Y and c_{XY}. When the constrained graph is tree shaped and the variable ordering is compatible with the father-child relation of the tree, DAC is enough to find the preference of an optimal solution [11]. Such an optimal preference is the best preference in the domain of the root variable after DAC and an optimal solution, can be found by a backtrack-free search which instantiates variables in the same order used for DAC. In our running example, if we choose the variable ordering $\langle X, Y, Z \rangle$, achieving DAC means first enforcing the property over the constraint on Y and Z and then over the constraint on X and Y. The first phase modifies the preference value of $Y = b$ to $max(min(f_Y(b), f_{Y,Z}(b, a), f_Z(a)), min(f_Y(b), f_{Y,Z}(b, b), f_Z(b))) = max(min(0.7, 0.2, 1), min(0.7, 0.5, 1)) = max(0.2, 0.5) = 0.5$. Similarly, the second phase sets the preference values of both $X = a$ and $X = b$ to 0.5. We also note that, by achieving DAC w.r.t. ordering o, we obtain a total order with ties over the values of the first variable in o, where each value is associated to the preference of the best solution having such a variable instantiated to such a value.

3   Problem Definition

The deliberation tasks we consider involve choosing an option consisting of a complete assignment to a set of variables, $X = \{X_1, \ldots, X_n\}$, where each variable X_i can take values from its domain $D(X_i) = \{\sigma_1, \ldots, \sigma_m\}$.

To represent the agent's preferences with respect to a particular attribute we use a fuzzy constraint satisfaction problem (FCSP) defined over the variables in X (see Sect. 2.2). Thus, we will have one FCSP for each attribute. Here, we consider FCSPs where the constraint-graph is tree-shaped. This, in addition to making preference propagation tractable, allows us to topologically sort the variables which is a property that will use.

As an example, consider a scenario where a user has to decide on what to have for dinner. Her preferences in terms of attributes *Taste* and *Health* on the available options are expressed by the two FCSPs depicted in Fig. 3. Both FCSPs are defined on same set of variables, that is, *Entree*, denoted by E, with domain $\{m, v\}$ for *meat* and *vegetables*, respectively, and *Drink*, denoted with D, with values $\{w, r\}$ for *white* wine and *red* wine. From Fig. 3 we see, for example, that the user, prefers the taste of meat to the taste of vegetables and prefers red wine with it. On the other hand, the user is also aware that vegetables are healthier than meat, and that red wine is healthier than white wine.

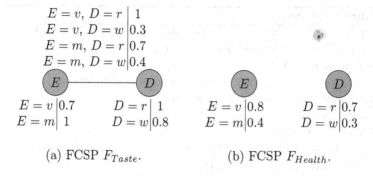

$E = v, D = r$ | 1
$E = v, D = w$ | 0.3
$E = m, D = r$ | 0.7
$E = m, D = w$ | 0.4

$\boxed{E}$────────$\boxed{D}$

$E = v$ | 0.7 $D = r$ | 1 $E = v$ | 0.8 $D = r$ | 0.7
$E = m$ | 1 $D = w$ | 0.8 $E = m$ | 0.4 $D = w$ | 0.3

(a) FCSP F_{Taste}.   (b) FCSP F_{Health}.

Fig. 3. *Taste* and *Health* preferences expressed as Fuzzy CSPs.

We assume that the decision on which value to assign to each variable X_i is taken by running one or more MDFT models and in what follows we describe three different approaches. In the first two methods each variable X_i has an associated MDFT model $\mathbf{Q}_{\mathbf{X_i}} = \langle \mathbf{M}_{\mathbf{X_i}}, \mathbf{p}_{\mathbf{X_i}} \rangle$ where the personal evaluations in matrix $\mathbf{M}_{\mathbf{X_i}}$ is obtained from the FCSPs. In the third approach we use a single MDFT model, $\mathbf{Q} = \langle \mathbf{M}, \mathbf{p} \rangle$, where the options are the complete assignments and the evaluations in the $\mathbf{M}$ matrix are the preferences associated to the assignments by the FCSPs corresponding top attributes.

4   Decision-Making Procedures

We now describe the three decision-making procedures.

4.1   Fixed-Order Sequential Decision-Making: Algorithm FO-Seq-DM

The first approach to deliberation we consider is to sequentially find a value for each variable X_i via a run of an MDFT associated with it following a fixed order O. Algorithm 1 shows the steps of this process which we call FO-Seq-DM.

We recall that our goal is to find a complete assignment to a set of n variables $X = \{X_1, \ldots, X_n\}$ based on preferences expressed in terms of J attributes. FO-Seq-DM takes in input one FCSP for each attribute and a total order O over X. In addition, it also takes in input probability distribution $\mathbf{p}$ which will serve as the default probability distribution for the attention weights in the MDFT models associated to the variables.

FO-Seq-DM returns a complete assignment to the variables in X which is stored in array of variable-value assignments *decisions* during the procedure. Array *decisions*. For each variable X_i in order O (line 2):

– We consider the constraint graph of each FCSP, F_j, with X_i as the root, and we return a topological order of X, denoted O_j^i, via function Topological-Order (line 3).

Algorithm 1. FO-Seq-DM

Input: $F = \{F_1, \ldots, F_J\}$: a set of tree-shaped FCSPs over variable set $X = \{X_1, \ldots, X_n\}$; //J *is the number of attributes*
O: a total order over X;
$\mathbf{p} = (p_1, \ldots, p_J)$: an probability distribution over J elements;
Output: An assignment to the variables in X;

 1: Var-value decisions = []
 2: **for** each X_i in order O **do**
 3:     O_j^i = Topological-order(X,F_j,X_i);
 4:     DAC(F_j,$r(O_j^i)$), $\forall j \in \{1, \ldots J\}$;
 5:     Evaluation Matrix $\mathbf{M_i}$ = Pref-extract(F, X_i);
 6:     Attention weights probability distribution $\mathbf{p_{X_i}} = \mathbf{p}$;
 7:     MDFT Model $\mathbf{Q_{X_i}} = \mathbf{Q}(\mathbf{M_{X_i}}, \mathbf{p_{X_i}})$;
 8:     $\sigma_i = \mathbf{Q_{X_i}}(D(X_i))$;
 9:     Append(decisions,$X_i = \sigma_i$);
10:     Update(F_j,$X_i = \sigma_i$), $\forall j \in \{1, \ldots J\}$;
11: **end for**
12: **return** decisions

- We then enforce DAC on each FCSP F_j following O_j^i in reverse order, denoted $r(O_j^i)$. This ensures that the preferences are propagated to the values in the domain of X_i.
- We define the evaluation matrix $\mathbf{M_{X_i}}$ for the MDFT model of variable X_i in line 5 via function Pref-extract. This function takes in input all of the FCSPs and defines a $|D(X_i)| \times J$ matrix where the j-th column contains the preferences over the domain of X_i, $D(X_i)$, according to FCSP F_j.
- In lines 6 and 7 we complete the initialization of MDFT model $\mathbf{Q_{X_i}}$ associated to X_i by setting the attention weights probability distribution to the default one, $\mathbf{p}$, given in input.
- At this point we run MDFT model $\mathbf{Q_{X_i}}$ to select a value σ_i in $D(X_i)$ to be assigned to X_i. This is denoted as $\mathbf{Q_{X_i}}(D(X_i))$ in line 8.
- In line 9 we add the new assignment to *decisions* and in line 10 we update the FCSPs with the new assignment by setting to 0 all the preferences of all domain values of X_i other than σ_i.
- After each variable is assigned a value as described above, the complete assignment to the variables in X is returned in line 12.

We observe that FO-Seq-DM could be easily generalized in several ways. For example, while we have assumed that the set of attributes is the same for all of the variables, we could easily adapt FO-Seq-DM to handle the case in which different attributes are involved in the decision of each variables. This could be implemented by allowing FCSPs to involve subsets of variables. Similarly, each variable could also have a specific attention weight probability distribution. The generalization to non tree-shaped FCSPs would be possible by replacing DAC with Full Arc consistency [11] which would come, however, at an additional computational cost.

Let us now take a closer look at how FO-Seq-DM works, considering the example depicted in Fig. 3 and described in Sect. 3. In line 4 of FO-Seq-DM, DAC is enforced on FCSPs F_{Health} and F_{Taste} with root E. FCSP F_{Taste} after DAC is shown in Fig. 4(a). The $Health$ FCSP is already DAC since there is no constraint between the two variables. The values in matrix $\mathbf{M_E}$ are defined using the four preferences associated to the domain values in $D(E)$ in F_{Taste} and F_{Health}, respectively. Let us assume, that running $Q_E(D(E))$ returns $E = v$. That is, by running $Q_E(D(E))$ we accumulate the preference of $E = m$ and $E = v$ alternating between considering taste and health and then choosing the option with highest preference after a number of iterations. This choice is implemented in both FCSPs by setting the preference of $E = m$ to 0.

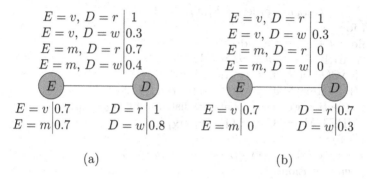

(a)   (b)

Fig. 4. (a) F_{Taste} after update with assignment $E = v$ and DAC. (b) F_{Taste} after update with assignment $E = v$ and DAC.

The next variable to be considered in order O is D. DAC propagates the effect of the assignment to E to the preferences of values in the domain of D. This step modifies the preference value of $D = r$ to 0.3. The outcome of this step is shown in Fig. 4(b). Finally, MDFT $\mathbf{Q_D}$ is applied to the domain of D to select an assignment for D. For example, if MDFT($\mathbf{Q_D}$) $= r$ then, overall, the choice is to have vegetables and red wine.

4.2 Distributed Synchronous Decision-Making: Algorithm Dist-Sync-DM

We now consider a setting where the deliberation on which value to assign the variables in X happens synchronously and the order of the decisions is not fixed but, rather, based on which process terminates first. For example, this method could model a setting where multiple users are in charge of deciding values for different variables. We assume that the MDFT process that reaches the preferential threshold first results in an assignment which is adopted and propagated before deliberation is restarted on all the remaining variables. Considering the example mentioned before, deliberation would start simultaneously on both the food and drink variables.

Algorithm 2 shows the steps of this procedure, which we call Dist-Sync-DM, for Distributed Synchronous Decision-Making.

Algorithm 2. Dist-Sync-DM

Output: $F = \{F_1, \ldots, F_J\}$: a set of tree-shaped FCSPs over variable set $X = \{X_1, \ldots, X_n\}$; $//J$ *is the number of attributes*
$\mathbf{p} = (p_1, \ldots, p_J)$: an probability distribution over J elements;
Output: An assignment to the variables in X;

1: var *decisions* = []
2: var set *notAssigned* = $\{X_1, \ldots, X_n\}$
3: **for** each X_i in *notAssigned* **do**
4:     O_j^i = Topological-order(X, F_j, X_i);
5: **end for**
6: **while** *notAssigned* $\neq \emptyset$ **do**
7:     **for** each X_i in *notAssigned* **do**
8:         DAC($F_j, r(O_j^i)$), $\forall j \in \{1, \ldots J\}$;
9:         Evaluation Matrix $\mathbf{M_i}$ = Pref-extract(F, X_i);
10:         Attention weights probability distribution $\mathbf{px_i} = \mathbf{p}$;
11:         MDFT Model $\mathbf{Q_{X_i}} = \mathbf{Q}(\mathbf{M_{X_i}}, \mathbf{px_i})$;
12:     **end for**
13:     σ_h = Sync-MDFT($\{\mathbf{Q}_{X_i} | X_i \in notAssigned\}$)
14:     append(decisions, $X_h = \sigma_h$);
15:     Update($F_j, X_h = \sigma_h$), $\forall j \in \{1, \ldots J\}$;
16:     *notAssigned* = *notAssigned* $\setminus \{X_h\}$;
17: **end while**
18: **return** decisions

In lines 3–4 we use function Topological-order described in Sect. 4.1 to compute the topological orders over the variables for every FCSP and for each variable considered as the root. These orders are used for preference propagation in line 8. In line 13 function Sync-MDFT launches all of the MDFT models, $\mathbf{Q}_{X_i}$, on the domains of the corresponding variables and returns the first value σ_h to be selected by any of the deliberation processes. Notice that we can designate any of the topological orders computed in line 4 for the purpose of breaking ties in case of concurrent termination. Once all variables have been assigned, array *decisions* is returned.

4.3   One-Shot Approach

Another possibility for making an overall decision is to run the deliberation process only once over the set of candidate options consisting of all complete assignments. Although this approach is cognitively less plausible as well as computationally intractable, we consider it as a means of comparison with the sequential and synchronous approaches.

We assume a single decision to be made over the combinatorial structure. In other words, each complete assignment to all variables in X is treated as an option, and its preference according to each FCSP is its evaluation with respect to the attribute represented by the FCSP. For instance, In our (small) running example, $(E = m, D = r)$ has preference 0.7 according to attribute Taste and 0.4 for attribute Health. We evaluate all of the complete assignments via the FCSPs and we use such preferences to populate matrix $\mathbf{M}$.

We call this approach One-shot-DM. The pseudo-code is omitted in the interest of space. We note that unlike FO-Seq-DM and Dist-Sync-DM, One-shot-DM does not require for the FCSPs in input to be tree-shaped. In fact, given a complete assignment to the variables of a FCSP, computing its preference is linear in the number of constraints regardless of the shape of the constraint graph [11]. However, we also note that the complexity of One-Shot-DM is already exponential in the number of variables as it assumes that the preferences of all possible complete assignments are computed with respect to all FCSPs.

5   Experimental Results

We have implemented the sequential, synchronous and one-shot decision-making approaches and we have tested them on randomly generated problems. We consider a setting with two attributes. Thus, each generated instance comprises of a pair of tree-shaped fuzzy problems, one for each attribute, defined over the same set of binary variables. We consider a number of variables ranging between 2 and 8 with increments of 2 and a constraint tightness of 20%, meaning that, in each constraint, 20% of the tuples are associated with preference 0. For both of the approaches, the values for the $\mathbf{C}$ and $\mathbf{S}$ matrices are defined as described in Sect. 2.1 and according to the literature [20]. In all of the experiments involving FO-Seq-DM and One-Shot-DM we stop deliberation after a pre-determined number of iterations. For Dist-Sync-DM, we use a threshold on the accumulated preferences to halt deliberation.

Choice Analysis. After running the fixed-order, synchronous and one-shot approaches the same number of times on a given instance, options would ideally be returned as deliberated with similar frequencies by all approaches. Not surprisingly, this is not the case. By decomposing the decision-making into a set of local deliberation steps, we lose some of the information and we incur in a situation similar to the discursive dilemma in judgment aggregation [8] and a similar effect observed in sequential voting [9,13].

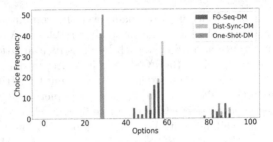

Fig. 5. Frequency with which options are returned as deliberated in one instance of eight variables.

Figure 5 shows the frequency with which solutions are returned in a problem with 8 variables after running both approaches 100 times. As we can see, out of the 2^8 possible choices, only 14 are returned by the fixed-order approach, 8 by the synchronous approach and 3 are returned by the one-shot one. We performed this same experiment on 100 different instances involving 8 variables and we observed a similar trend. The average size of the set containing solutions returned at least once by FO-Seq-DM and Dist-Sync-DM was 15.6 and 14.4 respectively, over a total of 256, while the size of the set returned by One-Shot-DM was 9. It is not surprising that the fixed-order and synchronous approaches have slightly more variability in their outputs. In fact, the uncertainty modeled by the probability distribution over the weights of the attributes affects the decision at each variable. Instead, in the case of the one-shot approach it only contributes once and at the global level. Nonetheless, all methods are capable of focusing only a few alternatives and manage to efficiently weed out unattractive candidates.

This is further corroborated by an analysis we have performed on the relationship between the options deliberated by the MDFT-based decision-making procedures and the optimal solutions of the FCSPs (that is, the most preferred complete assignments according to each attribute). The results, obtained from the same set of 100 instances, are shown in Fig. 6. In most cases the one-shot deliberation process returns an option which is optimal in at least one of the

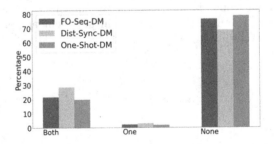

Fig. 6. Percentage of times the decision-making procedures return options that are not optimal or optimal in one or optimal in both FCSPs.

FCSPs. This suggests that the subjective preferences given in input by matrix **M** have the effect of focusing the deliberation process only on solutions which are highly ranked by at least one attribute. The results of the sequential approach, instead, reflect the decomposition of the decision process. In fact, running the deliberation process variable by variable implies applying the effect of both FCSPs variable by variable. The assignments selected for previous variables affect future deliberations on those which are connected to them via the constraints. We also computed the average difference between the preferences of the options returned by the procedures and the preference of the optimal solutions of the FCSPs. This difference was 0.14 for FO-Seq-DM, 0.11 for Dist-Sync-DM and 0.19 for One-Shot-DM, when averaged over 100 instances. The averages for the non-sequential approach are aligned with the fact that it often returns an option which is optimal in at least one FCSP. We note that the distance from optimal of the options deliberated by the fixed-order and synchronous approaches is also very small. Thus, it appears that the preference propagation obtained via DAC is sufficient to guarantee the selection of a solution which is of high quality for both attributes. Moreover, the fixed-order and synchronous approaches can be seen as better compromising between possible conflicting attributes.

Behavioral Effects. A key advantage of MDFT is its ability to replicate behavioral effects such as, the similarity, attraction and compromise effect (see Sect. 2.1), which are observed in human decision-making. An interesting question is whether using MDFT in a decoupled way, as in the sequential and synchronous approach can still replicate the behavioral effects over complete solutions. To this end we have further investigated the FO-Seq-DM. In fact, we note that a run of Dist-Sync-DM can be seen as running FO-Seq-DM with a specific variable ordering. Thus investigating FO-Seq-DM with different variable orderings suffices. In this set of experiments we consider problems with two attributes and three variables. Each generated instance comprises of a pair of tree-shaped FCSPs over the same set of variables. We consider variable domain sizes between 2 and 8 with increments of 2 and a number of deliberation iterations ranges between 20 and 100. We replicate the scenarios corresponding to the effects in our combinatorial setting. In particular, we consider settings in which the FCSPs have only up to 3 solutions as well as settings where the effects are observed on a subset of solutions among many others. In both cases we preferences in the constraints match the targeted behavior.

In all of the experiments we define two options A and B, each corresponding to a complete variable assignment, with asymmetric preferences in the two FCSPs (i.e., A is hugely preferred in one and disliked in the other FCSP and viceversa for B). A third option C is introduced in a separate set of instances with a preference defined by the constraints according to the effect that we want to observe. In the similarity case C's preferences are similar to A's in both FCSPs, in the attraction effect C's preferences are slightly below A's and in the compromise effect they are between A's and B's preferences in both FCSPs. As it can be seen, our model reproduces all three effects for the case of decision-making over complex structures. In Fig. 7[a], introducing C decreases the probability for A being chosen. In Fig. 7[b], C's introduction increases the probability for A being chosen and in Fig. 7[c], C is the favored compromising option.

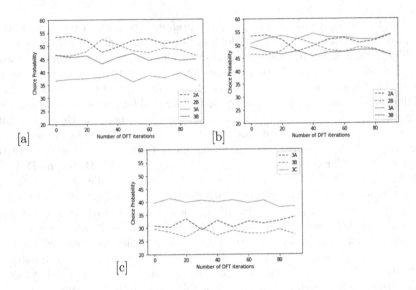

Fig. 7. [a] Similarity, [b] Attraction and [c] Compromise effect: choice probability over 100 runs as a function of the number of deliberation iterations. 2A/2B probability of choosing option A/B without C. 3A/3B probability of choosing A/B with C available.

Running Time. We conclude with a note on running time. Not surprisingly, both sequential and synchronous procedures are faster and comparable to each-other. The gap with the one-shot approach in terms of time becomes exponentially larger as the number of variables grows. For example, the average execution time of FO-Seq-DM and Dist-Sync-DM on problems with 8 variables is around 0.001 s while it is 2.7 for the one-shot approach[1].

6 Future Directions

In this work we investigated the task of modeling human choice over complex domains by leveraging the psychological cognitive model MDFT and fuzzy constraint satisfaction problems. In the future we plan to further validate our results with behavioral experiments as well as leverage our approach to learn local preferences from observing global choices.

References

1. Bartak, R., Morris, R.A., Venable, K.B.: An Introduction to Constraint-Based Temporal Reasoning. Synthesis Lectures on Artificial Intelligence and Machine Learning. Morgan & Claypool Publishers (2011)
2. Busemeyer, J.R., Diederich, A.: Survey of decision field theory. Math. Soc. Sci. **43**(3), 345–370 (2002)

---

[1] System Specifications: 2.3 GHz 18-Core Intel Xeon W, 256 GB 2666 MHz DDR4.

3. Busemeyer, J.R., Townsend, J.T.: Decision field theory: a dynamic-cognitive approach to decision making in an uncertain environment. Psychol. Rev. **100**(3), 432 (1993)
4. Busemeyer, J.R., Gluth, S., Rieskamp, J., Turner, B.M.: Cognitive and neural bases of multi-attribute, multi-alternative, value-based decisions. Trends Cogn. Sci. **23**(3), 251–263 (2019)
5. Gao, J., Lee, J.D.: Extending the decision field theory to model operators' reliance on automation in supervisory control situations. IEEE Trans. Syst. Man Cybern. Part A Syst. Hum. **36**(5), 943–959 (2006)
6. Hotaling, J.M., Busemeyer, J.R., Li, J.: Theoretical developments in decision field theory: comment on Tsetsos, Usher, and Chater (2010). Psychol. Rev. (2010)
7. Jannach, D., Zanker, M., Felfernig, A., Friedrich, G.: Recommender Systems: An Introduction, 1st edn. Cambridge University Press, New York (2010)
8. Kornhauser, L.A., Sager, L.G.: Unpacking the court. Yale Law J. **96**, 82–117 (1986)
9. Lang, J., Xia, L.: Sequential composition of voting rules in multi-issue domains. Math. Soc. Sci. **57**, 304–324 (2009)
10. Lee, S., Son, Y.-J., Jin, J.: Decision field theory extensions for behavior modeling in dynamic environment using Bayesian belief network. Inf. Sci. **178**(10), 2297–2314 (2008)
11. Meseguer, P., Rossi, F., Schiex, T.: Soft constraints. In: Rossi, F., Van Beek, P., Walsh, T. (eds.) Handbook of Constraint Programming. Elsevier (2005)
12. Phillips-Wren, G.: Ai tools in decision making support systems: a review. Int. J. Artif. Intell. Tools **21**, 04 (2012)
13. Dalla Pozza, G., Pini, M.S., Rossi, F., Venable, K.B.: Multi-agent soft constraint aggregation via sequential voting. In: Proceedings of IJCAI 2011, pp. 172–177 (2011)
14. Roe, R., Busemeyer, J.R., Townsend, J.T.: Multi-alternative decision field theory: a dynamic connectionist model of decision-making. Psychol. Rev. **108**, 370–392 (2001)
15. Rossi, F., Venable, K.B., Walsh, T.: A Short Introduction to Preferences: Between Artificial Intelligence and Social Choice. Morgan and Claypool (2011)
16. Sachan, S., Yang, J.-B., Xu, D.-L., Benavides, D.E., Li, Y.: An explainable AI decision-support-system to automate loan underwriting. Expert Syst. Appl. **144**, 113100 (2020)
17. Tenenbaum, J.B., Gershman, S.J., Malmaud, J.: Structured representations of utility in combinatorial domains. Am. Psychol. Assoc. **4**, 67–86 (2017)
18. Schiex, T.: Possibilistic constraint satisfaction problems or "how to handle soft constraints?" In: Proceedings of UAI 1992, pp. 268–275 (1992)
19. Shaikh, F., et al.: Artificial intelligence-based clinical decision support systems using advanced medical imaging and radiomics. Curr. Probl. Diagn. Radiol. **50**(2), 262–267 (2021)
20. Tsetsos, K., Usher, M., Chater, N.: Preference reversal in multiattribute choice. Psychol. Rev. **117**, 1275–1293 (2010)

User-Like Bots for Cognitive Automation: A Survey

Habtom Kahsay Gidey[1]([✉])(iD), Peter Hillmann[1](iD), Andreas Karcher[1], and Alois Knoll[2](iD)

[1] Universität der Bundeswehr München, Munich, Germany
{habtom.gidey,peter.hillmann,andreas.karcher}@unibw.de
[2] Technische Universität München, Munich, Germany
knoll@in.tum.de

Abstract. Software bots have attracted increasing interest and popularity in both research and society. Their contributions span automation, digital twins, game characters with conscious-like behavior, and social media. However, there is still a lack of intelligent bots that can adapt to the variability and dynamic nature of digital web environments. Unlike human users, they have difficulty understanding and exploiting the affordances across multiple virtual environments.

Despite the hype, bots with human user-like cognition do not currently exist. Chatbots, for instance, lack situational awareness on the digital platforms where they operate, preventing them from enacting meaningful and autonomous intelligent behavior similar to human users.

In this survey, we aim to explore the role of cognitive architectures in supporting efforts towards engineering software bots with advanced general intelligence. We discuss how cognitive architectures can contribute to creating intelligent software bots. Furthermore, we highlight key architectural recommendations for the future development of autonomous, user-like cognitive bots.

Keywords: software bot · cognitive architecture · cognitive automation

1 Introduction

Software bots are becoming an integral part of automation and social computing. Digital platforms, including software ecosystems and cyber-physical systems, are growing increasingly complex. The complexity of diverse digital systems can overwhelm even expert human users [1]. In such scenarios, software agents acting as autonomous users can assist in automating human user activities. As a result, there is a growing interest in augmenting bots into software-intensive business, social, or industrial environments for cognitive automation [2]. In Industry 4.0 (I4.0) and digital twins (DTs), software agents are playing a crucial role in enabling smart factories with higher flexibility, efficiency, and safety [3–5]. Additionally, in the service industry, robotic process automation (RPA) leverages bots to automate business processes [6].

G. Nicosia et al. (Eds.): LOD 2023, LNCS 14506, pp. 388–402, 2024.
https://doi.org/10.1007/978-3-031-53966-4_29

Software development bots, also known as DevBots, are making their mark in automated software engineering [7,8]. It has become common to see bots assisting in code review and bug-fixing on platforms like GitHub [8,9]. Wessel et al. [8] identified 48 bots used for this purpose. From committing code to coordinating open-source projects, bots are increasingly becoming a part of the software development life cycle [8,9]. Their impact on development can significantly affect how future digital innovation ecosystems are managed and governed [10–12]. Social platforms and games also serve as environments for social bots and virtual avatars [13,14]. Although claims of political intent and the influence of social bots on social media are exaggerated, bots are also prevalent on social platforms nowadays [15]. The diverse applications of bots highlight the desiderata and requirements for advanced cognitive agents [16].

However, the reality falls short of the expectations, and advanced social bots do not exist today [15,17]. First, the level of autonomy in industrial software agents and robotic process automation (RPA) is minimal [18,19]. Agents often have architectures that are tightly coupled with specific service platforms. As a result, bots are designed and optimized for these specific platforms, limiting their adaptability. They lack the ability to autonomously recognize the variability and then function effectively across diverse web or service environments. Consequently, they lack a sense of awareness and the ability to adapt to different contexts. Second, bots have very limited or no autonomy in their behavior. Addressing these challenges requires a focus on identifying and resolving architectural concerns [20–23]. It is important to note that bots in social media, games, industry, and business use cases may differ, but the software systems architecture and engineering challenges can span domains [24,25].

Thus, by reviewing and synthesizing existing works, this study aims to explore two architectural aspects of bots through two research questions. The first question investigates the role of cognitive architectures in bot behavior, while the second question examines the strict separation of bots and their operational environment. In our context, the second question is operationalized by evaluating *user-likeness* or similarity. A *user-like bot* refers to the level of similarity a bot possesses compared to a human user [26].

The subsequent sections are structured as follows: Sect. 2 provides the foundational background on bots and cognition, Sect. 3 describes the approach employed in surveying bots, Sect. 4 presents the evaluated results, Sect. 5 discusses the implications of the obtained results, and finally, Sect. 6 concludes the study.

2 Bots and Cognition

2.1 Software Bots

Software bot, or simply bot, is an umbrella term for diverse software agents [27]. The term is used loosely in domains such as RPA. RPA, robotic process automation, is an automation approach that employs software bots, sometimes referred to as digital workers [6]. RPAs and digital twins utilize various toolsets and development paradigms from agent-based systems [5,28].

Lebeuf [29] conducted a comprehensive study proposing a broad definition and general taxonomy of bots. She defines software bots as interfaces that connect users to software services and describes them as a *"new [user] interface paradigm"*. According to this perspective, users can access software services through a bot, where the user interface takes the form of a conversational interface. The user is typically human, although other programs and systems can also utilize the bot. Software services, in this context, refer to applications or digital platforms that provide additional functionalities. While these services are typically external, they can also be integrated as internal components of the bot.

Furthermore, Lebeuf [29] classifies bots based on their observable properties and behaviors. The taxonomy defines three dimensions: environment, intrinsic, and interactions. The environment dimension refers to the software service properties that bots operate on. Ideally, a software bot is separate from a specific platform and can operate on multiple platforms. However, most bots are tailored to a specific platform, such as a Twitter bot. The intrinsic dimension describes the internal abilities of the bot. Lebeuf also puts anthropomorphism as an intrinsic dimension, which specifies if the bot has human *user-like* features, such as name, visualization, and persona. The interaction dimension specifies how the bot accesses and interacts with the environment.

Lebeuf's taxonomy is based solely on observable properties and does not consider the architectural aspects and components of bots [29]. In our context, the interaction dimension concerns perception, action, and autonomy, which impact the system architecture and environment of the bot. Therefore, in this study, we introduce another term to restrict the scope to bots that are similar to users.

User-Like Software Bots. Today, the web is an operational environment for human users and software user agents [26]. Unlike virtual reality environments, where agents are typically manifested as conscious-like avatars, the web represents a different kind of mixed-reality environment. In this environment, both human users and software bots interact, tying virtual elements with real-world extensions, as exemplified by the Web of Things (WoT) [30–32]. A diverse set of software agents interact with the web as their environment [29,32].

In this study, we refer to these classes of mixed-reality software agents as *user-like bots* [26]. The term *user-like* implies that these bots exhibit similarities to human users. User-like bots use graphical user interfaces to perceive and act within a software service environment. They interact with keyboard and mouse operations.

As Gibson's theory of visual perception suggests, perception is not merely about passive observation; it's also about actively distinguishing the potential actions or *"affordances"* that an environment offers to an agent [33]. The environment, in this context, presents affordances [34]. Affordances refer to possibilities for interaction and action [26]. Platform services or features can be analogous to affordances in the real world. User-like bots are expected to understand or perceive these affordances and act on them similar to human users.

While the web is the primary context, a software service environment can also refer to any desktop application.

2.2 Cognitive Architectures

Research in engineering machine intelligence, particularly Artificial General Intelligence (AGI), aims to endow software systems with cognitive abilities that enable machines to think at or beyond the level of humans [23]. To achieve this goal, efforts to understand the brain from disciplines such as cognitive and neuroscience have led to various promising approaches. One such approach is the study of cognitive architectures, which focuses on designing high-level cognitive functions [26,35].

Cognitive architectures serve as essential architectural design foundations for artificial intelligence research. The ultimate objective of cognitive architectures is to enable software with cognitive abilities equal to or greater than human-like intelligence. Cognitive architectures can be described as a set of specifications or theories of cognition that outline the essential structural elements and capabilities of a cognitive system [36,37]. Metzler and Shea [38] compiled a list of cognitive functions as components that constitute a cognitive architecture, such as learning, reasoning, decision-making, perception, planning, and acting.

Cognitive architectures are designed to handle a broader set of cognitive tasks or cognitive functions. They can enable perpetual learning from the environment, adapting to changes, and reasoning based on available information. Due to this universal approach, an agent implementing a cognitive architecture may operate successively and simultaneously in various applications. While most cognitive architectures remain theoretical specifications, some, such as ACT-R, LIDA, and SOAR, have implementations and active communities [36]. Duch et al. [39] have conducted an in-depth comparison on a technical level of these architectures, among others.

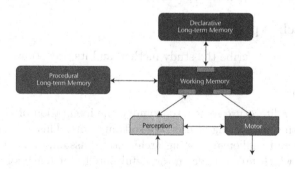

Fig. 1. Schematic Structure of the Standard Model of Cognition [40].

Currently, efforts are being made in the cognitive modeling research community to establish a comprehensive understanding of the architectural assumptions that define aspects of human-like cognition, whether natural or artificial [40]. Figure 1 shows a schematic structure of the Standard Model of Cognition [40]. The Standard Model of Cognition aims to consolidate knowledge from all existing cognitive architectures and establish a unified understanding of the generic aspects of cognition, such as perception and motor functions, common to all cognitive architectures.

2.3   Cognitive Automation

Traditional automation methods and approaches, such as process automation and RPA, have played crucial roles in automating repetitive tasks and workflows [2,41,42]. However, as organizations strive for increased agility, efficiency, and even hyper-automation, the increased complexity of cyber-physical systems necessitates additional layers of automation, i.e., cognitive automation [2].

Cognitive automation aims to advance the capabilities of traditional automation and RPA by combining them with other technologies in artificial intelligence. This integrated approach enables the automation of more complex and cognitive tasks that traditionally require human intervention. Cognitive automation aims to automate knowledge and service work that involves decision-making, problem-solving, and other cognitive activities [2,43]. It focuses on alleviating the burden of cognitive tasks on humans by automating their roles or enhancing mixed reality collaborations [2,42,44].

Similarly, advanced user-like bots or software agents can be employed to achieve generalizable intelligence or cognitive capabilities on digital platforms comparable to human users [26]. As autonomous users, software agents can then assist in automating human user activities. Consequently, one prominent application of cognitive automation can be utilizing user-like bots. Bots with generalizable intelligence, matching human users, are one way to address the augmentation of autonomous digital workers to cyber-physical systems. These bots can enhance efficiency in various domains, such as RPA, knowledge platforms, digital twins, and smart factories [5,26].

3   Research Approach

In this section, we describe the study method and its execution.

3.1   Planning

We conducted a literature review to examine the integration of cognitive architectures into the engineering of user-like software bots. This approach allows us to investigate and challenge existing architectural assumptions of software bot development, which fail to enact meaningful, intelligent behavior. The method also uses specific criteria to narrow the review's scope and establish a clear and replicable methodology. The survey is executed by systematically searching, selecting, and evaluating relevant works.

3.2   Initial Selection

Initially, we collected various bots, agents, software tools, and personal digital assistants to get an overview. Since terms such as agent, bot, software bot, user-agent, and chatbot are interchangeably used in literature, the initial efforts resulted in an extensive collection of sources. However, this initial process helped explore and map the software bots already developed throughout all time.

We then limited the collection to the bots that use some form of cognitive architecture or cognitive model. Furthermore, we determined the search to include works that distinctively show some similarity to a way a human user would access and with software service platforms, hence only user-like software bots.

First, we systematically selected relevant works from comprehensive databases, such as Scopus. To that end, specific keywords, including variations of "software bot," "cognitive architecture," and "user-like or user-agent," were devised and conducted. Next, we used forward and backward snowball sampling [45] to find citation chains of relevant studies that claim the development of bots with cognitive capabilities.

Our initial collection resulted in approximately 190 works. With a closer look, we found that many were unrelated to the distinct interest. The initial results are then narrowed to a representative selection of software bots for further investigation and evaluation.

3.3   Selection and Evaluation Criteria

The representative selection, also called candidates, is evaluated with four essential selection criteria. The four criteria used to evaluate the selection are derived to ensure the proper implementation of the interest and objective of the study. Each candidate bot selected is evaluated against all four criteria and the corresponding sub-criteria. Figure 2 depicts the four criteria and subcriteria of the first two: Software Bot and User Similarity.

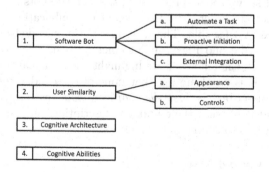

Fig. 2. Criteria used to evaluate the bots applying cognitive models.

(1) Software bot: this criterion evaluates bot capability. It helps exclude conversational interfaces or chatbots that do little or no autonomous action on service platforms. The criterion has three subcriteria: a. Automate a Task, b. Proactive Initiation, and c. External integration. One of the three aspects is the necessity that the bots automate a task in a digital environment (a. Automate a Task). In this context, the scope of the environment and the complexity of the task are neglected. To differentiate the candidate from other programs and scripts, as a second aspect, independence from the user is required (b. Proactive Initiation). Changes to other elements of the environment, combined with intrinsic motivations, should trigger the behavior and actions the bot performs. Indirect reactions to the actions of the user are permitted. For the third aspect, the candidate bot should operate externally in the targeted digital environment (c. External integration). Direct integration into the environment, as a fixed part of a system, is not considered user-like.

(2) User Similarity: Equal to the first criterion, this criterion stems from the distinction set regarding user-like software bots in Sect. 2. User-like is used to denote agents who either resemble human users or act on behalf of users. We focus on drawing a clear distinction between the bots that autonomously enact user-like behavior and others.

(3) Cognitive Architecture: While claiming the implementation of a cognitive architecture is relatively easy, its practical application poses a challenge.

(4) Cognitive Abilities: Implementing a cognitive architecture does not necessarily result in cognitive behavior. Consequently, this criterion unites various aspects regarding a candidate's cognitive abilities. The list aims at excluding narrowly set low-level heuristics from the selection while considering the desired high-level cognitive functions or generalized cognitive capabilities.

4 Results

This section presents the selected agents, personal assistants, and applied cognitive models. Their evaluation is based on the criteria outlined in Sect. 3. Furthermore, descriptions of notable selected results from the corresponding review are provided. Table 1 shows the main results of the software bots analyzed and evaluated. The entries are ordered by the year of the publication. However, other arrangements are also feasible. The meaning of the encoded column captions is provided in Sect. 3.3 and Fig. 2. The elements of the selection cover diverse research domains. In the following, we highlight the significant bases of the elements in Table 1. It is worth noting that none of the candidates in the related publications are described as *"user-like software bots"* as termed in this study. They were included because their implicit descriptions in the literature align with the previously set criteria and guidelines.

4.1   IDA and Virtual Mattie

The agent Virtual Mattie was developed by Franklin et al. in 1996 to perform clerical tasks related to the organization of seminar schedules [46]. Human seminar organizers are contacted via email. The primary cognitive abilities of Virtual

Table 1. Overview of selected bots, agents, or assistants, with evaluation results.

Study Ref.	Description	Software Bot			User Similarity		Cognitive Architecture	Cognitive Abilities
		a.	b.	c.	a.	b.		
Franklin et al. 1996 [46]	Virtual Mattie is a clerical agent that collects and shares weekly email information with a designated group.	✓	✓	✓	✓	✓	✓. Extended versions of Maes' behavior net [47] and Hofstadter and Mitchell's Copycat architecture [48].	X
Franklin et al. 1998 [49]	IDA: Intelligent distribution agent for US Navy sailor billet assignments via email.	✓	✓	✓	✓	✓	✓. Architecture based on global workspace theory. The basis for LIDA [50].	✓
Knoblock et al. 2003 [51]	A personal information agent that gathers information from sources on the internet and links them to the specific task.	✓	✓	✓	X	X	✓. Employs, in combination with others, the rule-based approach, Planing by Rewriting [52].	✓
Berry et al. 2004 [53]	PCalM is a personalized calendar agent that manages the calendar of an individual and coordinates as well as schedules meetings.	✓	✓	✓	X	X	✓. Utilizes the Open Agent Architecture to connect the cognitive components directly [54].	✓
Freed et al. 2008 [55]	RADAR is a personal assistant that reduces email overload by identifying task-relevant content and managing emails.	✓	✓	✓	X	X	X. Interconnected structure of task-specific, AI-enhanced modules/agents.	X
Berry et al. 2011 [56]	PTIME is a learning cognitive assistant agent that represents a personalized time manager.	✓	✓	✓	X	X	✓. Employs a preference module, which is connected to various schedulers.	✓
SRI International [57]	CALO is an extensive cognitive agent that learns and organizes. It is a predecessor of Apple Siri.	✓	✓	✓	X	X	✓. Cooperating cognitive agents including, for example, PTIME for language processing, see [58].	✓
Strain et al. 2014 [59]	Medical Agent X, an agent for clinical diagnostics.	✓	X	X	X	X	✓. Employs the cognitive architecture LIDA [50].	✓
Lebiere et al. 2015 [60]	An automated approach to determine the task of a piece of malware.	✓	X	X	X	X	✓. Partially employs the cognitive architecture ACT-R [61].	X
Shi et al. 2017 [62]	An experimental platform where agents learn to interact with web environments using only keyboard and mouse operations.	✓	✓	✓	X	✓	X. Training models based on supervised and reinforcement learning [62].	X
Wendt et al. 2018 [63]	A cognitive agent for building energy management	✓	✓	✓	X	X	✓. Employs the KORE cognitive architecture for building automation [64].	✓

Mattie are aimed at understanding the free-form and probably incomplete messages. While Virtual Mattie implements mechanisms for goals and attention, the small number of cognitive abilities does not meet the respective criteria. The other criteria are satisfied.

IDA, the intelligent distribution agent designed by Franklin et al. in 1998 for the US Navy, is the successor of Virtual Mattie and the predecessor of the cognitive architecture LIDA [65]. The agent assigns new long-term tasks to sailors who have finished their current ones. A notable feature of the architecture is its working memory, based on the global workspace theory [66]. In comparison to Virtual Mattie, IDA possesses more enhanced cognitive abilities. In addition to decision-making and attention modules, IDA also includes a module for emotions. Due to the communication by email and similar related access to human coworkers, IDA, at a basic level, satisfies the criteria set that evaluates the systemic architectural perspective of integrating behavior and user similarity.

4.2   The CALO Program (CALO, PTIME)

The Cognitive Assistant that Learns and Organizes (CALO) project brought together researchers from 22 organizations to advance research in cognitive software systems by developing a long-lasting, personalized cognitive agent [67]. The Defense Advanced Research Projects Agency (DARPA) funded it under the Perceptive Assistant that Learns (PAL) program. This project led to the CALO meeting assistant as well as a variety of related cognitive agents. One example developed to build on the foundation of CALO is PTIME [56]. With its ability to reason, plan schedules, and learn user preferences, PTIME meets the cognitive ability criteria set out in our methodology at a basic level. However, due to the implementation of the system with a user interface, it lacks the requested user similarity. The CALO meeting assistant integrates multiple previously developed cognitive concepts and possesses various cognitive abilities. It is the predecessor of Apple Siri. Nevertheless, in this study context, such systems are classified as conversational user interfaces rather than autonomous user agents.

4.3   Lebiere et al. 2015 and Wendt et al. 2018

Cognitive architectures can be directly applied to solve cognitive tasks previously performed by humans. In Lebiere et al. 2015, the cognitive architecture ACT-R is used to identify malware tasks [60]. The corresponding model was trained with historical malware data. While the agent system can perform its task by employing ACT-R, no other cognitive abilities were apparent. Additionally, the system must be initiated manually on the data, not meeting the respective criteria. Wendt et al. 2018 present a cognitive model capable of building energy management [63]. To achieve that, they employed the KORE cognitive architecture [64]. The agent is capable of being integrated externally into a building management interface. It processes data from physical sensors and establishes management rules based on that data. Like Lebiere et al.'s system, the agent does not employ all of the cognitive abilities in the criteria. However, some criteria, such as user similarity, are satisfied with the active perception of the environment and the decision-making.

4.4 World of Bits (WoB)

WoB is an experimental learning platform to train software bots in open-domain web environments [62]. Agents in WoB perceive the web environments in the form of the Document Object Model (DOM) and rendered pixels. Interestingly, though benchmark results are low compared to human users, the approach accomplished some web tasks by sending mouse and keyboard actions. Furthermore, agents are separated from the environments in which they interact. Tasks and activities are low-level operations and lack high-level cognitive capabilities. As a result, it falls short of the evaluation criteria.

5 Discussion

The study's results have important implications for the field of cognitive architectures and user-like software bots. Of the eleven agents and applied cognitive architectures examined in the previous section, Sect. 4, only one satisfied the established criteria. Interestingly, the agent, IDA by Franklin et al. [49], was developed a little over two decades ago. Except for the first two candidates, which both precede the year 2000, all the systems featured missed the second criterion of user similarity. The two systems satisfying the criteria were limited by their time and would have been implemented as standalone systems if they had been designed a few years later. Presumably, due to a lack of alternatives, they both use email to communicate with their environment. This limitation likely led to them satisfying the user similarity criteria, not necessarily through the designer's intent. The trend shown by the other candidates moved towards applications without user similarity.

5.1 Bots and Cognitive Architectures

One central question arising from the results is the absence of software bots currently employing cognitive architectures. To the best of our knowledge, it appears highly unlikely to find an agent or bot in current use that meets all the criteria established for this study. Despite significant advancements in other areas of artificial intelligence, the fact that only one bot, IDA by Franklin et al. [49], satisfied all the criteria suggests that the field of user-like intelligent bots is still nascent and in its early stages. These findings and observations align with existing literature, which indicates that integrating cognitive architectures into software bots is a complex and challenging endeavor [26,68].

However, these results also underscore the potential that cognitive architectures hold for enhancing the capabilities of software bots. From a simplified architectural perspective, the example of IDA, which met all the criteria, demonstrates how a cognitive architecture can enable advanced autonomous behavior and user similarity in software bots.

These findings also have practical implications. These findings highlight the need for developers and researchers to focus their efforts on integrating cognitive architectures into software bots. For industry professionals, the results can guide the development of more advanced software bots that align with the requirements of cognitive automation, enabling more effective automation of various tasks across different domains.

5.2   Autonomy and Behavior: Architectural Perspectives

The architectural aspects that can address the observed shortcomings in this study, particularly regarding the level of autonomy and generalized behavior, can be viewed from two perspectives: the strict separation of bots and their environment and the integration of a separate behavior model [26].

First, as mentioned earlier, the strict separation of agents and their environment facilitates architectural possibilities for autonomy [26]. By decoupling bots from a single environment, they can gain the ability to interact with multiple environments dynamically, achieving higher levels of variability and adaptability. This separation also enables the design of software bots with a user-like orientation, treating bots as if they were human users. Consequently, this orientation opens up possibilities for alternative architectural design patterns, allowing software bots with user-like characteristics to seamlessly integrate into existing user interfaces or operate on digital platforms without the need for APIs or integration protocols. The separation already places the bots in a position where they can potentially establish their own intentions, goals, and deliberate interactions, access, and actions.

Second, an agent's capacity for self-awareness, contextual awareness, and other high-level cognitive functions arises from holistic behavior models [26]. These behavior models are typically formulated using cognitive architectures or similar integrative models and architectures. Consequently, the componentization and integration of cognitive architectures into the system architecture of software bots become essential architectural considerations. The scientific understanding of machine intelligence, along with related models and principles, is still evolving and has not yet reached a crystallized state [17]. As a result, models of intelligent behavior will evolve over time, and older ones may require changes and modifications. Componentization can facilitate separation and change. Additionally, through this componentization, bots can potentially dynamically integrate separate behavior models in real-time.

Adopting these architectural perspectives can contribute to addressing the challenges related to autonomy and generalized behavior in bots, enabling them to operate with greater flexibility, adaptability, and user-like characteristics.

5.3   Study Limitations

Despite the study's important findings and insights into integrating cognitive architectures into user-like software bots, the study has limitations. The review was limited to publicly available literature and may not capture all existing software bots that employ cognitive architectures. Furthermore, the criteria for evaluating software bots may not encompass all possible features and capabilities. Future research could expand on our work by investigating databases further, employing different evaluation criteria, or examining the evolution of cognitive architectures in software bots over time.

6 Conclusion

The review highlights the importance of developing user-like software bots. These bots integrate cognitive architectures, enabling advanced autonomous behavior and user similarity. These requirements have significant implications for the engineering aspects of cognitive bots, particularly in cognitive automation.

Through the analysis of architectural recommendations and perspectives, this study has provided insights into achieving these goals at a design level. The distinctive view of bots and their environment, along with the dynamic integration of componentized behavior models, arise as key approaches to support both desiderata.

Consequently, implementing these architectural approaches can lead to increased autonomy and adaptability in software bots. This opens up new possibilities for developing autonomous user-like bots that can effectively perform a wide range of tasks. These findings contribute to the growing understanding of how to enhance the cognitive capabilities of software bots through architectural references.

Acknowledgement. We would like to thank Lorenz Bobber for his valuable support in the initial stages of the survey.

References

1. Jiang, J., et al.: A cognitive reliability model research for complex digital human-computer interface of industrial system. Saf. Sci. **108**, 196–202 (2018)
2. Engel, C., Ebel, P., Leimeister, J.M.: Cognitive automation. Electron. Mark. **32**(1), 339–350 (2022)
3. Leitao, P., Karnouskos, S., Ribeiro, L., Lee, J., Strasser, T., Colombo, A.W.: Smart agents in industrial cyber-physical systems. Proc. IEEE **104**(5), 1086–1101 (2016)
4. Lee, J., Davari, H., Singh, J., Pandhare, V.: Industrial artificial intelligence for industry 4.0-based manufacturing systems. Manuf. Lett. **18**, 20–23 (2018)
5. Karnouskos, S., Leitao, P., Ribeiro, L., Colombo, A.W.: Industrial agents as a key enabler for realizing industrial cyber-physical systems: multiagent systems entering industry 4.0. IEEE Ind. Electron. Mag. **14**(3), 18–32 (2020)
6. Ivančić, L., Suša Vugec, D., Bosilj Vukšić, V.: Robotic process automation: systematic literature review. In: Di Ciccio, C., et al. (eds.) BPM 2019. LNBIP, vol. 361, pp. 280–295. Springer, Cham (2019). https://doi.org/10.1007/978-3-030-30429-4_19
7. Erlenhov, L., de Oliveira Neto, F.G., Scandariato, R., Leitner, P.: Current and future bots in software development. In 2019 IEEE/ACM 1st International Workshop on Bots in Software Engineering (BotSE), pp. 7–11. IEEE (2019)

8. Wessel, M., et al.: The power of bots: characterizing and understanding bots in OSS projects. PACM:HCI **2**(CSCW), 1–19 (2018)

9. Monperrus, M.: Explainable software bot contributions: case study of automated bug fixes. In: 2019 IEEE/ACM 1st International Workshop on Bots in Software Engineering (BotSE), pp. 12–15. IEEE (2019)

10. Newton, O.B., Saadat, S., Song, J., Fiore, S.M., Sukthankar, G.: EveryBOTy counts: examining human-machine teams in open source software development. Top. Cogn. Sci (2022)

11. Platis, D.: Software development bot ecosystems (2021)

12. Maruping, L., Yang, Y.: Governance in digital open innovation platforms. In: Oxford Research Encyclopedia of Business and Management (2020)

13. Hendler, J., Berners-Lee, T.: From the semantic web to social machines: a research challenge for AI on the world wide web. Artif. Intell. **174**(2), 156–161 (2010)

14. Arrabales, R., Ledezma, A., Sanchis, A.: Towards conscious-like behavior in computer game characters. In: 2009 IEEE Symposium on Computational Intelligence and Games, pp. 217–224. IEEE (2009)

15. Gallwitz, F., Kreil, M.: The rise and fall of 'social bot' research. In: SSRN (2021)

16. McDonald, K.R., Pearson, J.M.: Cognitive bots and algorithmic humans: toward a shared understanding of social intelligence. Curr. Opin. Behav. Sci. **29**, 55–62 (2019)

17. Butlin, P., et al.: Consciousness in artificial intelligence: insights from the science of consciousness. arXiv preprint arXiv:2308.08708 (2023)

18. Kugele, S., Petrovska, A., Gerostathopoulos, I.: Towards a taxonomy of autonomous systems. In: Biffl, S., Navarro, E., Löwe, W., Sirjani, M., Mirandola, R., Weyns, D. (eds.) ECSA 2021. LNCS, vol. 12857, pp. 37–45. Springer, Cham (2021). https://doi.org/10.1007/978-3-030-86044-8_3

19. Vagia, M., Transeth, A.A., Fjerdingen, S.A.: A literature review on the levels of automation during the years. What are the different taxonomies that have been proposed? Appl. Ergon. **53**, 190–202 (2016)

20. Baldassarre, G., Santucci, V.G., Cartoni, E., Caligiore, D.: The architecture challenge: future artificial-intelligence systems will require sophisticated architectures, and knowledge of the brain might guide their construction. Behav. Brain Sci. **40**, e254 (2017)

21. Kraetzschmar, G.: Software engineering factors for cognitive robotics. Deliverable D-3.1 (2018)

22. Bosch, J., Olsson, H.H., Crnkovic, I.: Engineering AI systems: a research agenda. In: Artificial Intelligence Paradigms for Smart Cyber-Physical Systems, pp. 1–19 (2021)

23. Goertzel, B., Pennachin, C., Geisweiller, N.: Engineering General Intelligence, Part 1. Atlantis Thinking Machines, vol. 5 (2014)

24. Wooldridge, M., Jennings, N.R.: Pitfalls of agent-oriented development. In: Proceedings of the Second International Conference on Autonomous Agents, pp. 385–391 (1998)

25. Martínez-Fernández, S., et al.: Software engineering for ai-based systems: a survey. ACM Trans. Softw. Eng. Methodol. (TOSEM) **31**(2), 1–59 (2022)

26. Gidey, H.K., Hillmann, P., Karcher, A., Knoll, A.: Towards cognitive bots: architectural research challenges. In: Hammer, P., Alirezaie, M., Strannegård, C. (eds.) AGI 2023. LNCS, vol. 13921, pp. 105–114. Springer, Cham (2023). https://doi.org/10.1007/978-3-031-33469-6_11

27. Lebeuf, C., Storey, M.-A., Zagalsky, A.: Software bots. IEEE Softw. **35**(1), 18–23 (2017)

28. Wooldridge, M., Jennings, N.R.: Agent theories, architectures, and languages: a survey. In: Wooldridge, M.J., Jennings, N.R. (eds.) ATAL 1994. LNCS, vol. 890, pp. 1–39. Springer, Heidelberg (1995). https://doi.org/10.1007/3-540-58855-8_1
29. Lebeuf, C.R.: A taxonomy of software bots: towards a deeper understanding of software bot characteristics. Ph.D. thesis, UVic (2018)
30. Holz, T., Campbell, A.G., O'Hare, G.M.P., Stafford, J.W., Martin, A., Dragone, M.: MiRA-mixed reality agents. IJHC 69(4), 251–268 (2011)
31. Milgram, P., Kishino, F.: A taxonomy of mixed reality visual displays. IEICE Trans. Inf. Syst. 77(12), 1321–1329 (1994)
32. Charpenay, V., Käfer, T., Harth, A.: A unifying framework for agency in hypermedia environments. In: Alechina, N., Baldoni, M., Logan, B. (eds.) EMAS 2021. LNCS, vol. 13190, pp. 42–61. Springer, Cham (2022). https://doi.org/10.1007/978-3-030-97457-2_3
33. Gibson, J.J.: The Theory of Affordances. Hilldale, USA, vol. 1, no. 2, pp. 67–82 (1977)
34. Nye, B.D., Silverman, B.G.: Affordances in AI (2012)
35. Vernon, D.: Cognitive architectures. In: Cognitive Robotics. MIT Press (2022)
36. Vernon, D.: Artificial Cognitive Systems: A Primer. MIT Press, Cambridge (2014)
37. Kotseruba, I., Avella Gonzalez, O.J., Tsotsos, J.K.: A review of 40 years of cognitive architecture research: focus on perception, attention, learning and applications, pp. 1–74. arXiv (2016)
38. Metzler, T., Shea, K., et al.: Taxonomy of cognitive functions. In: DS 68-7: Proceedings of the 18th International Conference on Engineering Design (ICED 2011), Impacting Society through Engineering Design, vol. 7: Human Behaviour in Design. Lyngby/Copenhagen, Denmark, 15–19 August 2011, pp. 330–341 (2011)
39. Duch, W., Oentaryo, R.J., Pasquier, M.: Cognitive architectures: where do we go from here? In: AGI, vol. 171, pp. 122–136 (2008)
40. Laird, J.E., Lebiere, C., Rosenbloom, P.S.: A standard model of the mind: toward a common computational framework across artificial intelligence, cognitive science, neuroscience, and robotics. Magazine 38(4), 13–26 (2017)
41. Van Der Aalst, W.M.P., Van Hee, K.M.: Workflow Management: Models, Methods, and Systems (2004)
42. Van der Aalst, W.M.P., Bichler, M., Heinzl, A.: Robotic process automation (2018)
43. Bruckner, D., Zeilinger, H., Dietrich, D.: Cognitive automation-survey of novel artificial general intelligence methods for the automation of human technical environments. IEEE Trans. Industr. Inf. 8(2), 206–215 (2011)
44. Aguirre, S., Rodriguez, A.: Automation of a business process using robotic process automation (RPA): a case study. In: Figueroa-García, J.C., López-Santana, E.R., Villa-Ramírez, J.L., Ferro-Escobar, R. (eds.) WEA 2017. CCIS, vol. 742, pp. 65–71. Springer, Cham (2017). https://doi.org/10.1007/978-3-319-66963-2_7
45. Wohlin, C.: Guidelines for snowballing in systematic literature studies and a replication in software engineering. In: Proceedings of the 18th International Conference on Evaluation and Assessment in Software Engineering, pp. 1–10 (2014)
46. Franklin, S., Graesser, A., Olde, B., Song, H., Negatu, A.: Virtual Mattie - an intelligent clerical agent. In: AAAI Symposium on Embodied Cognition and Action, Cambridge, MA (1996)
47. Maes, P.: Situated agents can have goals. Robot. Auton. Syst. 6(1–2), 49–70 (1990)
48. Hofstadter, D.R., Mitchell, M.: The copycat project: a model of mental fluidity and analogy-making (1994)

49. Franklin, S., Kelemen, A., McCauley, L.: IDA: a cognitive agent architecture. In: SMC 1998 Conference Proceedings. IEEE Conference on Systems, Man, and Cybernetics, vol. 3, pp. 2646–2651. IEEE (1998)

50. Franklin, S., Strain, S., Snaider, J., McCall, R., Faghihi, U.: Global workspace theory, its LIDA model and the underlying neuroscience. Biol. Inspired Cogn. Arch. **1**, 32–43 (2012)

51. Knoblock, C.A.: Deploying information agents on the web. In: IJCAI, pp. 1580–1586. Citeseer (2003)

52. Ambite, J.L., Knoblock, C.A.: Planning by rewriting. J. Artif. Intell. Res. **15**, 207–261 (2001)

53. Berry, P., Gervasio, M., Uribe, T., Myers, K., Nitz, K.: A personalized calendar assistant. In: Working Notes of the AAAI Spring Symposium Series, vol. 76 (2004)

54. Cheyer, A., Martin, D.: The open agent architecture. Auton. Agent. Multi-Agent Syst. **4**(1), 143–148 (2001)

55. Freed, M., et al.: RADAR: a personal assistant that learns to reduce email overload. In: AAAI, vol. 8, pp. 1287–1293 (2008)

56. Berry, P.M., Gervasio, M., Peintner, B., Yorke-Smith, N.: PTIME: personalized assistance for calendaring. ACM Trans. Intell. Syst. Technol. (TIST) **2**(4), 1–22 (2011)

57. Apple Inc.: Siri Intelligent Personal Assistant. https://www.apple.com/uk/siri/. Accessed 18 June 2021

58. Tur, G., et al.: The CALO meeting assistant system. IEEE Trans. Audio Speech Lang. Process. **18**(6), 1601–1611 (2010)

59. Strain, S., Kugele, S., Franklin, S.: The learning intelligent distribution agent (LIDA) and Medical Agent X (MAX): computational intelligence for medical diagnosis. In: IEEE Symposium on Computational Intelligence for Human-Like Intelligence (CIHLI), pp. 1–8. IEEE (2014)

60. Lebiere, C., Bennati, S., Thomson, R., Shakarian, P., Nunes, E.: Functional cognitive models of malware identification. In: Proceedings of ICCM, ICCM 2015, pp. 9–11 (2015)

61. Ritter, F.E., Tehranchi, F., Oury, J.D.: ACT-R: a cognitive architecture for modeling cognition. Wiley Rev. Cogn. Sci. **10**(3), e1488 (2019)

62. Shi, T., Karpathy, A., Fan, L., Hernandez, J., Liang, P.: World of bits: an open-domain platform for web-based agents. In: ICML, pp. 3135–3144. PMLR (2017)

63. Wendt, A., Kollmann, S., Siafara, L., Biletskiy, Y.: Usage of cognitive architectures in the development of industrial applications. In: Proceedings of ICAART 2018 (2018)

64. Zucker, G., Wendt, A., Siafara, L., Schaat, S.: A cognitive architecture for building automation. In: 42nd Annual Conference of the IEEE Industrial Electronics Society, IECON 2016, pp. 6919–6924. IEEE (2016)

65. Franklin, S., Patterson, F.G., Jr.: The LIDA architecture: adding new modes of learning to an intelligent, autonomous, software agent. Pat **703**, 764–1004 (2006)

66. Baars, B.J.: Global workspace theory of consciousness: toward a cognitive neuroscience of human experience. Prog. Brain Res. **150**, 45–53 (2005)

67. Sri International: Artificial Intelligence Center - The CALO Project. https://www.ai.sri.com/project/CALO. Accessed 18 June 2021

68. Vernon, D.: The architect's dilemmas. In: Aldinhas Ferreira, M., Silva Sequeira, J., Ventura, R. (eds.) Cognitive Architectures. Intelligent Systems, Control and Automation: Science and Engineering, vol. 94. Springer, Cham (2019). https://doi.org/10.1007/978-3-319-97550-4_5

On Channel Selection for EEG-Based Mental Workload Classification

Kunjira Kingphai[ID] and Yashar Moshfeghi[✉][ID]

NeuraSearch Laboratory, Department of Computer and Information Sciences,
University of Strathclyde, Glasgow, Scotland
{kunjira.kingphai,yashar.moshfeghi}@strath.ac.uk

Abstract. Electroencephalogram (EEG) is a non-invasive technology
with high temporal resolution, widely used in Brain-Computer Interfaces (BCIs) for mental workload (MWL) classification. However, numerous EEG channels in current devices can make them bulky, uncomfort-
able, and time-consuming to operate in real-life scenarios. A Riemannian
geometry approach has gained attention for channel selection to address
this issue. In particular, Riemannian geometry employs covariance matrices of EEG signals to identify the optimal set of EEG channels, given
a specific covariance estimator and desired channel number. However,
previous studies have not thoroughly assessed the limitations of various
covariance estimators, which may influence the analysis results. In this
study, we aim to investigate the impact of different covariance estimators,
namely Empirical Covariance (EC), Shrunk Covariance (SC), Ledoit-
Wolf (LW), and Oracle Approximating Shrinkage (OAS), along with the
influence of channel numbers on the process of EEG channel selection.
We also examine the performance of selected channels using diverse deep
learning models, namely Stacked Gated Recurrent Unit (GRU), Bidirectional Gated Recurrent Unit (BGRU), and BGRU-GRU models, using a
publicly available MWL EEG dataset. Our findings show that although
no universally optimal channel number exists, employing as few as four
channels can achieve an accuracy of 0.940 (±0.036), enhancing practicality for real-world applications. In addition, we discover that the
BGRU model, when combined with OAS covariance estimators and a
32-channel configuration, demonstrates superior performance in MWL
classification tasks compared to other estimator combinations. Indeed,
this study provides insights into the effectiveness of various covariance
estimators and the optimal channel subsets for highly accurate MWL
classification. These findings can potentially advance the development of
EEG-based BCI applications.

Keywords: Mental workload classification · Machine learning · Deep
learning · Channel selection · Covariance estimator

1 Introduction

Mental workload (MWL) has been shown to impact human life in various ways,
such as attention disorders in children [18], driving fatigue [13], and reduced task

© The Author(s), under exclusive license to Springer Nature Switzerland AG 2024
G. Nicosia et al. (Eds.): LOD 2023, LNCS 14506, pp. 403–417, 2024.
https://doi.org/10.1007/978-3-031-53966-4_30

performances [32]. Neuroimaging techniques such as magnetoencephalography (MEG) [29], functional near-infrared spectroscopy (fNIRS) [11], and electroencephalography (EEG) [8] have been used to measure MWL levels by monitoring brain signal activity. Among these techniques, EEG is a preferred method for researchers due to its noninvasive nature and superior temporal resolution, enabling measurements on the millisecond scale [16]. Our study focuses on EEG signals, which have proven to be effective in measuring subjects' MWL levels. However, the development of models for MWL classification is a complex process [33]; moreover, multichannel recording techniques are often used to achieve high accuracy [30], which can be inconvenient in real-life applications and may capture redundant or irrelevant data [1]. Traditional channel selection strategies address these challenges [3] by eliminating duplicate channels [2], reducing computational times, and maintaining good classification performance [7]. These strategies also identify optimal channels for specific activities or applications, minimizing costs and improving user comfort [19]. Channel selection methods, including wrapping and filtering [28], differ in precision and processing demand. Wrapping approaches are accurate yet resource-intensive due to their classifier-dataset specificity, while filter approaches are faster, reliable, and classifier-independent [4]. Our study focuses on the filtering method to determine optimal EEG channels for MWL classification. This involves selecting only meaningful channels that increase the accuracy of the classification.

The Riemannian distance-based channel selection approach has become increasingly popular for reducing EEG channels while maintaining MWL classifier robustness and accuracy [10,23]. This approach involves using and manipulating EEG signal covariance matrices directly. However, the effectiveness of various covariance estimators used in this method has not been systematically evaluated. To address this gap, our study aims to investigate the impact of different covariance estimators on channel selection. Furthermore, we thoroughly study the effectiveness of the identified channels on a diverse range of state-of-the-art deep learning models for MWL level classification [30]. Specifically, we train models using features extracted from the selected channels, including Stacked Gated Recurrent Unit (GRU), Bidirectional Gated Recurrent Unit (BGRU), and BGRU-GRU. While deep learning techniques have demonstrated their ability to capture EEG signal variations accurately, they have drawbacks such as high computational demands, increased memory needs, and extended training times, especially with larger datasets [25]. To decrease computation time, we focus on an optimal set of channels providing the most relevant information. We evaluate our proposed algorithms using available datasets from a passive Brain-Computer Interfaces (BCIs) hackathon grand challenge [12].

2   Background

Effective channel selection methods for EEG data analysis are essential to enhance classification accuracy [14]. To perform channel selection, wrapper or filtering techniques can be used [3]. Wrapper techniques optimize a channel subset using classification accuracy as the primary measure [5]. For example, Mzurikwao et al. [22] used a wrapper strategy with a convolutional neural network (CNN) to select channels for decoding multiple motor imagery intention classes from four amputees. They achieved a classification accuracy of 99.7% with a CNN model trained on 64-channel EEG data, and channel selection based on weights extracted from the trained model resulted in 8-channel models with 91.5±% accuracy. Despite offering potentially high performance, these techniques can be computationally demanding, requiring the model to be retrained for each subset evaluated and carrying the risk of overfitting due to their inherent exhaustive search nature [5]. In contrast, filtering techniques evaluate subsets of channels a search algorithm generates using independent evaluation criteria, such as distance, dependency, or information measures, offer speed, independence from the classifier, and scalability [27]. These methods aim to maintain the accuracy achieved with all channels by training the model with an optimal channel set [4]. For instance, the mutual information maximization technique proposed in [17] ranks EEG channels based on their correlation with class labels, which lowers classification error. Similarly, the normalized mutual information technique proposed in [31] selects an optimal subset of EEG channels for emotion recognition, achieving high accuracy with a sliding window approach and short-time Fourier transform. The sparse common spatial pattern algorithm proposed in [3] optimizes channel selection under classification accuracy constraints and outperforms several other methods, achieving up to 10% improvements over three channels.

Recently, the Riemannian geometry approach has become a popular method for channel selection in EEG analysis. This technique utilizes the covariance matrices of EEG signals as features, which are then manipulated and classified directly. By examining the covariance properties of these features, researchers can determine the most meaningful channels for further analysis. For example, in the study by Barachant et al. [6], Riemannian geometry was used to select fewer electrodes for brain signal analysis. The method assessed how well different electrodes could distinguish between classes by measuring the Riemannian distance between their spatial covariance matrices. This method was applied to a two-class motor imagery paradigm, utilizing the sample spatial covariance matrix. Similarly, Qu et al. [23] employed Riemannian geometry to minimize information redundancy, extracting key features from the most relevant time-frequency bands of the selected channels to enhance decoding for BCIs. The EC estimator was utilised to analyse EEG signals in this binary classification problem, focusing on the left- and right-hand motor imagery tasks. This technique successfully reduced the number of electrodes from 61 to 18–32 using the LW estimator by sequentially pruning channels to maximize the Riemannian distance between the class-conditional covariance matrices [24].

Prior research primarily advanced binary classification. Our study extends Riemannian channel selection to more complex multiclass classification across easy, medium, and difficult MWL levels, requiring multiple class comparisons. Additionally, past studies utilized various covariance estimators without thoroughly exploring their advantages and drawbacks, potentially affecting technique effectiveness. Consequently, our primary aim is to assess different covariance estimators' impacts on channel selection and multiclass classification performance. By identifying the optimal number of channels for model accuracy, we aim to enhance the efficiency and effectiveness of channel selection in EEG analysis.

3   Methodology

Datasets Description. We investigated the effect of different covariance estimators on Riemannian channel selection using a publicly available EEG MWL dataset from a 2021 Neuroergonomics conference Passive BCI Hackathon[1] [12]. The dataset contains EEG signals from 15 participants, with 62 electrodes sampled at 500 Hz. Participants underwent two states: resting and testing. In resting, participants relaxed with open eyes for a minute while their EEG signals were recorded. In testing, they were subjected to the Multi-Attribute Task Battery II (MATB-II), a 15-min task with three 5-min blocks of different difficulty levels (i.e., easy, medium, or difficult MWL levels). The MATB-II, a NASA-developed software[2], is used to assess cognitive workload and performance [26].

Data Preprocessing. Before processing, artefacts in the EEG signals needed to be removed. Hackathon organisers preprocessed the dataset initially by splitting the data from the task and resting states of the complete EEG recording. Following this, the heart activity electrode was then removed and the data was segmented into two-second non-overlapping epochs. The dataset was then further subjected to high-pass filtering at 1 Hz and low-pass filtering at 40 Hz using FIR filters, resulting in the rejection of electrodes and noisy independent components from muscle, heart, and eye activity. Average re-referencing downsampled the signal to 250 Hz. To ensure data quality, we performed an automatic independent component analysis, ICA-ADJUST [21], to remove any remaining artefact components.

Feature Extraction. In machine learning, high-dimensional data can pose significant challenges, including time-consuming and computationally expensive calculations. We employed traditional feature extraction techniques to address this issue, capturing only relevant signal characteristics from EEG data. This strategy enables a compact, fast model tailored to our use case through customized features and interpretability. In this study, we calculated a set of features that are broadly classified into six groups as follows: 1) Frequency domain features include five Power Spectral Density (PSD) bands of delta (1–4 Hz), theta (4–8 Hz), alpha (8–12 Hz), beta (12–30 Hz), and gamma (30–40 Hz) band. 2) Statistical features that describe the signal's distribution include mean, variance, skewness, and kurtosis. 3) Morphological features include curve length, number of peaks, and average non-linear energy that characterize the signal's shape and form. These features are extracted in different frequency bands (theta,

---

[1] https://www.neuroergonomicsconference.um.ifi.lmu.de/pbci/.
[2] https://software.nasa.gov/software/LAR-17835-1.

alpha, beta, and low gamma) and from 1–40 Hz. 4) Time-frequency features are obtained by performing a wavelet transform on the signal. 5) Linear features, including the autoregressive coefficient (AR) with p = 2, describe time-varying processes. 6) Non-linear features include approximate entropy (ApEn) and Hurst exponent (H), which quantify the unpredictability of fluctuations and measure the self-similarity of the time series, respectively. In particular, we used a 2-s sliding window with a 2-s shift for feature extraction.

Deep Learning Models. Our research employs the Gated Recurrent Unit (GRU) family, including Stacked GRU, Bidirectional GRU (BGRU), and Bidirectional Stacked GRU (BGRU-GRU), to analyze sequential and time-series data effectively. These models have demonstrated efficacy in MWL classification [15,30]. The architectures of the models are: Stacked GRU (G128-G64-G40-D32-D3), BGRU (BG128-D32-D3), and BGRU-GRU (BG256-G128-G64-D32-D3), where G, BG, and D represent GRU, BGRU, and Dense layers, respectively. For instance, G128-G64-G40-D32-D3 implies a GRU layer with 128 units, GRU layers with 64 and 40 units, and Dense layers with 32 and 3 units. Model training involved the Adam optimizer with a learning rate of 1e-04, and to counter overfitting, we used early stopping after 30 epochs of no performance improvement.

Model Evaluation. Cross-validation is a crucial technique for evaluating a machine learning model's performance, and the choice of cross-validation method depends on the objectives of analysis [15]. The Hackathon organiser organised the EEG data into distinct folders, including easy, medium, and difficult levels. This rigorous technique resulted in the data's temporal sequence being reorganised. To mimic a real-world experiment, where each difficulty level would be performed randomly, we shuffled the features before splitting them into training and validation sets. We used stratified sampling to assign 80% of the data for model training and 20% for validation, thereby ensuring unbiased performance evaluation by equally representing labels in each class.

Evaluation Metrics. In our experiments, we evaluate the performance of the three-class classification using various metrics, including Accuracy, Sensitivity (Recall), Precision and F1-score, as follow:

$$Accuracy = \left[\sum_{i=1}^{3} \frac{tp_i + tn_i}{(tp_i + tn_i + fp_i + fn_i)} \right] /3, \tag{1}$$

$$Sensitivity(Recall) = \left[\sum_{i=1}^{3} \frac{tp_i}{(tp_i + fn_i)} \right] /3, \tag{2}$$

$$Precision = \left[\sum_{i=1}^{3} \frac{tp_i}{(tp_i + fp_i)} \right] /3, \tag{3}$$

$$F1 - score = \left[\sum_{i=1}^{3} \frac{2 * (precision * recall)}{(precision + recall)} \right] /3, \tag{4}$$

where true positives (tp_i) are cases correctly predicted as positive, and false negatives (fn_i) are cases wrongly predicted as negative. Conversely, true negatives (tn_i) are cases accurately predicted as negative, and false positives (fp_i) are cases incorrectly predicted as positive. Each MWL class L_i, with $i = 1, 2, 3$, is evaluated separately.

Covariance Estimation Techniques. While various channel selection strategies are currently in use, there remains a lack of comprehensive studies focusing on MWL level classification. To address this gap, we employed the Riemannian technique for channel selection, which involved calculating covariance matrices to examine the relationships between MWL levels. For those who are not familiar, the Riemannian distance measures the difference between two symmetric positive-definite matrices on a Riemannian manifold. In the context of EEG analysis, these covariance matrices represent the statistical relationships between channels. More specifically, the Riemannian distance is used to quantify the dissimilarity between matrices derived from different EEG conditions. In this paper, our investigation focuses on four distinct covariance estimators based on the Riemannian distance.

Empirical Covariance (EC). The traditional sample covariance matrix or EC is calculated using the following formula:

$$S = \frac{1}{n-1} \sum_{i=1}^{n} (x_i - \bar{x})(x_i - \bar{x})^T \tag{5}$$

where n is the number of observations, x_i is the i-th observation, and $\bar{x}$ is the sample mean. While widely used, this method has drawbacks like susceptibility to outliers, noise amplification, and multicollinearity tendency, possibly affecting the accuracy of subsequent analyses. Alternative covariance estimators such as SC, LW, and OAS have been developed to resolve these issues.

Shrunk Covariance (SC). The Shrunk Covariance matrix estimator addresses the limitations of the sample covariance matrix by combining it with a structured target matrix (T). The idea is to "shrink" the sample covariance matrix towards the target matrix to obtain a more stable and robust estimate. The SC matrix is computed as follows:

$$Shrunk\_Cov = \alpha T + (1 - \alpha)S \tag{6}$$

where S is the sample covariance matrix, T is the target matrix, and α is a shrinkage parameter between 0 and 1. The target matrix is typically an identity matrix or a diagonal matrix with the average of the variances on the diagonal. The choice of α can be made using cross-validation or by minimizing some criterion, such as the mean squared error.

Ledoit-Wolf (LW). The Ledoit-Wolf estimator, a shrinkage estimator, produces a more accurate estimate of the covariance matrix by minimizing the mean squared error between the true covariance matrix and the SC matrix.

$$Ledoit - Wolf = \beta I + (1 - \beta)S \tag{7}$$

where I is an identity matrix scaled by the average of the diagonal elements of the sample covariance matrix and β is the shrinkage factor. Similarly, the OAS estimator aims to find a shrinkage factor that minimizes the mean squared error in an oracle setting, where the true covariance matrix is known.

Oracle Approximating Shrinkage (OAS). The OAS estimator is calculated as

$$OAS = \gamma I + (1 - \gamma)S \tag{8}$$

where I is an identity matrix and γ is the shrinkage factor computed based on the trace and Frobenius norm of the sample covariance matrix.

Channel Number. In this study, we evaluated 4-, 8-, 16-, and 32-channel configurations to optimise the number of channels for each estimator. We aimed to identify the best covariance estimator and optimal channel number for high model performance. We also compared three neural network models: GRU, BGRU, and BGRU-GRU, to evaluate the effectiveness of these covariance estimators further.

We utilised the Riemannian distance within this framework, which quantifies the shortest distance between two points following a curved trajectory, as defined by Eq. 10. The Riemannian mean is expressed by Eq. 11.

$$\delta_R(C_1, C_2) = \log \|C_1^{-1}C_2\|_F = \left[\sum_{i=1}^{N} \log^2 \lambda_i \right]^{\frac{1}{2}} \tag{9}$$

where C_1, C_2 are two different covariance matrices respectively, λ_i denotes the th eigenvalue of $C_1^{-1}C_2$, $\| \cdot \|_F$ denotes the Frobenius norm, and $\log(\cdot)$ is the log-matrix operator.

$$\bar{C} = \arg \min_C \sum_{i=1}^{N} \delta_R^2(C, C_i), \tag{10}$$

$$\text{Crit} = \delta_R(\bar{C}_i, \bar{C}_j) = \|\log(\bar{C}_i^{-1}\bar{C}_j)\|_F \tag{11}$$

The pseudo-code employed for channel selection is shown in Algorithm 1.

To compute the sum of pairwise Riemannian distances for multiple classes, we calculated distances between Riemannian mean covariance matrices of each pair in our three MWL levels: $\bar{C}_1'$ and $\bar{C}_2'$, $\bar{C}_1'$ and $\bar{C}_3'$, and $\bar{C}_2'$ and $\bar{C}_3'$. The sum of these pairwise distances served as an overall measure of class dissimilarity in terms of Riemannian distances.

4   Results

Table 1 presents the performance results of three deep learning models: GRU, BGRU, and BGRU-GRU. These models were evaluated based on an optimal set of channels, selected using Riemannian distance with four covariance estimators. The channel configurations for the evaluations were set at 4, 8, 16, and 32.

Table 1. Average Performance of Deep Learning Models for MWL Level Classification based on four Covariance Estimators (CE), i.e. Empirical Covariance (EC), Shrunk Covariance (SC), Ledoit-Wolf (LW), Oracle Approximating Shrinkage (OAS), in Riemannian Distance-Based Channel Selection. The models' performance was assessed using accuracy, sensitivity, precision, and F1-score, with mean and standard deviation values reported for each metric. The abbreviation "NoC" represents the total number of channels involved in the analysis.

CE	NoC	Model	Accuracy	Sensitivity	Precision	F1-Score
EC	4	GRU	0.923 (±0.039)	0.885 (±0.058)	0.902 (±0.049)	0.882 (±0.060)
		BGRU	**0.930 (±0.041)**	0.895 (±0.061)	0.907 (±0.054)	0.894 (±0.063)
		BGRU-GRU	0.924 (±0.039)	0.886 (±0.058)	0.902 (±0.050)	0.884 (±0.059)
	8	GRU	0.923 (±0.040)	0.885 (±0.061)	0.908 (±0.048)	0.881 (±0.064)
		BGRU	**0.939 (±0.037)**	0.909 (±0.056)	0.922 (±0.048)	0.906 (±0.058)
		BGRU-GRU	0.930 (±0.038)	0.895 (±0.058)	0.914 (±0.048)	0.891 (±0.061)
	16	GRU	0.923 (±0.029)	0.885 (±0.044)	0.914 (±0.032)	0.879 (±0.046)
		BGRU	**0.950 (±0.032)**	0.925 (±0.048)	0.937 (±0.039)	0.923 (±0.050)
		BGRU-GRU	0.932 (±0.035)	0.898 (±0.053)	0.922 (±0.037)	0.893 (±0.057)
	32	GRU	0.934 (±0.028)	0.901 (±0.043)	0.925 (±0.030)	0.895 (±0.050)
		BGRU	**0.949 (±0.029)**	0.923 (±0.044)	0.938 (±0.033)	0.920 (±0.047)
		BGRU-GRU	0.925 (±0.035)	0.887 (±0.052)	0.916 (±0.034)	0.880 (±0.059)
SC	4	GRU	0.932 (±0.045)	0.898 (±0.067)	0.914 (±0.056)	0.897 (±0.069)
		BGRU	**0.940 (±0.036)**	0.910 (±0.054)	0.918 (±0.052)	0.908 (±0.055)
		BGRU-GRU	0.929 (±0.046)	0.893 (±0.068)	0.908 (±0.059)	0.890 (±0.072)
	8	GRU	0.946 (±0.032)	0.918 (±0.048)	0.931 (±0.041)	0.917 (±0.050)
		BGRU	**0.952 (±0.036)**	0.929 (±0.054)	0.939 (±0.043)	0.927 (±0.054)
		BGRU-GRU	0.943 (±0.035)	0.914 (±0.053)	0.928 (±0.045)	0.913 (±0.054)
	16	GRU	0.932 (±0.028)	0.898 (±0.041)	0.923 (±0.030)	0.893 (±0.045)
		BGRU	**0.950 (±0.031)**	0.924 (±0.046)	0.938 (±0.036)	0.922 (±0.048)
		BGRU-GRU	0.934 (±0.034)	0.901 (±0.051)	0.925 (±0.038)	0.894 (±0.056)
	32	GRU	0.928 (±0.041)	0.891 (±0.062)	0.919 (±0.037)	0.886 (±0.068)
		BGRU	**0.955 (±0.031)**	0.932 (±0.047)	0.943 (±0.037)	0.929 (±0.050)
		BGRU-GRU	0.927 (±0.034)	0.891 (±0.052)	0.921 (±0.037)	0.885 (±0.056)
LW	4	GRU	0.924 (±0.045)	0.886 (±0.068)	0.903 (±0.058)	0.882 (±0.072)
		BGRU	**0.931 (±0.044)**	0.896 (±0.066)	0.908 (±0.059)	0.894 (±0.068)
		BGRU-GRU	0.925 (±0.044)	0.887 (±0.066)	0.903 (±0.057)	0.885 (±0.068)
	8	GRU	0.940 (±0.033)	0.909 (±0.049)	0.924 (±0.041)	0.908 (±0.050)
		BGRU	**0.951 (±0.033)**	0.927 (±0.049)	0.936 (±0.042)	0.926 (±0.050)
		BGRU-GRU	0.938 (±0.042)	0.906 (±0.062)	0.925 (±0.047)	0.903 (±0.066)
	16	GRU	0.933 (±0.035)	0.899 (±0.052)	0.925 (±0.036)	0.894 (±0.056)
		BGRU	**0.952 (±0.026)**	0.928 (±0.038)	0.939 (±0.032)	0.926 (±0.040)
		BGRU-GRU	0.938 (±0.033)	0.907 (±0.049)	0.929 (±0.035)	0.903 (±0.053)
	32	GRU	0.936 (±0.029)	0.904 (±0.044)	0.927 (±0.031)	0.899 (±0.048)
		BGRU	**0.956 (±0.026)**	0.934 (±0.039)	0.946 (±0.031)	0.932 (±0.040)
		BGRU-GRU	0.930 (±0.033)	0.894 (±0.049)	0.921 (±0.035)	0.890 (±0.051)
OAS	4	GRU	0.924 (±0.048)	0.886 (±0.072)	0.903 (±0.058)	0.883 (±0.076)
		BGRU	**0.931 (±0.044)**	0.897 (±0.066)	0.910 (±0.057)	0.896 (±0.067)
		BGRU-GRU	0.928 (±0.041)	0.893 (±0.061)	0.907 (±0.054)	0.891 (±0.063)
	8	GRU	0.941 (±0.034)	0.912 (±0.051)	0.928 (±0.040)	0.910 (±0.055)
		BGRU	**0.953 (±0.031)**	0.929 (±0.047)	0.937 (±0.042)	0.928 (±0.048)
		BGRU-GRU	0.937 (±0.033)	0.906 (±0.049)	0.924 (±0.038)	0.904 (±0.050)
	16	GRU	0.938 (±0.032)	0.907 (±0.048)	0.929 (±0.033)	0.902 (±0.052)
		BGRU	**0.955 (±0.029)**	0.933 (±0.044)	0.943 (±0.034)	0.930 (±0.047)
		BGRU-GRU	0.942 (±0.029)	0.913 (±0.044)	0.931 (±0.035)	0.910 (±0.046)
	32	GRU	0.936 (±0.030)	0.903 (±0.045)	0.926 (±0.033)	0.899 (±0.047)
		BGRU	**0.958 (±0.027)**	0.937 (±0.041)	0.946 (±0.034)	0.935 (±0.042)
		BGRU-GRU	0.932 (±0.036)	0.898 (±0.054)	0.925 (±0.038)	0.892 (±0.058)
-	62	GRU	0.925 (±0.035)	0.887 (±0.053)	0.919 (±0.037)	0.880 (±0.058)
		BGRU	**0.958 (±0.026)**	0.937 (±0.039)	0.947 (±0.032)	0.935 (±0.042)
		BGRU-GRU	0.907 (±0.041)	0.861 (±0.062)	0.905 (±0.036)	0.852 (±0.070)

Algorithm 1. Pseudo code for Riemannian distance-based channel selection.

Input: The preprocessed N-channel EEG signals X_i, the number of selected channels N_{Ch}, Number of MWL levels N_{levels}

Output: N_{Ch} selected channel subset

1: **procedure** CHANNEL SELECTION
2:     Compute the covariance matrix C_i of X_i;
3:     Compute the Riemannian means of each level $\bar{C}_1, \bar{C}_2,...,\bar{C}_{N_{levels}}$;
4:     **for** n = 1: N_{Ch} **do**
5:         **for** k = 1: N **do**
6:             Remove k^{th} channel by reducing the kth row and column from matrix $\bar{C}_1, \bar{C}_2,...\bar{C}_{N_{levels}}$ to $\bar{C}'_1, \bar{C}'_2,...,\bar{C}'_{N_{levels}}$;
7:             Compute the sum of pairwise Riemannian distances between all classes' Riemannian means $D_{K_{sum}}$;
8:         **end for**
9:         Select the channel corresponding to a minimum $D_{K_{sum}}$ value;
10:     **end for**
11:     return N_{Ch} selected channels;
12: **end procedure**

Empirical Covariance (EC). For the 4-channel configuration, the BGRU model achieved the highest accuracy (0.930 ± 0.041), sensitivity (0.895 ± 0.061), and F1-score (0.894 ± 0.063) among the three models. The GRU and BGRU-GRU models had comparable performances, with slightly lower values across all metrics. When increasing the number of channels to 8, the BGRU model continued to outperform the other models, reaching an accuracy of 0.939 (±0.037), a sensitivity of 0.909 (±0.056), and an F1-score of 0.906 (±0.058). The BGRU-GRU model also showed improvement compared to the 4-channel configuration, but still, it was not as effective as the BGRU model. For the 16-channel configuration, the BGRU model again demonstrated the best performance, achieving an accuracy of 0.950 (±0.032), a sensitivity of 0.925 (±0.048), and an F1-score of 0.923 (±0.050). The BGRU-GRU model showed a slight increase in accuracy compared to the 8-channel configuration, but the BGRU model still outperformed it. Finally, in the 32-channel configuration, the BGRU model continued to demonstrate superior performance with an accuracy of 0.949 (±0.029), a sensitivity of 0.923 (±0.044), and an F1-score of 0.920 (±0.047). However, its performance was slightly lower than in the 16-channel configuration. On the other hand, the GRU model improved accuracy compared to the 16-channel configuration, but it still fell behind the BGRU model. The results indicate that the BGRU model consistently outperforms the GRU and BGRU-GRU models across all channel configurations in the EC estimator. Additionally, increasing the number of channels generally leads to better performance for all models, with the most improvements observed for the BGRU model. This suggests that the BGRU model is more effective at leveraging additional channels for improved classification performance.

Shrunk Covariance (SC). The BGRU model consistently demonstrated superior performance across various configurations. Specifically, for the 4-channel configuration, the BGRU model achieved the highest accuracy of 0.940 (±0.036), a sensitivity of 0.910 (±0.054), and an F1-score of 0.908 (±0.055), while the GRU and BGRU-GRU models had comparable performances with slightly lower values for all metrics. When the number of channels increased to 8, the BGRU model continued to outperform the other models, achieving an accuracy of 0.952 (±0.036), a sensitivity of 0.929 (±0.054), and an F1-score of 0.927 (±0.054), although the GRU model also improved substantially from the 4-channel configuration. In the 16-channel configuration, the BGRU model again demonstrated the best performance, achieving an accuracy of 0.950 (±0.031), a sensitivity of 0.924 (±0.046), and an F1-score of 0.922 (±0.048). However, the performance of each model decreased in this configuration. Finally, in the 32-channel configuration, the BGRU model showed the best performance, with an accuracy of 0.955 (±0.031), a sensitivity of 0.932 (±0.047), and an F1-score of 0.929 (±0.050). In contrast, the GRU and BGRU-GRU models showed a continuous decrease in accuracy compared to the 8- and 16-channel configurations.

Ledoit-Wolf (LW).   Results showed that in the 4-channel configuration, the BGRU model achieved an accuracy of 0.931 (±0.044), a sensitivity of 0.896 (±0.066), and an F1-score of 0.894 (±0.068), outperforming GRU and BGRU-GRU models. With 8 channels, the BGRU model improved to an accuracy of 0.951 (±0.033), a sensitivity of 0.927 (±0.049), and an F1-score of 0.926 (±0.050). The other models also improved, but BGRU remained superior. In the 16-channel configuration, the BGRU model excelled again with an accuracy of 0.952 (±0.026), a sensitivity of 0.928 (±0.038), and an F1-score of 0.926 (±0.040), while the GRU model decreased, and BGRU-GRU's accuracy was 0.938. In the 32-channel setup, BGRU maintained its lead with an accuracy of 0.956 (±0.026), a sensitivity of 0.934 (±0.039), and an F1-score of 0.932 (±0.040). The GRU model slightly improved, but BGRU-GRU performance decreased. Our results indicate that the BGRU consistently outperformed across configurations.

Oracle Approximating Shrinkage (OAS). In the 4-channel configuration, the BGRU model exhibited the best performance, with an accuracy of 0.931 (±0.044), sensitivity of 0.897 (±0.066), and F1-score of 0.896 (±0.067). The GRU model was the least performant. As the channel number increased to 8, the BGRU model maintained its superior performance with an accuracy of 0.953 (±0.031), sensitivity of 0.929 (±0.047), and F1-score of 0.928 (±0.048). While GRU and BGRU-GRU models improved, BGRU remained the best. In the 16-channel configuration, BGRU again led with an accuracy of 0.955 (±0.029), sensitivity of 0.933 (±0.044), and F1-score of 0.930 (±0.047). BGRU-GRU improved, but GRU's performance dropped. Lastly, with 32 channels, BGRU sustained its lead with an accuracy of 0.958 (±0.027), the sensitivity of 0.937 (±0.041), and F1-score of 0.935 (±0.042). Both GRU and BGRU-GRU performances declined. The findings show that BGRU consistently outperforms GRU and BGRU-GRU using the OAS estimator in all channel configurations.

Table 2. The mean values in the table represent the average performance of the deep learning models under each comparison.

Factor	Comparison	Mean	p-value
CE	EC vs SC	(0.932, 0.939)	0.027*
	EC vs LW	(0.932, 0.938)	0.072
	EC vs OAS	(0.932, 0.940)	0.012*
	SC vs LW	(0.939, 0.938)	0.669
	SC vs OAS	(0.939, 0.940)	0.819
	LW vs OAS	(0.938, 0.940)	0.522
NoC	4 vs 8	(0.928, 0.941)	0.001*
	4 vs 16	(0.928, 0.940)	0.027*
	4 vs 32	(0.928, 0.939)	0.032*
	4 vs 62	(0.928, 0.930)	0.922
	8 vs 16	(0.941, 0.940)	0.310
	8 vs 32	(0.941, 0.939)	0.265
	8 vs 62	(0.941, 0.930)	0.087
	16 vs 32	(0.940, 0.939)	0.916
	16 vs 62	(0.940, 0.930)	0.199
	32 vs 62	(0.939, 0.930)	0.250
Model	GRU vs BGRU	(0.932, 0.948)	0.000*
	GRU vs BGRU_GRU	(0.932, 0.931)	0.703
	BGRU vs BGRU_GRU	(0.948, 0.931)	0.000*

After examining the study's findings, we noticed that the BGRU model with 62 channels achieved comparable accuracy to the BGRU model using the OAS estimator and 32 channels. Nevertheless, in every covariance estimation technique, the GRU and BGRU-GRU models trained on 32-channel signals consistently outperformed those trained on 62-channel signals. This implies that an increased number of channels may cause overfitting issues, and a higher channel count does not necessarily guarantee improved performance in terms of channel configuration. Notably, the models maintained a commendable performance even when utilizing a limited number of channels, such as 4 or 8. This demonstrates the feasibility of creating effective EEG-based emotion recognition systems with reduced channels. This finding is valuable when the goal is to decrease complexity and computational demands while preserving a substantial overall model performance.

In this study, we also conducted a pairwise comparison using the Wilcoxon test to examine the influence of different factors on the accuracy of a model used for MWL level classification. The results, presented in Table 2, revealed significant performance differences between EC vs SC and EC vs OAS. In both these comparisons, EC underperforms, indicated by its lower mean value. The statistical significance of these differences is evident from the p-values being less than 0.05. For the other pairs evaluated - EC vs LW, SC vs LW, SC vs OAS, and LW vs OAS - no significant differences in model accuracy were found (p < 0.05). This underscores the specific impact of the EC covariance estimator relative to SC and OAS on model performance. In analyzing the impact of the NoC, it is observed that accuracy significantly improves when channels increase from 4 to 8, 16, and 32 (p < 0.05). However, this trend plateaus beyond 32 channels. When comparing configurations with 4 channels to 62 channels, the performance difference is not statistically significant (p > 0.05. Interestingly, increasing channels from 8 to 16, 32, 62, and from 16 to 32, 62, or from 32 to 62 does not yield significant improvements (p < 0.05). Furthermore, it is noted that an excessive number of channels may not contribute to enhanced performance. Instead, it could be counterproductive. In the Model factor, three pairwise comparisons are presented. The comparisons between GRU vs BGRU and BGRU vs BGRU-GRU show statistically significant differences (p < 0.0.5), indicating that the choice of model architecture significantly impacts the MWL level classification performance. However, the comparison between GRU vs BGRU-GRU does not exhibit a statistically significant difference (p > 0.05).

Overall, Table 2 provides valuable insights into the performance of deep learning models for MWL level classification based on different covariance estimators, number of channels, and model architectures. The statistical tests performed using the Mann-Whitney U test allow for rigorous comparisons and highlight significant differences in performance. These findings can aid researchers and practitioners in selecting the most suitable approach for MWL level classification in Riemannian Distance-Based Channel Selection.

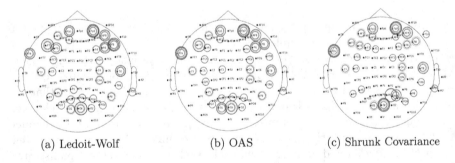

(a) Ledoit-Wolf   (b) OAS   (c) Shrunk Covariance

Fig. 1. Selected EEG channel using Riemannian Geometry with three different covariance estimator (Color figure online)

Figure 1 shows various channel configurations essential for analyzing EEG signals and MWL [20]. The 4-channel configuration, marked by red circles, spans the occipital to prefrontal regions, which is crucial for visual perception and cognitive functions. The 8-channel (blue circles) encompasses frontal to temporal regions, suitable for a wider MWL analysis. The 16-channel configuration, indicated by yellow circles, covers from frontal to occipital areas, while the 32-channel (green circles) spans five brain areas: frontal, central, parietal, occipital, and temporal, offering the most comprehensive MWL analysis.

Our research discerned key similarities in various configurations, revealing that channels Fp1 and AF8 were uniformly selected in 8, 16, and 32 configurations across the LW, OAS, and SC covariance estimators. Notably, Fp2 was a consistent choice across all configurations (4, 8, 16, and 32) for the LW, OAS, and SC estimators, while F7 was chosen in the 4, 16, and 32 configurations under the LW and OAS covariance estimators. The FT9 channel was versatile, being chosen in all configurations with OAS and SC estimators and in 8, 16, and 32 configurations with the LW estimator. Additionally, the Oz channel demonstrated significant applicability, being a common selection in 4, 16, and 32-channel configurations for LW and OAS estimators and was universally selected in all configurations when utilizing the SC estimator. These repeatedly selected channels highlight the importance of specific brain regions, regardless of the configuration size used. Channels AF8, Fp1, Fp2, F7, FT9 and F8, associated with the frontal and prefrontal regions, are important in influencing MWL [20]. In contrast, the Oz channel, which is linked to the occipital region, highlights the importance of visual perception in tasks related to MWL. Our findings support existing studies on the role of certain brain regions in mental effort [9], validating our results in exploring brain function and cognition.

5   Conclusion

Mental workload (MWL) is a cognitive construct that measures the mental effort needed to perform tasks. Assessing MWL is essential for optimizing human performance, decision-making, and designing efficient human-computer interactions. EEG has become popular for estimating MWL due to its high temporal resolution and non-invasiveness. However, current EEG devices are complex, involving many channels, making them unsuitable for practical use. Selecting the optimal number of channels is important, e.g., in BCI applications. This study evaluated different covariance estimators for Riemannian geometry-based channel selection and assessed their effectiveness with deep learning models to classify MWL levels. We examined four covariance estimators: EC, SC, LW, and OAS. The OAS estimator consistently delivered the best performance across all models and the covariance estimation technique. Our study showed that using as few as four channels can achieve an accuracy of 0.940 ($\pm$0.036), improving practicality for real-world applications. We also found that the BGRU model, combined with OAS covariance estimators and a 32-channel configuration, outperforms other estimators for MWL classification tasks. Our approach supports the development of user-friendly, efficient, and accurate brain-computer interfaces (BCIs) for various purposes, such as cognitive assessment and neurorehabilitation, by reducing the number of channels while retaining high classification accuracy. This has significant implications for enhancing EEG-based BCIs in real-world settings.

References

1. Alotaiby, T., El-Samie, F.E.A., Alshebeili, S.A., Ahmad, I.: A review of channel selection algorithms for EEG signal processing. EURASIP J. Adv. Signal Process. **2015**, 1–21 (2015)
2. Alyasseri, Z.A.A., Khader, A.T., Al-Betar, M.A., Alomari, O.A.: Person identification using EEG channel selection with hybrid flower pollination algorithm. Pattern Recogn. **105**, 107393 (2020)
3. Arvaneh, M., Guan, C., Ang, K.K., Quek, C.: Optimizing the channel selection and classification accuracy in EEG-based BCI. IEEE Trans. Biomed. Eng. **58**(6), 1865–1873 (2011)
4. Baig, M.Z., Aslam, N., Shum, H.P.: Filtering techniques for channel selection in motor imagery EEG applications: a survey. Artif. Intell. Rev. **53**, 1207–1232 (2020)
5. Baig, M.Z., Aslam, N., Shum, H.P., Zhang, L.: Differential evolution algorithm as a tool for optimal feature subset selection in motor imagery EEG. Expert Syst. Appl. **90**, 184–195 (2017)
6. Barachant, A., Bonnet, S.: Channel selection procedure using Riemannian distance for BCI applications. In: 2011 5th International IEEE/EMBS Conference on Neural Engineering, pp. 348–351. IEEE (2011)
7. Belakhdar, I., Kaaniche, W., Djemal, R., Ouni, B.: Single-channel-based automatic drowsiness detection architecture with a reduced number of EEG features. Microprocess. Microsyst. **58**, 13–23 (2018)
8. Borghini, G., Astolfi, L., Vecchiato, G., Mattia, D., Babiloni, F.: Measuring neurophysiological signals in aircraft pilots and car drivers for the assessment of mental workload, fatigue and drowsiness. Neurosci. Biobehav. Rev. **44**, 58–75 (2014)
9. Brouwer, A.M., Hogervorst, M.A., Van Erp, J.B., Heffelaar, T., Zimmerman, P.H., Oostenveld, R.: Estimating workload using EEG spectral power and ERPS in the n-back task. J. Neural Eng. **9**(4), 045008 (2012)
10. Chen, S., Sun, Y., Wang, H., Pang, Z.: Channel selection based similarity measurement for motor imagery classification. In: 2020 IEEE International Conference on Bioinformatics and Biomedicine (BIBM), pp. 542–548. IEEE (2020)
11. Herff, C., Heger, D., Fortmann, O., Hennrich, J., Putze, F., Schultz, T.: Mental workload during n-back task-quantified in the prefrontal cortex using fNIRS. Front. Hum. Neurosci. **7**, 935 (2014)
12. Hinss, M.F., et al.: An EEG dataset for cross-session mental workload estimation: passive BCI competition of the neuroergonomics conference 2021 (2021). https://doi.org/10.5281/zenodo.5055046. The project was validated by the local ethical committee of the University of Toulouse (CER number 2021-342)
13. Islam, M.R., Barua, S., Ahmed, M.U., Begum, S., Di Flumeri, G.: Deep learning for automatic EEG feature extraction: an application in drivers' mental workload classification. In: Longo, L., Leva, M.C. (eds.) H-WORKLOAD 2019. CCIS, vol. 1107, pp. 121–135. Springer, Cham (2019). https://doi.org/10.1007/978-3-030-32423-0_8
14. Jin, J., Miao, Y., Daly, I., Zuo, C., Hu, D., Cichocki, A.: Correlation-based channel selection and regularized feature optimization for mi-based BCI. Neural Netw. **118**, 262–270 (2019)
15. Kingphai, K., Moshfeghi, Y.: On time series cross-validation for deep learning classification model of mental workload levels based on EEG signals. In: Nicosia, G., et al. (eds.) LOD 2022, Part II. LNCS, vol. 13811, pp. 402–416. Springer, Cham (2023). https://doi.org/10.1007/978-3-031-25891-6_30

16. Lachaux, J.P., Axmacher, N., Mormann, F., Halgren, E., Crone, N.E.: High-frequency neural activity and human cognition: past, present and possible future of intracranial EEG research. Prog. Neurobiol. **98**(3), 279–301 (2012)

17. Lan, T., Erdogmus, D., Adami, A., Pavel, M., Mathan, S.: Salient EEG channel selection in brain computer interfaces by mutual information maximization. In: 2005 IEEE Engineering in Medicine and Biology 27th Annual Conference, pp. 7064–7067. IEEE (2006)

18. Lim, C.G., et al.: A brain-computer interface based attention training program for treating attention deficit hyperactivity disorder. PLoS ONE **7**(10), e46692 (2012)

19. Lin, B.S., Huang, Y.K., Lin, B.S.: Design of smart EEG cap. Comput. Methods Programs Biomed. **178**, 41–46 (2019)

20. Miller, E.K., Cohen, J.D.: An integrative theory of prefrontal cortex function. Annu. Rev. Neurosci. **24**(1), 167–202 (2001)

21. Mognon, A., Jovicich, J., Bruzzone, L., Buiatti, M.: ADJUST: an automatic EEG artifact detector based on the joint use of spatial and temporal features. Psychophysiology **48**(2), 229–240 (2011)

22. Mzurikwao, D., et al.: A channel selection approach based on convolutional neural network for multi-channel EEG motor imagery decoding. In: 2019 IEEE Second International Conference on Artificial Intelligence and Knowledge Engineering (AIKE), pp. 195–202. IEEE (2019)

23. Qu, T., Jin, J., Xu, R., Wang, X., Cichocki, A.: Riemannian distance based channel selection and feature extraction combining discriminative time-frequency bands and riemannian tangent space for mi-bcis. J. Neural Eng. **19**(5), 056025 (2022)

24. Roy, R.N., et al.: Retrospective on the first passive brain-computer interface competition on cross-session workload estimation. Front. Neuroergon. **3** (2022)

25. Roy, Y., Banville, H., Albuquerque, I., Gramfort, A., Falk, T.H., Faubert, J.: Deep learning-based electroencephalography analysis: a systematic review. J. Neural Eng. **16**(5), 051001 (2019)

26. Santiago-Espada, Y., Myer, R.R., Latorella, K.A., Comstock Jr., J.R.: The multi-attribute task battery II (MATB-II) software for human performance and workload research: a user's guide. Technical report (2011)

27. Shen, J., et al.: An optimal channel selection for EEG-based depression detection via kernel-target alignment. IEEE J. Biomed. Health Inform. **25**(7), 2545–2556 (2020)

28. Shi, B., Wang, Q., Yin, S., Yue, Z., Huai, Y., Wang, J.: A binary harmony search algorithm as channel selection method for motor imagery-based BCI. Neurocomputing **443**, 12–25 (2021)

29. Tanaka, M., Ishii, A., Watanabe, Y.: Neural effects of mental fatigue caused by continuous attention load: a magnetoencephalography study. Brain Res. **1561**, 60–66 (2014)

30. Varshney, A., Ghosh, S.K., Padhy, S., Tripathy, R.K., Acharya, U.R.: Automated classification of mental arithmetic tasks using recurrent neural network and entropy features obtained from multi-channel eeg signals. Electronics **10**(9), 1079 (2021)

31. Wang, Z.M., Hu, S.Y., Song, H.: Channel selection method for EEG emotion recognition using normalized mutual information. IEEE Access **7**, 143303–143311 (2019)

32. Yang, S., Yin, Z., Wang, Y., Zhang, W., Wang, Y., Zhang, J.: Assessing cognitive mental workload via EEG signals and an ensemble deep learning classifier based on denoising autoencoders. Comput. Biol. Med. **109**, 159–170 (2019)

33. Yin, Z., Zhang, J.: Cross-session classification of mental workload levels using EEG and an adaptive deep learning model. Biomed. Signal Process. Control **33**, 30–47 (2017)

What Song Am I Thinking Of?

Niall McGuire[(✉)] and Yashar Moshfeghi

NeuraSearch Laboratory, Department of Computer and Information Sciences,
University of Strathclyde, Glasgow, Scotland
{niall.mcguire,yashar.moshfeghi}@strath.ac.uk

Abstract. Information Need (IN) is a complex phenomenon due to the difficulty experienced when realising and formulating it into a query format. This leads to a semantic gap between the IN and its representation (e.g., the query). Studies have investigated techniques to bridge this gap by using neurophysiological features. Music Information Retrieval (MIR) is a sub-field of IR that could greatly benefit from bridging the gap between IN and query, as songs present an acute challenge for IR systems. A searcher may be able to recall/imagine a piece of music they wish to search for but still need to remember key pieces of information (title, artist, lyrics) used to formulate a query that an IR system can process. Although, if a MIR system could understand the imagined song, it may allow the searcher to satisfy their IN better. As such, in this study, we aim to investigate the possibility of detecting pieces from Electroencephalogram (EEG) signals captured while participants *"listen"* to or *"imagine"* songs. We employ six machine learning models on the publicly available data set, OpenMIIR. In the model training phase, we devised several experiment scenarios to explore the capabilities of the models to determine the potential effectiveness of Perceived and Imagined EEG song data in a MIR system. Our results show that, firstly, we can detect perceived songs using the recorded brain signals, with an accuracy of 62.0% (SD 5.4%). Furthermore, we classified imagined songs with an accuracy of 60.8% (SD 13.2%). Insightful results were also gained from several experiment scenarios presented within this paper. Overall, the encouraging results produced by this study are a crucial step towards information retrieval systems capable of interpreting INs from the brain, which can help alleviate the semantic gap's negative impact on information retrieval.

Keywords: Information systems · Information retrieval · Music Retrieval · Brain · EEG · Machine Learning

1 Introduction

All Information Retrieval (IR) systems aim to satisfy searchers' Information Needs (IN). Many IR systems rely on the searcher to assess their information requirements and convey them to the system, typically in the form of queries [58]. However, conventional queries may suffer from semantic gaps between the

G. Nicosia et al. (Eds.): LOD 2023, LNCS 14506, pp. 418–432, 2024.
https://doi.org/10.1007/978-3-031-53966-4_31

searcher's true IN and their formulated query [17,54]. Various studies [34] have tried to alleviate this problem by designing an IR system capable of understanding INs from biological features. Studies such as [36] focus on the use of Functional Magnetic Resonance Imaging (fMRI) to analyse the Blood Oxygenation Level Dependent (BOLD) signals to detect which areas of the brain are more active in the process of realising IN. Alternative methods to fMRI, such as Electroencephalograms (EEG), have been employed to understand further how INs are realised within a subject [28].

A subdivision of the IR field that may greatly benefit from the development of an IR system that can interpret INs from the brain is Music Information Retrieval (MIR) [46]. MIR is exploring and organising enormous amounts of music or music information based on its relevance to specific queries [46]. Recently this area has seen the use of EEG data to extract music-related information from subjects [15,30,49]. Prior studies have shown the ability to extract the tempo and meter of the music stimuli subjects are exposed to from their EEG data [50]. At the same time, others have successfully utilised EEG data that was recorded while a subject listened to a segment of music. To then determine which song the subject perceives solely from their brain signals.

Although these prior studies have demonstrated the potential of **Perceived** EEG song data for extracting music-related information [49,51], they have yet to effectively demonstrate the ability of **Imagined** EEG song data to extract song-related information for potential use within an IR system, which could then be consolidated into a MIR system, allowing for more efficient satisfaction of a searcher's IN. Thus, to start building towards a brain-driven MIR system, in this study, we aim to answer the following research questions:

- **RQ1:** *"Is it possible to classify a song from subjects' EEG data whilst Perceiving a given song?"*
- **RQ2:** *"Is it possible to classify a song from subjects' EEG data whilst Imagining a given song?"*
- **RQ3:** *"How similar are Perceived and Imagined EEG data? and can they be used interchangeably for training and testing of machine learning models"*
- **RQ4:** *"How much of a song would a subject have to Perceive/Imagine before the system can produce accurate results?"*

2   Background

2.1   Information Need Complexity

Information need (IN) is one of the key concepts in IR theory [24]. The subjective nature of IN makes it complex for IR systems to fully comprehend [44]. The source of this complexity lies in the IN's paradoxical nature, meaning that unlike many other essential human needs such as hunger or fatigue, the information that a subject requires to satisfy their needs is most likely unknown to the searcher. In prior studies, this has been explained by the idea that an IN is "intangible and visceral" and thus is "unknowable and non-specifiable", this creates a conundrum for the searcher such that if they are unable to create a query proficient enough for an IR system to understand and retrieve relevant documents, then they are unable to satisfy their IN [4,7]. An IN can link to a discrepancy in knowledge [2,3], uncertainty [26,61], anxiety [8], dissatisfaction [52], or doubt [54]. Many researchers have attempted to better understand this complex phenomenon in the past [2,26,53]. Searchers will realise an internal IN when presented with a gap within their Anomalous States of Knowledge (ASK) [4]; this, in turn, sets the search process into motion. A searcher would begin this process by transposing their IN into a suitable query processable by an IR system, this then prompts the IR systems to retrieve documents likely relevant to the initial query to fulfil the searchers IN [8], followed by the searcher reviewing the documents presented by the IR system and taking in any relevant information to satisfy their IN [36]. Unfortunately, it is common that when reviewing the output of the IR system, searchers will find that their IN is not sufficiently satisfied [43] firstly, since the IN was not formulated correctly. Secondly, due to the transformation of the IN into a query represented by keywords, which are considered to be noisy and uncertain [17] as they are limited as to how close they can be to the IN [54].

2.2   NeuraSearch

Recently, studies have begun utilising neurophysiological features from searchers in an attempt to understand IN information, this area of research has gone under the term NeuraSearch [28,31]. The widely utilised neuro-imaging modalities for this purpose include EEG [1,13,18,20,32–35,37,41,57], Functional Magnetic Resonance Imaging (fMRI) [32,34,37,39,40], Magnetoencephalography (MEG) [23], Functional Near Infrared Spectroscopy (fNIRS) [27,29]. Among these powerful techniques, EEG has drawn more attention from researchers because of its high temporal resolution (millisecond scale) and unobtrusiveness. EEG can be deployed in both clinical settings, such as to predict patients' epileptic seizure [56], stroke [19], or brain tumour [22], and non-clinical settings, including detecting fatigue [11], or performing a recognition task [9] or a realisation of IN [10]. Research conducted by [6,21,28] saw the utilisation of EEG, which has a high temporal resolution that allows for real-time capture of brain signals and Event-Related Potentials (ERP) to observe the brain whilst participants developed an IN. It can facilitate information acquisition performance [60]. The findings of this study [59] showed that the realisation of an IN is developed within the brain before the consciousness of the subject observes it. With the development of the understanding and processing of INs from neurophysiological features [48,55], the possibility of using said features within practical applications to overcome the issue of noisy and ill-defined IN queries becomes even more feasible [1,28]. A core area within IR that suffers substantially from the use of keyword queries is that of MIR. This is because current MIR systems require users to present their query using metadata such as the song title, lyrics, or the artist's name [46]. However, if the IN needs to be defined and the searcher cannot recall these pieces of information, they may have to imitate the desired song through singing/humming [51]. As a result, MIR may benefit significantly from an IR system capable of processing internally imagined song information as a query and obtaining relevant documents to the imagined song query [42].

2.3 Neuroscience and MIR

Recent works have begun incorporating neurophysiological data with Music Information Retrieval (MIR) applications. An early example of using EEG within MIR can be found in [45], where researchers had 10 participants listen to 7 short melody segments ranging from 3.26 to 4.36 s. A logistic regression classifier was utilised to classify the ERP of each trial. This study showed an above random classified accuracy that, when applied to individual subjects, varied from 25% to 70%, whereas when applied across all subjects, it ranged from 35% to 53%. Within works by [51], the researchers created the OpenMIIR data set, which contains the EEG data of participants who were recorded while exposed to several short music segments, as well as the recordings of those same participants asked to then imagine the song segment. This data set was then used in a follow-up study by [50], in which the researchers attempted to classify which song the participants were perceiving/imagining. A Support Vector Machine (SVM) and a Neural Network (NN) were used to classify the music from the perceived EEG data. The results showed that the SVM had a classification accuracy of 27.59% for the perceived data, and the NN had a classification accuracy of 27.22%, which is significantly higher than random classification (8%). In a more recent study by [38], the authors use the perceived EEG song data from the OpenMIIR data set to train various machine learning models on 9 out of the ten subjects, where the last subject was used for evaluating the models. The results from their study presented a song classification rate varying from 52.5% to 24%. Although these previous studies have achieved accuracy scores substantially higher than random classification (8%), there is still room for significant improvement in overall classification accuracy and prediction variance. As well as this, there has been no reporting of the above random classification of the Imagined song segments. This is arguably the most significant part of creating a MIR system capable of interpreting IN directly from a subject's brain hence why the findings presented within this paper are of great significance to tackling this problem.

3   Methodology

The main steps of this study are EEG signal acquisition, artefact removal, feature engineering, model evaluation and EEG music classification. Details of each step will be described in the following sections.

3.1   Data Set

Within this study, the OpenMIIR dataset was used to provide subject EEG data [51]. The dataset contains the EEG signals of 10 subjects recorded using 64 + 2 EEG electrodes sampled at a frequency of 512 Hz. As well as this, the horizontal and vertical EOG channels were recorded to capture the eye movements of the subject. The dataset was produced by having each subject perceive and imagine 12 short pieces of music 5 times to create repetitions. Each song ranged between 7 to 16 s in length whilst their EEG data was recorded. The short music segments are taken from a variety of musical pieces that varied across musical genres, songs that were recorded without lyrics, songs where the lyrics have been removed and purely instrumental pieces. Although the original dataset was captured under four various conditions, we only make use of one of those four conditions for this study, this being the recording of the subject's EEG data whilst they are perceiving a short music segment.

3.2   EEG Preprocessing

During EEG recording it is commonplace for electrical activities caused by the actions of the individual to affect the data causing artefacts to form. These are unwanted signal fragments that can negatively impact measurements and skew results. Thus, it is a crucial step to remove these artefacts to the best of our ability, to achieve this we made use of the MNE python toolbox [12]. Firstly, "bad" channels that had previously been marked by the dataset creator were removed from the raw dataset. This was then followed by performing channel interpolation.

Often during EEG recording the skin-electrolyte-electrode interface produces a drift in voltage which can be picked up by the EEG electrodes [16], to account for this, we filtered out these drifts using EEG using a bandpass filter with a frequency ranging from 0.5 Hz to 40 Hz. Another artefact that may be present in the data is the electrical interference caused by the subject blinking their eye, to handle this, we applied the Independent Component Analysis (ICA) technique which computes the independent components within the raw dataset allowing us to remove any components that share a high correlation with the EOG channels. The 64 EEG channels were then reconstructed from the remaining independent components.

3.3   EEG Feature Extraction

To capture significant EEG signal characteristics, we generated features from the processed EEG data. The details of features in their categories are discussed below:

1. **Morphological features:** the number of peaks, average non-linear energy and the curve length were extracted. Features and three morphological features were extracted in two different ways. Firstly, they were calculated at four frequency bands theta (4–8 Hz), alpha (8–13 Hz), beta (13–25 Hz), and low gamma (25–40 Hz).
2. **Statistical features:** skewness, mean, and kurtosis were computed to identify the distribution of the signal. [25] was calculated to describe time-varying processes

3. **non-linear features:** The approximate entropy (ApEn) and Hurst exponent (H) were used to quantify the unpredictability of fluctuations over a time series as well as measure the self-similarity of the time series, respectively.

3.4 Models

Machine learning techniques, particularly advanced models such as deep learning, have been employed to accurately extract variance characteristics in EEG data for song identification [47]. Therefore, to investigate the research questions, we adopted the six machine learning models: GNB, SVM, GRU, BGRU, LSTM, and BLSTM, in our analysis. These models have been widely used in the EEG signal for classification [5, 25]. GRU, BGRU, LSTM, and BLSTM are beneficial for learning sequential data with long-term dependencies [62].

Table 1. Deep learning model architectures

Model	Layers/Nodes
GRU	G256-G128-D12
BGRU	BG256-G128-D12
LSTM	L256-L128-D12
BLSTM	BL256-L128-D12

The architectures of the deep learning models are shown in Table 1. G, BG, L, BL, and D correspond to GRU, BGRU, LSTM, BLSTM, and Dense layer, respectively. For example, BL256-L128-D12 implies that there is a BLSTM layer with 256 units, an LSTM layer with 128 units and since there are twelve classes in our analysis, a Dense layer with twelve outputs for each model. Softmax activation was implemented in the last layer. This work implemented a dropout rate of 0.2, and Adam was used to train all deep-learning models. Additionally, early stopping was utilised to prevent an over-fitting issue. Furthermore, we stopped training after the model's performance stopped increasing after 200 epochs.

3.5 Classification Scenarios

To address each research question, we created several experimental scenarios, which are detailed as follows:

Scenario 1: Perceived/Imagined Song Classification. To investigate **RQ1** and **RQ2**, our model performs classification utilising the EEG signals of all subjects, bar one subjects data, that was left out of the training set and used for testing purposes, creating a 10-fold leave one out cross validation training loop (LOOCV). The task was performed twice using two distinct groups of data;

perception (**RQ1**) and imagination (**RQ2**), and classification was performed for each repetition in each stage to determine the difference in model performance (**RQ3**).

Scenario 2: Interchanging of Perceived and Imagined EEG Music Classification. To investigate **RQ3** and determine whether imagined and perceived songs can be classified interchangeably. Our models were trained with EEG signals from the perception stage and tested with EEG signals from the imagination stage, and vice versa. Like the first experiment condition, we again rotated the subject whose data was used to evaluate the model for both the perceived and imagined states creating a 10-fold LOOCV loop.

Scenario 3: Train/Test Data Limitation. To investigate **RQ4**, we trained the models the same way as in Scenario 1, but how we test our models is different. After completing the model training procedure, we test our model utilising n% of the initial portion of the data for each song repetition. The sizes of the testing data set varied between 10%, 25%, 50%, 75%, and 100%.

3.6   Metrics

This study evaluates the model performance using accuracy, precision, and recall.

Accuracy is the sum of the number of true positives and true negatives divided by the total number of examples.

Precision refers to the ratio of correct positive examples to the number of actual positive examples.

Recall is calculated as the sum of true positives across all classes divided by the sum of true positives and false negatives across all classes.

4   Results

In this section, we use a box plot to display the results for the models in several experimental scenarios. Over the ten cross-validation sets, each box plot presents five crucial pieces of information: the minimum, first, second (median), third, and maximum quarterlies [14], where the red dot in the box represents the mean of accuracy for each model.

4.1   Perceived/Imagined Song Classification

The results from training the models on the Perceived and Imagined datasets are presented in Fig. 1 and Fig. 2. Figure 1 details that the highest performing model trained on Perceived EEG song data is the LSTM at the fourth repetition with an average accuracy score of 62.0% (SD 5.4%). This result helps to address **RQ1**, demonstrating the ability of the techniques employed within this study to provide a higher classification accuracy than that achieved by prior studies.

As shown in Fig. 2, the highest performing model is again the LSTM on the fifth repetition with an average accuracy score of 60.8% (SD 13.2%). The result achieved by the LSTM trained on Imagined EEG song data helps to satisfy **RQ2**. Firstly, by showing that the techniques employed within this study can classify songs solely from Imagined music EEG data. Secondly, the model achieved a significantly higher classification score than random classification (8%). As well as, this Imagined classification was within 2% of Perceived classification, demonstrating the comparable performance of Imagined EEG data with Perceived EEG data. Further observations of both figures show that each model achieves similar results on both Perceived and Imagined data with slight variations. GRU, LSTM, BRGU, and BLSTM's accuracy appears to increase with the incremental addition of each repetition into the training data set, except GNB and SVM.

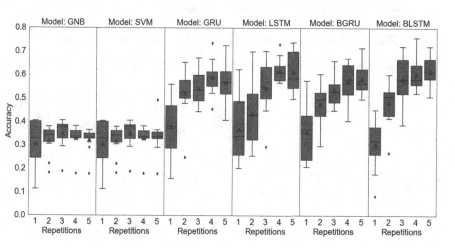

Fig. 1. Average results of models trained on **Perceived** song data

4.2   Testing Size Breakdown

The results of this experiment are presented in Fig. 3 and Fig. 4. The results illustrated in both figures highlight that the highest performing models at the smallest test sample size of 2 (which translates to 10% of the comprehensive testing data) are the Perceived GNB/SVM achieving an accuracy of 30.3% (SD 5.6%) and the Imagined GNB/SVM both scoring 30.5% (SD 5.2%). However, their accuracy falls behind the other models after the initial testing segment. Interestingly, when observing the performance of GRU, LSTM, BGRU, and BLSTM across both data types. We can see that segment sizes 10 (50% of testing) and 15 (75% of testing) share similar accuracy scores; for example, the highest Perceived model at segments 10 and 15 is the BLSTM with 57.3% (SD 5.5%) and 56.5% (SD 5.4%) respectively. Similarly, the highest Imagined model

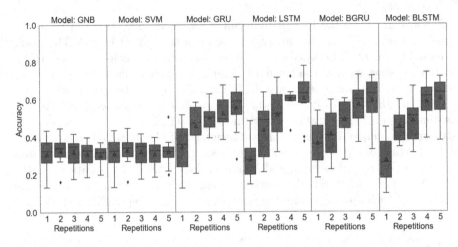

Fig. 2. Average results of models trained on **Imagined** song data

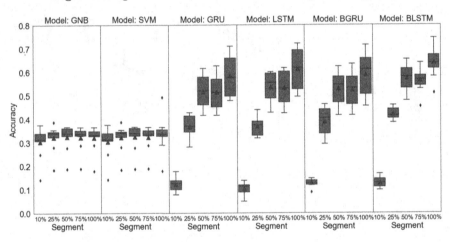

Fig. 3. Average results of models tested on the varied **Perceived** data set

at segments 10 and 15 is the LSTM which achieved 52.5% (SD 8.2%) and 53.4% (SD 7.3%). As well as this, the increase from segment size 10 to segment size 20 (100% of testing) does increase the accuracy significantly, with the Perceived BLSTM scoring accuracy of 64.1% (SD 6.9%) and the Imagined LSTM scoring 61.8% (SD 8.7%) on segment 20 (100% of the testing set). An important consideration when viewing these results is that the song segments used for training already average around 7's in length. That means if we use 50% of the test set and achieve a classification close to 50%, we are achieving that by using only 3.5's of an entire song. These results help to address **RQ4**.

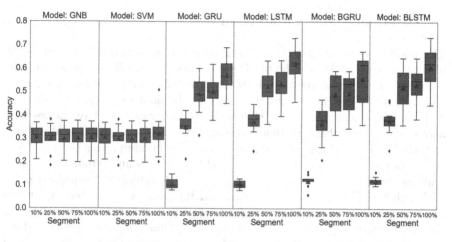

Fig. 4. Average results of models tested on the varied **Imagined** data set

Table 2. Interchanging of Perceived and Imagined for train/test

Trained on/Tested on	Model	Mean Precision (SD)	Mean Recall (SD)	Mean Accuracy (SD)
Perceived/Imagined	GNB	31.8%, (5.9%)	32.0%, (6.8%)	32.0%, (6.8%)
	SVM	43.7%, (12.7%)	43.7%, (12.4%)	43.7%, (12.4%)
	GRU	59.2%, (7.4%)	60.0%, (9.0%)	60.0%, (9.0%)
	LSTM	62.1%, (8.5%)	63.4%, (9.4%)	63.4%, (9.4%)
	BGRU	59.7%, (10.6%)	59.9%, (10.9%)	59.9%, (10.9%)
	BLSTM	62.1%, (8.5%)	63.6%, (8.8%)	63.6%, (8.8%)
Imagined/Perceived	GNB	25.6%, (6.2%)	19.1%, (5.6%)	19.1%, (5.6%)
	SVM	34.5%, (4.3%)	34.0%, (4.6%)	34.0%, (4.6%)
	GRU	57.4%, (8.8%)	57.5%, (8.9%)	57.5%, (8.9%)
	LSTM	61.2%, (8.3%)	61.4%, (7.5%)	61.4%, (7.5%)
	BGRU	54.6%, (7.9%)	56.3%, (8.7%)	56.3%, (8.7%)
	BLSTM	61.2%, (8.3%)	61.2%, (8.2%)	61.2%, (8.2%)

4.3   Changing of Perceived and Imagined for Train/test

The results of this scenario are detailed in Table 2 for the initial case of training the models on the perceived data set and evaluating them on the imagined data. We can observe that the highest performing model for this case is the BLSTM with a score of 63.6% (SD 8.8%). Furthermore, in the case where we trained on the imagined data and evaluated the perceived data. The model with the highest accuracy is the LSTM, with a score of 63.4% (SD 7.5%). The results highlight the comparable nature of Perceived and Imagined EEG song data. The findings of this experiment scenario help to directly address **RQ3**.

5   Discussion and Conclusion

The findings from our first experiment scenario, as seen in Fig. 1 and Fig. 2, directly address **RQ1** and **RQ2**. These results give significant weight to the points made earlier in the introduction on the effectiveness of Imagined EEG data to extract IN-related data that can then be used to provide the searcher with documents capable of satisfying their IN. In our case, we have successfully extracted a potential IN (an Imagined song) directly from neurological features (recorded EEG data) and created a system capable of interpreting those features into data that a standard IR system could then understand. From these initial findings, we can also see the similar nature of Perceived and Imagined data. Although the EEG data for Perceived and Imagined are collected under two different conditions, the models appear to share performance across both types, this could imply that the process of Perceiving and Imagining a song follows a similar process within the brain and produces a pattern that the machine learning models can identify.

The second experiment scenario was designed to address **RQ3**. The results from this experiment were displayed in Table 2. We observe that the models can classify songs from a data type (e.g. Perceived or Imagined) that it had yet to be trained on, at a similar classification rate across both data types. The models produced classification results comparable to those presented in the first experiment scenario when the models were trained and evaluated on the same data type, this adds further weight to the point that Perceiving and Imagining a song can produce similar EEG outputs that the models can interpret. This implies that for a real-world brain-driven MIR system, that no matter what data type the searcher presents to the system, it will likely be capable of classifying the song correctly.

Experiment scenario three results are displayed in Figs. 3 and 4. The findings from this experiment highlighted the ability of the models to detect the songs with 50% accuracy when given only 3.5 s of the music segment, this has significant implications for a brain-driven MIR system as it would be the system's goal to limit the time a searcher would have to Perceive/Imagine a song while producing the most optimal classification results.

In conclusion, this study investigated the use of neurophysiological features to represent searchers IN better. To help bridge the semantic gap between searchers IN. In particular, we focused on using neurophysiological characteristics within the MIR domain. Furthermore, we examined the EEG recordings of Perceived and Imagined music stimuli to assess their capabilities within a brain-driven MIR system. The findings produced by this study demonstrate the ability of our techniques to achieve Perceived song classification significantly higher than that achieved by prior studies. As well as this, we are the first to report above high accuracy classification of Imagined song stimuli from EEG data.

References

1. Allegretti, M., Moshfeghi, Y., Hadjigeorgieva, M., Pollick, F.E., Jose, J.M., Pasi, G.: When relevance judgement is happening? An EEG-based study. In: Proceedings of the 38th International ACM SIGIR Conference on Research and Development in Information Retrieval, pp. 719–722 (2015)
2. Belkin, N.J.: Anomalous states of knowledge as a basis for information retrieval. Can. J. Inf. Sci. **5**(1), 133–143 (1980)
3. Belkin, N.J., Oddy, R.N., Brooks, H.M.: Ask for information retrieval: Part II. Results of a design study. J. Doc. (1982)
4. Belkin, N.J., Oddy, R.N., Brooks, H.M.: Ask for information retrieval: Part I. Background and theory. J. Doc. **38**, 61–71 (1997)
5. Chen, J.X., Jiang, D.M., Zhang, Y.N.: A hierarchical bidirectional GRU model with attention for EEG-based emotion classification. IEEE Access **7**, 118530–118540 (2019). https://doi.org/10.1109/ACCESS.2019.2936817
6. Chen, X., et al.: Web search via an efficient and effective brain-machine interface. In: Proceedings of the Fifteenth ACM International Conference on Web Search and Data Mining, pp. 1569–1572 (2022)
7. Cole, C.: A theory of information need for information retrieval that connects information to knowledge. J. Am. Soc. Inf. Sci. Technol. **62**(7), 1216–1231 (2011). https://doi.org/10.1002/asi.21541, https://onlinelibrary.wiley.com/doi/abs/10.1002/asi.21541
8. Cole, C.: A theory of information need for information retrieval that connects information to knowledge. J. Am. Soc. Inf. Sci. Technol. **62**(7), 1216–1231 (2011)
9. Davis III, K.M., Kangassalo, L., Spapé, M., Ruotsalo, T.: BrainSourcing: crowdsourcing recognition tasks via collaborative brain-computer interfacing. In: Proceedings of the 2020 CHI Conference on Human Factors in Computing Systems, pp. 1–14 (2020)
10. Eugster, M.J., et al.: Predicting term-relevance from brain signals. In: Proceedings of the 37th International ACM SIGIR Conference on Research & Development in Information Retrieval, SIGIR 2014, pp. 425–434. Association for Computing Machinery, New York (2014). https://doi.org/10.1145/2600428.2609594
11. Gao, Z.K., Li, Y.L., Yang, Y.X., Ma, C.: A recurrence network-based convolutional neural network for fatigue driving detection from EEG. Chaos: Interdisc. J. Nonlinear Sci. **29**(11), 113126 (2019)
12. Gramfort, A., et al.: MEG and EEG data analysis with MNE-python. Front. Neurosci. 267 (2013)
13. Gwizdka, J., Hosseini, R., Cole, M., Wang, S.: Temporal dynamics of eye-tracking and EEG during reading and relevance decisions. J. Am. Soc. Inf. Sci. **68**(10), 2299–2312 (2017)
14. Hofmann, T.: Collaborative filtering via gaussian probabilistic latent semantic analysis. In: Proceedings of the 26th Annual International ACM SIGIR Conference on Research and Development in Information Retrieval, pp. 259–266 (2003)
15. Hsu, J.L., Zhen, Y.L., Lin, T.C., Chiu, Y.S.: Personalized music emotion recognition using electroencephalography (EEG). In: 2014 IEEE International Symposium on Multimedia, pp. 277–278 (2014). https://doi.org/10.1109/ISM.2014.19
16. Huigen, E., Peper, A., Grimbergen, C.: Investigation into the origin of the noise of surface electrodes. Med. Biol. Eng. Compu. **40**(3), 332–338 (2002)
17. Ingwersen, P., Järvelin, K.: The Turn: Integration of Information Seeking and Retrieval in Context. Springer, Heidelberg (2005). https://doi.org/10.1007/1-4020-3851-8

18. Jacucci, G., et al.: Integrating neurophysiologic relevance feedback in intent modeling for information retrieval. J. Am. Soc. Inf. Sci. **70**(9), 917–930 (2019)

19. Jordan, K.G.: Emergency EEG and continuous EEG monitoring in acute ischemic stroke. J. Clin. Neurophysiol. **21**(5), 341–352 (2004)

20. Kangassalo, L., Spapé, M., Jacucci, G., Ruotsalo, T.: Why do users issue good queries? neural correlates of term specificity. In: Proceedings of the 42nd International ACM SIGIR Conference on Research and Development in Information Retrieval, pp. 375–384 (2019)

21. Kangassalo, L., Spapé, M., Ravaja, N., Ruotsalo, T.: Information gain modulates brain activity evoked by reading. Sci. Rep. **10**(1), 1–10 (2020)

22. Karameh, F.N., Dahleh, M.A.: Automated classification of EEG signals in brain tumor diagnostics. In: Proceedings of the 2000 American Control Conference. ACC (IEEE Cat. No. 00CH36334), vol. 6, pp. 4169–4173. IEEE (2000)

23. Kauppi, J.P., et al.: Towards brain-activity-controlled information retrieval: decoding image relevance from meg signals. Neuroimage **112**, 288–298 (2015)

24. Keshavarz, H.: Human information behaviour and design, development and evaluation of information retrieval systems. Program **42**(4), 391–401 (2008)

25. Kingphai, K., Moshfeghi, Y.: On EEG preprocessing role in deep learning effectiveness for mental workload classification. In: Longo, L., Leva, M.C. (eds.) H-WORKLOAD 2021. CCIS, vol. 1493, pp. 81–98. Springer, Cham (2021). https://doi.org/10.1007/978-3-030-91408-0_6

26. Kuhlthau, C.C.: Inside the search process: information seeking from the user's perspective. J. Am. Soc. Inf. Sci. **42**(5), 361–371 (1991)

27. Maior, H.A., Ramchurn, R., Martindale, S., Cai, M., Wilson, M.L., Benford, S.: fNIRS and neurocinematics. In: Extended Abstracts of the 2019 CHI Conference on Human Factors in Computing Systems, pp. 1–6 (2019)

28. Michalkova, D., Parra-Rodriguez, M., Moshfeghi, Y.: Information need awareness: an EEG study. In: Proceedings of the 45th International ACM SIGIR Conference on Research and Development in Information Retrieval, SIGIR 2022, pp. 610–621. Association for Computing Machinery, New York (2022). https://doi.org/10.1145/3477495.3531999

29. Midha, S., Maior, H.A., Wilson, M.L., Sharples, S.: Measuring mental workload variations in office work tasks using fNIRS. Int. J. Hum Comput Stud. **147**, 102580 (2021)

30. Morita, Y., Huang, H.H., Kawagoe, K.: Towards music information retrieval driven by EEG signals: architecture and preliminary experiments. In: 2013 IEEE/ACIS 12th International Conference on Computer and Information Science (ICIS), pp. 213–217 (2013). https://doi.org/10.1109/ICIS.2013.6607843

31. Moshfeghi, Y.: NeuraSearch: neuroscience and information retrieval. In: CEUR Workshop Proceedings, vol. 2950, pp. 193–194 (2021)

32. Moshfeghi, Y., Pinto, L.R., Pollick, F.E., Jose, J.M.: Understanding relevance: an fMRI study. In: Serdyukov, P., et al. (eds.) ECIR 2013. LNCS, vol. 7814, pp. 14–25. Springer, Heidelberg (2013). https://doi.org/10.1007/978-3-642-36973-5_2

33. Moshfeghi, Y., Pollick, F.E.: Search process as transitions between neural states In: Proceedings of the 2018 World Wide Web Conference, pp. 1683–1692 (2018)

34. Moshfeghi, Y., Pollick, F.E.: Neuropsychological model of the realization of information need. J. Am. Soc. Inf. Sci. **70**(9), 954–967 (2019)

35. Moshfeghi, Y., Triantafillou, P., Pollick, F.: Towards predicting a realisation of an information need based on brain signals. In: The World Wide Web Conference, pp. 1300–1309 (2019)

36. Moshfeghi, Y., Triantafillou, P., Pollick, F.E.: Understanding information need: an fMRI study. In: Proceedings of the 39th International ACM SIGIR Conference on Research and Development in Information Retrieval, SIGIR 2016, pp. 335–344. Association for Computing Machinery, New York (2016). https://doi.org/10.1145/2911451.2911534

37. Moshfeghi, Y., Triantafillou, P., Pollick, F.E.: Understanding information need: an fMRI study. In: Proceedings of the 39th International ACM SIGIR Conference on Research and Development in Information Retrieval, pp. 335–344 (2016)

38. Ntalampiras, S., Potamitis, I.: A statistical inference framework for understanding music-related brain activity. IEEE J. Sel. Top. Signal Process. **13**(2), 275–284 (2019). https://doi.org/10.1109/JSTSP.2019.2905431

39. Paisalnan, S., Moshfeghi, Y., Pollick, F.: Neural correlates of realisation of satisfaction in a successful search process. Proc. Assoc. Inf. Sci. Technol. **58**(1), 282–291 (2021)

40. Paisalnan, S., Pollick, F., Moshfeghi, Y.: Towards understanding neuroscience of realisation of information need in light of relevance and satisfaction judgement. In: Nicosia, G., et al. (eds.) LOD 2021. LNCS, vol. 13163, pp. 41–56. Springer, Cham (2022). https://doi.org/10.1007/978-3-030-95467-3_3

41. Pang, S., Hu, X., Cai, Z., Gong, J., Zhang, M.: Building change detection from bi-temporal dense-matching point clouds and aerial images. Sensors **18**(4), 966 (2018)

42. Ras, Z.W., Wieczorkowska, A.: Advances in Music Information Retrieval, vol. 274. Springer, Heidelberg (2010). https://doi.org/10.1007/978-3-642-11674-2

43. van Rijsbergen, C.J.: (Invited paper) a new theoretical framework for information retrieval. In: Proceedings of the 9th Annual International ACM SIGIR Conference on Research and Development in Information Retrieval, SIGIR 1986, pp. 194–200. Association for Computing Machinery, New York (1986). https://doi.org/10.1145/253168.253208

44. Savolainen, R.: Information need as trigger and driver of information seeking: a conceptual analysis. Aslib J. Inf. Manage. **69**, 2–21 (2017)

45. Schaefer, R.S., Farquhar, J., Blokland, Y., Sadakata, M., Desain, P.: Name that tune: decoding music from the listening brain. NeuroImage **56**(2), 843–849 (2011). https://doi.org/10.1016/j.neuroimage.2010.05.084, https://www.sciencedirect.com/science/article/pii/S1053811910008402, multivariate Decoding and Brain Reading

46. Schedl, M., Gómez, E., Urbano, J., et al.: Music information retrieval: recent developments and applications. Found. Trends® Inf. Retrieval **8**(2–3), 127–261 (2014)

47. Sonawane, D., Miyapuram, K.P., Rs, B., Lomas, D.J.: GuessTheMusic: song identification from electroencephalography response. In: Proceedings of the 3rd ACM India Joint International Conference on Data Science & Management of Data (8th ACM IKDD CODS & 26th COMAD), pp. 154–162 (2021)

48. Spape, M., Davis, K., Kangassalo, L., Ravaja, N., Sovijarvi-Spape, Z., Ruotsalo, T.: Brain-computer interface for generating personally attractive images. IEEE Trans. Affect. Comput. **1**(1) (2021)

49. Stober, S.: Toward studying music cognition with information retrieval techniques: Lessons learned from the OpenMIIR initiative. Front. Psychol. **8**, 1255 (2017)

50. Stober, S.: Toward studying music cognition with information retrieval techniques: lessons learned from the OpenMIIR initiative. Front. Psychol. **8** (2017). https://doi.org/10.3389/fpsyg.2017.01255, https://www.frontiersin.org/articles/10.3389/fpsyg.2017.01255

51. Stober, S., Sternin, A., Owen, A.M., Grahn, J.A.: Towards music imagery information retrieval: introducing the OpenMIIR dataset of EEG recordings from music perception and imagination. In: International Society for Music Information Retrieval Conference (2015)

52. Taylor, R.S.: The process of asking questions. Am. Doc. **13**(4), 391–396 (1962)

53. Taylor, R.S.: Question-negotiation an information-seeking in libraries. Technical report, Lehigh Univ Bethlehem PA Center for Information Science (1967)

54. Taylor, R.S.: Question-negotiation and information seeking in libraries. Coll. Res. Libr. **76**, 251–267 (1968)

55. de la Torre-Ortiz, C., Spapé, M.M., Kangassalo, L., Ruotsalo, T.: Brain relevance feedback for interactive image generation. In: Proceedings of the 33rd Annual ACM Symposium on User Interface Software and Technology, pp. 1060–1070 (2020)

56. Tzallas, A.T., Tsipouras, M.G., Fotiadis, D.I.: Epileptic seizure detection in EEGs using time-frequency analysis. IEEE Trans. Inf Technol. Biomed. **13**(5), 703–710 (2009)

57. van der Veen, V., dutt-Sharma, N., Cavallaro, L., Bos, H.: Memory errors: the past, the present, and the future. In: Balzarotti, D., Stolfo, S.J., Cova, M. (eds.) RAID 2012. LNCS, vol. 7462, pp. 86–106. Springer, Heidelberg (2012). https://doi.org/10.1007/978-3-642-33338-5_5

58. Wissbrock, F.: Information need assessment in information retrieval; beyond lists and queries. In: Proceedings of the 27th German Conference on Artificial Intelligence (2004)

59. Ye, Z., et al.: Brain topography adaptive network for satisfaction modeling in interactive information access system. In: Proceedings of the 30th ACM International Conference on Multimedia, pp. 90–100 (2022)

60. Ye, Z., et al.: Towards a better understanding of human reading comprehension with brain signals. In: Proceedings of the ACM Web Conference 2022, pp. 380–391 (2022)

61. Zhang, J., et al.: Global or local: constructing personalized click models for web search. In: Proceedings of the ACM Web Conference 2022, pp. 213–223 (2022)

62. Zhao, R., Yan, R., Wang, J., Mao, K.: Learning to monitor machine health with convolutional bi-directional LSTM networks. Sensors **17**(2), 273 (2017)

Path-Weights and Layer-Wise Relevance Propagation for Explainability of ANNs with fMRI Data

José Diogo Marques dos Santos[1,2] (iD) and José Paulo Marques dos Santos[3,4,5(✉)] (iD)

[1] Faculty of Engineering, University of Porto, R. Dr Roberto Frias, 4200-465 Porto, Portugal
[2] Abel Salazar Biomedical Sciences Institute, University of Porto, R. Jorge de Viterbo Ferreira, 4050-313 Porto, Portugal
[3] University of Maia, Av. Carlos de Oliveira Campos, 4475-690 Maia, Portugal
jpsantos@umaia.pt
[4] LIACC - Artificial Intelligence and Computer Science Laboratory, University of Porto, R. Dr Roberto Frias, 4200-465 Porto, Portugal
[5] Unit of Experimental Biology, Faculty of Medicine, University of Porto, Alameda Prof. Hernâni Monteiro, 4200-319 Porto, Portugal

Abstract. The application of artificial neural networks (ANNs) to functional magnetic resonance imaging (fMRI) data has recently gained renewed attention for signal analysis, modeling the underlying processes, and knowledge extraction. Although adequately trained ANNs characterize by high predictive performance, the intrinsic models tend to be inscrutable due to their complex architectures. Still, explainable artificial intelligence (xAI) looks to find methods that can help to delve into ANNs' structures and reveal which inputs most contribute to correct predictions and how the networks unroll calculations until the final decision.

Several methods have been proposed to explain the black-box ANNs' decisions, with layer-wise relevance propagation (LRP) being the current state-of-the-art. This study aims to investigate the consistency between LRP-based and path-weight-based analysis and how the network's pruning and retraining processes affect each method in the context of fMRI data analysis.

The procedure is tested with fMRI data obtained in a motor paradigm. Both methods were applied to a fully connected ANN, and to pruned and retrained versions. The results show that both methods agree on the most relevant inputs for each stimulus. The pruning process did not lead to major disagreements. Retraining affected both methods similarly, exacerbating the changes initially observed in the pruning process. Notably, the inputs retained for the ultimate ANN are in accordance with the established neuroscientific literature concerning motor action in the brain, validating the procedure and explaining methods. Therefore, both methods can yield valuable insights for understanding the original fMRI data and extracting knowledge.

Keywords: Artificial neural networks (ANN) · Explainable artificial intelligence (XAI) · Layer-wise relevance propagation (LRP) · Functional magnetic resonance imaging (fMRI)

G. Nicosia et al. (Eds.): LOD 2023, LNCS 14506, pp. 433–448, 2024.
https://doi.org/10.1007/978-3-031-53966-4_32

1 Introduction

Although functional magnetic resonance imaging (fMRI) data analysis with artificial neural networks (ANNs) is more than one decade old [1–5], it has received renewed recent interest [6, 7]. Inherently noisy and highly correlated data (spatially and temporally), extreme unbalance between inputs (magnitude of hundreds of thousands) and training epochs (magnitude of hundreds), difficulty in finding features pertinent to the study's goal, figure among the common hurdles, although the recognized advantages in ANNs' modeling abilities, which may be used for decoding brain states, for example [8]. An additional difficulty in ANNs is understanding how they make predictions [9, 10]. Models that deliver high prediction accuracies are helpful. However, if they are transparent, allowing one to understand which inputs contribute more to the correct predictions (explain) and understand how the progress of calculation leads to the prediction (interpret), they would be even more helpful, for example, because such models may contribute to improving knowledge about the process. Therefore, explainable and interpretable artificial intelligence (XAI) in ANNs is needed in neuroscience.

Simply put, an fMRI data file is a 4D collection of values, where there is a 3D array containing voxels, i.e., "pixel with volume", and the signal in each voxel varies in time, the other 1D. Thus, the fMRI data is a 3D array whose cells contain signals that fluctuate in time. A typical acquisition involves around 150,000 voxels, acquired 200 to 400 times in a study involving around 30 subjects. This is the source of the extreme unbalance between inputs and training epochs if each voxel is considered an input. The result would be a subtrained network, even for shallow neural networks, due to the excess of connections taking into account the available training epochs.

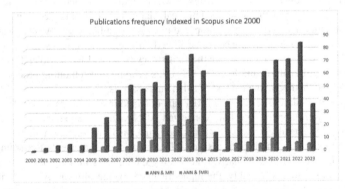

Fig. 1. Publications frequency indexed in Scopus since 2000, using the keywords combination "ANN & MRI" (blue) and "ANN & fMRI" (orange) as of 2023-06-07. (Color figure online)

To overcome the unbalancing issue, authors have been limiting the application of ANNs to parts of the brain [1] or to voxel clusters [2], to identify the participation of resting state networks at the individual level [5], using PCA to reduce data dimensionality [3], or ICA for the same purpose [4, 11]. Despite the initial impetus, it is remarked that few studies use ANNs for the analysis of fMRI data, at least when compared to studies that use ANNs for the analysis of anatomical MRI images (cf. Fig. 1).

Some causes for the dejection observed in the late 2010s, following the somewhat hype witnessed at the beginning of the decade, may be those already pointed out in the first paragraph. Among them, the extreme unbalance between inputs and training epochs may be the most challenging hurdle. Nonetheless, some solutions have been proposed with consequent interesting results [3, 4, 11], which means that ANN-based models for whole-brain fMRI data analysis were already built and are predicting correctly well above the chance level. This means that modeling has extracted and screened useful information from data. However, neither PCA nor ICA selects the sources of information most relevant for the process. A drawback in ANN's modeling is its "black box" nature [12].

A recent development has been the introduction of explainable artificial intelligence (XAI), mainly in health-related models, aiming in helping to understand and comprehend how the ANN works to achieve high prediction rates [9, 10]. Explaining an ANN means selecting the sources of information most relevant to the process. In the case of the fMRI data analysis, that means selecting the inputs that encompass information pertinent for the appropriate modeling and consequent correct predictions. If ICA is used for dimensionality reduction, that means selecting the most important independent components (ICs).

Dense networks are challenging to explain and interpret. Therefore, it is common to prune the network in the XAI process, which facilitates its explanation. Pruned networks are less dense, although such operation tends to reduce their prediction performance. One common further step is to retrain the network aiming to recover the prediction rates [13, 14].

Addressing the explainability of ANN-built models of fMRI data has been recently tackled [15–17]. One computational model for such purpose is layer-wise relevance propagation (LRP) [13, 18], which was already applied in ANNs [6, 19]. The purpose of the present study is to compare LRP and the path-weights concept suggested in [15, 16], answering the following questions:

Is the path-weights-based analysis in accordance with the state-of-the-art ANN interpretability method of LRP-based analysis regarding input importance for the network's prediction?

Does the pruning process affect the path-weights-based analysis and the LRP-based analysis similarly?

Does the retraining process affect the path-weights-based analysis and the LRP-based analysis similarly?

To test the two explaining procedures, a publicly accessible dataset containing fMRI acquisitions of a motor paradigm is used. Choosing a motor paradigm is the first step in the process of validation. Motor action in the brain has been studied [20]. Thus, it may be used as a reference framework for validation purposes, i.e., an explaining method must achieve the targets predicted by established neuroscientific knowledge to validate

2   Method

The functional magnetic resonance imaging (fMRI) data processing stages are represented in Fig. 2. It encompasses three stages: firstly, the raw data processing, which includes extracting the data from the publicly accessible dataset and preparing it, split the data into two groups, train and test, reduce train data's dimensionality with independent component analysis (ICA) and generate masks for test data screening, and, finally, extract features from the two datasets; secondly, the two datasets, train and test, are used to build the model, which is a fully connected artificial neural network (ANN); thirdly, the model is refined, passing through a pruning stage and subsequent model retraining. The processing yields three network models: one fully connected, another just pruned, and another one that is pruned and retrained. These models are explained with path-weights and layer-wise relevance propagation (LRP), whose metrics are addressed at the end of this section.

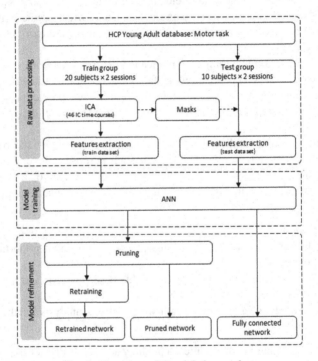

Fig. 2. Flowchart of the global procedure.

2.1 Raw Data Processing

The first stage involves preliminary raw data processing aiming at adequate data for the ANN training and testing. Data is split into two groups, train and test, and there is no further mixing between the two datasets until the end. By the end of this stage, features are extracted in both datasets.

FMRI Data. Raw data is obtained from the Young Adults database of the Human Connectome Project (HCP), specifically from the motor paradigm in the 100 Unrelated Subjects subset, and its characteristics are detailed elsewhere [16]. In this paradigm, subjects are asked to perform five tasks [21, 22]:

- LF: squeeze their left foot;
- LH: tap their left-hand fingers;
- RF: squeeze their right foot;
- RH: tap their right-hand fingers;
- T: move their tongue.

Train and Test Datasets. The 30 acquisitions were randomly distributed by the two groups, 20 for the train group and 10 for the test group. Because the two sessions have different sequences, we considered them independently. Therefore, the train group has $20 \times 2 = 40$ files, and the test group has $10 \times 2 = 20$ files. There is no mixing of subjects' data between the two groups.

Dimensionality Reduction with ICA. The procedure for dimensionality reduction has already been described elsewhere [16]. The output is a set of 46 ICs' timecourses.

Feature Extraction. The feature extraction procedure is the same for train and test groups, although data is never mixed. We assume the peak of the haemodynamic response is the feature that represents the functional participation of the brain in the task. To that end, the seventh, eighth, and ninth time points after stimulus onset are averaged, aiming to maximize the capture of the haemodynamic response peak. On the contrary, we assume the haemodynamic response's baseline to be non-task relevant. Hopefully, the differences between the haemodynamic response's peaks and baselines are proxies of the brain's participation in the tasks.

The mean time difference between the seventh, eighth, and ninth time points and stimulus onset is 5.285 s, which is expected to be near the peak [23]. After averaging the selected time points and standardization, the result for the train group dataset is a matrix with 400 rows (20 subjects $\times$ 2 sessions $\times$ 5 stimuli $\times$ 2 stimulus/session) and 46 columns (each corresponding to an IC), and a matrix with 200 rows (10 subjects $\times$ 2 sessions $\times$ 5 stimuli $\times$ 2 stimulus/session) and 46 columns (each corresponding to an IC mask) for the test group. Each row corresponds to a training or testing epoch.

2.2 ANN Architecture, Training, and Testing

The AMORE package v. 0.2–15 [24] implemented in R v. 4.2.3 [25] and RStudio v. 2023.03.0 Build 386 is used to design and perform the necessary calculations of the backpropagation feedforward shallow neural network (ANN).

The ANN has one hidden layer composed of 10 hidden nodes. The intention is to start with a network as simple as possible. Inputs are 46, each corresponding to an IC, and the outputs are five, each corresponding to a task (LF, LH, RF, RH, and T). The ANN is fully connected. The hidden nodes' activation function is hyperbolic tangent ("tansig"), and the output nodes' activation function is sigmoid. The training method is the adaptive gradient descent with momentum (ADAPTGDwm), and the neural network training error criteria is the least mean squares error (LMS).

The R script is run 50,000 times, and the network with the highest global correct prediction is chosen (the "best network"). Its global accuracy is 83.0%.

2.3   Model Refinement

Although the "best network" is considered in the path-weights-based and LRP-based analysis, two more versions are considered in parallel: one pruned, and another pruned and retrained. Pruned, and pruned and retrained networks are simpler than the "best network", which is fully connected, which may facilitate the network's interpretability and explainability.

Pruning. Network pruning may be performed by two strategies: pruning nodes or pruning connection weights. It was found that pruning nodes is preferable [26]. However, removing a hidden node in the present shallow neural network would represent removing 51 connection weights at once (46 afferent weights from the inputs plus 5 efferent weights to the outputs), which is 10% of the network's connections. Pursuing a finer approach to pruning, nonetheless, the option here is for pruning connection weights.

Following some pruning criteria for fMRI data discussed and tested recently [16, 17], it is considered the path-weights ranking [16]. It is possible to do the pruning globally, i.e., firstly, rank all the possible path-weights (in this case $46 \times 10 \times 5 = 2300$), then, threshold, and, finally, remove the path-weights below the threshold, as well as per output, i.e., firstly, rank the path-weights per output (in this case $46 \times 10 = 460$), then, threshold, and, finally, remove the path-weights below the threshold in each input. We found to be preferable to prune per output (unpublished data).

"Best network's" 460 path-weights per output (46 weights from the inputs to one hidden node $\times$ 10 hidden nodes) were ranked. The top 10 higher ranked path-weights per output were kept. All the weights that did not participate in one of these kept path-weights at least were pruned, resulting in a lightened network, named the "pruned network".

Retraining. The same R package, AMORE, is used in the network retraining stage with the same parameters except for the learning rate and momentum. A new exploration found that 0.900 is the best for the learning rate and 0.65 for momentum. The "pruned network" biases and weights are used as initial values, and the original train dataset is presented 500 times for retraining purposes. The yielded network has the same structure as the "pruned network" but with updated biases and weights. It is named the "retrained network".

2.4   Layer-Wise Relevance Propagation (LRP) Analysis

Layer-Wise Relevance Propagation (LRP) is a state-of-the-art method for interpreting an ANN's predictions introduced by Bach, Binder, Montavon, Klauschen, Müller and Samek [18]. LRP analysis works by computing relevance scores for each input feature by propagating the prediction through the network in reverse order, i.e., from the output layer to the input, allowing the user to identify input features' importance in the final prediction. The LRP value is calculated according to the basic rule (LRP-0) as presented in [13]:

$$R_j = \sum_k \frac{a_j w_{jk}}{\sum_{0,j} a_j w_{jk}} R_k \tag{1}$$

where R_j is the relevance of node j, a_j is the activation of node j, w_{jk} is the weight of the connection between nodes j and k, and R_k is the relevance of node k. According to the LRP-0 rule, the relevance value of an output node is the prediction score of the output node before applying the activation function.

Although LRP is usually applied to deep neural networks, in order to help explain their black-box nature and, thus, enabling the extraction of pertinent information and useful knowledge from such models, it has already been used to explain neural networks' decisions, for instance, regarding EEG motor-imagery data [19].

LRP computation is implemented using R's library innsight [27], version 0.2.0. As innsight is based on torch, it only accepts models in torch, Keras, or neuralnet formats. Therefore, as our model was created, trained, pruned, and retrained in AMORE, it was manually translated to the neuralnet package syntax. In AMORE, the ANN uses a tansig activation function in the hidden nodes of the form:

$$tansig(x) = 1.7159 \times \tanh(\frac{2}{3}x) \tag{2}$$

However, such activation function does not exist in neuralnet, and the innsight package does not currently allow custom activation functions, so a simple tanh function is used instead. Furthermore, as LRP is a method based on the calculation of backward propagation, it is computed over several epochs. In this study, it was computed over the entirety of the train group dataset for a total of 400 epochs.

2.5   Grand-Weight (GW) and Grand-Relevance (GR) Computation

In the path-weights-based analysis, there is a need for a metric that allows for the direct comparison of individual ICs between themselves for a given stimulus. In this way, the metric Grand-weight (GR) is the sum of the absolute values of the path-weights from each IC to a specific output, according to the formula:

$$GW_{ik} = \sum_j \left| path - weight_{ijk} \right| \tag{3}$$

where GW_{ik} is the Grand-weight from input i to output k, and $path\text{-}weight_{ijk}$ is the path-weight that goes from input i to output k through hidden node j (the weight from input i to hidden node j times the weight from hidden node j to output k).

For the LRP-based analysis, each IC has a relevance value per epoch of calculation and per output. Therefore, to keep consistency with the path-weights-based analysis, the absolute values of the relevancy score for each IC for a given output are summed, obtaining the metric Grand-relevance (GR), which allows the direct comparison between ICs regarding their relevance. GR formula is:

$$GR_{ik} = \sum_l |R_{ikl}| \tag{4}$$

where GR_{ik} is the Grand-relevance from input i to output k, and R_{ikl} is the relevance score for input i and output k on computational epoch l.

In essence, a similar hurdle that impedes the direct comparison of either path-weights or relevance values is present in both approaches. So, to enable comparisons between methods as well, the metrics of GW and GR are calculated, as GW allows the comparison between ICs' weights and GR allows the comparison between ICs' relevance for a given output. Furthermore, as both yield one value per IC per output, conclusions may be extracted from the relative positioning of the ICs according to each method, even though the values may not be on a comparable scale.

3　Results

The results are presented in two sections: the first refers to the network's performance, followed by the interpretability and explainability based on LRP.

3.1　Network Performance

Table 1. Confusion matrix of the "best network" (fully connected) predictions based on the test data, including the partial and global accuracies and precisions (LF: left foot; LH: left hand; RF: right foot; RH: right hand; T: tongue).

Stimulus		Prediction					Total
		LF	LH	RF	RH	T	
Input	LF	27	1	6	5	1	40
	LH	3	36	0	1	0	40
	RF	8	0	31	1	0	40
	RH	0	2	1	37	0	40
	T	4	0	1	0	35	40
Total		42	39	39	44	36	200
Accuracy (%)		67.5	90.0	77.5	92.5	87.5	83.0
Precision (%)		64.3	92.3	79.5	84.1	97.2	

Networks' partial and global accuracies and precisions are represented, respectively, in Table 1 for the "best network", Table 2 for the "pruned network", and Table 3 for the "retrained network". It is remarkable that, after an initial loss in the global accuracy of the "pruned network", its retraining recovered to the original level (83.0% to 64.0% to 81.0%). However, networks' architectures are notably different. While the "best network" encompasses 510 connection weights, the "retrained network" is only 36 (cf. Table 4).

Table 2. Confusion matrix of the "pruned network" predictions, including the partial and global accuracies and precisions (LF: left foot; LH: left hand; RF: right foot; RH: right hand; T: tongue).

Stimulus		Prediction					Total
		LF	LH	RF	RH	T	
Input	LF	19	1	9	2	9	40
	LH	1	31	4	4	0	40
	RF	5	0	17	8	10	40
	RH	0	1	2	36	1	40
	T	7	1	3	4	25	40
Total		32	34	35	54	45	200
Accuracy (%)		47.5	77.5	42.5	90.0	62.5	64.0
Precision (%)		59.4	91.2	48.6	66.7	55.6	

Table 3. Confusion matrix of the "retrained network" predictions based on the test data, including the partial and global accuracies and precisions (LF: left foot; LH: left hand; RF: right foot; RH: right hand; T: tongue).

Stimulus		Prediction					Total
		LF	LH	RF	RH	T	
Input	LF	32	0	5	1	2	40
	LH	0	29	0	10	1	40
	RF	8	0	31	1	0	40
	RH	0	1	2	37	0	40
	T	3	0	4	0	33	40
Total		43	30	42	49	36	200
Accuracy (%)		80.0	72.5	77.5	92.5	82.5	81.0
Precision (%)		74.4	96.7	73.8	75.5	91.7	

Table 4. Accuracies, partial and global, and structures' complexity (number of connections and correct predictions per connection) of the three types of networks.

Network	Conn	Hits	Hits per conn	Accuracies (%)					
				LF	LH	RF	RH	T	Global
Best network	510	166	0.325	67.5	90.0	77.5	92.5	87.5	83.0
Pruned network	36	128	3.556	47.5	77.5	42.5	90.0	62.5	64.0
Retrained network	36	162	4.500	80.0	72.5	77.5	92.5	82.5	81.0

Assuming that the number of correct hits per connection represents effectiveness, then the "retrained network" is more effective, increasing this index to 4.500 when compared with the "pruned network" (3.556) and the "best network" (0.325) (cf. Table 4).

3.2   Path-Weights-Based and LRP-Based Interpretability and Explainability

Figure 3 depicts the Grand-weights calculated for the fully connected "best network". According to the path-weights-based analysis, the most relevant ICs for all the stimuli are IC 5, IC 7, IC 11, and IC 12, although the absolute value is not constant across all stimuli. Figure 4 depicts the Grand-relevance calculated for the fully connected "best network". According to the LRP-based analysis, the most relevant ICs are not as constant across all the stimuli. For LF, IC 11 has a higher Grand-relevance than the rest, with ICs 5, 7, and 12 also achieving high values. For LH, it is IC 7. In the case of RF, there are two ICs with a high GR comparatively to the rest, ICs 7 and 12. Lastly, for T, the values are overall lower than for the other stimulus, and the peaks are not as pronounced, but IC 5 still achieves a relatively high value.

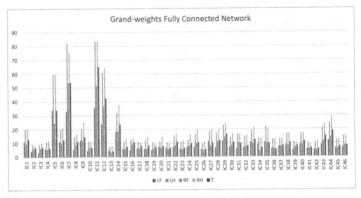

Fig. 3. Grand-weights for the "best network" (fully connected).

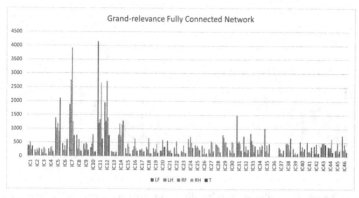

Fig. 4. Grand-relevance for the "best network" (fully connected).

Figure 5 depicts the path-weights, and LRP obtained for the "pruned network" based on the top 10 path-weights per input. For the path-weights-based analysis, as the value is calculated by summing the absolute values of all the path-weights for each IC, the expected change after pruning the network is a change in the value but for the IC ranking to keep somewhat stable. It is possible to see that the obtained results follow expected behavior except for IC 7 in LF and IC 5 in T, gaining a more relative GR. As for the LRP-based analysis, the same behavior is expected and observed, except for a significant drop in the GR value of IC 7 for RF.

Figure 6 depicts the path-weights, and LRP obtained for the pruned ANN based on the top 10 path-weights per input after the retraining step. It is important to note that the scale of the vertical axis on the graphs of this figure is different from the two previous figures. This is because during the retraining step, the ANN adjusts its weights and, as there is less input signal than in the fully connected ANN (only 6 ICs survived the pruning process), the ANN compensates for the fall in signal strength by having higher values for the remaining weights. Thus, the absolute value of the path-weights and LRP is higher. However, for input ranking, the relative value of path-weights and LRP for each IC is more important than the absolute value of either metric. For the path-weights-based analysis, the relative ranking of the ICs is the same as before the retraining step, except for the RH stimulus, where IC 5 goes from fourth to second most relevant IC. For the LRP-based analysis, the relative order of the ICs changes mostly in accordance with the trends observed when going from the fully connected "best network" to the "pruned network" being exacerbated. For the LF stimulus, IC 12 continues to grow relative to IC 1, overtaking it in the order of relevancy, and IC 7 achieves a higher LRP value than IC 5. For LH and RF, IC 11 achieves higher relative relevancy than IC 12. For RH, the relevancy of ICs 5 and 11 is even more pronounced. Lastly, for T, the relative relevancy of IC 5 diminishes as ICs 7 and 11 have closer LRP values than before the retraining step.

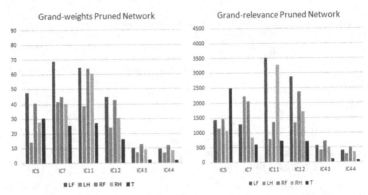

Fig. 5. Grand-weights (left column) and Grand-relevance (right column) of the "pruned network".

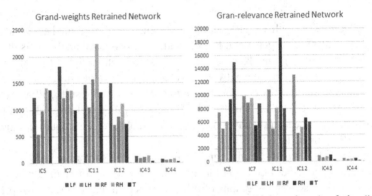

Fig. 6. Grand-weights (left column) and Grand-relevance (right column) of the "retrained network".

4   Discussion

The three main questions this study aims to answer are:

- Is the path-weights-based analysis in accordance with the state-of-the-art ANN interpretability method of LRP-based analysis regarding input importance for the network's prediction?
- Does the pruning process affect the path-weights-based analysis and the LRP-based analysis similarly?
- Does the retraining process affect the path-weights-based analysis and the LRP-based analysis similarly?

To answer the first question, the most relevant figure to analyze in this case is Fig. 3, as the model has all the inputs/ICs present, allowing for a better comparison between methods. Overall, both methods yield the same ICs as the most important, mainly ICs 5, 7, 11, and 12. Thus, the path-weights-based analysis is congruent with the state-of-the-art method, the LRP-based analysis.

It is important to note that, even though IC 14 appears as more relevant for the model's decision than ICs 43 and 44, the latter survived the pruning process, and the former did not, as the pruning process was based on the top 10 absolute path-weight values for each output and not the sum of the absolute values of the path-weights for each IC per output to attain a sparser ANN after the pruning process.

Answering the second question, the change observed in the path-weights-based analysis is expected, as during the pruning process, the connections with less weight associated with them are removed, all the changes that occur from Fig. 3, 4 and Fig. 5 are a diminishing of the obtained values. As for the LRP-based analysis, the changes are overall similar, but, in some cases, there is an increase in the obtained values, so, although the pruning process affects both methods differently, the results still seem mostly in accordance with both methods.

To answer the third question, the changes from Fig. 5 and Fig. 4 to Fig. 6 mostly represent an exacerbation of the changes observed from Fig. 3 and Fig. 4 to Fig. 5. The scale of the y-axis had to be changed for both methods, as there are fewer connections, the obtained values at each node will be lower, so during the retraining process the ANN compensates by having higher weights, as both path-weights and LRP are dependent on the weights between layers this compensation leads to significantly higher values for the new path-weights and LRP. So, the effect of retraining the pruned ANN seems similar for both methods. However, it is dependent on the changes observed during the pruning process.

Hence, it is possible to conclude that the path-weight-based procedure explains the ANN model in coherence with the layer-wise relevance propagation-based method. The pruned and retrained network recovers previous levels of correct predictions despite being a much lighter model, thus, less computationally intensive. In addition, pruning and retraining do not affect explainability in either case, continuing to point to the same inputs as the responsible for the predictions, i.e., both methods are useful for extracting pertinent knowledge from the fMRI data.

Compared with established neuroscientific knowledge concerning motor action execution [20, 22], both methods are correctly selective. The input IC 5 encompasses significant activations in areas known to activate head muscles, tongue included, bordering the Sylvian fissure, and a minor activation for feet in the longitudinal fissure. IC 7 has a more extensive activation for feet than IC 5, an activation for the tongue, and a prominent activation for hands in the right hemisphere. Right hemisphere hand control activates the contralateral hand, i.e., the left one. IC 11 mirrors IC 7. Activations for feet and tongue exist, but the activation for hand is at the left hemisphere, which means right-hand movements. Finally, IC 12 encompasses an extensive activation in the longitudinal fissure for the feet and an activation for the tongue. It also includes two large deactivations, one in each hemisphere in the hand areas.

Considering the anatomical components of each IC, one may consider they are rough. That derives from the method used for dimensionality reduction, ICA, which outputs brain networks whose time courses are coherent, independently of eventual anatomical boundaries. Other methods may be used in this stage in the future, more sensible to anatomical segmentation if one finds this is important for their work.

Another limitation of the present work is the dataset considered, which is about motor action only. Further work should extend the application to cognitive processes, aiming to generalize the procedure.

Although these limitations, one may conclude that both methods are coherent when explaining the ANN model, but both are coherent with established neuroscientific knowledge because both selected the same inputs as those containing more pertinent information for correct classification. Those inputs are expected from the literature, where more invasive and causal methods were employed for tracking the connections between the brain and muscles. In conclusion, the procedure depicted in Fig. 2 is suitable for analyzing fMRI data, here included building a model that accurately represents the process, the model has a high predicting ability, and is sufficiently explainable for knowledge extraction from data and human understanding.

Acknowledgments. This work was partially financially supported by Base Funding - UIDB/00027/2020 of the Artificial Intelligence and Computer Science Laboratory – LIACC - funded by national funds through the FCT/MCTES (PIDDAC).

References

1. Hanson, S.J., Matsuka, T., Haxby, J.V.: Combinatorial codes in ventral temporal lobe for object recognition: Haxby (2001) revisited: is there a "face" area? Neuroimage **23**, 156–166 (2004). https://doi.org/10.1016/j.neuroimage.2004.05.020
2. Sona, D., Veeramachaneni, S., Olivetti, E., Avesani, P.: Inferring cognition from fMRI brain images. In: de Sá, J.M., Alexandre, L.A., Duch, W., Mandic, D. (eds.) ICANN 2007. LNCS, vol. 4669, pp. 869–878. Springer, Heidelberg (2007). https://doi.org/10.1007/978-3-540-74695-9_89
3. do Espírito Santo, R., Sato, J.R., Martin, M.G.M.: Discriminating brain activated area and predicting the stimuli performed using artificial neural network. Exacta **5**, 311–320 (2007). https://doi.org/10.5585/exacta.v5i2.1180
4. Santos, J.P., Moutinho, L.: Tackling the cognitive processes that underlie brands' assessments using artificial neural networks and whole brain fMRI acquisitions. In: 2011 IEEE International Workshop on Pattern Recognition in NeuroImaging (PRNI), Seoul, Republic of Korea, pp. 9–12. IEEE Computer Society (2011)
5. Hacker, C.D., et al.: Resting state network estimation in individual subjects. Neuroimage **82**, 616–633 (2013). https://doi.org/10.1016/j.neuroimage.2013.05.108
6. Thomas, A.W., Heekeren, H.R., Müller, K.-R., Samek, W.: Analyzing neuroimaging data through recurrent deep learning models. Front. Neurosci. **13** (2019). https://doi.org/10.3389/fnins.2019.01321
7. Liu, M., Amey, R.C., Backer, R.A., Simon, J.P., Forbes, C.E.: Behavioral studies using large-scale brain networks – methods and validations. Front. Hum. Neurosci. **16** (2022). https://doi.org/10.3389/fnhum.2022.875201
8. Haynes, J.-D., Rees, G.: Decoding mental states from brain activity in humans. Nat. Rev. Neurosci. **7**, 523–534 (2006). https://doi.org/10.1038/nrn1931
9. Samek, W., Müller, K.-R.: Towards explainable artificial intelligence. In: Samek, W., Montavon, G., Vedaldi, A., Hansen, L.K., Müller, K.-R. (eds.) Explainable AI: Interpreting, Explaining and Visualizing Deep Learning. LNCS (LNAI), vol. 11700, pp. 5–22. Springer Cham (2019). https://doi.org/10.1007/978-3-030-28954-6_1

0. Adadi, A., Berrada, M.: Peeking inside the black-box: a survey on explainable artificial intelligence (XAI). IEEE Access **6**, 52138–52160 (2018). https://doi.org/10.1109/ACCESS.2018.2870052

1. Marques dos Santos, J.P., Moutinho, L., Castelo-Branco, M.: 'Mind reading': hitting cognition by using ANNs to analyze fMRI data in a paradigm exempted from motor responses. In: International Workshop on Artificial Neural Networks and Intelligent Information Processing (ANNIIP 2014), Vienna, Austria, pp. 45–52. Scitepress (Science and Technology Publications, Lda.) (2014)

2. de Oña, J., Garrido, C.: Extracting the contribution of independent variables in neural network models: a new approach to handle instability. Neural Comput. Appl. **25**, 859–869 (2014). https://doi.org/10.1007/s00521-014-1573-5

3. Montavon, G., Binder, A., Lapuschkin, S., Samek, W., Müller, K.-R.: Layer-wise relevance propagation: an overview. In: Samek, W., Montavon, G., Vedaldi, A., Hansen, L.K., Müller, K.-R. (eds.) Explainable AI: Interpreting, Explaining and Visualizing Deep Learning. LNCS (LNAI), vol. 11700, pp. 193–209. Springer, Cham (2019). https://doi.org/10.1007/978-3-030-28954-6_10

4. Clark, P., Matwin, S.: Using qualitative models to guide inductive learning. In: Utgoff, P. (ed.) Proceedings of the Tenth International Conference on International Conference on Machine Learning, ICML 1993, pp. 49–56. Morgan Kaufmann Publishers Inc., University of Massachusetts, Amherst (1993)

5. Marques dos Santos, J.D., Marques dos Santos, J.P.: Towards XAI: interpretable shallow neural network used to model HCP's fMRI motor paradigm data. In: Rojas, I., Valenzuela, O., Rojas, F., Herrera, L.J., Ortuño, F. (eds.) IWBBIO 2022. LNCS, vol. 13347, pp. 260–274. Springer, Cham (2022). https://doi.org/10.1007/978-3-031-07802-6_22

6. Marques dos Santos, J.D., Marques dos Santos, J.P.: Path weights analyses in a shallow neural network to reach Explainable Artificial Intelligence (XAI) of fMRI data. In: Nicosia, G., et al. (eds.) Machine Learning, Optimization, and Data Science. LNCS, vol. 13811, pp. 417–431. Springer, Cham (2023). https://doi.org/10.1007/978-3-031-25891-6_31

7. Thomas, A.W., Ré, C., Poldrack, R.A.: Benchmarking explanation methods for mental state decoding with deep learning models. Neuroimage **273**, 120109 (2023). https://doi.org/10.1016/j.neuroimage.2023.120109

8. Bach, S., Binder, A., Montavon, G., Klauschen, F., Müller, K.-R., Samek, W.: On pixel-wise explanations for non-linear classifier decisions by layer-wise relevance propagation. PLoS ONE **10**, e0130140 (2015). https://doi.org/10.1371/journal.pone.0130140

9. Sturm, I., Lapuschkin, S., Samek, W., Müller, K.-R.: Interpretable deep neural networks for single-trial EEG classification. J. Neurosci. Methods **274**, 141–145 (2016). https://doi.org/10.1016/j.jneumeth.2016.10.008

0. Penfield, W., Boldrey, E.: Somatic motor and sensory representation in the cerebral cortex of man as studied by electrical stimulation. Brain **60**, 389–443 (1937). https://doi.org/10.1093/brain/60.4.389

1. Buckner, R.L., Krienen, F.M., Castellanos, A., Diaz, J.C., Yeo, B.T.T.: The organization of the human cerebellum estimated by intrinsic functional connectivity. J. Neurophysiol. **106**, 2322–2345 (2011). https://doi.org/10.1152/jn.00339.2011

2. Yeo, B.T.T., et al.: The organization of the human cerebral cortex estimated by intrinsic functional connectivity. J. Neurophysiol. **106**, 1125–1165 (2011). https://doi.org/10.1152/jn.00338.2011

3. Buckner, R.L.: Event-related fMRI and the hemodynamic response. Hum. Brain Mapp. **6**, 373–377 (1998). https://doi.org/10.1002/(SICI)1097-0193(1998)6:5/6<373::AID-HBM8>3.0.CO;2-P

4. Limas, M.C., et al.: AMORE: A MORE flexible neural network package (0.2-15). León (2014)

25. R Development Core Team: R: A Language and Environment for Statistical Computing. R Foundation for Statistical Computing, Vienna (2010)
26. Bondarenko, A., Borisov, A., Alekseeva, L.: Neurons vs weights pruning in artificial neural networks. In: 10th International Scientific and Practical Conference on Environment. Technologies. Resources, vol. 3, pp. 22–28. Rēzekne Academy of Technologies, Rēzekne (2015)
27. Koenen, N., Baudeu, R.: Innsight: Get the Insights of your Neural Network (0.2.0) (2023)

Sensitivity Analysis for Feature Importance in Predicting Alzheimer's Disease

Akhila Atmakuru[1] , Giuseppe Di Fatta[2] , Giuseppe Nicosia[3]([✉]) ,
and Ali Varzandian[1] , and Atta Badii[1]

[1] University of Reading, Berkshire, UK
atta.badii@reading.ac.uk
[2] Free University of Bozen-Bolzano, Bolzano, Italy
giuseppe.difatta@unibz.it
[3] University of Catania, Catania, Italy
giuseppe.nicosia.1@gmail.com

Abstract. Artificial Intelligence (AI) classifier models based on Deep Neural Networks (DNN) have demonstrated superior performance in medical diagnostics. However, DNN models are regarded as "black boxes" as they are not intrinsically interpretable and, thus, are reluctantly considered for deployment in healthcare and other safety-critical domains. In such domains explainability is considered a fundamental requisite to foster trust and acceptability of automatic decision-making processes based on data-driven machine learning models. To overcome this limitation, DNN models require additional and careful post-processing analysis and evaluation to generate suitable explainability of their predictions. This paper analyses a DNN model developed for predicting Alzheimer's Disease to generate and assess explainability analysis of the predictions based on feature importance scores computed using sensitivity analysis techniques. In this study, a high dimensional dataset was obtained from Magnetic Resonance Imaging of the brain for healthy subjects and for Alzheimer's Disease patients. The dataset was annotated with two labels, Alzheimer's Disease (AD) and Cognitively Normal (CN), which were used to build and test a DNN model for binary classification. Three Global Sensitivity Analysis (G-SA) methodologies (Sobol, Morris, and FAST) as well as the SHapley Additive exPlanations (SHAP) were used to compute feature importance scores. The results from these methods were evaluated for their usefulness to explain the classification behaviour of the DNN model. The feature importance scores from sensitivity analysis methods were assessed and combined based on similarity for robustness. The results indicated that features related to specific brain regions (e.g., the hippocampal sub-regions, the temporal horn of the lateral ventricle) can be considered very important in predicting Alzheimer's Disease. The findings are consistent with earlier results from the relevant specialised literature on Alzheimer's Disease. The proposed explainability approach can facilitate the adoption of black-box classifiers, such as DNN, in medical and other application domains.

Keywords: Sensitivity Analysis · Explainability · Neural Network · predicting Alzheimer's · Feature Importance · SHAP · Sobol · Morris · Fast and High Dimensional Dataset

The Author(s), under exclusive license to Springer Nature Switzerland AG 2024
. Nicosia et al. (Eds.): LOD 2023, LNCS 14506, pp. 449–465, 2024.
tps://doi.org/10.1007/978-3-031-53966-4_33

1 Introduction

In recent times, there has been a surge in the usage of AI tools in the field of healthcare. These AI tools, including Machine Learning (ML) and Deep Neural Network (DNN) based models, are being used for the diagnosis and prediction of degenerative neurological disorders such as Alzheimer's Disease (AD), Parkinson's Disease (PD), and Mild Cognitive Impairment (MCI). In situations that involve critical healthcare, it is essential that the AI models output should be explainable, interpretable, accurate, reliable, and trustworthy. However, the DNN models are often perceived to lack transparency and re-adaptability, and their black-box nature and high level of complexity make for poor explainability and interpretability of the basis of their inferences. Therefore, it is imperative to evaluate and enhance their explainability to improve their reliability and acceptability. Sensitivity analysis is an effective approach for assessing and improving the explainability of the model.

Sensitivity Analysis (SA) examines the effects of variations in independent input variables and model parameters on the output of the model. This analysis is useful for researchers to gain a deeper understanding of the internal state vectors of the model and the rationale behind the predictions. Additionally, this study can facilitate the refinement or modification of the model, focusing on significant aspects or addressing any underlying defects. Consequently, SA can contribute to the valuation and enhancement of the interpretability and explainability of the model, ultimately leading to greater reliability, transparency, and reduced complexity. While various interpretations of interpretability and explainability exist in the literature, this analysis adopts the definition in [1].

Interpretability refers to the investigation of the model parameters and their impact on the output. The process requires comprehending the internal mechanics of the model and applying that understanding to make forecasts. This enhances the credibility of the model and ensures equitable predictions. Machine Learning (ML) models are highly interpretable as they are simple, use fewer parameters, and adhere to a set of operational principles. Therefore, ML models are highly interpretable. Conversely, DNNs are essentially black boxes by nature and exhibit intricate structure with numerous layers and complex operations, which makes it difficult to understand their inner workings and grasp the rationale behind their predictions.

Explainability refers to a detailed examination of the model input to evaluate its impact on the output. It refers to the ability to clearly explain predictions and to pinpoint the variables that affected them. Explainability leads to a clear and intuitive explanation for determining the most essential or important input features for the developed model and its predictions. This also enables understanding of the faults and biases in the model that impact its performance. While achieving interpretability for DNNs is challenging, explainability can be utilised to comprehend the most influential input features of the model and the rationale behind predictions.

Sensitivity Analysis [2] encompasses two complementary techniques: Global Sensitivity Analysis (GSA) and Local Sensitivity Analysis (LSA). Although both methods can be used for interpretability and explainability, this study primarily focuses on explainability. GSA examines variations in the model output behaviour with respect to the entire input space. The method involves varying the value of all input features at once and measuring the resulting output changes, with the objective of identifying the input eatures having the most influence on the output. LSA examines how variations in one ndependent input variable impact the model output. The method involves altering the alue of one input feature at a time while observing the output changes. Therefore, GSA s used to understand key features across the entire dataset, while LSA is used to understand essential features in a single case. The global sensitivity offers a comprehensive xplanation of the model. Two popular GSA methodologies and tools that are useful or assessing explainability are SHAP and SALib python libraries consisting of Sobol, Morris, and FAST methods.

Sobol and FAST are variance-based methods, which provide a quantifiable measure f the relative importance of each variable and its interactions by breaking down the utput of the model variation into contributions from each input variable and their interctions. Sobol is computationally demanding for large datasets because of the enormous umber of model assessments that must be conducted. Whereas, FAST computes sentivity indices quickly using the Fast Fourier Transform, making it an efficient tool for igh- dimensional datasets.

The Morris method uses a screening technique, Morris elementary effects, to estinate the influence of each input variable on the output by changing one variable at a time nd estimating the change in output. The Morris method generates sample space using different sampling strategy, and various strategies yield different sensitivity analyses. or the Morris method to provide sample data that is typical of actual data, the input ariables are required to be independent in nature however, the method reveals interactions between the input features as an output. SHAP is a game theory-based method that enerates feature significance as a Shapley value, which indicates the average contribution of the feature to the model output. SHAP delivers consistent results and clearly xplains complicated interactions.

The present paper describes assessment of explainability of DNN models developed for detecting Alzheimer's Disease, based on feature importance scores determined sing global sensitivity analysis. The classification was performed on a high-dimensional lzheimer's dataset with two labels: Alzheimer's Disease and Cognitively Normal. Two atasets were used: one that comprised all feature metrics gathered from the FreeSurfer nd the other that included a subset with fewer metrics. The GSA methodologies or ools, namely SHAP and SALib libraries containing Sobol, Morris, and FAST methods, ere used to identify the feature importance scores that are significant for explaining e model. The first approach used the SALib python package and implemented Sobol, Morris, and FAST methods for determining the feature importance scores for the two atasets and evaluating the three methods and their outcomes. The second approach omputed the feature importance scores using SHAP. Finally, the results obtained from e sensitivity analysis methods were analysed to determine similarity among them and e resultant methods were combined to form the final list of the most important features.

The remainder of this article is structured as follows: Sect. 2 presents a brief literature survey, Sect. 3 describes the datasets utilised for analysis, Sect. 4 elaborates the nplementation of diverse global sensitivity analysis methodologies for ascertaining the ature importance, while Sect. 5 provides the obtained results and ensuing discussions. astly, Sect. 6 presents some conclusions.

2 Literature Review

This literature review is focused on different research approaches utilised for identifying feature importance using sensitivity analysis, the use of sensitivity analysis in the context of deep neural networks and some general aspects of the pre-diagnostics for the detection of Alzheimer's Disease.

The words interpretability and explainability are frequently used interchangeably in the broad literature, although it is more appropriate to adopt different meanings in specialised work. A distinction between interpretability and explainability has been used in [1], which correspond to a standard definition for the terms that is widely accepted. The interpretable models typically offer a justification inherently, i.e. they provide a way to understand the logic behind their output in terms of semantics of the target problem and the input variables. Interpretability is typically implicitly provided by the model itself. On the other hand, the explainability of a model refers to a posthoc justification that is explicitly created after the model is built for the specific purpose to understand a model that typically is not interpretable on its own. Explainability is typically aimed at linking predictions and the input variables that influenced them by building a supplemental model for that purpose.

Some of the popular methodologies used for sensitivity analysis are Sobol, FAST and Morris. Shapely Additive exPlanations (SHAP) is a popular method used to generate posthoc explanations for black-box machine learning models. SHAP is based on the Shapley values from game theory and measures the contributions of each input feature to the predictions. In order to do that, SHAP compares the results of two scenarios, one which includes a feature and another without that feature. This comparison is performed for all possible combinations of the features. The obtained Shapley values indicate feature importance scores, which are computed as the average marginal contribution of the feature when included in the feature subset and when excluded from the subset. SHAP was found to be more robust and consistent with results [3], when compared with LIME (Local Interpretable Model-Agnostic Explanations), another explainability method. However, another study found that SHAP could be computationally very expensive for large numbers of input variables [4].

Sobol is a variance-based GSA method which quantifies the contribution of each input feature to the variance in the output of a model. Sobol analysis involves computing two values, the first order Sobol index and the total effect Sobol index. The first-order Sobol index is used to compute the contribution of each input feature to the variation of the output, whereas the total-effect Sobol index is used to compute the overall contribution of an input feature and its interactions with other features [5]. The method assumes the features are independent and uncorrelated however second order indices which is interaction between the input features is given as an output [6].

The Morris approach is a variance-based GSA method whereby each input feature is perturbed (changed) one at a time and the corresponding output variance is measured. A scaled variance of a uniform distribution determines the amount of required perturbation. Each component is perturbed numerous times to produce a series of basic effects. The average of these Morris elementary effects absolute values offers an approximation of the most important of those input features [7]. However, this approach requires a large number of samples for computing the feature importance. Also, it requires the input data to be independent in nature for it to generate a sample space which is similar to the input data. The methodology produces an infinity in the generated samples for the inputs that are not meeting the expectations of the Morris approach [8].

The Fourier Amplitude Sensitivity Test (FAST) [9] method involves variance based global sensitivity analysis. The approach employs Fourier analysis which involves producing a set of Fourier coefficients based on the input features and utilising these coefficients to estimate the variation of the output of the model. The Fourier coefficients are then used to determine the relative contribution of the variation of each input feature and rank the features depending on their contribution to the output variance. The approach is quite fast and efficient even while dealing with a high dimensional dataset as it uses Fourier transform techniques.

In [10], the SHAP methodology was applied along with other sensitivity analysis for identifying the feature importance of multiple subsets of Alzheimer's Disease dataset from ADNI. The procedures carried out by the study were feature selection, model training, and sensitivity analysis using machine learning models. The most significant features found in one of the datasets were the hippocampal and the cortical thickness.

Alzheimer's Disease (AD) is a neurodegenerative disease resulting in permanent damage to the neurological system of the brain. AD is externally characterised by behavioural changes such as language and short-term memory issues, trouble in executing tasks that require cognition, changes in personality, and loss of social and interpersonal skills, which act as early indicators of the disease (Radiology.org) [11]. The onset and the progress of the disease can be detected from observable changes in the physiological structures in the brain, which may be obtained from radiological images. An accurate diagnostic tool should be able to select the already established physiological changes with high accuracy. Research shows that AD is directly linked to structural changes in the hippocampus and in the entorhinal cortex located in the medial temporal lobe of the brain [12].

3 Datasets

The present study utilized a data set that contained brain T1-weighted structural MRI scans with a slice thickness of 1.5 mm obtained from 1901 research participants. The data for this study was sourced from three publicly available data repositories, namely the Australian Imaging Biomarker & Lifestyle Flagship Study of Ageing (AIBL), Alzheimer's Disease Neuroimaging Initiative (ADNI), and Information eXtraction from images (IXI).

The main purpose of this study was to develop a tool to support improvement in the diagnosis of AD. It was noted that the ADNI library is composed of multiple images of the same subject taken from various studies. To ensure consistency, screening and baseline scans were selected as the adopted photos since they were the earliest accessible scans of a subject. When there were multiple images of the same subject the image with the greatest contrast-to-noise ratio (CNR) was chosen.

To facilitate operations such as skull stripping, image registration, cortical and subcortical segmentation, hippocampal subfields segmentation, and calculation of cortical thickness, surface, and volume, all images were pre-processed using FreeSurfer version 6.0 [13].

A significant number of files containing numerical measurements connected to specific regions of interest were produced during the pre-processing stage (ROI). KNIME [14] and its extension KSurfer [15] were used to extract, filter, and clean the data created by the pre-processing.

A total of 446 traits were obtained from the data generated by FreeSurfer. The utilisation of ICV normalisation and the estimated total intracranial volume (ICV) were not included. Inaccurate or duplicate features were discarded during the data cleansing phase, resulting in the elimination of 42 traits. The numerical measurements of the brain were referred to as the feature set $F = \{fi\}$, with $|F| = 404$. The selection of traits was not based on specific domain expertise.

The raw dataset included information from both the right and left-brain hemispheres. Within each hemisphere, five distinct types of metrics were calculated from MRI scan images, comprising volume, thickness, thickness standard deviation, mean curvature, and area. Our concentration for feature selection was specifically on AD and CN as targets to perform experiments. Dataset 1 consisted of 401 features for classes AD and CN that encompassed all of the metrics excluding Gender, Age and Label. Dataset 2 consisted of 265 features for classes AD and CN that encompassed all of the metrics excluding Gender, Age and Label.

4 Methodologies

This section outlines the two methodologies implemented for conducting sensitivity analysis on two distinct Deep Neural Network (DNN) models coded in Python. The sensitivity analysis leveraged two Python method libraries, namely SHAP and the SALib for Sobol, Morris, and FAST methods. The goal of the analysis methodologies was to execute sensitivity analysis on a specified DNN model for a given input dataset and produce output predictions utilising different methods. The analysis outcome comprises of a set of features and their corresponding feature importance scores, specific to a designated DNN model, input dataset, and analysis methodology. Subsequently, the resulting feature importance scores across all methods are evaluated to determine the most significant features.

The sensitivity analysis was conducted on two distinct DNN models, which differ by the input dataset and the internal architecture of the model. As previously discussed in Sect. 3 Datasets, there exist two distinct datasets of features. The dataset 1 consists of 401 features, excluding the Label, whereas the 2 dataset 2 consists of a subset of 265 features, excluding the Label.

The DNN Model 1 meant for analysing 401 features and is composed of an initial layer that utilises the 'ReLU' activation function, 3 sets of hidden layers along with dropout layers, and a last output layer that uses a Sigmoid activation function. The input layer contained neurons tailored for 401 features with the activation of 'ReLU'. Each hidden layer, which contained the 'ReLU' activation function, was followed by a dropout layer with a dropout rate of 30%. The initial hidden layer set comprises 200 neurons, the second hidden layer set comprises 100 neurons, and the last hidden layer set comprises 50 neurons. The ultimate output layer included one neuron that employed a sigmoid activation function for binary classification. Prediction accuracy of the DNN Model 01 obtained with Dataset 1 using Stratified 10 Cross validation was approx. 91% with 3% standard deviation.

The DNN Model 2 meant for analysing 265 features. It featured an input layer that utilised the 'ReLU' activation function, 3 sets of hidden layers along with dropout layers, and a final output layer that employed the Sigmoid activation function. The input layer contained neurons specifically designed for 265 features and used the 'ReLU' activation function. Each hidden layer started with the 'ReLU' activation function and ended with a dropout layer that has a 30% dropout rate. The first hidden layer had 150 neurons, the second had 75, and the last hidden layer had 30 neurons. The final output layer consisted of a single neuron that was equipped with a sigmoid activation function for binary classification. Prediction accuracy of the DNN Model 2 obtained with Dataset 2 using Stratified 10 Cross validation was approx. 91.4% with 4.4 as standard deviation.

4.1   Methodology 1

This section describes the Python implementation of the SALib library for the Sobol, Morris, and FAST methods. The implementation made use of two distinct datasets, namely Dataset 1 and Dataset 2. Individual DNN models were created for each dataset. The objective of this methodology is to perform various sensitivity analyses on the two DNN models and to identify the most important features in predicting Alzheimer's Disease.

The implementation procedure begins with the preparation of the dataset for model training. Datasets 1 and 2 are used to define training features and the target label, the training dataset is then scaled in preparation for normalisation. The DNN Models 1 and 2 were trained using Dataset 1 and Dataset 2, respectively. Using the trained DNN models and corresponding datasets, the Sobol, Morris, and FAST methods were utilised for prediction.

For each of the methods, a definition of the original data was provided which includes the data mean, standard deviation, and distribution as parameters for generating the sample dataset. The sample size parameter was also provided which determined the resultant sample size of the dataset. The resultant sample dataset was created by following the perturbation techniques of the respective specified methods which involved adding small amounts of noise to the original dataset and increasing the dataset size.

The generated resultant sample dataset was used for prediction. The predicted target and the definition of the dataset was provided to the 'analyse' function of specified methods. The output from the 'analyse' function provided the results of sensitivity analysis containing feature importance scores.

To obtain stable, and reliable results, the above procedure was run a number of times by training the DNN model with a specified dataset, then prediction using the resultant sample dataset generated by the specified method and subsequent analysis for feature importance.

The DNN models are usually randomly initialised with weights and biases which can lead to a slightly different output for each run. While training, stochastic optimisation techniques such as ADAM were used. This leads to a selection of different subsets of training data resulting in different outputs each time. Carrying out the training and prediction in the DNN models for multiple iterations will produce robust, reliable estimates of the performance of the model as well as the feature importance score.

The resultant feature importance scores from all the methodologies namely Sobol, Morris and FAST require post-processing of the data to identify the most important features for each method.

Figure 1 is a schematic flow chart representation of the Methodology 1. The same procedure was followed for Sobol, Morris, and FAST methods.

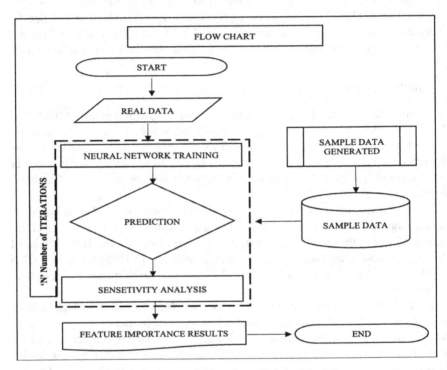

Fig. 1. Schematic Flowchart of Methodology 1

4.2 Methodology 2

This section describes the implementation of the SHAP method using python library. The implementation uses DNN model 1 and DNN model 2 which were created for dataset 1 and 2 respectively as described in the previous section. The objective of this methodology was to perform SHAP sensitivity analyses on the two DNN models and to identify the most important features in predicting Alzheimer's Disease.

The implementation procedure begins with the preparation of the dataset for model training. The dataset 1 and 2 was used to define training features and the target label. The training dataset was then scaled in preparation for normalisation. The corresponding datasets were used to train both DNN Model 1 and Model 2.

In order to carry out Shapley analysis, the SHAP explainer function was initialised using the input data and the trained DNN model. For generating Shapley values. The analysis focus was to explicate the effect of the input data on the output for the respective model.

The procedure followed for Methodology 2 comprised of multiple runs of the DNN models to produce robust and reliable estimates of feature importance. The resultant feature importance scores from SHAP require post processing to identify the most important features for the methodology.

Figure 2 is a schematic flow chart representation of the Methodology 2 using the SHAP method.

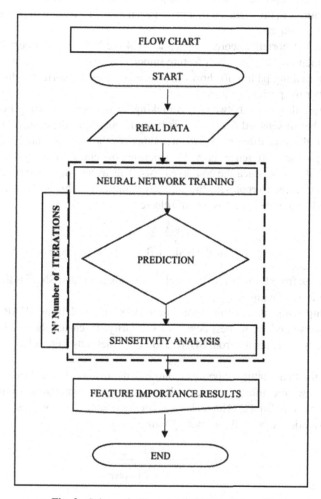

Fig. 2. Schematic Flowchart of Methodology 2

5 Results and Discussions

Explainability of DNN classifier models developed for detection of Alzheimer's Disease was assessed using compatible Alzheimer's datasets and sensitivity analysis methods, namely SHAP and SALib containing Sobol, Morris and FAST.

The two DNN models 1 and 2 were developed and trained using datasets 1 and 2 respectively. The developed models were analysed using SHAP, Sobol, Morris, and FAST methods. Under Methodology 1, Sobol, Morris and FAST methods were used to analyse both DNN models. Under Methodology 2, SHAP method was used to analyse both DNN models. In each analysis, the chosen method was run 500 times for Dataset 1 with 401 features and 300 times for Dataset 2 having 265 features so as to average out any fluctuations in the obtained outputs. In each analysis, feature importance scores were obtained as outputs.

The features importance score obtained using 4 methods was thoroughly analysed to determine their similarities. Each feature importance score list was converted into a corresponding ranking pattern to show difference between the corresponding rankings over the number of specified features.

The absolute difference between two rankings was computed using Eq. (1) for a specified number of selected features. This formula quantifies the relative discrepancy by dividing the absolute difference between rankings by the specified number of features. The averaging of rank differences provides a measure of central tendency that represents the overall similarity between the compared lists. This averaging of the results yields a single aggregate value to represent the collective similarity. This analysis as shown in Fig. 3 was performed over various sets of selected features values

$$\frac{abs(A-B)}{SNSF}$$
$$\text{if } A \leq SNSF \text{ or } B \leq SNSF \tag{1}$$

where A is a rank from Rank list A, B is rank from rank list B and SNSF is the Specified Number of Selected Features.

Upon comparison, it was found that the results obtained from SHAP and Sobol methods demonstrated the highest degree of similarity. To ensure the reliability of the methodology, only the results from SHAP and Sobol methods were selected for further analysis.

To create the final feature importance ranking, the results obtained from SHAP and Sobol methods were combined using the Rank position (reciprocal rank) method [16] which is illustrated in Eq. (2). The rank score for document 'i' is computed, utilising its position information across all retrieval systems (j = 1... n).

$$r(d_i) = \frac{1}{\sum_j \frac{1}{position(d_{ij})}} \tag{2}$$

The Rank Position score was calculated for each document to be combined. These scores were then used to sort the documents into a non-decreasing order. The scores derived from this method were subsequently ranked to form the ultimate feature importance ranking.

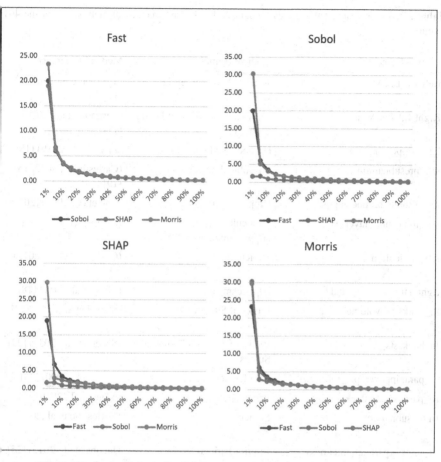

Fig. 3. Similarity analysis for 4 different approaches and 401 features dataset

Table 1 describes the 20 most important features recognised by the rank recipro-
al method out of the 401 features. The results were obtained from DNN model 01
sing SHAP and Sobol methods. The table contains a list of features along with their
orresponding medical terminologies and references to medical literature.

The results in the Tables 1 and 2 highlight the significance of the brain regions
adicated by the selected features in early detection of Alzheimer's Disease.

Table 2 outlines the 20 most important features as identified by the reciprocal rank-
ig method out of the 265 features 60% of which are identical to the Table 1. The table
ndings have adopted the DNN model 02 for SHAP and Sobol along with their cor-
esponding medical terminologies. Also, references to medical literature are provided,
mphasising the importance of these features or brain regions in the early detection of
lzheimer's Disease.

The comparison between Tables 1 and 2 highlights the abiding significance of spe-
ific brain regions in the early detection of Alzheimer's Disease. The fact that both tables

Table 1. 20 most important features recognised by the rank reciprocal method out of the 401 features.

Feature Name	Medical Names	Medical Reference
Left-Inf-Lat-Vent	Temporal horn of left lateral ventricle	(Vernooij et al., 2020) [17]
Right-Inf-Lat-Vent	Temporal horn of right lateral ventricle	(Vernooij et al., 2020) [17]
left_Hippocampal_tail	Hippocampal tail	(Zhao et al., 2019) [18]
left_presubiculum	Pre subiculum	(Carlesimo et al., 2015) [19]
left_Whole_hippocampus	Hippocampus	(Rao et al., 2022) [20]
left_molecular_layer_HP	Molecular Layer Hippocampus	(Stephen et al.,1996) [21]
left_subiculum	Subiculum	(Carlesimo et al., 2015) [19]
right_Hippocampal_tail	Hippocampal tail	(Zhao et al.,2019) [18]
lh_bankssts_volume	Banks of Superior Temporal Sulcus	(Sacchi et al.,2023) [22]
lh_bankssts_thicknessstd	Banks of Superior Temporal Sulcus	(Sacchi et al., 2023) [22]
lh_parahippocampal_thickness	Para Hippocampal	(Van et al., 2000) [23]
rh_paracentral_thicknessstd	Paracentral	(Yang et al., 2019) [24]
right_subiculum	Subiculum	(Carlesimo et al.,2015) [19]
rh_inferiorparietal_thickness	Inferior Parietal	(Jacobs et al., 2012) [25]
lh_transversetemporal_meancurv	Transverse Temporal	(Peters et al.,2009) [26]
Left-Amygdala	Amygdala	(Poulin et al.,2011) [27]
left_hippocampal fissure	Hippocampal Sulcus	(Bastos et al.,2006) [28]
left_GC-ML-DG	Granule Cell (GC) and Molecular Layer (ML) of the Dentate Gyrus (DG)	(Ohm et al., 2007) [29]
Right-Amygdala	Amygdala	(Poulin et al., 2011) [27]
rh_inferiortemporal_volume	Inferior Temporal	(Scheff et al., 2011) [30]

demonstrate a significant 60% overlap in characteristics, while using distinct models for validation, serves to highlight the resolute nature of these common traits. This prevailing trend as also confirmed by findings reported in relevant medical literature, serves to underscore the paramount importance of the specific brain regions in the field of Alzheimer's research.

Table 2. 20 most important features as identified by the reciprocal ranking method out of the 265 features.

Feature Name	Medical Names	Medical Reference
Left-Inf-Lat-Vent	Temporal horn of left lateral ventricle	(Vernooij et al., 2020) [17]
Right-Inf-Lat-Vent	Temporal horn of right lateral ventricle	(Vernooij et al., 2020) [17]
right_Hippocampal_tail	Hippocampal tail	(Zhao et al., 2019) [18]
left_presubiculum	Presubiculum	(Carlesimo et al., 2015) [19]
left_subiculum	Subiculum	(Carlesimo et al., 2015) [19]
left_Hippocampal_tail	Hippocampal tail	(Zhao et al., 2019) [18]
left_hippocampal-fissure	Hippocampal Sulcus	(Bastos et al., 2006) [28]
lh_parahippocampal_thickness	Para Hippocampal	(Van Hoesen et al., 2000) [23]
left_molecular_layer_HP	Molecular Layer Hippocampus	(Stephen et al., 1996) [21]
rh_entorhinal_thickness	Entorhinal	(Van Hoesen et al., 1991) [31]
rh_rostralmiddlefrontal_thickness	Rostral Middle Frontal	(Vasconcelos et al., 2014) [32]
rh_inferiorparietal_thickness	Inferior Parietal	(Greene SJ et al., 2010) [33]
left_Whole_hippocampus	Hippocampus	(Rao et al.,2022) [20]
rh_precuneus_thickness	Precuneus	(Giacomo et al., 2022) [34]
Left-Amygdala	Amygdala	(Poulin et al., 2011) [27]
Optic-Chiasm	Optic-Chiasm	(Sadun AA et al., 1990) [35]
Right-Pallidum	Pallidum	(Miklossy J et al., 2011) [36]
rh_entorhinal_volume	Entorhinal	(Van Hoesen et al.,1991) [31]
right_presubiculum	Pre-Subiculum	(Carlesimo et al.,2015) [19]
Left-Pallidum	Pallidum	(Miklossy J et al., 2011) [36]

The trustworthiness of medical diagnostic systems relies heavily on their repeatability. Utilising a vast number of input features, the DNN-based classifier accurately predicts outcomes, particularly with a select subset of features. To ensure the explainability of the DNN classifier, sensitivity analysis methods are employed to identify the most significant features. The repeatability of these crucial features signifies the consistency of the diagnosis. For evaluating the developed model, repeatability is assessed by analysing the important features obtained through SHAP and Sobol analyses. In SHAP analysis, the output of each iteration is scrutinised to determine the input feature that yields the highest score. By analysing the output of 500 iterations, it was possible to determine how frequently an input feature appeared with the maximum score. The results are presented as histograms in Fig. 4 below which shows the similar outputs obtained through Sobol analysis. The recurrence of these features instils confidence in the accuracy of the diagnosis.

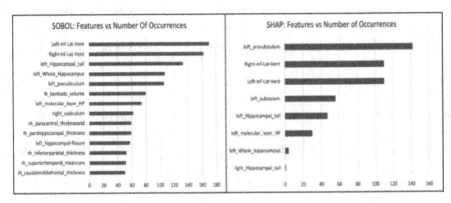

Fig. 4. Repeatability analysis for 2 different approaches with 401 features dataset

6   Conclusions

Classification models based on Deep Neural Networks (DNN) were developed for predicting Alzheimer's Disease and were studied using sensitivity analysis techniques to assess model explainability based on feature importance scores. An Alzheimer's dataset with two labels, Alzheimer's Disease and Cognitively Normal, was utilised for the classification task. Two DNN models were developed to analyse two datasets of different sizes. The analysis aimed at assessing the extent of the explainability and used two approaches based, respectively, on SHAP and on Global Sensitivity Analysis (G-SA) techniques (Sobol, Morris and FAST) from the python library SALib.

In the study of Alzheimer's Disease diagnosis, numerous significant characteristics have been recognised as noteworthy in terms of their correlation with the evolution and diagnosis of the disease. These characteristics incorporate the temporal horn of the left and right lateral ventricles, the hippocampal tail, the pre-subiculum, the whole hippocampus, the molecular layer of the hippocampus, the subiculum, the banks of the superior temporal sulcus, the para hippocampal region, the paracentral area, the inferior parietal region, the transverse temporal area, the amygdala, the hippocampal sulcus, and the inferior temporal area. These features have been extensively scrutinized and linked with various aspects of Alzheimer's Disease, such as neuroimaging, structural modifications, atrophy, and clinical correlations. The assessment of the significance of these features has contributed to enhancing our understanding of the use of sensitivity analysis techniques for the explainability of machine learning models and for feature selection. A subset of important features was effectively utilised by the DNN-based classifier to accurately forecast outcomes.

By using sensitivity analysis techniques significant features were identified. The repeatability of the results of a medical diagnostic system is a crucial factor in determining its reliability. The consistent presence of some features in multiple analyses serves to enhance the credibility of the diagnostic model.

This research has made notable contributions to the field of computer science by advancing the use of machine learning models for Alzheimer's Disease prediction and addressing the critical aspect of explainability. By developing and analysing the results from Deep Neural Network (DNN) models, the study has contributed to insights into their performance and the extent of explainability. The research has focused on assessing the explainability of these models through sensitivity analysis techniques, shedding light on the significant variables. Furthermore, the study has conducted a comprehensive feature analysis, exploring key features such as the temporal horn of the lateral ventricles, hippocampal regions, and other relevant areas, providing a deep understanding of the complex dynamics of the disease. Additionally, by presenting a comparative analysis of SHAP and sensitivity analysis techniques, the research has provided valuable insights into their performance and suitability for feature importance assessment, guiding researchers in selecting the most effective approach for predictive models.

References

1. Rudin, C.: Stop explaining black box machine learning models for high stakes decisions and use interpretable models instead. Nat. Mach. Intell. **1**(5), 206–215 (2019)
2. Razavi, S., et al.: The future of sensitivity analysis: An essential discipline for systems modelling and policy support. Environ. Model. Softw. **137**, 104954 (2021)
3. Lundberg, S.M., Lee, S.I.: A unified approach to interpreting model predictions. In: Advances in Neural Information Processing Systems, vol. 30 (2017)
4. Jia, R., et al.: Towards efficient data valuation based on the Shapley value. In: The 22nd International Conference on Artificial Intelligence and Statistics, pp. 1167–1176. PMLR (2019)
5. Sobol, I.M.: Global sensitivity indices for nonlinear mathematical models and their Monte Carlo estimates. Math. Comput. Simul. **55**(1–3), 271–280 (2001)
6. Helton, J.C., Davis, F.J.: Latin hypercube sampling and the propagation of uncertainty in analyses of complex systems. Reliab. Eng. Syst. Saf. **81**(1), 23–69 (2003)
7. Morris, M.D.: Factorial sampling plans for preliminary computational experiments. Technometrics **33**(2), 161–174 (1991)
8. Borgonovo, E., Plischke, E.: Sensitivity analysis: a review of recent advances. Eur. J. Oper. Res. **248**(3), 869–887 (2016)

9. Cukier, R.I., Levine, H.B., Shuler, K.E.: Nonlinear sensitivity analysis of multiparameter model systems. J. Comput. Phys. **26**(1), 1–42 (1978)

10. Bloch, L., Friedrich, C.M., Alzheimer's disease neuroimaging initiative, machine learning workflow to explain black-box models for early Alzheimer's disease classification evaluated for multiple datasets. SN Comput. Sci. **3**(6), 509 (2022)

11. Radiology for patients. https://www.radiologyinfo.org/en/info/alzheimers

12. Raji, C.A., Lopez, O.L., Kuller, L.H., Carmichael, O.T., Becker, J.T.: Age, Alzheimer disease, and brain structure. Neurology **73**(22), 1899–1905 (2009)

13. Fischl, B.: FreeSurfer. Neuroimage **62**(2), 774–781 (2012)

14. Berthold, M., et al.: "KNIME: the Konstanz Information Miner", workshop on multi-agent systems and simulation (MAS&S). In: 4th Annual Industrial Simulation Conference (ISC), Palermo, Italy, June 5–7 2006, pp.58–61 (2006)

15. Sarica, A., Di Fatta, G., Cannataro, M.: K-Surfer: a KNIME extension for the management and analysis of human brain MRI FreeSurfer/FSL data. In: Ślęzak, D., Tan, AH., Peters, J.F., Schwabe, L. (eds.) Brain Informatics and Health. Lecture Notes in Computer Science(), vol. 8609, pp. 481–492. Springer, Cham (2014). https://doi.org/10.1007/978-3-319-09891-3_44

16. Nuray-Turan, R., Can, F.: Automatic ranking of retrieval systems using fusion data. Inf. Process. Manage. **42**, 595–614 (2006). https://doi.org/10.1016/j.ipm.2005.03.023

17. Vernooij, M.W., van Buchem, M.A.: Neuroimaging in dementia. In: Hodler, J., Kubik-Huch, R., von Schulthess, G. (eds.) Diseases of the Brain, Head and Neck, Spine 2020–2023. IDKD Springer Series. Springer, Cham (2020). https://doi.org/10.1007/978-3-030-38490-6_11

18. Zhao, W., Wang, X., Yin, C., He, M., Li, S., Han, Y.: Trajectories of the hippocampal subfields atrophy in the Alzheimer's disease: a structural imaging Study. Front. Neuroinform. **22**(13), 13 (2019). https://doi.org/10.3389/fninf.2019.00013. PMID:30983985; PMCID:PMC6450438

19. Carlesimo, G.A., Piras, F., Orfei, M.D., Iorio, M., Caltagirone, C., Spalletta, G.: Atrophy of pre-subiculum and subiculum is the earliest hippocampal anatomical marker of Alzheimer's disease. Alzheimer's Dementia: Diagn., Assess. Dis. Monit. **1**, 24–32 (2015). https://doi.org/10.1016/j.dadm.2014.12.001

20. Rao, Y.L., Ganaraja, B., Murlimanju, B.V., Joy, T., Krishnamurthy, A., Agrawal, A.: Hippocampus and its involvement in Alzheimer's disease: a review 3 Biotech. **12**(2), 55 (2022). https://doi.org/10.1007/s13205-022-03123-4. Epub 2022 Feb 1. PMID: 35116217; PMCID: PMC8807768

21. Scheff, S., Sparks, D.L., Price, D.: Quantitative assessment of synaptic density in the outer molecular layer of the hippocampal dentate Gyrus in Alzheimer's disease. Dementia **7**(4) 226–232 (1996). https://doi.org/10.1159/000106884

22. Sacchi, L., et al.: Banks of the superior temporal sulcus in Alzheimer's disease: a pilot quantitative susceptibility mapping study. J Alzheimer's Dis. **93**(3), 1125–1134 (2023). https://doi.org/10.3233/JAD-230095. PMID: 37182885

23. Van Hoesen, G.W., Augustinack, J.C., Dierking, J., Redman, S.J., Thangavel, R.: The parahippocampal Gyrus in Alzheimer's disease: clinical and preclinical neuroanatomical correlates Ann. N. Y. Acad. Sci. **911**, 254–274 (2000). https://doi.org/10.1111/j.1749-6632.2000.tb06731.x

24. Yang, H., et al.: Study of brain morphology change in Alzheimer's disease and amnestic mild cognitive impairment compared with normal controls. Gen Psychiatr. **32**(2), e100005 (2019) https://doi.org/10.1136/gpsych-2018-100005. PMID:31179429; PMCID:PMC6551438

25. Jacobs, H.I., Van Boxtel, M.P., Jolles, J., Verhey, F.R., Uylings, H.B.: Parietal cortex matters in Alzheimer's disease: an overview of structural, functional and metabolic findings Neurosci. Biobehav. Rev. **36**(1), 297–309 (2012). https://doi.org/10.1016/j.neubiorev.2011.06.009. Epub 2011 Jun 30 PMID: 21741401

26. Peters, F., Collette, F., Degueldre, C., Sterpenich, V., Majerus, S., Salmon, E.: The neural correlates of verbal short-term memory in Alzheimer's disease: an fMRI study. Brain **132**(Pt 7), 1833–1846 (2009). https://doi.org/10.1093/brain/awp075. Epub 2009 May 11 PMID: 19433442

27. Poulin, S.P., Dautoff, R., Morris, J.C., Barrett, L.F., Dickerson, B.C.: Alzheimer's disease neuroimaging initiative. Amygdala atrophy is prominent in early Alzheimer's disease and relates to symptom severity. Psychiatry Res. **194**(1), 7–13 (2011). https://doi.org/10.1016/j.pscychresns.2011.06.014. Epub 2011 Sep 14. PMID: 21920712; PMCID: PMC3185127

28. Bastos-Leite, A.J., van Waesberghe, J.H., Oen, A.L., van der Flier, W.M., Scheltens, P., Barkhof, F.: Hippocampal sulcus width and cavities: comparison between patients with Alzheimer disease and nondemented elderly subjects. AJNR Am J Neuroradiol. **27**(10), 2141–5 (2006). PMID: 17110684; PMCID: PMC7977199

29. Ohm, T.G.: The dentate Gyrus in Alzheimer's disease. Prog. Brain Res. **163**, 723–740 (2007). https://doi.org/10.1016/S0079-6123(07)63039-8. PMID: 17765747

30. Scheff, S.W., Price, D.A., Schmitt, F.A., Scheff, M.A., Mufson, E.J.: Synaptic loss in the inferior temporal Gyrus in mild cognitive impairment and Alzheimer's disease. J Alzheimer's Dis. **24**(3), 547–557 (2011). https://doi.org/10.3233/JAD-2011-101782. PMID:21297265; PMCID:PMC3098316

31. Van Hoesen, G.W., Hyman, B.T., Damasio, A.R.: Entorhinal cortex pathology in Alzheimer's disease. Hippocampus **1**(1), 1–8 (1991). https://doi.org/10.1002/hipo.450010102. PMID: 1669339

32. Vasconcelos Lde, G., et al.: The thickness of posterior cortical areas is related to executive dysfunction in Alzheimer's disease. Clinics (Sao Paulo) **69**(1), 28–37 (2014). https://doi.org/10.6061/clinics/2014(01)05. PMID:24473557; PMCID:PMC3870310

33. Greene, S.J., Killiany, R.J.: Alzheimer's disease Neuroimaging Initiative. Subregions of the inferior parietal lobule are affected in the progression to Alzheimer's disease. Neurobiol. Aging **31**(8), 1304–11 (2010). https://doi.org/10.1016/j.neurobiolaging.2010.04.026. Epub 2010 Jun 8. PMID: 20570398; PMCID: PMC2907057

34. Koch, G., et al.: Precuneus magnetic stimulation for Alzheimer's disease: a randomized, sham-controlled trial. Brain **145**(11), 3776–3786 (2022). https://doi.org/10.1093/brain/awac285

35. Sadun, A.A., Bassi, C.J.: Optic nerve damage in Alzheimer's disease. Ophthalmology **97**(1), 9–17 (1990). https://doi.org/10.1016/s0161-6420(90)32621-0. PMID: 2314849

36. Miklossy, J.: Alzheimer's disease - a neurospirochetosis. Analysis of the evidence following Koch's and hill's criteria. J. Neuroinflammation **8**, 90 (2011). https://doi.org/10.1186/1742-2094-8-90. PMID: 21816039; PMCID: PMC3171359

A Radically New Theory of How the Brain Represents and Computes with Probabilities

Gerard Rinkus[✉] [iD]

Neurithmic Systems, Newton, MA 02465, USA
rod@neurithmicsystems.com

Abstract. It is widely believed that the brain implements probabilistic reasoning and that it represents information via some form of population (distributed) code. Most prior probabilistic population coding (PPC) theories share basic properties: 1) continuous-valued units; 2) fully/densely distributed codes; 3) graded synapses; 4) rate coding; 5) units have innate low-complexity, usually unimodal, tuning functions (TFs); and 6) units are intrinsically noisy and noise is generally considered harmful. I describe a radically different theory that assumes: 1) binary units; 2) sparse distributed codes (SDC); 3) *functionally* binary synapses; 4) a novel, *atemporal*, combinatorial spike code; 5) units initially have flat TFs (all weights zero); and 6) noise is a controlled resource used to cause similar inputs to be mapped to similar codes. The theory, Sparsey, was introduced 25 + years ago as: a) an explanation of the physical/computational relationship of episodic and semantic memory for the spatiotemporal (sequential) pattern domain; and b) a canonical, mesoscale cortical probabilistic circuit/algorithm possessing fixed-time, unsupervised, single-trial, non-optimization-based, unsupervised learning and fixed-time best-match (approximate) retrieval; but was not described in terms of probabilistic computation. Here, we show that: a) the active SDC in a Sparsey coding field (CF) simultaneously represents not only the likelihood of the single most likely input but the likelihoods of all hypotheses stored in the CF; and b) that entire explicit distribution can be transmitted, e.g., to a downstream CF, via a set of simultaneous single spikes from the neurons comprising the active SDC.

Keywords: Sparse distributed representations · probabilistic population coding · cell assemblies · canonical cortical circuit/algorithm

1 Introduction

It is widely believed that the brain implements some form of probabilistic reasoning to deal with uncertainty in the world [1], but exactly how the brain represents probabilities/likelihoods remains unknown [2, 3]. It is also widely agreed that the brain represents information with some form of distributed—a.k.a. population, cell-assembly ensemble—code [see [4] for relevant review]. Several population-based probabilistic coding theories (PPC) have been put forth in recent decades including those in which the state of all neurons comprising the population, i.e., the *population code*, is viewed as representing: a) the single most likely/probable input value/feature [5]; or b) the entire

© The Author(s), under exclusive license to Springer Nature Switzerland AG 2024
G. Nicosia et al. (Eds.): LOD 2023, LNCS 14506, pp. 466–480, 2024.
https://doi.org/10.1007/978-3-031-53966-4_34

probability/likelihood distribution over features [6–10]. Despite their differences, these approaches share fundamental properties. (1) Neural activation is continuous (graded). (2) *All* neurons in the coding field (CF) formally participate in the active code whether it represents a single hypothesis or a distribution over all hypotheses. Such a representation is referred to as a *fully distributed* representation. (3) Synapse strength is continuous. (4) They are typically formulated in terms of rate-coding [11]. (5) They assume *a priori* that *tuning functions* (TFs) of the neurons are unimodal, e.g., bell-shaped, over any one dimension, and consequently do not explain how such TFs might naturally emerge, e.g., through a learning process. (6) Individual neurons are assumed to be intrinsically noisy, e.g., firing with Poisson variability, and noise is viewed primarily as a problem to be dealt with, e.g., reducing noise correlation by averaging.

At a deeper level, it is clear that despite being framed as population models, they are really based on an underlying localist interpretation, specifically, that an individual neuron's firing rate can be taken as a perhaps noisy estimate of the probability that a single preferred feature (or preferred value of a feature) is present in its receptive field [12], i.e., consistent with the "Neuron Doctrine". While these models entail some method of combining the outputs of individual neurons, e.g., averaging, each neuron is viewed as providing its own individual, i.e., localist, estimate of the input feature. For example, this can be seen quite clearly in Fig. 1 of [9] wherein the first layer cells (sensory neurons) are unimodal and therefore can be viewed as detectors of the value at their modes (preferred stimulus) and the pooling cells are also in 1-to-1 correspondence with directions. This localist view is present in the other PPC models referenced above as well.

However, there are compelling arguments against such localistically rooted conceptions. From an experimental standpoint, a growing body of research suggests that individual cell TFs are far more heterogeneous than classically conceived [13–20], also described as having "mixed selectivity" [21], and more generally, that sets (populations, ensembles) of cells, i.e., "cells assemblies" [22], constitute the fundamental representational units in the brain [23, 24]. And, the greater the fidelity with which the heterogeneity of TFs is modeled, the less neuronal response variation that needs to be attributed to noise, leading some to question the appropriateness of the traditional concept of a single neuron I/O function as an invariant TF plus noise [25]. From a computational standpoint, a clear limitation is that the maximum number of features/concepts, e.g., oriented edges, directions of movement, that can be stored in a localist coding field of N units is N. More importantly, as explained here, the efficiency, in terms of time and energy, with which features/concepts can be stored (learned) and retrieved/transmitted is far greater if items of information (memories, hypotheses) are represented with *sparse distributed codes* (SDCs) rather than localistically [26–28].

The theory described herein, Sparsey [26–28], constitutes a radically new way of representing and computing with probabilities, diverging from most existing PPC theories in many fundamental ways, including: (1) The representational units (principal cells) comprising a CF need only be binary. (2) Individual items (hypotheses) are represented by fixed-size, sparsely chosen subsets of the CF's units, referred to as modular sparse distributed codes (MSDCs), or simply "codes" if unambiguous. (3) Decoding (read-out) not only of the most likely hypothesis but of the whole distribution, i.e., the likelihoods of all hypotheses stored in a CF, by downstream computations, requires only binary synapses. (4) The whole distribution, is sent via a wave of effectively simultaneous (i.e., occurring within some small window, e.g., at some phase of a local gamma cycle [29–32]) single spikes from the units comprising an active code to a downstream (possibly recurrently to the source) CF. (5) The initial weights of all afferent synapses to a CF are zero, i.e., the TFs are completely flat. The classical, roughly unimodal TFs [as would be revealed by low-complexity probes, e.g., oriented bars spanning a cell's receptive field (RF), cf. [33]] emerge as a side-effect of the model's single/few-trial learning process of storing MSDCs in superposition. (6) Neurons are not assumed to be intrinsically noisy. However, the canonical, mesoscale (i.e., the cell assembly scale) circuit normatively uses noise as a resource during learning. Specifically, noise, presumably mediated by neuromodulators, e.g., ACh [34], NE [35], proportional to input novelty, is explicitly injected into the code selection process to achieve the specific goal of (statistically, approximately) mapping more similar inputs to more similar MSDCs, where the similarity measure for MSDCs is intersection size. In this approach, patterns of correlation that emerge amongst principal cells are simply artifacts of this learning process.

2 The Model

Figure 1a shows a small Sparsey model instance with an 8x8 binary units (pixel) input field that is fully connected, via binary weights, all initially zero, to a *modular sparse distributed coding* (MSDC) coding field (CF). The CF consists of Q winner-take-all (WTA) *competitive modules* (CMs), each consisting of K binary neurons. Here, $Q = 7$ and $K = 7$. Thus, all codes have exactly Q active neurons and there are K^Q possible codes. We refer to the input field as the CF's receptive field (RF). Figure 1b shows an input, A, which has been associated with a code, $\phi(A)$ (black units); lines from active pixels to active coding units indicate the bundle [cf. "Synapsemble", [29]] of weights that would be increased from 0 to 1 to store this association (memory trace).

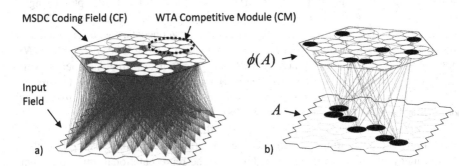

Fig. 1. The *modular sparse distributed code* (MSDC) coding field (CF). See text.

Figure 2 illustrates MSDC's key property that: *whenever any one code is fully active in a CF, i.e., all Q of its units are active, all codes stored in the CF will simultaneously be active (in superposition) in proportion to the sizes of their intersections with the single* *maximally active code.* Figure 2 shows five hypothetical inputs, A-E, which have been learned, i.e., associated with codes, φ(A) - φ(E). These codes were manually chosen to illustrate the principle that similar inputs should map to similar codes ("SISC"). That is, inputs B to E have progressively smaller overlaps with A and therefore codes φ(B) to φ(E) have progressively smaller intersections with φ(A). Although these codes were manually chosen, Sparsey's Code Selection Algorithm (CSA), described shortly, has been shown to statistically enforce SISC for both the spatial and spatiotemporal (sequential) input domains [26–28, 36]: a simulation-backed example, further demonstrating this for the spatial domain, is given in the Results section.

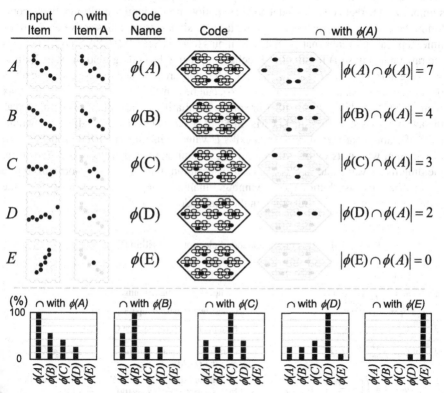

Fig. 2. The probability/likelihood of a feature can be represented by the fraction of its code that is active. When φ(A) is fully active, the hypothesis that feature A is present can be considered maximally probable. Because the similarities of the other features to the most probable feature, A, correlate with their codes' overlaps with φ(A), their probabilities/likelihoods are represented by the fractions of their codes that are active. In "∩" columns, black units are those intersecting with the input A and with its code, φ(A); gray indicates non-intersecting units.

For input spaces for which it is plausible to assume that input similarity correlates with probability/likelihood, i.e., for vast regions of natural input spaces, the single active code can therefore also be viewed as a probability/likelihood distribution over all stored inputs. This is shown in the lower part of Fig. 2. The leftmost panel at the bottom of

Fig. 2 shows that when $\phi(A)$ is 100% active, the other codes are partially active in proportions that reflect the similarities of their corresponding inputs to A, and thus the probabilities/likelihoods of the inputs they represent. The remaining four panels show input similarity (probability/likelihood) approximately correlating with code overlap when each of the four other stored codes is maximally active.

2.1   The Learning Algorithm

A simplified version of Sparsey's Code Selection Algorithm (CSA), sufficient for this paper's examples involving only purely spatial inputs, is given in Table 1 and we briefly summarize it here. [The full model handles spatiotemporal inputs, multiplicatively combining bottom-up, top-down, and horizontal (i.e., signals from codes active on the prior time step via recurrent synaptic matrices) inputs to a CF. See [28]] CSA Step 1 computes the raw input sums (u) for all $Q \times K$ cells comprising the coding field. In Step 2, these sums are normalized to U values, in [0,1]. All inputs are assumed to have the same number of active pixels, thus the normalizer, π_U, can be constant. In Step 3, we find the max U in each CM and in Step 4, a measure of the familiarity of the input, G, is computed as the average max U across the Q CMs. In Steps 5 and 6, G is used to adjust the parameters of a nonlinear transform from a cell's U value to its unnormalized probability, μ, of winning, within its own CM. In Step 7, each unit applies that "U-to-μ" transform, yielding the μ value and in Step 8, the μ distribution in each CM is renormalized to a total probability distribution (ρ) of winning. Finally, in Step 9, a draw is made from the ρ distribution in each CM resulting in the final code.

Table 1. Simplified Code Selection Algorithm (CSA)

	Equation	Short Description
1	$u_i = \sum_{j \in RF_U} x(j)w(j,i)$	Compute raw input (u) sums
2	$U_i = u_i / \pi_U \, w_{max}$	Compute normalized input sums. π_U
3	$\hat{U}_q = \max_{i \in CM_q} U_i$	Find the max U, $\hat{U}_q$, in each CM, CM_q
4	$G = \sum_{q=1}^{Q} \hat{U}_k \Big/ Q$	Compute the input's *familiarity*, G, as average $\hat{U}$ value over the Q CMs
5	$\eta = 1 + \left(\left[\frac{G-G^-}{1-G^-}\right]^+\right)^\gamma \times \chi \times K$	Determine expansivity (η) of U-to-μ sigmoid function. In this paper, $\gamma = 1$, $\chi = 100$, $G^- = 0.1$
6	$\sigma_1 = \frac{((\eta-1)/0.001)^{1/\sigma_4} - 1}{e^{\sigma_2 \sigma_3}}$	Sets σ_1 so that the overall sigmoid shape is preserved over full η range. $\sigma_2 = 8$, $\sigma_3 = 0.5$, $\sigma_4 = 8.5$
7	$\mu_i = \frac{(\eta-1)}{(1+\sigma_1 e^{-\sigma_2(U_i - \sigma_3)})^{\sigma_4}} + 1$	To each cell, apply sigmoid function, which collapses to constant fn, $\mu_I = 1$, when $G \leq G^-$
8	$\rho_i = \mu_i / \sum_{k \in CM} \mu_k$	In each CM, normalize relative (μ) to final (ρ) probabilities of winning
9	Select a final winner in each CM according to the ρ distribution in that CM	

G's influence on the "U-to-μ" transform, and thus on the ρ distributions can be summarized as follows.

a) When high global familiarity is detected ($G{\approx}1$), those distributions are exaggerated to bias the choice in favor of cells that have high input summations, and thus, high *local* familiarities (U), which acts to increase correlation.

b) When low global familiarity is detected ($G{\approx}0$), those distributions are flattened so as to reduce bias due to local familiarity, which acts to increase the expected Hamming distance between the selected code and previously stored codes, i.e., to decrease correlation (increase code separation).

Since the U values represent *signal*, exaggerating the U distribution in a CM increases signal whereas flattening it increases noise. The above behavior (and its smooth interpolation over the range, $G = 1$ to $G = 0$) is the means by which Sparsey achieves SISC. And, it is the enforcement (statistically) of SISC during learning, which ultimately makes possible the immediate, i.e., fixed time, retrieval of the best-matching (most likely, most relevant) hypothesis. By "fixed time", we mean that the number of algorithmic steps needed to do the retrieval remains constant as the number of stored codes (inputs) increases.

3 Results

The simulation-backed example of this section demonstrates that the CSA achieves the property, i.e., statistical (approximate) preservation of similarity from inputs to codes, qualitatively described in Fig. 2. In the experiment, the six inputs, I_1 to I_6, at top of Fig. 3a, were presented once each and assigned to the codes, $\phi(I_1)$ to $\phi(I_6)$ (not shown), via execution of the CSA (Table 1). The six inputs are disjoint only for simplicity of exposition. The input field (receptive field, RF) is a 12x12 binary pixel array and all inputs are of the same size, 12 active pixels. Since all inputs have exactly 12 active pixels, input similarity is simply $sim(I_x, I_y) = |I_x \cap I_y|/12$, shown as decimals under inputs. The CF consists of $Q = 19$ WTA CMs, each having $K = 8$ binary cells. The second row of Fig. 3a shows a novel stimulus, I_7, and its varying overlaps (gray pixels) with I_1 to I_6. Figure 3b shows the code, $\phi(I_7)$, activated (by the CSA) in response to presentation of I_7. Black indicates cells that also won for I_1, gray indicates active cells that did not win for I_1. Figure 3c shows (using the same color interpretations) the detailed values of all relevant variables (u, U, μ, and ρ) computed by the CSA when I_7 presents, and the winners drawn from the ρ distribution (black/gray bars) in each of the $Q = 19$ CMs.

If we consider presentation of I_7 to be a retrieval test, then the desired result is that the code of the most similar stored input, I_1, should be retrieved (reactivated). In this case, the gray cells in a given CM can be viewed as errors, i.e., in most CMs having a gray bar, the corresponding cell did not have the max U value in the CM, but since the final winner is a draw, occasionally a cell with a (possibly much) lower U (and thus, ρ) value wins, e.g., in CMs, 2, 7, 10. However, these are sub-symbolic scale errors, not errors at the scale of whole inputs (hypotheses), as a whole input is collectively represented by the entire MSDC code (entire cell assembly). In this example, appropriate threshold settings in downstream computations, would allow the model as a whole to return the correct answer given that 11 out of 19 cells of I_1's code, $\phi(I_1)$, are activated, similar to thresholding schemes in other associative memory models [37, 38].

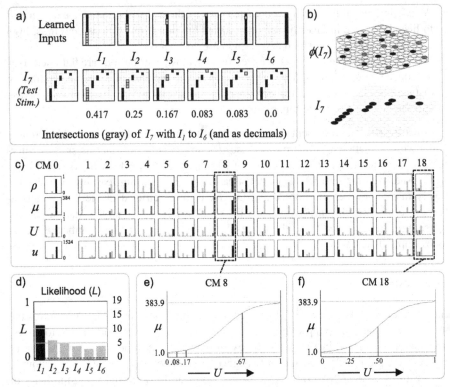

Fig. 3. In response to a novel input, I_7, the codes for the six previously learned (stored) inputs I_1 to I_6, i.e., hypotheses, are activated with strength approximately correlated with the similarity (pixel overlap) of I_7 input and those stored inputs. Test input I_7 is most similar to learned input, I_1, shown by the intersections (gray pixels) in panel a. Thus, the code with the largest fraction of active cells is $\phi(I_1)$ (11/19≈58%) (black bar in panel d). The codes of the other inputs are active in rough proportion to their similarities with I_7 (gray bars). (c) Raw (u) and normalized (U) input summations to all cells in all CMs. Note: all weights are effectively binary, though "1" is represented with 127 and "0" with 0. Hence, the max u value possible in any cell when I_7 is presented is 12x127 = 1524. The U values are transformed to un-normalized win probabilities (μ) in each CM via a sigmoid transform whose properties, e.g., max value of 383.9, depend on G and other parameters. μ values are normalized to true probabilities (ρ) and one winner is chosen in each CM (indicated by black or dark gray bars: black: winner for I_7 that also won for I_1; dark gray: winner for I_7 that did not win I_1. (e, f) Details for CMs, 8 and 18. In CM 8, cell 7 wins. It has $u = 1,016$ (thus, $U = 0.67$) meaning it has max weight synapses from 8 of the 12 active pixels in I_7, which in turn means that it was active not only as part of $\phi(I_1)$ but also in one or more of the other codes as well. CM18's ρ distribution is more compressed. Cell 1 (dark gray bar) has non-maximal ρ, but ends up winning.

More generally, when I_7 Is presented, we would like *all* of the stored inputs to be reactivated in proportion to their similarities to the test probe, I_7, as approximately occurs for the single presentation of I_7 shown here (Fig. 3d). Thus, the black bar in Fig. 3d represents the fact that the code, $\phi(I_1)$, for the best matching stored input, I_1

has the highest active code fraction, 57% (11 out 19, the black cells in Fig. 3b) of the cells of $\phi(I_1)$ are active in $\phi(I_7)$. The gray bar for the next closest matching stored input, I_2, indicates that 6 out of 19 of the cells of $\phi(I_2)$ (code note shown) are active in $\phi(I_7)$. In general, some of these 6 may be common to the 11 cells in $\{\phi(I_7) \cap \phi(I_1)\}$. And similarly for the other stored hypotheses. [Note that even the code for I_6 which has zero intersection with I_7 has four cells in common with $\phi(I_7)$. In general, the expected code intersection for the zero input intersection condition is not zero, but chance, since in that case, the winners are chosen from the uniform distribution in each CM, in which case the expected intersection is Q/K.] While Fig. 3 shows the results of a single presentation of I_7, we presented I_7 10 times and computed the average intersection of $\phi(I_7)$ with $\phi(I_1)$ to $\phi(I_6)$ across those 10 trials. The average code intersections were 88% (Pearson) correlated with the input pattern (pixel) intersections, I_7 with I_1 to I_6.

If, instead of viewing presentation of I_7 as a retrieval test, we view it as a learning trial, we want the sizes of intersection of the code, $\phi(I_7)$, activated in response, with the six previously stored codes, $\phi(I_1)$ to $\phi(I_6)$, to approximately correlate with the similarities of I_7 to inputs, I_1 to I_6. But again, this is what Fig. 3d shows. As noted earlier, we assume that the similarity of a stored input I_x to the current input can be taken as a measure of I_x's probability/likelihood. And, since all codes are of size Q, we can divide code intersection size by Q, yielding a measure normalized to [0,1]: $L(I_1) = |\phi(I_7) \cap \phi(I_1)|/Q$. Thus, this result shows that the CSA, a single-trial, unsupervised, non-optimization-based, and *most importantly, fixed time,* algorithm statistically enforces SISC. In this case, the gray cells in Fig. 3b would not be considered errors: they would just be part of a new code, $\phi(I_7)$, being assigned to represent a novel input, I_7, in a way that respects similarity with previously stored inputs. Crucially, because all codes are stored in *superposition* and because, when each one is stored, it is stored in a way respecting similarities with all previously stored codes, the patterns of intersection amongst the set of stored codes reflects not simply the *pairwise* similarity structure over the inputs, but, in principle, the similarity structure *of all orders present* in the input set. This is similar in spirit to another neural probabilistic model [2, 39], proposing that overlaps of distributed codes (and recursively, overlaps of overlaps), encode the domain's latent variables (their identities and valuednessess), cf. "anonymous latent variables" [40].

A cell's U value represents the total *local evidence*, i.e., its normalized input summation, that it should be activated. However, rather than simply picking the max U cell in each CM as winner (i.e., hard max), which would amount to executing only steps 1–3 of the CSA, the remaining CSA steps, 4–9, are executed, in which the U distributions are transformed as described earlier and winners are chosen as draws from the ρ distributions in each CM. Thus, an extremely cheap-to-compute (CSA Step 4) *global function* of the whole CF, G, is used to influence the *local* decision process in each CM. We repeat for emphasis that no part of the CSA explicitly operates on, i.e., iterates over, stored hypotheses (codes); indeed, there are no explicit (localist) representations of stored hypotheses on which to operate.

To further demonstrate this paper's primary result, Fig. 4 shows that presentation of different novel inputs to the model that has previously stored inputs, I_1 to I_6, yields different likelihood distributions that correlate approximately with input similarity. Input I_8 (Fig. 4a) has highest pixel intersection with I_2 and a different pattern of intersections

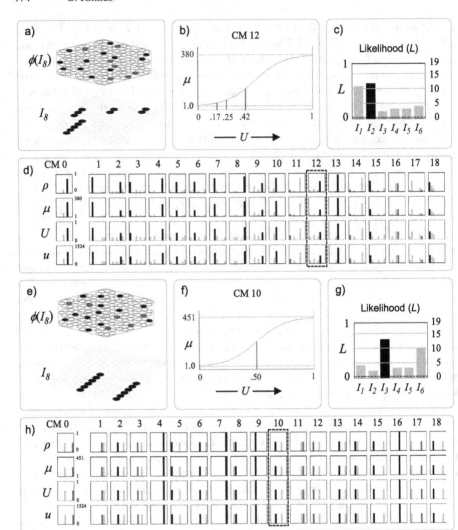

Fig. 4. Details of presenting other novel inputs, I_8 (panels a-d) and I_9 (panels e-h). In both cases the resulting likelihood distributions (panels c,g) correlate closely with the input overlap patterns Panels b and f show details of one example CM (dashed boxes in panels d and h) for each input.

with the other learned inputs as well (refer to Fig. 3a). Figure 4c shows that the codes of the stored inputs become active in approximate proportion to their similarities with I_8 i.e., their likelihoods are simultaneously physically represented by the fractions of their codes which are active. The G value in this case, 0.526, yields, via CSA steps 5–7, the U-to-μ transform shown in Fig. 4b, which is applied in all CMs. Its range is [1,380] and given the particular U distributions shown in Fig. 4d, the cell with the max U in each CM ends up being strongly favored in most CMs. The dashed gray box shows the u, U, μ, and ρ distribution for CM 12. Thus, cell 5 has $U = 0.42$ which maps to approximately $\mu \approx$

150 whereas cell 2 has $U = 0.25$ and cells 0 and 4 have $U = 0.17$. The effect of pushing the U values through the transform squashes the final probabilities (ρ) of these other cells relative to that of cell 5. Similar statistical conditions exist in many other CMs. Overall, presentation of I_8 activates a code $\phi(I_8)$ that has 12 out of 19 cells in common with $\phi(I_2)$ manifesting the high likelihood estimate for I_2. We presented I_8 10 times and computed the average intersections of $\phi(I_8)$ with $\phi(I_1)$ to $\phi(I_6)$. The average code intersections were 91% (Pearson) correlated with the input pattern (pixel) intersections, I_8 with I_1 to I_6.

Finally, Fig. 4e shows presentation of a more ambiguous input, I_9, having half its pixels in common with I_3 and the other half with I_6. Figure 4g shows that the codes for I_3 and I_6 have both become approximately equally (with some statistical variance) active and are both more active than any of the other codes. Thus, the model is representing that these two hypotheses are the most likely and approximately equally likely. The remaining hypotheses' likelihoods also approximately correlate with their pixelwise intersections with I_9. The qualitative difference between presenting I_8 and I_9 is readily seen by comparing the U rows of Fig. 4d and 4h and seeing that for the latter, a tied max U condition exists in almost all the CMs, reflecting the equal similarity of I_9 with I_3 and I_6. In approximately half of these CMs the cell that wins intersects with $\phi(I_3)$ and in the other half, the winner intersects with $\phi(I_6)$. In Fig. 4h, the four CMs in which there is a single black bar, CMs 4, 7, 9, and 16, indicates that the codes, $\phi(I_3)$ and $\phi(I_6)$, intersect in these three CMs. We presented I_9 10 times and computed the average intersections of $\phi(I_9)$ with $\phi(I_1)$ to $\phi(I_6)$. The average code intersections were 99% (Pearson) correlated with the input pattern (pixel) intersections, I_9 with I_1 to I_6.

3.1 A MSDC Simultaneously Transmits the Full Likelihood Distribution via an Atemporal Combinatorial Spike Code

The use of MSDC allows the likelihoods of *all* hypotheses stored in the distribution, to be transmitted via a set of simultaneous single spikes from the neurons comprising the active MSDC. This is shown in the example given in Fig. 5e, which, at the same time, compares this fundamentally new *atemporal, combinatorial spike code*, with temporal spike codes and one prior (in principle) atemporal code. For a single source neuron, two types of spike code are possible, rate (frequency) (Fig. 5b), and latency (e.g., of spike(s) relative to an event, e.g., phase of gamma) (Fig. 5c). Both are fundamentally temporal and have the crucial limitation that only one value (item) represented by the source neuron can be sent at a time. Most prior population-based codes also remain fundamentally temporal: the signal depends on spike *rates* of the afferent axons, e.g., [1–4] (not shown in Fig. 5).

Figure 5d illustrates an (effectively) atemporal population code [5] in which the *fraction* of active neurons in a source field carries the message, coded as the number of simultaneously arriving spikes to a target neuron (shown next to the target neuron for each of the four signals values). This *variable-size* population (a.k.a. "thermometer") code has the benefit that all signals are sent in the same, short time, but it is not combinatorial in nature, and has limitations, including: a) the max number of representable values (items/concepts) is the number (N) of units comprising the source CF; and b) as for the temporal codes defined with respect to a single source neuron, any single message sent can represent *only one* item, e.g., a single value of a scalar variable, i.e., implying that any one message carries only $\log_2 N$ bits.

In contrast, consider the fixed-size MSDC code of Fig. 5e. The source CF consists of $Q = 5$ CMs, each with $K = 4$ binary units. Thus, all codes, are of the same fixed size, $Q = 5$. As done in Fig. 2, the codes for this example were manually chosen to reflect the similarity structure of scalar values (Col. a) (the prior section has already demonstrated that the CSA statistically preserves similarity). As suggested by charts at right of Fig. 5, any single MSDC, ϕ_i, represents (encodes) the similarity distribution over all items (values) stored in the field. Note: gray denotes active units not in the intersection with ϕ_1. We're assuming that input (e.g., scalar value) similarity correlates with likelihood, which again, is reasonable for vast portions of input spaces having natural statistics.

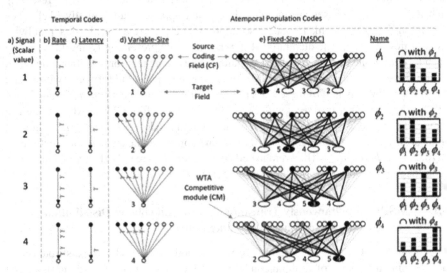

Fig. 5. Temporal vs. atemporal spike coding concepts. The fixed-size MSDC code has the advantage of being able to send the entire distribution, i.e., the likelihoods of *all* codes (hypotheses) stored in the source CF, with a set of simultaneous single spikes from $Q = 5$ units comprising an active MSDC code. See text for details.

Since any one MSDC, ϕ_i, encodes the full likelihood distribution, the set of single spikes sent from it simultaneously transmits that full distribution, encoded as the instantaneous sums at the target cells. Note: when any MSDC, ϕ_i, is active, 20 wts (axons) will be active (black), thus, all four target cells will have $Q = 5$ active inputs. Thus, due to the *combinatorial* nature of the MSDC code, the specific values of the binary weights are essential to describing the code (unlike the other codes where we can assume all wts are 1). Thus, for the example of Fig. 5e, we assume: a) all wts are initially 0; b) the four associations, $\phi_1 \rightarrow$ target cell 1, $\phi_2 \rightarrow$ target cell 2, etc., were previously stored (learned) with single trials; and c) on those learning trials, coactive pre-post synapses were increased to $w = 1$. Thus, if ϕ_1 is reactivated, target cell 1's input sum will be 5 and other cells' sums will be as shown (to left of target cells). If ϕ_2 is reactivated, target cell 2's input sum will be 5, etc. [Black line: active $w = 1$; dotted line: active $w = 0$; gray line: $w = 0$.] As described in Fig. 3 of [7], the four target cells could be embedded in a recurrent field with inhibitory infrastructure allowing sequential read out in descending input sum order, implying that the full similarity (likelihood) order information over all four stored items is sent in each of the four cases. Since there are 4! orderings of the four items, each such message, each a set of 20 simultaneous spikes sent from five active CF units, sends $\log_2(4!) = 4.58$ bits. I suggest this marriage of fixed-size MSDCs and an atemporal spike code is a crucial advance beyond prior population-based models, i.e., the "distributional encoding" models (see [8, 9] for reviews), and may be key to explaining the speed and efficiency of probabilistic computation in the brain.

4   Discussion

We described a radically different theory, from prevailing probabilistic population coding (PPC) theories, for how the brain represents and computes with probabilities. This theory, Sparsey, avails itself only in the context of *modular sparse distributed coding* (MSDC), as opposed to the fully distributed coding context in which the PPC models have been developed (or a localist context). Sparsey, was originally described as a model of the canonical cortical circuit and a computationally efficient explanation of episodic and semantic memory for sequences, but its interpretation as a way of representing and computing with probabilities was not emphasized. The PPC models [5, 7–11, 39] share several fundamental properties: 1) continuous neurons; 2) full/dense coding; 3) due to 1 and 2, synapses must either be continuous or rate coding must be used to allow decoding; 4) they generally assume rate coding; 5) individual neurons are generally assumed to have unimodal, e.g., bell-shaped, tuning functions (TFs); 6) individual neurons are assumed to be noisy, and noise is generally viewed as degrading computation, thus, needing to be mitigated, e.g., averaged out.

In contrast to these PPC properties/assumptions, Sparsey assumes: 1) binary neurons; 2) items of information are represented by small (relative to whole CF) sets of neurons (MSDCs); 3) only effectively binary synapses; 4) signaling via waves of simultaneous single (e.g., first) spikes from a source MSDC; 5) all weights are initially zero, i.e., the TFs are initially completely flat, and emerge via single/few-trial, unsupervised learning to reflect a neuron's specific history of inclusion in MSDCs; 6) rather than being viewed as a problem imposed by externalities (e.g., common input, intrinsically noisy cell firing), noise functions as a resource, controlled usage of which yields the valuable property that similar inputs are mapped to similar codes (SISC).

The CSA's algorithmic efficiency, i.e., both learning (storage) and best-match retrieval are fixed time operations, has not been shown for any other computational method, including hashing methods, either neurally-relevant [41–43], or more generally reviewed in [44]]. Although time complexity considerations like these have generally not been discussed in the PPC literature, they are essential for evaluating the overall plausibility of models of biological cognition, for while it is uncontentious that the brain computes probabilistically, we also need to explain the extreme speed with which these computations, over potentially quite large hypothesis spaces, occur.

One key to Sparsey's computational speed is its extremely efficient method of computing the *global* familiarity, G, simply as the average of the max U values of the Q

CMS. In particular, computing G *does not require* explicitly comparing the new input to every stored input (nor to a log number of inputs as for tree-based methods). G is then used to adjust, in the same way, the transfer functions of all neurons in a CF. This dynamic, and fast timescale (e.g., 10 ms), modulation of the transfer function, based on the *local* (to the CF) measure, G, is a strongly distinguishing property of Sparsey: in most models, the transfer function is static. While there has been much discussion about the nature, causes, and uses of correlations and noise in cortical activity (see [45–47] for reviews), the G-based titration of the amount of noise present in the code selection process, to achieve the specific goal of approximately preserving similarity (SISC) is a novel contribution to the discussion.

Enforcing SISC in the context of an MSDC CF realizes a balance between:

a) maximizing the storage capacity of the CF, and
b) embedding the similarity structure of the input space in the set of stored codes, which in turn enables fixed-time best-match retrieval.

In exploring the implications of shifting focus from information theory to coding theory viz. Theoretical neuroscience, [48] pointed to this same tradeoff, though their treatment uses error rate (coding accuracy) instead of storage capacity. Understanding how neural correlation ultimately affects things like storage capacity is considered largely unknown and an active area of research [49]. Our approach implies a straightforward answer. Minimizing correlation, i.e., maximizing average Hamming distance over the set of codes stored in an MSDC CF, maximizes storage capacity. Increases of any correlations of pairs, triples, or subsets of any order, of the CF's units increases the strength of embedding of statistical (similarity) relations in the input space.

References

1. Pouget, A., et al.: Probabilistic brains: knowns and unknowns. Nat. Neurosci. **16**(9), 1170–1178 (2013)
2. Pitkow, X., Angelaki, D.E.: How the brain might work: statistics flow in redundant population codes. (submitted) (2016)
3. Ma, W.J., Jazayeri, M.: Neural coding of uncertainty and probability. Annu. Rev. Neurosci **37**(1), 205–220 (2014)
4. Barth, A.L., Poulet, J.F.A.: Experimental evidence for sparse firing in the neocortex. Trends Neurosci. **35**(6), 345–355 (2012)
5. Georgopoulos, A., et al.: On the relations between the direction of two-dimensional arm movements and cell discharge in primate motor cortex. J. Neurosci. **2**(11), 1527–1537 (1982)
6. Pouget, A., Dayan, P., Zemel, R.: Information processing with population codes. Nat. Rev Neurosci. **1**(2), 125–132 (2000)
7. Pouget, A., Dayan, P., Zemel, R.S.: Inference and computation with population codes. Annu Rev. Neurosci. **26**(1), 381–410 (2003)
8. Zemel, R., Dayan, P., Pouget, A.: Probabilistic interpretation of population codes. Neura Comput. **10**, 403–430 (1998)
9. Jazayeri, M., Movshon, J.A.: Optimal representation of sensory information by neura populations. Nat. Neurosci. **9**(5), 690–696 (2006)
10. Ma, W.J., et al.: Bayesian inference with probabilistic population codes. Nat. Neurosci. **9**(11) 1432–1438 (2006)

11. Sanger, T.D.: Neural population codes. Curr. Opin. Neurobiol. **13**(2), 238–249 (2003)
12. Barlow, H.: Single units and sensation: a neuron doctrine for perceptual psychology. Perception **1**(4), 371–394 (1972)
13. Cox, D.D., DiCarlo, J.J.: Does learned shape selectivity in inferior temporal cortex automatically generalize across retinal position? J. Neurosci. **28**(40), 10045–10055 (2008)
14. Nandy, A.S., et al.: The fine structure of shape tuning in area V4. Neuron **78**(6), 1102–1115 (2013)
15. Mante, V., et al.: Context-dependent computation by recurrent dynamics in prefrontal cortex. Nature **503**(7474), 78–84 (2013)
16. Nandy, A.S., et al.: Neurons in macaque area V4 are tuned for complex spatio-temporal patterns. Neuron **91**(4), 920–930 (2016)
17. Bonin, V., et al.: Local diversity and fine-scale organization of receptive fields in mouse visual cortex. J. Neurosci. **31**(50), 18506–18521 (2011)
18. Yen, S.-C., Baker, J., Gray, C.M.: Heterogeneity in the responses of adjacent neurons to natural stimuli in cat striate cortex. J. Neurophys. **97**(2), 1326–1341 (2007)
19. Smith, S.L., Häusser, M.: Parallel processing of visual space by neighboring neurons in mouse visual cortex. Nat. Neurosci. **13**(9), 1144–1149 (2010)
20. Herikstad, R., et al.: Natural movies evoke spike trains with low spike time variability in cat primary visual cortex. J. Neurosci. **31**(44), 15844–15860 (2011)
21. Fusi, S., Miller, E.K., Rigotti, M.: Why neurons mix: high dimensionality for higher cognition. Curr. Opin. Neurobiol. **37**, 66–74 (2016)
22. Hebb, D.O.: The Organization of Behavior; A Neuropsychological Theory. Wiley, NY (1949)
23. Yuste, R.: From the neuron doctrine to neural networks. Nat. Rev. Neurosci. **16**(8), 487–497 (2015)
24. Saxena, S., Cunningham, J.P.: Towards the neural population doctrine. Curr. Opin. Neurobiol. **55**, 103–111 (2019)
25. Deneve, S., Chalk, M.: Efficiency turns the table on neural encoding, decoding and noise. Curr. Opin. Neurobiol. **37**, 141–148 (2016)
26. Rinkus, G.: A combinatorial neural network exhibiting episodic and semantic memory properties for Spatio-temporal patterns, in cognitive & neural systems. Boston U.: Boston (1996)
27. Rinkus, G.: A cortical sparse distributed coding model linking mini- and macrocolumn-scale functionality. Front. Neuroanat. **4**, 1235 (2010)
28. Rinkus, G.J.: Sparsey^TM: spatiotemporal event recognition via deep hierarchical sparse distributed codes. Front. Comput. Neurosci. **8**, 116453 (2014)
29. Buzsáki, G.: Neural syntax: cell assemblies, synapsembles, and readers. Neuron **68**(3), 362–385 (2010)
30. Watrous, A.J., et al.: More than spikes: common oscillatory mechanisms for content specific neural representations during perception and memory. Curr. Opin. Neurobiol. **31**, 33–39 (2015)
31. Igarashi, K.M., et al.: Coordination of entorhinal-hippocampal ensemble activity during associative learning. Nature **510**(7503), 143–147 (2014)
32. Fries, P.: Neuronal gamma-band synchronization as a fundamental process in cortical computation. Annu. Rev. Neurosci. **32**(1), 209–224 (2009)
33. Hubel, D.H., Wiesel, T.N.: Receptive fields, binocular interaction and functional architecture in the cat's visual cortex. J. Physiol. **160**(1), 106–154 (1962)
34. McCormick, D.A., Prince, D.A.: Mechanisms of action of acetylcholine in the guinea-pig cerebral cortex in vitro. J. Physiol. **375**, 169–194 (1986)
35. Sara, S.J., Vankov, A., Hervé, A.: Locus coeruleus-evoked responses in behaving rats: a clue to the role of noradrenaline in memory. Brain Res. Bull. **35**(5–6), 457–465 (1994)

36. Rinkus, G.: A cortical theory of super-efficient probabilistic inference based on sparse distributed representations. In: CNS 2013, Paris (2013)

37. Willshaw, D.J., Buneman, O.P., Longuet-Higgins, H.C.: Non holographic associative memory. Nature **222**, 960–962 (1969)

38. Marr, D.: A theory of cerebellar cortex. J. Physiol. **202**(2), 437–470 (1969)

39. Rajkumar, V., Pitkow, X.: Inference by reparameterization in neural population codes (2016)

40. Bengio, Y.: Deep learning of representations: looking forward. In: Dediu, AH., Martín-Vide, C., Mitkov, R., Truthe, B. (eds.) Statistical Language and Speech Processing. SLSP 2013. Lecture Notes in Computer Science(), vol. 7978, pp. 1–37. Springer, Berlin, Heidelberg (2013). https://doi.org/10.1007/978-3-642-39593-2_1

41. Salakhutdinov, R., Hinton, G.: Semantic Hashing. In: SIGIR workshop on Information Retrieval and Applications of Graphical Models (2007)

42. Salakhutdinov, R., Hinton, G.: Semantic hashing. Int. J. Approximate Reasoning **50**(7), 969–978 (2009)

43. Grauman, K., Fergus, R.: Learning binary hash codes for large-scale image search. In: Cipolla, R., Battiato, S., Farinella, G. (eds.) Machine Learning for Computer Vision. Studies in Computational Intelligence, vol. 411, pp. 49–87. Springer, Berlin, Heidelberg (2013). https://doi.org/10.1007/978-3-642-28661-2_3

44. Wang, J., et al.: Learning to hash for indexing big data - a survey. Proc. IEEE **104**(1), 34–57 (2016)

45. Kohn, A., et al.: Correlations and neuronal population information. Annu. Rev. Neurosci. **39**(1), 237–256 (2016)

46. Cohen, M.R., Kohn, A.: Measuring and interpreting neuronal correlations. Nat. Neurosci. **14**(7), 811–819 (2011)

47. Schneidman, E.: Towards the design principles of neural population codes. Curr. Opin. Neurobiol. **37**, 133–140 (2016)

48. Curto, C., et al.: Combinatorial neural codes from a mathematical coding theory perspective. Neural Comput. **25**(7), 1891–1925 (2013)

49. Latham, P.E.: Correlations demystified. Nat. Neurosci. **20**(1), 6–8 (2017)

Author Index

G. Nicosia et al. (Eds.): LOD 2023, LNCS 14506, pp. 481–483, 2024.
https://doi.org/10.1007/978-3-031-53966-4

Printed in the United States
by Baker & Taylor Publisher Services